Made in the USA
Monee, IL
13 September 2026

62978986R10155

THE ROUTLEDGE COMPANION TO THEATRE, PERFORMANCE AND COGNITIVE SCIENCE

The Routledge Companion to Theatre, Performance and Cognitive Science integrates key findings from the cognitive sciences (cognitive psychology, neuroscience, evolutionary studies and relevant social sciences) with insights from theatre and performance studies. This rapidly expanding interdisciplinary field dynamically advances critical and theoretical knowledge, as well as driving innovation in practice. The anthology includes 30 specially commissioned chapters, many written by authors who have been at the cutting-edge of research and practice in the field over the last 15 years. These authors offer many empirical answers to four significant questions:

- How can performances in theatre, dance and other media achieve more emotional and social impact?
- How can we become more adept teachers and learners of performance both within and outside of classrooms?
- What can the cognitive sciences reveal about the nature of drama and human nature in general?
- How can knowledge transfer, from a synthesis of science and performance, assist professionals such as nurses, care-givers, therapists and emergency workers in their jobs?

A wide-ranging and authoritative guide, *The Routledge Companion to Theatre, Performance and Cognitive Science* is an accessible tool for not only students, but practitioners and researchers in the arts and sciences as well.

Rick Kemp is Professor of Theatre and Head of Acting and Directing at Indiana University of Pennsylvania, USA. An actor, director and Fellow of the Salzburg Global Seminar on Neuroscience and Art, his publications include *Embodied Acting: What Neuroscience Tells Us About Performance* (2012) and *The Routledge Companion to Jacques Lecoq* (2016).

Bruce McConachie, Emeritus Professor at the University of Pittsburgh, USA, has published widely in theatre history and cognitive studies. His scholarship includes *Engaging Audiences* (2008), *Evolution, Cognition, and Performance* (2015) and chapters in *Theatre Histories: An Introduction* (3rd edition, 2016). A former president of the American Society for Theatre Research, McConachie also acts and directs.

ROUTLEDGE THEATRE AND PERFORMANCE COMPANIONS

THE ROUTLEDGE COMPANION TO PUPPETRY AND MATERIAL PERFORMANCE
Edited by Dassia N. Posner, Claudia Orenstein, and John Bell

THE ROUTLEDGE COMPANION TO DRAMATURGY
Edited by Magda Romanska

THE ROUTLEDGE COMPANION TO COMMEDIA DELL'ARTE
Edited by Judith Chaffee and Olly Crick

THE ROUTLEDGE COMPANION TO MICHAEL CHEKHOV
Edited by Marie Christine Autant Mathieu and Yana Meerzon

THE ROUTLEDGE COMPANION TO JACQUES LECOQ
Edited by Mark Evans and Rick Kemp

THE ROUTLEDGE COMPANION TO SCENOGRAPHY
Edited by Arnold Aronson

THE ROUTLEDGE COMPANION TO ADAPTATION
Edited by Dennis Cutchins

THE ROUTLEDGE COMPANION TO BUTOH PERFORMANCE
Edited by Bruce Baird and Rosemary Candelario

THE ROUTLEDGE COMPANION TO THEATRE, PERFORMANCE AND COGNITIVE SCIENCE
Edited by Rick Kemp and Bruce McConachie

For more information about this series, please visit: https://www.routledge.com/handbooks/products/SCAR30

THE ROUTLEDGE COMPANION TO THEATRE, PERFORMANCE AND COGNITIVE SCIENCE

Edited by Rick Kemp and Bruce McConachie

Routledge
Taylor & Francis Group
LONDON AND NEW YORK

First published 2019
by Routledge
2 Park Square, Milton Park, Abingdon, Oxon OX14 4RN

and by Routledge
605 Third Avenue, New York, NY 10017

First issued in paperback 2021

Routledge is an imprint of the Taylor & Francis Group, an informa business

British Library Cataloguing-in-Publication Data
A catalogue record for this book is available from the British Library

Library of Congress Cataloging-in-Publication Data
A catalog record has been requested for this book

ISBN 13: 978-1-03-209486-1 (pbk)
ISBN 13: 978-1-138-04889-8 (hbk)

Typeset in Bembo
by codeMantra

This book is dedicated to our students

CONTENTS

FIGURES AND TABLES

Figures

Tables

NOTES ON CONTRIBUTORS

Ana Margarida Abrantes studied German and English at the Universities of Aveiro, Essen and Innsbruck. She completed her master's degree in cognitive linguistics in 2001, with a dissertation on the semantic and pragmatic dimensions of euphemism in the press. In 2008, she received her doctoral degree in German language and literature from the Catholic University of Portugal, with a thesis on a cognitive poetic approach to selected works by Peter Weiss. Between 2006 and 2009, she was visiting scholar at the Center for Semiotics of Aarhus University and at the Department of Cognitive Science of Case Western Reserve University in Cleveland, USA. In 2006, she joined the Research Center for Communication and Culture Studies at the Catholic University of Portugal in Lisbon, where she is currently senior researcher in the research line Culture, Translation and Cognition. She is Professor of Languages and Linguistics at the same university. Her publications include *Meaning and Mind. A Cognitive Approach to Peter Weiss' Prose Work* (2010) and *Cognition and Culture. An Interdisciplinary Dialogue* (co-edited with Peter Hanenberg, 2011).

Rhonda Blair is Professor of Theatre at Southern Methodist University in Dallas. Publications include *Theatre, Performance and Cognition: Languages, Bodies and Ecologies* (2016, co-edited with Amy Cook); *The Actor, Image, and Action: Acting and Cognitive Neuroscience* (2008); editing an edition of Richard Boleslavsky's *Acting: The First Six Lessons* that includes documents from the American Laboratory Theatre (2010); articles in the anthologies *Affective Performance and Cognitive Science*, *Routledge Companion to Stanislavsky*, and *Performance and Cognition: Theatre Studies and the Cognitive Turn*, and in the *Journal of Dramatic Theory and Criticism*, *Theatre Survey*, *TDR: The Drama Review*, and *Theatre Topics* among others. Board member, Centre for Kinesthetics, Cognition and Performance. Editorial board member, JDTC, Theatre Topics, and Theatre, Dance and Performance Training. She has keynoted or given featured talks on applications of cognitive science to acting, theatre and performance in Paris, Rome, Kent, Wroclaw and Zurich, as well as in various U.S. cities. She directs and creates performances pieces. She was President of the American Society for Theatre Research, 2009–2012.

Shelby Brewster is Doctoral Student in Theatre and Performance Studies at the University of Pittsburgh, where she received the 2015–2016 Provost's Humanities Predoctoral

Fellowship. She received her MA in Theatre History, Literature, and Criticism from The Ohio State University in 2015. Her current research explores how theatre and performance artists use speculative strategies to critique the relationship between humans and their environments and imagine new ways of being human. Her work has been published in *The Journal of American Drama* and *Theatre and Foundation: The International Journal of Science Fiction*.

Experience Bryon, PhD, trained as an actor and opera singer before becoming interested in interdisciplinary performance practice. She developed the Integrative Performance Theory for her PhD from Monash University in Australia, subsequently refined into Integrative Performance Practice. She is senior lecturer and course leader of the MA/MFA in Performance Practice as Research at the Royal Central School of Speech and Drama, and has delivered keynotes at the Moscow Art Theatre's International Movement Conference, the Embodied Cognition Symposium as part of the Society of Artificial Intelligence and Simulation of Behaviour at Goldsmith University and the Interdisciplinary Voice Studies Centre at Plymouth University. Her publications include *Integrative Performance: Practice and Theory for the Interdisciplinary Performer* (monograph, 2014), *Performing Interdisciplinarity: Working Across Disciplinary Boundaries Through an Active Aesthetic* (2017) and *Embodied Cognition, Acting and Performance* (co-edited, 2017). Her research interests include Practice as Research, interdisciplinary performance practice(s), physical/vocal praxis and the emergence of transdisciplines as performance engages with other disciplines.

Sharon Marie Carnicke is Professor of Dramatic Arts and Slavic Languages and Literatures at the University of Southern California and a widely published author on performance in Russia, acting history and methods and film performance. She is internationally known for her groundbreaking book *Stanislavsky in Focus*, now in its second edition. Her other books include *Anton Chekhov: 4 Plays and 3 Jokes*, *Checking out Chekhov*, *The Theatrical Instinct: The Work of Nikolai Evreinov*, and with Cynthia Baron, *Reframing Screen Performance*. Her translation of Chekhov's *The Seagull* won a Kennedy Center award. Her honours include grants and awards from the Rockefeller Foundation, the American Council of Learned Societies, the NEH and the National Science Foundation. Additionally, she is a director and master teacher of the rehearsal technique, Active Analysis, as created by Stanislavsky and developed by Maria Knebel. Most recently, she directed Chekhov's *Uncle Vanya* at the Norwegian National Academy of the Arts and taught an intensive workshop in Los Angeles for the National Alliance for Acting Teachers. Her work with scientists has leveraged active analysis for research on the expression of human emotion and interactive digital storytelling.

Amy Cook's most recent book, *Building Character: The Art and Science of Casting* (2018), is called a 'masterpiece that lies at the intersection of the humanities and cognitive science' by Mark Turner. Cook is an associate professor in English and Theatre Arts at Stony Brook University, NY, and specialises in the intersection of cognitive science and theories of performance, with particular attention to Shakespeare and contemporary performance. She has published *Shakespearean Neuroplay* (2010) and essays in *Theatre Journal*, *TDR*, *SubStance*, *Journal of Dramatic Theory and Criticism* and several edited volumes, including *Bloom's Modern Critical Interpretations*, *The Oxford Handbook of Dance and Theater* (2015) and *The Oxford Handbook of 4E Cognition* (2018).

Rachel DeSoto-Jackson is Assistant Professor of Applied Theater, director of SPATE (Simulated Patient/Applied Theater Ensemble) and managing/education director of Footlight Players youth theatre at Indiana University of Pennsylvania. She holds an MFA in Performance Pedagogy, MA in Theatre and Performance Studies and an MA Certificate in Film Studies from the University of Pittsburgh. DeSoto-Jackson currently serves as a Board Member for the Pedagogy and Theatre of the Oppressed organisation and on the Editorial Board of the organisation's journal. She also serves on the Steering Committee of the Latinx Theatre Commons. As a Latina educator, scholar and theatre-maker, her work focuses on promoting equity, diversity and inclusion through Applied Theater methods. Her artistic practice, scholarship and teaching explore these practices within non-traditional settings and with interdisciplinary fields including healthcare, food and nutrition, psychology and criminology. DeSoto-Jackson has developed workshops, presentations, lectures and course designs on various applications of Applied Theatre. She currently teaches the course, *The Performance of Caring*, which builds skills in empathy, active listening and verbal/non-verbal communication to improve trust and patient/client interaction. She has been a grant recipient for her beginning research in simulated patient training.

Robin Dixon is Honorary Associate of the Department of Theatre and Performance Studies, Sydney University. His primary area of research interest is the stagecraft and performance of ancient Roman theatre, but he has taught widely on the history of Western theatre, and pursues a range of interdisciplinary research interests. His PhD thesis on the spatial dramaturgy of Plautine comedy was submitted in 2011, and since then he has taught at Sydney University and the National Institute of Dramatic Art, where he was Convener of Performance Practices 2015–2017. He is currently a research assistant on an ARC project examining historical rehearsal practices and an associate investigator on a project on Shakespearean performance funded through the Centre for the History of Emotions.

Stanton B. Garner, Jr., is Professor of English and Adjunct Professor of Theatre at the University of Tennessee. He is the author of *The Absent Voice: Narrative Comprehension in the Theater* (1989), *Bodied Spaces: Phenomenology and Performance in Contemporary Drama* (1994, nominated for the Joe A. Callaway Prize and the Barnard Hewitt Award) and *Trevor Griffiths: Politics, Drama, History* (1999). His current research explores the kinaesthetic dynamics of theatrical spectatorship — how we respond to the movements we observe by vicariously enacting them — from phenomenological and cognitive perspectives. His book on this subject — *Kinesthetic Spectatorship in the Theater: Phenomenology, Cognition, Movement* – will be published in 2018 as part of its series Cognitive Approaches to Literature and Performance. Professor Garner's essays and reviews have appeared in numerous journals, including *Shakespeare Survey*, *Studies in Philology*, *Medieval and Renaissance Drama in England*, *Comparative Drama*, *Essays in Theatre/Études théâtrales*, *Degrés*, *Theatre Journal*, *Journal of Dramatic Theory and Criticism*, *Theatre Survey*, *Theatre Research International* and *Modern Drama*. With J. Ellen Gainor and Martin Puchner, he is co-editor of the *Norton Anthology of Drama* (3rd ed., 2017). He received his PhD from Princeton University.

Thalia R. Goldstein, PhD, is Assistant Professor of Applied Development Psychology at George Mason University and the director of the Social Skills, Imagination and Theatre Lab. She studies how children participate in and create fictional worlds, how actors construct characters onstage and the effects of these activities on empathy, theory of mind, emotion regulation, compassion and altruism. Her other work focuses on how children and adults understand social categorisation at the fiction/reality border, and how children react to watching fictional worlds. Her work has been supported by the National Endowment of the Arts, The John Templeton Foundation, Arts Connection, the National Science Foundation, American Psychological Foundation and the Department of Homeland Security. She has won awards from the Society for Research in Child Development, American Psychological Association and the International Society for the Empirical Study of Literature and Media. She is editor of the journal *Psychology of Aesthetics, Creativity and the* Arts (APA Division 10). Dr. Goldstein earned her B.A. from Cornell University in Theatre and Psychology, her PhD from Boston College in Developmental Psychology and completed a postdoctoral fellowship Yale University. She spent several years as a professional actress and dancer in New York City.

James Hamilton, Professor of Philosophy at Kansas State University in Manhattan, Kansas, USA, works on issues in aesthetics and the philosophy of art. An empirically and formally oriented philosopher, he works on topics having to do with the nature of theatrical enactment, how to model the reception of performances in theater, music, and dance and our interactions with puppets and other animated objects. He strives to make work that is informed by decision theory, formal learning theories, cognitive neuroscience, psychology, anthropology and history. He is author of the book, *The Art of Theater.* He has published articles in *Journal of Dramatic Theory and Criticism, Journal of Aesthetics and Art Criticism, Organon F, Proceedings of the European Society of Aesthetics, British Journal of Aesthetics, Routledge Companion to Shakespeare and Philosophy, Performance Research, Puppet Notebook, Revue internationale de philosophie, Routledge Companion to Aesthetics, Blackwell's Guide to Aesthetics, Encyclopedia of Aesthetics, Narratologia,* as well as in a number of edited books. He is a member of the *Theater and Performance Research Association,* the *American Society of Theater Research, Performance Philosophy, The British Society of Aesthetics, The American Society for Aesthetics,* the *Dance Scholars Association* and the *International Federation of Theater Research.* hamilton@ksu.edu

Pil Hansen is an Assistant Professor in performing arts at the University of Calgary, a founding member of Vertical City Performance and a dance/devising dramaturg. Her empirical and PaR experiments examine cognitive dynamics of memory, perception and learning in creative processes. She developed the tool-set 'Perceptual Dramaturgy' and, with Bruce Barton, the interdisciplinary research model 'Research-Based Practice.' She has dramaturged 27 premieres and remounts, including award winning and both nationally and internationally touring works. Hansen's scholarly research is published in *TDR: The Drama Review, Performance Research, Journal of Dramatic Theory and Criticism, Connection Science, Theatre Topics, Canadian Theatre Review, Peripiti, Koreografisk Journal, MAPA D3* and 13 essay collections on dramaturgy, PaR, cognitive performance studies and research methods. Hansen is primary editor of the essay collections *Dance Dramaturgy: Modes of Agency, Awareness and Engagement* (2015) and *Performing the Remembered Present: The Cognition of Memory in Dance, Theatre and Music* (2017) and first editor on three issues of *Canadian Theatre Review* and *Performance Research.* Hansen furthermore chairs the PSi working group on Performance and Dramaturgy.

Rick Kemp has received awards on both sides of the Atlantic for his work as an actor and director, including the Institut Français award for theatre, the British Telecom Innovations Award and the 2004 Heinz Endowments Creative Heights Award. He has worked with leading British companies and theatres such as The Almeida, 1982 Co., Complicité,

The Oxford Playhouse, Riverside Studios, Tricycle Theatre, and Traverse Theatre, and was the co-founder and joint Artistic Director of Commotion. In the USA, he has worked with Unseam'd Shakespeare, the Pittsburgh Playhouse, Quantum Theatre, New York's Perry Street Theatre, Squonk Opera, and internationally at Toronto's Harbourfront Theatre, Warsaw's Teatr Polski, Madrid's Festival de Otono and at Peter Brook's Bouffes du Nord theatre in Paris. Rick holds an MA from Oxford University, an MFA in Performance Pedagogy, and a PhD in Theatre and Performance Studies. He is a Fellow of the Salzburg Global Seminar on Neuroscience and Art, Professor of Theatre and the Head of Acting and Directing in the Department of Theater and Dance at Indiana University of Pennsylvania. Publications include *Embodied Acting: What Neuroscience Tells Us About Performance* (2012) and *The Routledge Companion to Jacques Lecoq* (2016).

Nancy Kindelan is Professor in the Department of Theatre at Northeastern University where she teaches courses in both the theatre department and the honours programme. Her PhD is from the University of Wisconsin-Madison in Dramatic Literature and Criticism. She has published widely, including her first book *Shadows of Realism: Dramaturgy and the Theories and Practices of Modernism*, chapters in *Experiential Education: Making the Most of Learning Outside the Classroom* and *Course-Based Undergraduate Research: Educational Equity and High-Impact Practice* as well as essays in *Liberal Education*, *The Journal of General Education*, *Theatre Topics*, *Journal of Dramatic Theory and Criticism*, and *New England Theatre Journal*. Her second book, *Artistic Literacy: Theatre Studies and a Contemporary Liberal Education*, received a Special Honorable Mention for the Outstanding Book Award from the Association for Theatre in Higher Education (ATHE) in 2013. She is the founding editor for the book series, *The Arts in Higher Education*. In addition, she directs plays and speaks widely about how the signature pedagogy of theatre studies encourages transformative lifelong learning through creative thought, research and practice.

Jeanne Klein, Associate Professor Emerita, taught theatre for young audiences, elementary and secondary theatre education, child media psychology and US theatre history, among other courses, for 30 years at the University of Kansas. Over the course of her 40-year academic career, she directed over 50 productions for and with young people of all ages and led drama classes and workshops with elementary students. Her reception studies with child audiences earned several awards from the American Alliance for Theatre and Education, a grant from the Children's Theatre Foundation of America and international recognition at the Netherlands Theatre Institute in Amsterdam. As a long-time member of TYA/USA (the US chapter of ASSITEJ International), she attended many international theatre festivals for young audiences. In addition to book chapters, she has published in *Youth Theatre Journal*, *Theatre Topics*, *The Lion and the Unicorn*, *Journal of the History of Childhood and Youth* and *Journal of American Drama and Theatre*. In retirement, she remains active as an independent scholar, journal and book reviewer, and master gardener.

Gunter Lösel holds a PhD in Theater Studies as well as a diploma in psychology and is working as an improvisational actor with his own theatre groups "Improtheater Bremen" and "The Stupid Lovers." Since 2014, he is heading the Research Focus Performative Practice at the Zurich University of the Arts. He has been publishing on the theme of improvisation since 2004 (with two books being available in English: *The Play of Archetypes – Basic Forms of Human Encounter* (2012) and *Impro Talks* (2017). His main fields of research are improvisation, creativity, embodiment and collaboration. Currently, he is the main applicant of the research project 'Research Video' (funded by the Swiss National Foundation), exploring the use of annotated video in the publication of artistic research. In parallel, he is doing research on artificial intelligences on the stage.

www.gunterloesel.theater

www.improtheater-bremen.de

www.stupidlovers.de

www.zhdk.ch/3700.

John Lutterbie is Professor at Stony Brook University where he has a joint appointment in the Department of Theatre Arts and the Department of Art. He serves as co-director of the Center for Embodied Cognition, Creativity and Performance. Lutterbie is currently working on two monographs, one on time-based aesthetics that explores the cognitive foundations of the embodied experience of art, and the other a textbook that engages students at the intersection of theatre/performance studies and cognitive science. He is the author of *Toward a General Theory of Acting: Cognitive Science and Performance* and *Hearing Voices: Modern Drama and the Problem of Subjectivity*. Articles have been published with *Theatre Journal*, *Journal of Dramatic Theory and Criticism*, *Performance Research*, and other periodicals and books. Lutterbie is a co-editor with Nicola Shaughnessy of the series Science and Performance with Methuen. He can be contacted at john.lutterbie@gmail.com

David Mason is Chair of Theatre and director of Asian Studies at Rhodes College, Editor-in-Chief of the journal *Ecumenica: Performance and Religion* and a board member of the Association for Asian Performance. He is the author of *The Performative Ground of Religion and Theatre* (2018), *Brigham Young: Sovereign In America* (2014) and *Theatre and Religion on Krishna's Stage* (2009). He has published essays on Sanskrit drama in *Theatre Research International* and *New Literary History*, as well as essays on performance theory in journals including the *Journal of Dramatic Theory and Criticism*, *Journal of Popular Culture* and *Studies in South Asian Film and Media*. He earned an MA in South Asian Studies and a PhD in Theatre Research at the University of Wisconsin-Madison. He has been a Fulbright Fellow and a George A. and Eliza Gardner Howard Fellow.

Bruce McConachie recently became Emeritus Professor in Theatre and Performance Studies at the University of Pittsburgh. His scholarship ranges among theatre history, historiography, cognitive criticism and the evolution of performance. Major works include *Interpreting the Theatrical Past: Essays in the Historiography of Performance* (with Thomas Postlewait, 1989), *Melodramatic Formations: American Theatre and Society, 1820–1870* (1992), *American Theater in the Culture of the Cold War* (2003), *Engaging Audiences: A Cognitive Approach to Spectating in the Theatre* (2008), *Evolution, Cognition, and Performance* (2015) and three editions of *Theatre Histories: An Introduction* (co-written with others, 2006, 2010, 2016). For his scholarly contributions, the American Society for Theatre Research awarded McConachie the Barnard Hewitt and the Distinguished Scholar awards. In addition to teaching, directing and acting at the College of William and Mary and the University of Pittsburgh,

McConachie has held visiting positions at Northwestern, Warsaw, Helsinki and Queens (Belfast) universities. He served as president of ASTR from 2001 to 2003, and continues to co-edit two publication series, *Cultural Histories of Theatre and Performance* (with Claire Cochrane) for Bloomsbury Methuen and *Cognitive Studies in Literature and Performance* (with Blakey Vermeule). E-mail: bamcco@pitt.edu.

Erin B. Mee is the author of Theatre of Roots: Redirecting the Modern Indian Stage, co-editor of Antigone on the Contemporary World Stage, editor of Drama Contemporary: India, and co-editor of Modern Asian Theatre and Performance 1900–2000. She has written numerous articles for *TDR, Theatre Journal, Performance Research* and other journals and books. Her born-digital Scalar article 'Hearing the Music of the Hemispheres' won the ATHE-ASTR Award for Best Digital Article in 2016. She has directed at the Public Theater, NYTW, the Guthrie, the Magic Theatre and elsewhere in the USA, and with Sopanam in India. She is the Founding Artistic Director of This Is Not A Theatre Company with whom she has directed Pool Play 2.0, Café Play, Versailles 2015, Readymade Cabaret and numerous other productions. She is assistant arts professor, Department of Drama, Tisch, NYU. www.erinbmee.com

Peter Meineck holds the endowed chair of Professor of Classics in the Modern World at New York University. He was the founding director of Aquila Theatre where he produced and directed over 50 shows throughout the world and developed several national Arts and Humanities public programmes. He has published widely on ancient drama, most recently a monograph, *Theatrocracy: Greek Drama, Cognition and the Imperative for Theatre* (2017).

Vladimir Mirodan, PhD, FRSA, is Emeritus Professor of Theatre, University of the Arts London. Trained on the Directors Course at Drama Centre London, he has directed over 50 productions in the UK as well as internationally, and has taught and directed in most leading drama schools in the UK. He was vice-principal and director of Drama at the Royal Conservatoire of Scotland, principal of Drama Centre London and director of Development and Research Leader, Drama and Performance, Central Saint Martins. He is a former chairman of the Conference of Drama Schools and a deputy chair of the National Council for Drama Training. He is currently the chair of the Directors Guild of Great Britain Trust and of the Directors Charitable Foundation. Professor Mirodan's research interests revolve around issues of acting psychology, in particular as this relates to the neuropsychology of gesture and posture. He has published on these topics and is currently completing *The Actor and the Character*, a book on the psychology of transformation in acting, to be published in 2018. Together with neuroscientists from University College London, Professor Mirodan is engaged in a research project on emotional contagion in acting, funded by the Leverhulme Foundation.

Ariel Nereson is Assistant Professor of Dance Studies at the University at Buffalo – SUNY. Her research focuses on embodiment and identity in movement-based performance across dance, theatre and musical theater. She is also a choreographer and dramaturg. Her book project, *Democracy Moving: The Lincoln Dances of Bill T. Jones/Arnie Zane Dance Company*, thinks across dance, performance, cognitive and American studies to analyse how these dances theorise and enact corporeality as historiographical strategy. Her essays can be found in *Theatre Journal, Theatre Survey, Studies in Musical Theatre, American Quarterly, Critical Stages* and *The Journal of American Drama and Theatre*, amongst others.

Helga Noice is Professor Emerita of Psychology at Elmhurst College; she earned her PhD in cognition at Rutgers University. During her early teaching years, Helga developed a great interest in specialised mental processes. She hypothesised that many older adults who were experiencing memory problems might benefit by learning the strategies of professional actors who routinely remember long scripts verbatim. As a result of this original insight, the Noices started investigating the connections between psychology and theatre and have pursued this interest for over 20 years. Their joint research, which combines theory and application, has resulted in a unique evidence-based theatrical intervention that has been successfully performed in dozens of retirement homes, hospital senior centres and select university programmes. Before and after taking a brief acting course, the older adults are given a battery of cognitive assessment tests. The dramatic results have revealed statistically significant improvements in memory, comprehension, creativity and problem solving. This work has been funded by the National Science Foundation (NSF), Pew Charitable Trust, Schweizer Nationalfond, Elizabeth Morse Charitable Trust and the National Institutes of Health (NIH). The latter has supported the Noices' research with four consecutive multi-year grants. In addition to the dozens of peer-reviewed articles in professional books and journals, this theatrical intervention has been featured on over 25 national and international media outlets, including the NBC Nightly News with Brian Williams, PBS-TV, ABC-TV, New York Times, the Chicago Tribune, the Los Angeles Times and the popular German magazine, *Der Spiegel.*

Tony Noice is currently Professor of Theatre at Elmhurst College, IL, and earned his PhD at Wayne State University in Michigan. He splits his time between the academic and professional worlds. In terms of the latter, he has performed in over a 100 productions under union contracts (Equity, SAG-AFTRA).

David Palmer taught philosophy and literature in the Humanities Department at the Massachusetts Maritime Academy until his retirement in June 2017. His interests in philosophy of mind, theories of the self and ethics led him to develop the course *The Brain, Narrative, and the Self: Evolutionary Foundations of Tragedy* and to explore the plays of Samuel Beckett, Arthur Miller and Eugene O'Neill. He has published on each of these dramatists and on Plato's moral theory. In spring 2015, he was a Travis Bogard Fellow at Tao House, the Eugene O'Neill National Historic Site in Danville, California. He is president of the Arthur Miller Society, a board member of the Eugene O'Neill Society and the editor of *Visions of Tragedy in Modern American Drama: From O'Neill to the Twenty-First Century* (2018). His degrees are from Harvard, The New School for Social Research and Columbia. Email: dpalmer@maritime.edu.

Esther Pascual (PhD, VU Amsterdam, 2003) is currently Senior Researcher and Assistant Professor at Zhejiang University in Hangzhou under the 'Hundred Talents' programme. She is mainly working on what she has labelled 'fictive interaction.' She has published extensively on the topic in indexed journals, such as *Cognitive Linguistics*, *Review of Cognitive Linguistics*, *Text & Talk* and *Pragmatics*, as well as in her 2014 monograph *Fictive Interaction: The Conversation Frame in Thought, Language, and Discourse* and her 2016 co-edited volume *The Conversation Frame: Forms and Functions of Fictive Interaction*. Dr. Pascual has received various prestigious research grants to study the phenomenon from the Fulbright foundation and the Netherlands Organisation for Scientific Research. She has been a visiting scholar at the University of California, San Diego, The University of Ghent, Belgium, the University of

California, Berkeley, and Case Western Reserve University in Ohio. She is the vice-president of the International Association for Cognitive Semiotics, an elected International Organizing Committee of the Language, Culture, and Mind conference series, and a permanent member of the scientific committee of the Spanish Association of Cognitive Linguistics. She is also a founding co-editor-in-chief of the peer-reviewed international journal *Language Under Discussion*.

Nicola Shaughnessy is Professor of Performance at the University of Kent where she founded the Centre for Cognition, Kinaesthetics and Performance. Her research and teaching interests are in the areas of contemporary performance, applied and socially engaged theatre, identity, autobiography and neurodiversity. Her interdisciplinary collaborations explore the intersections between cognitive and affective neuroscience and theatre through creative and participatory research methods. She was Principal Investigator for the AHRC funded project 'Imagining Autism: Drama, Performance and Intermediality as Interventions for Autism.' Her publications are wide ranging with essays in *Interdisciplinary Science Reviews* (2013), the *Wiley Handbook of Developmental Psychopathology* (2017) and *The Cognitive Humanities* (2016), as well as her contributions to theatre and performance studies. She is the author of *Applying Performance: Live Art, Socially Engaged Theatre and Affective Practice* (2012) and the edited collection *Affective Performance and Cognitive Science: Body, Brain and Being* (2013). She is series editor (with Professor John Lutterbie) for Methuen's Performance and Science volumes for which she is contributing a new collection (co-edited with Philip Barnard: *Performing Psychologies: Imagination, Creativity and Dramas of the Mind* (2018).

Gabriele Sofia is Associate Professor in Performing Arts at the Université Grenoble Alpes (France) and member of the Laboratory Litt&Arts (UMR 5316-CNRS). A PhD in Tecnologie digitali per la ricerca sullo spettacolo and Esthétique, Sciences et Technologies des Arts, he worked under the co-supervision of the Sapienza Università di Roma and the Laboratoire d'Ethnoscénologie (Université Paris 8). Since 2006, he has been carrying out a transdisciplinary research project on the neurophysiology of the actor and the spectator. From 2009 to 2013, he promoted and organised in Rome five editions of the International Conference 'Dialogues between Theatre and Neuroscience.' From 2014 to 2016, he organised in France three International Conferences on 'Cognitive Sciences and Performing Arts.' He edited the special issue 'Theatre and Neuroscience' for the *Brazilian Journal on Presence Studies* (vol. 4 n. 2, 2014) and the book *Theatre and Cognitive Neuroscience* (2016 with Clelia Falletti and Victor Jacono). He is the author of the book *Le acrobazie dello spettatore. Dal teatro alle neuroscienze e ritorno* [The spectator's acrobatics. From theatre to neuroscience and back] (2013). From 2012 to the present, he has been part of the Editorial Board of *Teatro e Storia*. For further information, see www.gabrielesofia.it.

Claire Syler is Assistant Professor in the Department of Theatre at the University of Missouri, USA. She holds a PhD in Theatre & Performance Studies (with an emphasis in the Learning Sciences) from the University of Pittsburgh, and a MFA in Directing from the University of Memphis. A scholar and artist, Claire has published articles in *Theatre, Dance and Performance Training*, *Theatre Topics*, *Youth Theatre Journal*, and is a former Education Director of the Nashville Shakespeare Festival in Nashville, Tennessee. Her dissertation, 'Actor Coaching: Talking Performance into Being,' was awarded the honourable mention for the 2017 American Alliance of Theatre Education's Distinguished Dissertation Award. Email: sylerc@missouri.edu

Evelyn Tribble is Professor of English at the University of Connecticut. Her books include *Margins and Marginality: The Printed Page in Early Modern England* (1993), *Cognitive Ecologies and the History of Remembering* (with Nicholas Keene, 2011), *Cognition in the Globe: Attention and Memory in Shakespeare's Theatre* (2011) and *Early Modern Actors and Shakespeare's Theatre: Thinking with the Body* (2017). She has also published articles in *Shakespeare Quarterly*, *Shakespeare*, *Shakespeare Survey*, *Shakespeare Studies*, *Textual Practice*, and *ELH*, among others. Email: evelyn_tribble@mac.com

Melissa Trimingham is Senior Lecturer in Drama and Director of the Centre for Cognition, Kinaesthetics and Performance at the University of Kent. She specialises in cognitive approaches to scenography, puppetry and costume; working with autistic people in performance; and modernist aesthetics. As co-investigator on the Arts and Humanities Research Council project 'Imagining Autism: Drama, Performance and Intermediality as Interventions for Autistic Spectrum Conditions' (2011–2014), she designed and built drama environments for children on the autistic spectrum using puppetry, masks, costumes, sound, light and projection, and she has since developed Imagining Autism into training work with families and professionals in health and public services. Her monograph *The Theatre of the Bauhaus: the Modern and Postmodern Stage of Oskar Schlemmer* was published in 2011. Publications include a seminal article on 'The Methodology of Practice as Research' (2002); 'Touched by Meaning: Haptic Affect in Autism' in *Affective Performance and Cognitive Science, Body, Brain and Being*, (2013); 'Ecologies of Autism: Vibrant Space in Imagining Autism' in *Scenography Expanded* (2017); 'Surprised by Beauty' in *Anthropology and Beauty: From Aesthetics to Creativity* (2017) and jointly with Nicola Shaughnessy 'Material Voices: Intermediality and autism' *Research in Drama Education* (*RiDE*) (2016: 21.3, 293–308).

Darren Tunstall read English at Cambridge and trained as an actor at Bristol Old Vic Theatre School, then later with Philippe Gaulier, Theatre de Complicite and David Glass. He was a professional actor, director, writer and movement director for over 20 years, working for the Royal Shakespeare Company, the National Theatre, BBC, ITV, Film Four and many other theatre and television companies. During the 1990s and early 2000s, he was heavily involved in the 'physical theatre' movement in Britain, collaborating with such companies as The Right Size, Told by an Idiot, Theatre de Complicite, Peepolykus and Scarlet Theatre. In 2007, he moved full-time into teaching, and since 2015, he has been a lecturer in Acting at the Guildford School of Acting, University of Surrey. Publications include chapters in *The Routledge Companion to Actors' Shakespeare, The Routledge Companion to Jacques Lecoq* and *Time, Temporality and Performer Training*, and articles in *Shakespeare Bulletin* and *Journal of Adaptation in Film and Performance*. His first book *Shakespeare and Gesture in Practice* was published in 2016, and he is currently working on his second, *The Evolutionary Imperative in Theatre and Film*.

GENERAL INTRODUCTION

Bruce McConachie

The Routledge Companion to Theatre, Performance and Cognitive Science puts cognitive scientific disciplines in conversation with theatre and performance studies. Under the general heading of 'cognitive science,' we include the methods and insights of neuroscience, cognitive psychology, evolutionary studies and those social scientific approaches that have been influenced by these three disciplines. Cognitive scientists are committed to discovering and validating empirical evidence about perception, imagination, empathy, the emotions, movement, consciousness and other areas of human behaviour that can provide reliable insights across the varieties of our species' cultures, histories and languages. Theatre and performance scholars have long understood the general relevance of these areas of knowledge to acting, spectating, training, criticism, history and other parts of our field, but have mostly relied on traditional concepts of mind and body to investigate them. In the past 20 years, scholars and scientists have begun to systematically apply the empirical knowledge derived from the cognitive sciences to how we practise and think about varieties of performance. We are happy that several of the leading investigators in this emerging interdiscipline have contributed chapters to our *Routledge Companion*.

Many of our authors have significantly advanced the field of theatre, performance and cognitive science. Several of the early books in our field primarily applied insights from cognitive studies to theatre theory, history and dramatic criticism. These included James Hamilton's *The Art of Theater* (2007), my own *American Theater in the Culture of the Cold War* (2003), Amy Cook's *Shakespearean Neuroplay* (2010) and Evelyn Tribble's *Cognition in the Globe* (2011). Tony and Helga Noice were the first cognitive investigators to focus on acting with their *The Nature of Expertise in Professional Acting: A Cognitive View* in 1997. Joining that book were Rhonda Blair's *The Actor, Image, and Action* (2010), John Lutterbie's *Toward a General Theory of Acting* (2011), Rick Kemp's *Embodied Acting* (2012), Gunter Lösel's *Impro Talks* (2017) and Vladimir Mirodan's *The Actor and the Character* (2018). Several anthologies edited or co-edited by our writers contained essays that made significant contributions to the interdisciplinary conversation. These included *Cognition and Culture: An Interdisciplinary Dialogue*, (2011), ed. by Ana Margarida Abrantes, *Affective Performance and Cognitive Science* (2013), ed. by Nicola Shaughnessy, *Fictive Interactions: The Conversation Frame of Thought, Language, and Discourse* (2016), ed. by Esther Pascual, *Dance Dramaturgy: Modes of Agency, Awareness, and Engagement* (2015), ed. by Pil Hansen and *Theatre and Cognitive Neuroscience* (2016), ed. by Gabriele Sofia.

As the dates of these titles suggest, our field has grown substantially in the last five years. There are now two book series specialising in publishing monographs in performance and cognitive studies – *Science and Performance*, at Bloomsbury Metheun, and *Cognitive Studies in Literature and Performance* at Palgrave Macmillan. In addition to possibilities for presentations at several academic conferences, the field now has an annual international conference, Cognitive Futures in the Arts and Humanities, which frequently features papers in theatre and performance. A sampling of the titles of published monographs in the last few years gives a good indication of the continuing vigour and broadening range of the field: *Integrative Performance: Practice and Theory for the Interdisciplinary Performer* (2014), *Evolution, Cognition, and Performance* (2015), *Theatrocracy: Greek Drama, Cognition, and the Imperative for Theatre* (2017), *Early Modern Actors and Shakespeare's Theatre* (2017), *Building Character: The Art and Science of Casting* (2018), *Kinesthetic Spectatorship in the Theater: Phenomenology, Cognition, Movement* (2018) and *The Performative Ground of Religion and Theatre* (2018).

In the midst of gathering essays for this book, we soon realised that discrete divisions among the topics we wanted to cover in *The Companion* would be impossible to carve out. Many of our contributors are practicing artists as well as scholars, for instance, and the same insight from cognitive science could as easily serve to justify an artistic choice as a scholarly point. Consequently, we decided to welcome the inevitability of overlap among the major parts of our anthology. We have divided our 30 essays into four major parts, each containing an introduction and from five to nine of our chapter essays: I Artistry, II Learning, III Scholarship and IV Translational Applications. In the hope that our readers will treat these part titles as general aims rather than watertight categories, we have placed those chapters oriented towards the goal of improving the artistry of performances in Part I, located essays investigating the evolution and psychology of learning – both in and out of school – in Part II and plunked pieces primarily intended to assist scholarly investigations in Part III. Our title for Part IV, Translational Applications, nods to the scientific concept of 'translational research' and is intended to demonstrate how the confluence of knowledge from performance and the cognitive sciences may facilitate action in such areas as therapy, emergency services and political change.

Given the scientific difficulty of some of our terminology and our desire to reach as many readers as possible, we have also included a lengthy Glossary at the back of the book. For the rest of this General Introduction and for all four of the part introductions, we will emphasise in **bold print** those terms that that also appear in the Glossary. Please look them up if you wish a richer understanding of their meaning than what we can provide in our introductions.

The remainder of this introduction will begin with a summary of the major paradigms that have dominated studies in the cognitive sciences from the 1940s until 2000, and then provide a discussion of enactive, embodied and distributed cognition, three contemporary fields that are having a major impact on practice and scholarship in theatre and performance. Recognising that the science behind this work will be new to many of our readers, we will introduce some of the key scientific concepts you will encounter in several of the chapters, along with some salient examples, so that reading the essays may flow more smoothly. Our part introductions will use these concepts and categories to connect the science to the specific areas of theatre, performance and related areas of research discussed in the part essays to come.

An overview of major cognitive paradigms

In the middle decades of the twentieth century, **behaviouralism** dominated the study of human psychology and sociology and also shaped many scientific findings about performance and the theatre. Behavioural scientists confined themselves to observable behaviour and excluded mental and emotional processes. In general, the behaviouralists believed that the brain, in addition to controlling the internal workings of the body, generated individual and social behaviour by responding to external stimuli. Because they assumed that stimulus-response operations provided the basis for most behaviour, the behaviouralists believed that cognition, per se, was relatively unimportant. Also, because people could exercise little control over their goals and intentions, they could be conditioned to do almost anything; there were no predispositions and constraints in their minds and bodies as the result of their evolution that encouraged or prevented them from responding in ways that society had programmed them to do. Behaviouralism influenced significant practices in theatre studies, from Vsevolod Meyerhold's biomechanical ideas about performing to the method of acting of Lee Strasberg.

After some psychologists demonstrated that mental processes did matter for behaviour, **cognitivism** emerged in the late 1950s to challenge behaviouralism. Partly because psychologist George A. Miller was the first to show that early computers could perform some of the same kinds of information processing as the brain, the cognitivists looked to the computer as a possible model of the mind. Jerry Fodor advanced one of the most influential versions of the **Computational Theory of Mind** (CTM) in the 1970s. In Fodor's theory, a person's mind uses mental representations corresponding to the realities of the internal and external world in order to facilitate her (or his) cognition. Later critics of CTM would point out that Fodor could not explain how the person's sensory experiences of the world could be translated into the symbolic representations needed for his (or her) mental computations.

When CTM had difficulty accounting for a growing range of cognitive phenomena, **connectionism** emerged as the next major approach in cognitive science in the 1980s. Connectionism was less a fully realised paradigm than a means of explaining the speed and fluidity of parallel processing in the brain by the mathematical modelling of cognition as the changing organisation of neurons and neuronal networks. The human brain has approximately 100 billion neurons, which can 'wire together and fire together' in a nearly infinite number of networks through changes in electrical impulses, chemical signaling, fluctuations in blood flow and shifts in the utilisation of oxygen and glucose. Connectionism arose partly to explain the results of functional Magnetic Resonance Imaging (fMRI), the technology that locates neuronal activation by charting increased cerebral blood flow when activities involving the mind are performed. FMRI studies are valuable because all thought involves neural processing that remains beneath conscious awareness. The limitation of such studies is that correlating the primary brain location with external activity does not always point to causation. Speaking, for example, generally activates blood flow to Broca's area in the brain, located in the prefrontal cortex, but other areas also 'light up' when we speak. While the overall findings from brain imaging studies have been helpful, they also demonstrate further complexities still to be explained in our neural mechanisms.

When Francisco Varela, Evan Thompson and Eleanor Rosch proposed **enactivism** in their book, *The Embodied Mind: Cognitive Science and Human Experience* (1991), they knew that scientists had not yet found enough evidence to support enaction as an accepted scientific paradigm. Even in 2010, after many enactivists had substantiated several of the claims of enaction and clearly distinguished it from previous cognitive paradigms, philosopher of science John Stewart called enaction a 'proto-paradigm' (2010, 1). Despite progress since then, enaction remains a paradigm-in-waiting for most mainstream cognitive scientists in 2018. Accordingly, we will discuss enactive cognition in one of the following three parts of this introduction that deal with significant contemporary fields of investigation.

Since the turn of the century, the meanings of 'cognition' have escaped from the brain, moved to the rest of the body, animated intentional action and even colonised the environment. For the majority of theatre and performance artists and scholars working in the cognitive sciences, cognition is 'embodied,' 'enactive' and 'distributed.' We shall examine salient examples in these categories and summarise the science behind each of them in turn.

Embodied cognition

Not surprisingly, perhaps, getting cognition 'out of the head' and 'into the body' has proven to be the most accessible arena of cognitive science for many academics in theatre and performance studies. Most cognitive scientists working within a non-representational version of connectionism also accept many of the major insights of **embodied cognition**.

Minding is embodied not simply because the brain needs a body to function, but because the active body shapes the kinds of concepts that the mind produces. We know, for example, what the action of grasping a stick, a handle or another person's arm entails. But why do cultures around the world that have a word meaning 'to grasp' also use the same word to denote understanding – 'grasping' – an idea? Because ideas cannot literally be grasped – to speak of 'grasping an idea' is a metaphor. Many metaphors extend our embodied notions of movement and space to make sense of our experience. When we say, 'Life is a journey,' for example, we map the notion of walking from one place to another onto the duration of a person's natural life, thereby gaining a physical sense of what it means to live. Cognitive linguist George Lakoff and cognitive philosopher Mark Johnson investigated the embodied nature of many metaphors in their book *Metaphors We Live By* in 1980, and much of their work since then builds upon and extends these ideas.

Insights into embodied cognition that started with cognitive linguistics also flourished in other areas of the field. Cognitive scientists influenced by the philosophical tradition of phenomenology have drawn attention to the need to distinguish between the body as an object of study and the body as the centre of personal experience. As a result of **proprioception**, the awareness we have of our body when it moves, as well as the visual and tactile knowledge we understand through movement, we know that embodiment centres much of the knowledge we gain through our own actions. In *How the Body Shapes the Mind* (2005), cognitive phenomenologist Shaun Gallagher usefully distinguishes between body image and body schema. A person's body image is largely defined by how the person feels about her or his body when looking in the mirror; personal health, local weather and social pressures shape body image. In contrast, a person's body schema is typically less conscious; it results from the accumulated knowledge over a lifetime of what 'my body' is able to do. 'I can walk, run, grasp, climb a tree, play tennis, enact a role, etc.' Clearly, actors and other performers can benefit from a heightened awareness of both their body image and body schema.

The importance of embodied approaches in the cognitive sciences is reshaping how we understand social cognition, those interactions among people in the same society that facilitate the creation and enforcement of new social norms and that encourage social cohorts to cooperate on projects that can be mutually beneficial. Since the 1970s, many psychologists have assumed that people needed a 'theory of mind' (ToM) – some notion of how the beliefs, intentions and desires of others would likely affect their behaviour – before they could successfully interact with others. While there are different conceptions about how ToM operates, most psychologists agreed that successful social interaction depended upon it. Many proponents of embodied cognition, however, now question the need for a mental theory about how other people's minds work as a prerequisite for effective sociality. Daniel

Hutto (2008) and other advocates of embodied cognition, for example, point to primary and secondary modes of intersubjectivity that occur in the normal development of all infants and children as providing the basis for later social interactions. Infants learn to track eye movements and imitate facial expressions, and one-year olds can already read the emotions of adult care-givers. This early basis for social interaction occurs before infants have any interest in or ability to develop a theory of mind. Building on this base, the growing child watches the actions, observes the emotions and may even construct imagined narratives about others to figure out effective ways of interacting in society.

Finally, embodiment plays several key roles in the cognitive operations of **memory**. The senses of the body often provide mnemonic registers of past events; smells, sights, sounds and tastes bring back specific memories and emotions. We learn and recall procedural memories through physical repetition and refinement; how to ride a bike or play the violin are good examples. Learning skills in apprenticeship situations often involves procedural memory as well and, especially in traditional cultures, has historically been an important part of the socialisation of teens into adulthood. Lawrence Barsalou proposes an embodied understanding for what is usually termed semantic memory – the memory of arbitrary, decontextualised categories in language, such as the word 'dog.' According to Barsalou, we do not recall dogs as a general category; instead, we remember experiences with specific dogs in concrete situations, and we simulate one or several of these embodied experiences (remembering and reenacting specific sounds, sights, and interactive emotions) every time we hear someone say, 'Dog.' Even our understanding of dictionary-like definitions of general categories, in other words, may work through embodiment.

Enactive cognition

The enactive approach to cognition builds upon many of the assumptions of embodiment. In terms of past cognitive paradigms, it does not rule out most versions of connectionism, but it does challenge the **representationalism** of computational ToM. Called 'enactivism' because its proponents believe that all animals, including individual humans, effectively 'enact' (rather than represent) the world in which they live to enable their flourishing, this paradigm links perception to action. In brief, we perceive the (social, material, spiritual, etc.) world in ways that are useful to us and act within that perception (often unconsciously) to ensure our survival. Our minds do not manipulate symbolic representations of what is 'out there' in an attempt to arrive at an objective perception of reality, but use our direct experience of memory and the senses to search for what is necessary to predict events and survive in the future.

Five intertwined principles constitute enactive cognition: autonomy, sense-making, emergence, embodiment and experience. All living systems possess **autonomy**, in the sense that they interact with other systems as self-governing agents; they do not simply respond to external stimuli or internal demands. Enactive agents also use their senses to understand their social and physical environments. Further, meaning emerges for people in the course of taking action. According to several enactivists, '[The **emergence** of] meaning is not to be found in elements belonging to the environment or in the internal dynamics of the agent, but belongs to the relational domain established between the two' (DiPaulo, Rohde, and DeJaegher 2010, 40). Spectators at a play, for example, do not find meanings in the performance or in themselves, but in the experience – 'the relational domain' – established between the spectator and the performance. For enactivists, the body is much more than a functional way of taking action in the world. All animate bodies allow for varying degrees of openness to the world through the senses and shape how cognition occurs in the mind. Regarding experience, enactivists emphasise that experiences transform our bodies and brains over time. In younger animals especially, new experiences can link memory to learning to enable new capabilities, interactions and identities.

Our bodies and brains dynamically engage all of the principles of enaction to survive and thrive in the world. Cognitive scientists call this complex system the perception-action cycle and use the mathematical modeling of **dynamical systems theory** (DST) to describe it. In *How Brains Make Up Their Minds* (2000), Walter J. Freeman describes the major stages and components of the perception-action cycle:

> Our actions emerge through a continuous loop that we can divide into three stages. The first stage is the emergence and elaboration within our brains of goals concerning future states. The goals are in nested layers, ranging from what we do in the next few seconds to our ultimate survival and enjoyment of life. The second stage of the loop involves acting and receiving the sensory consequences of actions and constructing their meanings. In the third stage, we modify our brains by learning, which guides each successive pattern.
> *(2000, 91–2)*

Then the loop repeats itself, leading to more goal-setting, acting and learning. Freeman notes, as well, that the entire process is suffused with **emotion**, which regulates the system throughout.

Given the importance of emotions in the perception-action cycle, it is not surprising that enactive approaches to cognition have challenged traditional understandings of emotional experience. Scientists have long understood the neuronal and chemical basis of our emotional lives. Traditionally, however, most scientists held that our emotions require a cognitive 'appraisal' of a situation in order to trigger an emotional response. For example: see the bear, appraise bears as dangerous, respond with the emotion of fear. When linked with the Theory of Basic Emotions, this assumption about cognitive appraisal preserved the age-old distinction between reason (i.e., appraisal) and emotion and ensured that episodes of anger, surprise, disgust and so forth would be over after they ran their course and individuals could return to a kind of bodily homeostasis that allowed for 'normal' functioning. Scientists like Giovanna Colombetti, however, in her *The Feeling Body: Affective Science Meets the Enactive Mind* (2014), have found persuasive evidence that our emotions always pervade our perceptions of the world; there is no separate 'time out' for appraisal that comes between perception and response. Further, Colombetti emphasises the action-based reality of our emotional-cognitive lives. Emotions exert pressure on our behaviour all the time because they carry with them specific action tendencies.

Empathy provides a window into the emotional lives of others. According to enactivist Evan Thompson's *Mind in Life* (2007), empathy begins in the mirror neuron systems of our bodies. Networks of mirror neurons effectively 'mirror' intentional motor activity produced by another person and perceived by the empathiser. If a spectator watches an aerialist take a step on a high wire, for example, the same group of neurons is activated in the observer's brain as in the aerialist's; it is almost as if the observer had taken the step himself. Through 'mirroring' the action of another, an empathiser can attune him or herself to the emotions of the other person. As early as nine months old, an empathiser can begin to assume the perspective of another – to see the world through the other's eyes. Toddlers can take the next step, the practice of reiterated empathy, by which, as Thompson explains, 'I can empathetically imagine your empathetic experience of me and you can empathetically imagine my empathetic experience of you' (2007, 398). Thompson's final level of empathy, usually attained when a child is around six years old, is 'moral perception,' the capacity to experience 'other-regarding feelings of concern' (401). For Thompson, empathy gradually gains in complexity from affective attunement to the possibility of ethical response.

In addition to providing an important basis for moral development, empathy helps people to predict the emotions and beliefs of others. In his *Surfing Uncertainty: Prediction, Action, and the Embodied Mind* (2017), philosopher of cognitive science Andy Clark joins connectionist modelling to aspects of embodied and enactive cognitive science to advance a theory about predictive processing. Clark is interested in understanding how ordinary people think about what will likely happen to them and their friends and relatives in both the near and distant future. His model expands the complexity of the perception-action cycle, noted earlier, to include how memory and expectation join with immediate information from our senses to create recurring loops of processing that gradually refine our predictions. As Clark notes near the end of his book, 'The upshot is a hugely complex cognitive-emotional-action-oriented [psychological] economy whose fundamental guiding principles are simple and consistent: the multilevel, multi-area, flow of prediction, inflected at every stage by changing estimations of our own uncertainty' (2016, 237).

Distributed cognition

Also known as 'situated' or 'embedded,' and occasionally as 'extended' cognition, distributed cognition moves human cognitive operations beyond the brain and body and into the environment. In most uses, however, extended cognition differs from distributed cognition. Andy Clark and David Chalmers, a scientist who has investigated the puzzle of consciousness, pushed for an extended understanding of cognition in 1998 when they claimed that a note written on a piece of paper and used as an aid to memory ought to be considered a part of cognition just as surely as a brain inside of a body. In his article on 'Extended Cognition' (2014), however, cognitive scientist Ken Aizawa has questioned whether such aids to memory actually contain the properties or undergo the processes of cognition. Many advocates of embodied and enactive cognition have also found problems in the definitional logic of Clark and Chalmers' claim. The consensus among scientists following these approaches generally restricts cognition to living, autonomous animals. Nonetheless, the proponents of extended cognition, sometimes aligned with scientists who assert that computer-based robots can think, continue to press for an understanding of extended cognition that would include the possible cognitive properties of inanimate objects.

In contrast, advocates of **distributed cognition** are interested in how animate species use objects in their environment to assist them with tasks that require cognitive effort, but they do not attribute cognitive properties to the objects themselves. Clearly, we off-load cognitive tasks all the time to clocks, cars and computers; long before we had these tools, our ancestors used sticks and stones for hunting, fighting and other tasks that required careful planning and execution. In *Cognition in the Wild* (1995), cognitive scientist Edwin Hutchins initially coined the term **distributed cognition** to denote the many kinds of cognitive activities needed to sail an ocean-going ship. As Hutchins pointed out, the cognition necessary for this task was distributed among sailors with different kinds of know-how, involved tools ranging from charts and maps to heavy machinery and required a high level of training and social organisation. The same levels of cognitive networking and cooperation, of course, are necessary for many theatrical productions. In short, even more so than ants, buffalo, chimpanzees and other social animals, the ability to build complex networks of distributed cognition is a special talent of our own ultra-social species. Hutchins used the term 'cognitive ecology' to describe this approach (Hutchins 2010a, 1) because it emphasises the interplay that occurs between animals with cognition and their social and physical environments.

The American perceptual psychologist James J. Gibson (1904–1979) pioneered an ecological approach to cognition. He used the term 'affordance' to describe the spatial 'fit' between animals, including humans, and specific objects and areas in their biological niches. Certain kinds of caves 'afford' bears with an opportunity for hibernation, for example, just as the height and dimensions of specific rocks and downed tree trunks afford humans with a seat for sitting. The material features of all environments, in other words, both enable certain kinds of actions and constrain others. Gibson's ecological sensibility has been an important asset to theatre and stage designers, who must take account of the potential affordances of their spaces for the movements of spectators and actors.

Hutchins extends Gibson's ecological sense of relational affordances into the arenas of social and technological networking. Applying Gibson's notion to Hutchins' case study of navigation, one could say that the social and educational training of the ship's crew, together with all the technologies that enable navigation, 'afford' the captain of the ship the ability to navigate successfully in the midst of nearly all contingencies. Hutchins emphasises that the proper unit for the analysis of cognitive systems should not stop with individual brains and bodies. In some systems, he states, 'high-level cognitive functions such as memory, planning, decision-making, reasoning, error detection and correction, computation, learning, and so on can be identified and analyzed in the culturally organized activities of groups of people in interaction with one another and with technology' (2010b, 426). Not surprisingly, Hutchins' analysis of distributed cognition networks has influenced scholars of systems' management operations and historians interested in understanding how past groups of artisans or professionals were able to work together effectively on massive projects that required extensive networking and coordination.

References

Aizawa, Ken. 2014. "Extended Cognition." In *The Routledge Handbook of Embodied Cognition*, edited by Lawrence Shapiro, 31–8. London and New York: Routledge.

Barsalou, Lawrence W. 2009. Chapter 14: "Situating Concepts." In *The Cambridge Handbook of Situated Cognition*, edited by Philip Robbins and Murat Aydede, 236–63. Cambridge: Cambridge University Press.

Clark, Andy. 2016. *Surfing Uncertainty: Prediction, Action, and the Embodied Mind*. Oxford: Oxford University Press.

Colombetti, Giovanna. 2014. *The Feeling Body: Affective Science Meets the Enactive Mind*. Cambridge, MA: MIT Press.

DiPaolo, Ezequiel, Marieke Rohde, and Hanne DeJaegher. 2010. "Horzions for the Enactive Mind: Values, Social Interaction, and Play." In *Enaction: Toward a New Paradigm for Cognitive Science*, edited by John Stewart, Olivier Gapenne, and Ezequiel A. DiPaolo, 33–88. Cambridge, MA: MIT Press.

Freeman, Walter J. 2000. *How Brains Make Up Their Minds*. New York: Columbia University Press.

Gallagher, Shaun. 2005. *How the Body Shapes the Mind*. Oxford: Oxford University Press.

Hutchins, Edwin. 1995. *Cognition in the Wild*. Cambridge, MA: MIT Press.

Hutchins, Edwin 2010a. "Cognitive Ecology." *Topics in Cognitive Science* (1): 1–11.

Hutchins, Edwin 2010b. "Enaction, Imagination, and Insight." In *Enaction: Toward a New Paradigm for Cognitive Science*, edited by John Stewart, Olivier Gapenne, and Ezequiel A. DiPaolo, 425–50. Cambridge, MA: MIT Press.

Hutto, Daniel D. 2008. *Folk Psychological Narratives: The Sociocultural Basis of Understanding Reasons*. Cambridge, MA: MIT Press.

Lakoff, George and Mark Johnson. 1980. *Metaphors We Live By*. Chicago, IL and London: University of Chicago Press.

McConachie, Bruce. 2015. *Evolution, Cognition, and Performance*. Cambridge: Cambridge University Press.

Stewart, John. 2010. "Foundational Issues in Enaction as a Paradigm for Cognitive Science: From the Origins of Life to Consciousness and Writing." In *Enaction: Toward a New Paradigm for Cognitive Science*, edited by John Stewart, Olivier Gapenne, and Ezequiel A. DiPaolo, 1–32. Cambridge, MA: MIT Press.

Thompson, Evan. 2007. *Mind in Life: Biology, Phenomenology, and the Sciences of Mind*. Cambridge, MA: Harvard University Press.

Varela, Francisco, Evan Thompson, and Eleanor Rosch. 1991. *The Embodied Mind: Cognitive Science and Human Experience*. Cambridge, MA: MIT Press.

PART I: ARTISTRY

Introduction

Rick Kemp

This part considers how insights from **cognitive neuroscience** can inform the practice of various genres of performance. Information from this field has informed the work of both practitioners and scholars of theatre and performance since the early 2000s. Neuroscience investigates the organisation and functions of the nervous system, seeking to better understand the brain's relationship to behaviour and cognitive functions. These include, for example, perceiving, thinking, imagining, remembering, speaking, gesturing, planning and doing. Evidently, all of these activities occur in the creation and performance of theatre, and in its reception by audiences. However, much of this cognitive activity happens below the level of conscious awareness, so it is difficult to analyse. The discoveries of neuroscience allow theatre and performance practitioners to better understand both their subject matter – human experience and behaviour – and the often intuitive processes involved in the preparation and presentation of performance in multiple genres. Researchers in cognitive neuroscience combine data from experimental studies with discipline-specific knowledge to build theories that are consistent with empirical evidence of the brain's functions. The ensuing body of knowledge has led to significant changes in the ways in which we understand phenomena such as perception, language and gesture, emotion, imagination, memory, empathy – even the idea of self. These changes challenge many of the traditional concepts of the mind, brain and self that underlie the practices, training and scholarship of Western theatre. Consequently, an increasing number of practitioners and scholars have been inspired to apply this new knowledge to their art form. There are, however, challenges involved in doing this, as cognitive neuroscience is a broad multi-disciplinary field with many competing theories. Theatre and performance practitioners and scholars have tended to focus on the '**embodied cognition**' view of the brain's functions, which proposes that what people perceive, how they conceive and what they do all develop interactively and are tied to their environment. Even within this broad understanding, there are variances of opinion that will appear in the chapters of this part. We hope that these variances, rather than appearing contradictory, will inspire readers of this book to go deeper into the field to determine which perspective accords with their own experience.

Given the widespread influence of Stanislavsky on Western theatre in the twentieth century, we begin this part with a cognitive perspective on his approach by Sharon Carnicke, author of the highly influential *Stanislavsky in Focus* (1998, 2nd ed. 2009). Carnicke has argued that Stanislavsky's techniques can continue to be productive in the twenty-first century and has applied Active Analysis (AA, his last rehearsal technique) to both new media and non-Aristotelian plays, as well as utilizing it in a scientific study on the emotional expressivity of physical gesture. In these activities, she continues the work that she began in *Stanislavsky in Focus*, of freeing Stanislavsky's ideas from misinterpretation. One of the most significant misrepresentations has been the associating of his work with behaviourism – a process that originated with the Soviet insistence on materialism. Behaviourism held that there was a simple correlation between stimulus and behaviour, confining itself to the study of observable behaviour. It was within this psychological orientation that Stanislavsky's System was adopted by Lee Strasberg in the USA. Carnicke connects Stanislavsky's work with neuroscientist Antonio Damasio's research on human consciousness and demonstrates how AA can be used experimentally in conjunction with **Theory of Mind** to generate new scientific knowledge.

This chapter is followed by one on improvisation by psychologist and theatre practitioner Gunter Lösel that considers many of the different manifestations of this activity through the lens of cognitive science. The juxtaposition of these two chapters addresses the traditional idea of Western theatre history that there are two basic categories of theatrical practice – scripted and improvised drama. Cognitive science offers a more sophisticated understanding of meaning and interpretation in theatre through identifying varying parameters within which creators of performance can make expressive choices. Lösel addresses the fact that while the skills of acting and improvisers overlap, there seem to be specific abilities and cognitive functions that are necessary for improvisational theatre. As he points out, improvisers have a greater variety of options than actors, since they are creating content as they perform it. Lösel examines how it is that improvisers can make choices without hesitation or apparent 'thinking.' He describes three approaches to identifying the skills of improvisation by considering, firstly, the vocabulary of improvising practice, secondly, the idea that improvisation is a form of problem solving and, thirdly, the application of theories of embodiment and social cognition.

This perspective involves the model of **dynamic systems** theory (addressed more fully in Part III), which also informs my chapter on devising. A dynamic system is a model of interaction that describes the flow of relationships among the components of a whole phenomenon, acknowledging the real-time 'circular causality' of elements within a system. As such, it is a useful way to describe the ways in which people, environment and actions simultaneously affect one another in the context of devising a dramatic performance, as well as presenting that performance. While devised theatre has become a familiar feature of Western performance, little has been written about what connects its many disparate expressions. In this chapter I use theories from embodied cognition to identify underlying principles that may be applied across varied devising processes. In doing so, I hope to provide practitioners, researchers and students with some useful concepts that can frame the processes of devising. Among these is the fact that devising is inherently creative, and I suggest that its multi-modality is a significant cause of this by describing how embodied multi-modality leads to both metaphorical thinking and ideational combination.

Devising is a form that uses improvisation but arrives at a fixed 'performance score,' thus disrupting the traditional binary of scripted and improvisatory theatre that I mentioned earlier. This disruption is furthered by the examination of some historical forms of theatre. In the next two chapters, Darren Tunstall and Peter Meineck trace ideas of embodied cognition through performance of Shakespeare and Greek theatre, respectively. The recognition that Shakespeare's actors rehearsed with cue scripts has been a significant factor in understanding the processes of performance in his time. A cue-script shows only one's own speeches and the lines that cue them. Reading these without the benefit of seeing the whole play written out meant that actors had to be singularly alert for their cues, resulting in a degree of spontaneity in their speech and action. In contemporary rehearsal processes, the cue-script approach is employed by only a few companies, with a variety of methods used by others. Tunstall reviews some different strategies for rehearsing Shakespearean text modelled upon some influential theatre practitioners, and investigates how an understanding of embodied cognition might enhance these processes. Peter Meineck was the founding director of Aquila Theatre and also holds the endowed chair of Professor of Classics in the Modern World at New York University. He thus combines the understanding of a practitioner with his scholarly research. Meineck points out that the study of Greek drama has tended to be either textual or archaeological, with little attention paid to the embodied experience of performance and reception. In his chapter, he explores how the application of cognitive theory to ancient drama can help us better understand how these plays functioned in antiquity. His approach applies research from the affective sciences, cognitive theory and social psychology to examine the extra-textual elements of the performance. Here he focuses on environment, masks and movement, and identifies cognitive processes that support historical evidence that suggests that the experience of watching a play in the fifth century BCE was highly emotional. His analysis offers valuable information that can inform the ways in which these ancient plays are understood and presented by modern practitioners.

The next two chapters also demonstrate how cognitive science can influence contemporary practice in different genres – one improvisatory and one scripted. Pil Hansen explores interdisciplinary studies of cognitive science and dance improvisation, while Jeanne Klein applies basic cognitive-affective principles of child development to the field of Theatre for Young Audiences (TYA). Hansen explains that in dance for the stage, improvisation can be used to create new material that is then 'set' in choreography or incorporated into a system or score in performance. She goes on to examine the concept of presence and how this is challenged by some contemporary improvisers. Employing the concept of '**constraints**' from dynamic systems theory, she discusses how the incorporation of explicit constraints may give dancers the ability to actively stimulate levels of memory and planning that are otherwise implicit. The contrast of explicit and implicit cognitive processes is also a feature of Jeanne Klein's chapter. As a TYA practitioner, she has long argued that productions need to take account of children's developmental cognitive-affective perspectives in order to optimise child spectators' aesthetic experiences. In her view, optimal experiences occur when audience members recognise, articulate and apply artists' intended meanings to themselves and society. This is more likely to occur when practitioners appropriately match their material to one of the four age-related epistemological orientations in their audiences.

Rhonda Blair was one of the first scholar/practitioners to apply cognitive science to acting practice in *The Actor, Image, and Action: Acting and Cognitive Neuroscience* (2008). In her chapter for this volume, she describes how she has applied theories of embodied cognition to her directing. Using case studies of two productions, she gives examples of how an understanding of cognitive principles can enhance the interaction of text, research and embodiment. She proposes that this can help us better understand both theatrical practice and aspects of human cognition, since rehearsal processes and performances provide discrete models of cognitive ecologies. The term **cognitive ecology** describes mutually interdependent elements in a cognitive system, and represents a shift in cognitive theory from defining units of analysis by their inherent properties to units defined in terms of dynamic patterns of correlation across elements. Blair uses her examples of preparing and presenting Thornton Wilder's *Our Town*, and Caryl Churchill's *Light Shining in Buckinghamshire* as demonstrations of the ways in which both rehearsal and performance can be recognized as cognitive systems. Thus, theatrical activity provides a frame that integrates text, individuals and material items in a coalescence of meanings.

In the final chapter in this part, Vladimir Mirodan addresses the following common question: 'are stage emotions actually experienced or are actors only giving the appearance of emotion-driven behaviour?' There is no simple answer, since cognitive neuroscience offers competing explanations of emotion. Mirodan first outlines the principal theories regarding the arousal, perception and transmission of emotion, and then looks at ways in which certain widely used acting approaches relate to them. The so-called 'read-out' theories of emotion, supported by neuroscientists such as Antonio Damasio and Joseph LeDoux, suggest that a stimulus provokes physiological reactions that, when registered in consciousness, are recognised as emotions. Neuroscientist Jaak Panksepp proposes an alternative explanation, that there is state of 'affective consciousness' that is unreflective and unthinking. In other words, we do not need to consciously 'know' what we are feeling in order to feel. Cognitive scientist and philosopher Giovanna Colombetti suggests that scientists have neglected the actual subjective (or phenomenological) feeling of emotion in their studies – that in order to fully understand emotion we need to know not only *what* is experienced, but *how* it is experienced. She also states that emotion is an intrinsic part of cognition – that one cannot separate the two. Mirodan adroitly describes these positions before going on to examine how cognitive explanations of emotion – while still partial and uncertain – might begin to influence acting practices.

The rich variety of the chapters in this part indicates how lively and exploratory the field of cognitive theatre and performance studies has become in its application of scientific theories to the practices of performance. While there are varied opinions within the field, there is a widespread recognition that the Cartesian separation of mind and body that was implicit in earlier explanations of practice is no longer valid or helpful. Holistic and situated descriptions of cognition provide a fertile ground for the linking of performance theory and practice that looks likely to continue to grow.

1

STANISLAVSKY'S PRESCIENCE

The conscious self in the system and Active Analysis as a theory of mind

Sharon Marie Carnicke

This chapter interrogates the continuing value of Konstantin Stanislavsky's legacy to acting in the wake of twenty-first-century discoveries in cognitive science, specifically Theory of Mind. During the first three decades of the twentieth century, Stanislavsky conducted a broad search through many fields of knowledge, including the science and psychologies of his era, for ways to verbalize his tacit experience with acting. His effort was deeply experimental, resulting in a globally influential acting System that has been successively transmitted and transformed by actors from generation to generation. While his System in its many, often distorted, forms continues to hold exceptional authority in actor training programs and rehearsal halls worldwide, that authority is also being challenged by rapid technological and dramaturgical changes that impact actors' work. Stanislavsky himself acknowledged the inevitability of change, when in the opening pages of his autobiography he recalls everyday items from his nineteenth-century Russian childhood that his students no longer knew—candles made from lard, the horse-drawn *tarantas*, flint-lock muskets, and the like (Stanislavskii 1988–1999, vol. 1: 53).[1] By extension, neither he nor his students could have imagined the world in which twenty-first-century actors live.

As I have argued elsewhere, Stanislavsky's pragmatic techniques can constitute a living legacy as long as they can accommodate production and dramaturgical conditions that post-date his life. To test such adaptability, I have experimented with applying his last rehearsal technique, Active Analysis (AA), to new media and non-Aristotelian plays. For example, I have used AA at the University of Southern California to generate motion capture data for a scientific study on the emotional expressivity of physical gesture and in my private Studio to rehearse post-dramatic plays that do not require actors to impersonate fictional characters (Carnicke 2016b). In these cases, I have found that AA provides actors with exactly the kind of flexibility they need for such twenty-first-century work.

Discoveries in cognitive science that outdate the psychologies of Stanislavsky's era can also be counted among the current challenges to his authority. In this chapter, I join other performance scholars and actors/directors, including Rhonda Blair (2008), John Lutterbie (2011) and Rick Kemp (2012), in arguing that Stanislavsky's System can not only withstand this challenge, but also be newly validated by twenty-first-century scientists. Actor/director and teacher Joelle Ré Arp-Dunham succinctly expresses the pragmatic importance of our argument, when she observes that, 'looking through the lens of recent research in the field of cognitive science may help us fine-tune which of Stanislavski's[2] ideas and techniques should be emphasized in the classroom and in the rehearsal hall' (2017: 68).

A full examination of the ways in which the dynamic field of cognitive science might reinvigorate Stanislavsky's complex theories on acting is far beyond the scope of this little chapter. Therefore, I confine myself to the following: (1) a brief survey of the scholarship on Stanislavsky's science, (2) how Antonio Damasio's research on human consciousness suggests the value of revisiting the System through the lens of neuroscience and (3) how AA can be used experimentally in conjunction with Theory of Mind to generate new scientific knowledge.

Studies of Stanislavsky's science

A brief survey of how Stanislavsky's science has been studied reveals that a new scholarly methodology is emerging in the wake of twenty-first-century discoveries in cognitive science. At base, Stanislavsky's impulse to understand acting through science was not unique or particularly original. In *The Player's Passion*, performance scholar Joseph Roach demonstrates that 'conceptions of the human body drawn from physiology and psychology have dominated theories of acting from antiquity to the present' (1985: 11). Furthermore, by detailing the specific ways that actors have mirrored science's changing views of human nature over the centuries, Roach observes that each new paradigm shift in science has generated new acting theories and practices.

Consequently, studies on the sources of science in the System have also dominated scholarship on Stanislavsky's scientific thinking. Natalie Crohn Schmitt identifies his 'science of acting [as] based on a perception of nature very like Aristotle's' that assumes 'the principles underlying all art, nature and life are the same' (1990: 94–5). Despite having no access to the Russian texts that prove Stanislavsky's familiarity with Aristotle, Crohn Schmitt accurately connects the two. Moreover, she concludes that the System's 'success' largely rests upon his incorporation of widely accepted 'Aristotelian precepts' (1990: 94).

Jonathan Pitches understands the System as 'fundamentally Newtonian' with 'a view of the world which is objectively measurable and material' (1999: 4). Pitches further argues that this 'mechanistic thinking [...] prov[ed] particularly popular in post-revolutionary Russia' where Soviet behaviourists, like Ivan Pavlov and Ivan Sechenov, stressed bodily, hence material, aspects of psychology (1999: 15).

In this comment, Pitches points to a particularly sad chapter in Stanislavsky studies, when Soviet cultural policies and Stalinist censorship banned all but the materialist approach to psychology. Within this cultural and political context, Stanislavsky had no choice but to bring his System into conformance with Soviet behaviourism. Yet, the record shows scant evidence that Stanislavsky did little more than pay lip-service to it (Carnicke 2009: 162–3). Nonetheless, by 1962, Soviet scientist Pavel Simonov had published *The Stanislavsky Method and the Physiology of Emotion* positioning the work of Pavlov and Sechenov at the System's heart. By 1963, the Russian actor Aleksei Popov had fully accepted the Soviet view of Stanislavsky as 'a natural materialist [who] walked hand in hand with the great physiologist I.P. Pavlov' (Vinogradskaia 2003, vol. 4: 465). The overstated link between Stanislavsky's System and behaviourism proved so persuasive that American Method actors also accepted the Soviet scientific heritage as primary in the System. Emphasizing the role of self-conditioning in actor training, Method guru Lee Strasberg said, 'That's how we're trained, not from Freud, but from Pavlov' (Munk 1966: 198). Some twenty-first-century scholars continue to find value in this Soviet view when aligning the System with the biological underpinnings of neuroscience. Ysabel Clare notes the irony in the fact that 'even the "mystical" aspects of the system' might be conceived as 'material,' especially 'if recent findings in the field of neuroscience resituate human experience in what turns out to be materialist terms' (2016: 101).

Stanislavsky actually preferred the writings of Théodule Armand Ribot to those of the Soviet behaviourists. Founder of French experimental psychology, Ribot, studied emotion from a more holistic perspective than that of the Soviets. Thus, Ribot better supports Stanislavsky's simultaneous interests in yoga and art's ability to convey transcendent aspects of human experience (Carnicke 2009: 170–84). Ribot understood emotion as a monistic phenomenon that embraces a total psychophysical experience. He therefore treats objective and subjective human experience as two sides of the same coin. 'An emotion which does not vibrate through the whole body is nothing but a purely intellectual state' (1897: 163). For Ribot, the tendency in Western civilisation to presume a separation between body and mind was entirely beside the point. In fact, he criticised the French language for forcing him to make 'an arbitrary distinction' between the internal (or 'organic') and external (or 'motor') functions of emotion merely 'for the sake of clearness in exposition' (1897: 113). In writing about emotion in acting, Stanislavsky emulates Ribot's monism by choosing to use the Russian word, *chuvstva*, which simultaneously refers to the physical senses and psychological feelings. 'Once you can grow pale or blush at the memory of something you have experienced,' Stanislavsky writes, 'once you are frightened to think about something unhappy that you lived through long ago, you have a memory for *chuvstva* (senses, feelings)' (Stanislavskii 1988–1999, vol. 2: 281).[3]

Rose Whyman meticulously places the System into the science of Stanislavsky's era by examining the Russian language texts and translations to which he had direct access. Her studies on 'Darwinian ideas of the physiological basis of emotion' and Russian associationist philosophy with its 'pre-Freudian concepts of the unconscious' and its methodology 'based on introspection and observation' are especially enlightening (2008: 8, 4). While Whyman sees these threads as anticipating behaviourism, she also takes account of the constraints that Soviet censorship places upon a full study of Stanislavsky's science by acknowledging that Soviet views of 'what was "scientific" had become more circumscribed' than his private thinking. She interprets his now famous disavowal of science in the prologue to his 1936 acting manual (Stanislavskii 1988–1999, vol. 2: 41–2) as his way of sidestepping Soviet censorship in order to study human nature as broadly as possible (Whyman 2008: 8).

These studies of Stanislavsky's science are invaluable in understanding the evolution of his System. But, studies of his scientific sources alone are insufficient to assure contemporary actors of the System's continuing relevance. In accord with Roach's thesis, twenty-first-century discoveries in cognitive science are prompting actors to rethink their acting theories. Nonetheless, in his detailed study of the System, Roach also holds the door open for the possibility that Stanislavsky's theories may be more flexible than the limits of his era's science. Roach explains that Stanislavsky's 'relationship to science evolved over several decades,' continually gaining 'depth and complexity,' and always grounded by an 'intense, even obsessive interest in the inner psychological context of an action, [and] in the nature of consciousness itself.' Thus, his System 'def[ies] tidy summary, because [his theories] take into account the complexity of higher organisms, including the phenomenon of double or multiple consciousness' (1985: 204–6).

Many scholars have already walked through this door, exchanging the traditional methodology of studying Stanislavsky's scientific sources for an inverted method that evaluates his insights by weighing them against cutting-edge developments in neuroscience. Among these scholars are the editors of and contributors to this anthology, as well as the actor/

director Rhonda Blair, who observed in 2000 that 'cognitive scientists, neurophysiologists and psychologists are proving that Stanislavsky, seventy-five years ago, began intuiting something fundamental about how we, as human beings and as actors, work' (2000: 204). Blair's work implicitly proves Roach's thesis—that acting responds to paradigm shifts in science—when she describes her overall aim as bringing acting 'into the next generation' through 'a transformed way of thinking about imagination and action, based in knowledge of the neurocognitive ground of memory, feeling, and imagery' (2008: 52).

This inverted method of studying, not Stanislavsky's understanding of science but the science in his System, depends, however, on a highly volatile field where much new information is often as provisional as it is tantalising. As Blair acknowledges, 'scientists disagree about their work at least as intensely as we do about ours' (2008: 6). All the same, measuring the System by means of new science is 'exciting,' to quote Ré Arp-Dunham, because it promises to provide 'strong scientific validation' in future for Stanislavsky's acting theories and techniques (2017: 68).

Damasio and Stanislavsky on the conscious self

Antonio Damasio's research on human consciousness can test the efficacy of studying Stanislavsky's theories through the lens of cognitive science. Born in Lisbon, Damasio is a neurobiologist who heads the Brain and Creativity Institute at the University of Southern California. He studies the biological processes of the brain in close collaboration with his wife, Hanna Damasio, an expert in neuroimaging. Their findings are changing how scientists envision the interdependence of reason and emotion and the mysteries of consciousness, subjects which Stanislavsky saw as directly pertinent to actor training. Moreover, Damasio is also changing cultural discourse though his vivid and accessible writings. His 1994 book, *Descartes' Error,* on the importance of emotion to social cognition and decision-making, was translated into 30 languages and became a finalist for the *Los Angeles Times* book award. His 1999 book, *The Feeling of What Happens: Body and Emotion in the Making of Consciousness*, found an even broader readership, winning countless book awards in the United States and globally. In short, at the end of the twentieth century, Damasio offered one new way to envision emotion that, together with the work of other cognitive scientists, is prompting a paradigm shift that directly affects how we understand ourselves as humans and how actors understand acting.

For the purposes of this chapter, Damasio's potential to impact acting theories is most relevant. He stands among the neuroscientists who are oft-cited by performance scholars. Indeed, at times, Damasio seems to speak directly to acting. Consider the opening of *The Feeling of What Happens*:

> I have always been intrigued by the specific moment when, as we sit waiting in the audience, the door to the stage opens and a performer steps into the light; or, to take the other perspective, the moment when a performer who waits in semidarkness sees the same door open, revealing the lights, the stage, and the audience. [...] I sense that stepping into the light is also a powerful metaphor for consciousness, for the birth of the knowing mind, for the simple and yet momentous coming of the sense of self into the world of the mental.
>
> *(1999: 3)*

Given Stanislavsky's obsessive quest to understand what happens in actors' minds during performance, Damasio's theatrical metaphor is especially apt. However, his reference to 'sense of self' is even more striking in regard to the System. Damasio explains that in studying consciousness, he 'had come up against the obstacle of self, for something like a sense of self was needed to make the signals that constitute the feeling of emotion known to the organism having the emotion' (1999: 8). Stanislavsky too insists that the actor needs 'a sense of self' (*samochuvstvie*) in order to create as an actor. 'You can never lose yourself on stage.[...] There's no walking away from yourself' (Stanislavskii 1988–1999, vol. 2: 294).

'Sense of self' marks only the first in a series of insights that Stanislavsky and Damasio share. 'Images' that create a 'movie-in-the-brain,' the inner/outer split in consciousness and the mind's multi-layered operations are three additional points of comparison. I will take each of these points in turn.

First, a cinematic metaphor helps both men describe mental processes. Damasio represents consciousness as 'images' in the mind that play out as 'a movie-in-the brain' (1999: 9). Stanislavsky exactly anticipates Damasio's formulation when he teaches the actor to create during performance a 'filmstrip' (*kinolenta*) of 'images' (*videniia*) that mirror the character's thoughts in order to focus the actor's mind (Stanislavskii 1988–1999, vol. 2: 130).[4] He trains the actor's ability to create this imaginary 'filmstrip' by adapting from yoga a visualisation exercise that works systematically through the senses. The best-known instance of this exercise can be found in the first volume of Stanislavsky's acting manual, when the fictional teacher, Tortsov, invites his students to imagine themselves as trees. They begin by envisioning the specific species, the shape and colour of the leaves, the texture of the bark and so forth. They then progressively add more imaginary senses: the tactile feel of the roots buried deep into the earth and the branches reaching towards the sky; the sap's flavour; the sounds and smells of the specific place in which the tree grows. Finally, the students add a sixth sense, emotion, by imagining a story that unfolds at the base of the tree: a battle ensues; a romantic tryst takes place; a family enjoys a picnic (Stanislavskii 1988–1999, vol. 2: 133–6). Significantly, the sensory aspect of Stanislavsky's exercise mirrors Damasio's 'movie-in-the-brain' which enters the mind through 'as many sensory tracks as our nervous system has sensory portals—sight, sound, taste, and olfaction, touch, inner senses, and so on' (Damasio 1999: 9).

Of course, Stanislavsky's assumption that emotion operates as a sixth sense stands outside Damasio's science. Yet, Damasio would not mock Stanislavsky for this or any other of his unscientific insights. Consider, for example, the fact that Stanislavsky defines the actor's creative state as 'that which occurs in the artist's soul (*dusha*) during the creating and preparing' of a role (Stanislavskii 1988–99, vol. 2: 410).[5] In *Descartes' Error,* Damasio explicitly honours the notion of 'soul':

> To discover that a particular feeling depends on activity in a number of specific brain systems interacting with a number of body organs does not diminish the state of that feeling as a human phenomenon. [...] Precisely the opposite should be true: Our sense of wonder should increase before the intricate mechanisms that make such magic possible. Feelings form the base for what humans have described for millennia as the human soul or spirit.
>
> *(1994: xx)*

Second, the split between inner consciousness and the external world of things is fundamental to the study of both neurobiology and acting. For both men, the subjective mind's access to the objective world depends upon the creation of mental images of 'objects.' For Stanislavsky, 'objects of attention' name anything that demands the actor's focus during performance, whether partner or prop (Stanislavskii 1988–1999, vol. 2: 149). Damasio also uses the term 'object' to name different kinds of things. In asking 'how the brain inside the human organism engenders the mental patterns we call, for lack of a better term, the images of an object,' Damasio explains that 'by *object* I mean entities as diverse as a person, a play, a melody, a toothache, a state of bliss' (1999: 9). Even more startling is how both men posit the relationship between inner perception and outer objects as provoking action. Stanislavsky teaches that action defines drama, often citing the word's Greek etymology, to prove his point (Vinogradskaia 2000: 496). Moreover, in AA, he emphasises that drama results from the interactions between the self in performance and the objects of attention on stage. Damasio seems to restate Stanislavsky's basic premise—that action undergirds drama—in the following passage:

> I began seeing the problem of consciousness in terms of two key players, the *organism* and the *object*, and in terms of the *relationships* those players hold in the course of their natural interactions. [...] Seen in this perspective, consciousness consists of constructing knowledge about two facts: that the organism is involved in relating to some object, and that the object in the relation causes a change in the organism.[6]
>
> *(1999: 19–20)*

Third, both men understand mind as multi-levelled and complex. Stanislavsky divides the mind into the 'conscious' and the 'unconscious,' which in turn comprise a 'subconscious,' which works automatically underneath our awareness, and a 'superconscious,' a term from yoga that refers to transcendent, spiritual aspects of human experience (Stanislavskii 1988–1999, vol. 4: 142). Damasio explains that 'consciousness is not a monolith' and distinguishes 'core consciousness' that deals automatically with 'the here and now' from 'extended consciousness' that 'provides the organism with an elaborate sense of self [...] and places that person at a point in individual historical time, richly aware of the lived past and of the anticipated future, and keenly cognizant of the world beside it' (1999: 16). Viewed through Damasio's terminology, Stanislavsky's 'subconscious' would be akin to 'core' and his 'superconscious' akin to 'extended consciousness.' Moreover, when Damasio then identifies within 'extended consciousness' a 'supersense'—his term for the awareness of being aware—he even more directly brings to mind Stanislavsky's notion of the 'superconscious.' Just as Damasio sees the 'supersense' as that which 'eventually brings a full construction of being into the light [and...] permit[s] human creativity' (1999: 17), Stanislavsky understands the 'superconscious' as that which 'most of all elevates a person's soul, and thus most of all must be valued and preserved in our art' (Stanislavskii 1988–1999, vol. 4: 140).

One must certainly acknowledge that Damasio and Stanislavsky part company in their overall goals. As a scientist, Damasio seeks to 'discover[...] the biological underpinnings for the curious ability we humans have of constructing, not just the mental patterns of an object, [...] but also the mental patterns which convey, automatically and naturally, the sense of self in the act of knowing' (1999: 11). As an artist, Stanislavsky seeks knowledge for the purpose of practicing his art. Damasio recognises this difference of goal when he writes:

> We [humans] can control, in part, the expression of some emotions—suppress our anger, mask our sadness—but most of us are not very good at it and that is one reason why we pay a lot to see good actors who are skilled at controlling the expression of their emotions.
>
> *(1999: 48)*

This look at Stanislavsky through the prism of Damasio's neuroscience, however cursory, suggests the value of measuring the System against twenty-first-century paradigms. Additionally, it suggests that Ribot's research and yoga may have fostered Stanislavsky's prescience more effectively than links in his System to behaviourism, because Ribot and yogis share a more insistently holistic conception of the mind, which better aligns with the complexities in twenty-first-century studies on human consciousness, like Damasio's.

AA as a theory of mind

Another way to test Stanislavsky's theories for the twenty-first century is to use them in experimental projects that seek to further scientific knowledge. As a theatre scholar and practitioner, known for my research on acting theories and Stanislavsky's System, I have had opportunities to consult on a number of such projects, including an investigation on how robots and intelligent virtual agents can interact with humans more credibly (Carnicke 2005), and the extended study, referenced earlier, on the bodily expression of emotion through motion capture (Carnicke 2012).

To conclude this chapter, I report on the scientific project in which I am engaged, while writing. The scientific team is based at Northeastern University and headed by Stacy Marsella, a widely published professor of computer science and psychology, and Magy Seif El-Nasr, associate professor of computer science with special expertise in arts, media and design. We are joined by two Ph.D. candidates, Dan Feng and Elin Carstendottir. Our project, which is funded by the US National Science Foundation, links AA to interactive storytelling in digital media.

The field of interactive storytelling investigates how digital games and artificial intelligence can be utilised, not only for entertainment, but also to address a wide variety of social and psychological needs. During the 2010s, a great number of interactive training programs were developed to help people cope with all sorts of social interactions, ranging from communication between doctors and patients to bullying. One particularly strong area of research has been the use of virtual reality to treat veterans with post-traumatic stress disorder.[7] However, the interactive narratives created for such training systems are often hampered by too few alternative choices for the humans who use the programmes and by overly simple stories.

Our team seeks to overcome both these difficulties. First, 'our approach begins with a paradigm shift that re-conceptualizes social skills simulation as rehearsing and improvising roles instead of performing a role' (Feng et al. 2016: 157). More specifically, we use the rehearsal techniques of AA to model how users can replay narratives in order to generate variations on the story. Second, relying upon 'the human skill to have and use beliefs about the mental processes and states of others, commonly called Theory of Mind (ToM),' we adapt the theoretical principles of AA to digital simulations that 'embed ToM training in the [interactive] experience to support better learning outcomes' (Feng et al. 2016: 157). This adaptation further aims at complicating and enriching the narrative content for interactive training programmes.

As this overview makes clear, our project brings Stanislavsky's AA into collaboration with scientific studies on Theory of Mind. Therefore, the histories of AA and ToM provide fuller contexts within which to understand our team's efforts.

Stanislavsky developed AA in his last Studio, the Opera-Dramatic (1934–1938), while he was under virtual house arrest by Stalin for artistic ideas that violated Soviet policies on Socialist Realism. Stanislavsky worked out the principles of AA by rehearsing plays in a variety of performance styles and genres with a select group of actors behind closed doors. Given the censorship of the times, he never wrote publicly about his new rehearsal approach. Moreover, the Soviet press and Soviet scholars promoted a politically cleansed version of his experiments, called the Method of Physical Actions, in order to bring his late experimentation into alignment with Marxist materialism and realism (Carnicke 2010b, 2009: 183–94). With the Soviet 'thaw' in the arts during the 1960s, Stanislavsky's most trusted actors began to speak and write openly about his hidden experiments. His chief proponent was Maria Knebel, who named his late work AA in order to distinguish it from the Soviet Method of Physical Actions. She promoted AA until her death in 1985 through her influential directing, teaching and prolific writings (Carnicke 2010a: 99–116). While widely used in contemporary Russia, AA is still largely misunderstood elsewhere (Carnicke 2016a).

ToM began as an issue in philosophy, but moved into psychology in 1944 when social psychologists Fritz Heider and Marianne Simmel showed animated cartoons of geometric shapes to observers, who readily attributed emotional states and purposeful intentions to the moving shapes. Heider's seminal book, *The Psychology of Interpersonal Relations* (1958), moved the common-sense psychology of ToM into the realm of sophisticated, experimental science. Over the decades, ToM has generated many different mental models that now also impact studies in neuroscience and artificial intelligence. As Damasio explains, 'the traditional worlds of philosophy and psychology have gradually joined forces with the world of biology and created an odd but productive alliance' (1999: 13). From the 1980s onwards, Robert Gordon's and Jane Heal's model for ToM, called Simulation Theory, has become widely accepted among cognitive scientists. This model posits a process by which an observer puts himself or herself imaginatively in the shoes of the other in order to understand how that person might be thinking and planning (see Gordon 1995). Bruce McConachie, co-editor of this volume, has observed that such simulation is 'very close to the notion of empathy put forward by Konstantin Stanislavsky,' who likewise asks actors to imagine themselves in the given circumstances of their characters (McConachie 2006: 55).

Our scientific project uses both the rehearsal practices and theoretical principles of AA to design an interactive training system that teaches ToM skills through interactive stories. In terms of rehearsal practices, AA replaces Stanislavsky's earlier process of analysing plays through lengthy discussions with actors around a table with AA's experimental process of embodied analysis. AA operates through a series of guided improvisations that require actors to explore the interactive possibilities in a dramatic event by working them out on their feet. Actors read a scene in order to make a hypothesis about the interpersonal dynamics implied by the text. They then test their hypothesis through an improvised performance (called an 'etude' from the French word for study). Etudes can proceed either silently through physical behaviour alone, or verbally, with actors using their own words to explore the scene. Following each etude, the actors assess how closely their improvisation came to embodying the text as written. This evaluation leads to adjustments in the group's hypothesis, then to the next etude, then to the next evaluation—a three-step process that is repeated until the full complexity of the scene emerges. In short, the actors actively and collaboratively interrogate the full range of possibilities in a scene by repeatedly replaying it through improvisatory etudes. This iterative process provides a productive model for the replay with variations that our team believes will create richer narratives than the field of digital storytelling generally supports.

In terms of theory, AA provides a clearly articulated structure, which actors can use reliably to attribute mental processes and inner states to their acting partners. This theoretical structure conceives of performance as a chain of events, with each event resulting from the collision between an impelling action and a counteraction. Vector analysis from physics, which assesses the direction and intensity of forces, has always afforded me a convenient analogy to explain the underlying principles of AA to actors. If I throw a baseball, I initiate an action; gravity now becomes the baseball's partner, exerting a counteraction that changes the trajectory of the baseball, curving its path down toward the ground. When the baseball finally falls, an event occurs and allows for a new impelling action. Perhaps a dog comes along, picks up the ball and starts running. If I chase him (a new counteraction), the next scene ensues and moves towards another event, and so on. This AA structure seems to be precisely what the performance scholar Patrice Pavis seeks when he calls for 'a theory of vectors that group together and dynamize entire moments of performance' (2003: 23).

In practice, AA's dynamic structure also prompts ToM skills because it assists actors in reading, reacting and adjusting to what they presume is going on in the minds of their acting partners. Actors are trained to pay close and sustained attention to how the external behaviours of their partners express their inner actions or counteractions. Damasio explains that, while consciousness is 'entirely private,' it is also 'closely tied to the external behaviours that can be observed by third persons,' and thus 'we know quite well how [first-person mind and third-person behaviour] are intercorrelated, first because of our own self-analysis, second because of our natural propensity to analyse others' (1999: 13).

If I were playing a scene in which my impelling action is to ask my boss for a raise, I would continuously assess her counteraction by attending to every nod of the head, every shift in the chair, every lapse in conversation which allows me to speculate about her attitudes, intentions, thoughts and feelings. As I asses this behavioural information, I modify my own behaviour, changing the strategies and tactics I use within the scene to persuade her that I deserve a raise. My partner simultaneously pays me the same close attention and adjusts to the information she observes in my behaviour. By the end of the scene, an event occurs between us when I get the raise or leave the office without it. Thus, our etude develops as an interactive dance in which we use the principles of action, counteraction and event as pragmatic tools to read each other's minds.

Yet, knowing who impels the action and who counteracts is not enough for actors to create a fully satisfying performance. Greater specificity comes when actors state their actions and counteractions as strong, active verbs that name their inner intentions in regard to one another. Using these verbs to advance their actions and counteractions in the etudes better allows actors to create credibly complex interactions that unfold moment to moment during performance. When directing, I begin with a brief discussion in which each actor determines whether he or she impels the scene or resists through a counteraction.[8] The actors then choose specific verbs that they believe might help them embody their actions and counteractions. By committing fully to their chosen verbs in their etudes, they often discover the scene's unanticipated possibilities, which they can test in future etudes.

Using AA as the rehearsal method for our social training project has opened my eyes to many premises in ToM that shed light on familiar aspects of acting. I trace three such premises here. First, in studying the developmental stages of ToM, scientists have found that children's ability to engage in pretend play develops at age two and is directly contingent upon knowing how to behave 'as if' something not true were true, for example, to use a banana as if it were a telephone receiver (Perner et al. 1994). Stanislavsky anticipates Perner by calling one of the System's foundational techniques 'magic if' and by observing that 'the actor must treat fiction exactly as if it were reality' (Stanislavskii 1986, vol. 2: 265). Additionally, studies in ToM find that the ability to attribute a 'false belief' to another develops in children at age three or four, when they can recognise that beliefs sometimes diverge from reality. In 1983, Heinz Wimmer and Josef Perner developed the now familiar 'false-belief task' to test this ability. A puppet, named Sally, puts a toy in a location and leaves the room. Another puppet, named Anne, comes in and moves the toy to a different location. The child-observer is then asked to predict where Sally will look for the toy when she re-enters. The child who predicts that Sally will first look in the place where she last left the toy passes the test. The child who expects Sally to look for the toy where Anne put it fails the test. My readers will surely associate ToM's 'false belief' with the theatrical notion of 'the illusion of the first time.' Actors are, after all, trained to be masters of the 'false-belief task' and thus able to behave during performances, as if they do not know how the stories, which they have thoroughly rehearsed, will end.

Second, in studying the human capacity to recognise behaviour as expressive of a plan, psychologist Bertram F. Malle observes that ToM 'frames and interprets perceptions of human behaviour [...] as perceptions of agents who can act intentionally and who have feelings, desires, and beliefs that guide their actions' (2005: 227). This insight supports Stanislavsky's demand that actors engage in purposeful action driven by precisely the same trio, feelings (*chuvstva*), desire (*khotenie*) and belief (*vera*). In one of the most famous passages from the first volume of his acting manual, Stanislavsky's fictional teacher, Tortsov, asks a beginning student to go on stage and sit there, waiting for further instructions. After a few moments, he tells her that she has completed her work well. She is stunned, because, in her view, she has done nothing. But, as Torstov explains, the very fact that she sat 'for a specific purpose, even so simple a one as waiting for something to happen' was enough to take her into 'the realm of living art' (Stanislavskii 1988–1999, vol. 2: 88–9).

In order to infer purpose, however, people must recognise a series of actions, performed in sequence, as an intentional plan. In the classic 1978 study on plan recognition, Charles F. Schmidt and his team set up an experiment in which a man, Steve, walks to a cabinet, opens it, takes out a record, removes the record jacket and then drops the record. The team found that observers made sense of this sequence by creating narratives that could logically account for Steve's intent—perhaps he wanted to play the record but accidentally breaks it, or perhaps he wanted to smash it because he associates it with a bad relationship. The scientists concluded that 'the problem of plan recognition is to take as input a sequence of actions performed by an actor and to infer the goal pursued by the actor and also to organize the action sequence in terms of a plan structure' (Schmidt et al. 1978: 52). Actors will readily connect this experiment to Stanislavsky's advice on developing a 'score of actions' that clarifies the logic of a role's story.

Such plan recognition allows an observer to predict future outcomes, in other words, to reason about imaginary possibilities. Schmidt and Marsella, who head our project, explain:

> By definition, planning to achieve some goal involves the use of and representation of states of affairs that are not true of the current world. And plan recognition involves attributing to some other acting agent belief about past, present, and possible future states of affairs as well as an intent to bring about some future state of affairs.
>
> *(1991: 109)*

In short, imagination is as necessary to ToM as to Stanislavsky, who, like Damasio, associates it with the 'images' of 'objects' that make up the imaginary 'filmstrip' in the actor's mind while performing (Stanislavskii 1988–1999, vol. 2: 129–31).

Third, scientists value ToM for its fostering of those social skills, such as empathy, cooperation and collaboration, which define human culture and are the target for training programmes in the field of interactive digital storytelling. As Malle passionately writes:

> The ability to represent, conceptualize and reason about mental states is one of the greatest achievements of human evolution.[...In fact,] recent theories and evidence suggest that the evolutionary emergence of a genuine theory of mind occurred after the hominid-line split off and may thus be uniquely human.
>
> *(2005: 225)*

Stanislavsky might have said much the same of acting, which in his opinion is an art that relies upon human empathy and cooperation through collaborative work.

In the first case, Stanislavsky uses the Russian linguistic connection between human feelings (*chuvstva*) and empathy (*sochuvstvie*)[9] to explain how actors create their roles through 'a process of analogy that occurs when recollections arise from both reading and listening to stories about other people' (Stanislavskii 1988–1999, vol. 2: 312). He continues:

> When we first encounter a dramatic work, we usually, with rare exception, start with empathy for the characters in the story. Then in rehearsing the play, we, who are humans and actors, transform this empathy into our own authentic feelings.
>
> *(Ibid)*

In the second case, Stanislavsky stresses the need for actors to work together cooperatively as an 'ensemble,' which he defines as 'a union formed for the sake of a single collective goal' (Stanislavskii 1986, vol. 1: 377). Moreover, he explains:

> This union must never be forced by a director, but free. Directors should attain their goal, not through fear or compulsion but by inspiring actors' fantasies and free-wills.
>
> *(Ibid)*

The digital project, in which I am engaged, began with a formal experiment that collected alternative interactions for a two-person scene through crowdsourcing—an Internet-based methodology that allows a large number of people to contribute data. This experiment yielded promising results that were presented by Dan Feng at the Ninth International Conference on Interactive Digital Storytelling in November 2016.

Based upon my providing a primer on AA, the scientists developed a three-step social skills training system: (1) *Framing* sets the parameters of the scene. (2) *Improvisation* then takes place between a human participant and virtual actors, with the participant writing short sentences that trace the development of the interaction by using verbs to express the actions and counteractions. (3) *Performance Analysis* concludes the process by providing feedback on how the participant understood the beliefs and intentions of the virtual character (Feng et al. 2016: 158–9). At the conclusion of this three-step process, a new piece of information about the dramatic situation or the characters would be added and the process repeated. 'This design is to encourage each [participant] to re-evaluate the same scenario but from different perspectives, much like AA rehearsal directors do with their actors' (Feng et al. 2016: 161).

This first experiment focused on one scene from a longer scenario, *Mistaken Guilt on a Train*, that the scientists wrote for the project. Following the erroneous arrest of an African-American good Samaritan, who intervenes to protect his fellow passengers when a passenger on a train begins attacking people, a reporter *(R)* goes to interview the police chief *(PC)* at his house, but is blocked from entry by a guard *(G)* (Feng et al. 2016: 158). Crowd workers, who functioned as human actors, generated 108 interactions based upon this scene, with each interaction averaging six lines of text. These interactions proved to be 'very rich in terms of character actions and the intention of each action,' despite the fact that the intentions tended to remain 'fairly static' over the course of the scene (Feng et al. 2016: 161). Here are two examples of the interactions that were collected:

> *Bribe:* *R* asks *G* to see *PC*. *G* declines her request. *R* tries to bribe *G*. *G* declines *R*'s bribe. *R* adds $50.00 more to the bribe. *G* accepts the bribe.
>
> *Manipulation*: *R* flirts with *G*. *G* tries to ignore *R*. *R* compliments *G* a lot. *G* begins to flirt with *R*. *R* tells *G* that she just needs a few teeny minutes with *PC*. *G* becomes wary and tells her to leave.
>
> *(Feng et al. 2016: 161)*

A different group of 120 crowd workers annotated how the verbs in the collected data 'altered the beliefs, goals and attitudes of the characters.' They also assessed the dramatic coherence, elicited by the improvised interactions of the players (Feng et al. 2016: 162).

In assessing our first experiment, Feng reported that 'the results of using AA and ToM as theoretical foundations show the promise of such a framework to collect, annotate, and generate interactive narratives, broader in scope and greater in richness, than those currently available in social skills training' (Feng et al. 2016: 166).[10] Moreover, during the Ninth International Conference on Digital Storytelling, we were delighted to meet another group from Italy, who are also using AA to develop a computational model for creating and analysing drama (Albert et al. 2016). This meeting further convinced me of Stanislavsky's value to science.

Conclusion

Can Stanislavsky's twentieth-century theories on acting continue to inspire contemporary actors when cognitive science promises radically new paradigms through which to understand humanity and, in turn, acting? In this chapter, I argue that Stanislavsky can indeed continue to speak to twenty-first-century actors. On the one hand, my case study on Stanislavsky and neurobiologist Damasio demonstrates the efficacy of the emerging methodology among scholars and actors/directors to study Stanislavsky's work through the lens of new science. On the other hand, my collaborative work with scientists, who are applying Stanislavsky's AA to the generation of new knowledge, positions his work as a living legacy in acting that not only continues, but also evolves with the times.

Notes

1 All translations from Russian language sources, including Stanislavsky's writings, are mine.
2 I transliterate the Russian name 'Stanislavskii' with a final 'y' in accord with standard transliteration practice in the Slavic languages field in which I hold a Ph.D. However, in this chapter I retain the alternate transliteration with a final 'i' as used in cited sources and quotations.
3 For more on the influences of Soviet behaviourism and Ribot on the System, see Carnicke (2009: chapters 8 and 9).
4 Blair finds the concept of 'image' especially salient as its prominence in the title of her 2008 book demonstrates.

5 While Stanislavsky fought with the Soviet censors to retain 'soul' (*dusha*) and 'spiritual' (*dushevnyi*) in his publications, Jean Benedetti erases both words in his 2008 translation by consistently translating *dusha* as 'mind' (*um*) and *dushevnyi* as 'mental' (*umnyi*). For example, with regard to my quotation here, Benedetti translates Stanislavsky's definition of the creative state as 'what happens in an actor's mind as he is performing and rehearsing' (Stanislavski 2008: 205).

6 The italics are Damasio's.

7 Leading the way in this work is Albert Rizzo, Director for Medical Virtual Reality at the Institute for Creative Technologies at the University of Southern California. He won the American Psychological Association's 2010 Award for Outstanding Contributions to the Treatment of Trauma.

8 If there are more than two actors in a scene, alliances are created in which actors assist with either the action or the counteraction. Such alliances ensure that ensemble scenes maintain clear focus and that all actors in the scene work collectively.

9 Jean Benedetti translates *sochuvstvie* as 'fellow feeling' (Stanislavski 2008: 224).

10 We have initiated the next phase of our work. The scientists will observe my directing of the project's scenario with live actors in order to design an intelligent virtual agent (IVA) that can, like a human director, assist participants to expand their imaginative choices as they prepare for improvisations. The scientists hope that designing this 'director agent' will allow them 'to incorporate more aspects of AA into the social training experience' (Feng et al. 2016: 166). My modelling for an IVA boggles my mind, but brings a smile to the face of my long-term colleague, Stacy Marsella. Only time will tell if such an avenue of research will prove productive.

Works Cited

Albert, Giacomo, Pizzo, A., Lombardo, V., Damiano, R., and Terzulli, C. (2016) 'Bringing Authoritative Models to Computational Drama (Encoding Knebel's Action Analysis),' *Interactive Storytelling: International Conference on Digital Storytelling*, Berlin: Spinger (LNAI 10045), 285–97.

Blair, Rhonda (2000) 'The Method and the Computational Theory of Mind,' in D. Krasner (ed.), *Method Acting Reconsidered: Theory, Practice, Future*, New York: St. Martin's Press, 201–18.

——— (2008) *The Actor, Image, and Action: Acting and Cognitive Neuroscience*, London: Routledge.

Carnicke, Sharon Marie (2005) 'Emotional Expression in Virtual and Robotic Actors,' *National Conference on Robotics*, Los Angeles.

——— (2009) *Stanislavsky in Focus: An Acting Master for the Twenty-First Century* (2nd ed.), London: Routledge.

——— (2010a) 'The Knebel Technique: Active Analysis in Practice,' in A. Hodge (ed.), *Actor Training*, London: Routledge, 99–116.

——— (2010b) 'Stanislavsky and Politics: Active Analysis and the American Legacy of Soviet Oppression,' in L. Tyler-Renaud and E. Margolin (eds), *The Politics of American Actor Training*, London: Routledge, 15–30.

——— (2012) 'Emotional Expressivity in Motion Picture Capture Technology,' in J. Sternagel, D. Levitt, and D. Mersch (eds), *Acting and Performance in Moving Image Culture: Bodies, Screens, Renderings*, Berlin: Verlag für Kommunikation, Kultur und soziale Praxis, 321–38.

——— (2016a) 'Active Analysis,' in G. Cody and M. Cheng (eds), *Reading Contemporary Performance: Theatricality Across Genres*, London: Routledge, 98–9.

——— (2016b) 'Stanislavsky's Active Analysis for Twenty-First Century Actors: Be Flexible,' lecture at Rose Bruford, Sidcup UK. Available on Digital Theatre Plus, www.digitaltheatreplus.com (accessed 12/30/2016).

Clare, Ysabel (2016) 'A System behind the System: But Is It Stanislavski?,' *Stanislavski Studies*, vol. 4, no. 2: 89–109.

Crohn Schmitt, Natalie (1990) *Actors and Onlookers: Theater and Twentieth Century Scientific Views of Nature*. Evanston, IL: Northwestern University Press.

Damasio, Antonio (1994) *Descartes' Error: Emotion, Reason and the Human Brain*, New York: Penguin.

——— (1999) *The Feeling of What Happens in the Body: Body and Emotion in the Making of Consciousness*, New York: Harcourt.

Feng, Dan, Carstensdóttir, E., Carnicke, S. M., Seif-El-Nasr, M., and Marsella, S. C. (2016) 'An Active Analysis and Crowd Sourced Approach to Social Training,' *Interactive Storytelling: International Conference on Digital Storytelling*, Berlin: Spinger (LNAI 10045), 156–67.

Gordon, Robert M. (1995) 'Folk Psychology as Simulation', in M. Davies and T. Stone (eds), *Mental Simulation: Evaluations and Applications*, Oxford: Basil Blackwell, 53–67.

Kemp, Rick (2012) *Embodied Acting: What Neuroscience Tells Us about Performance*, New York: Routledge.

Lutterbie, John (2011) *Toward a General Theory of Acting: Cognitive Science and Performance*, New York: Palgrave.

Malle, Bertram F. (2005) 'Folk Theory of Mind: Conceptual Foundations of Social Cognition,' in R. R. Hassan, J. S. Ullman, and J. A. Bergh (eds), *The New Unconscious*, New York: Oxford University Press, 225–55.

McConachie, Bruce (2006) 'Cognitive Studies and Epistemic Competence in Cultural History: Moving beyond Freud and Alcan,' in B. McConachie and F. E. Hart (eds), *Performance and Cognition: Theatre Studies and the Cognitive Turn*, London: Routledge, 52–75.

Pavis, Patrice (2003) *Analyzing Performance: Theater, Dance, and Film*, Ann Arbor: University of Michigan Press.

Perner, Josef, Baker, S., and Hutton, D. (1994) 'Relief: The Conceptual Origins of Belief and Pretense,' in C. Lewis and P. Mitchell (eds), *Children's Early Understanding of Mind*, Hillsdale, NJ: Erlbaum, 261–86.

Pitches, Jonathan (1999) *Science and the Stanislavsky Tradition*, London: Routledge.

Ré Arp-Dunham, Joelle (2017) 'The Cognitive Stanislavski in the Rehearsal Hall,' *Stanislavski Studies*, vol. 5, no. 1: 67–74.

Ribot, Théodule (1897) *The Psychology of Emotions*, London: Walter Scott, Ltd.

Roach, Joseph (1985) *The Player's Passion: Studies in the Science of Acting*, Newark: University of Delaware.

Schmidt, Charles F. and Marsella, Stacy C. (1991) 'Planning and Plan Recognition from a Computational Point of View,' in A. Whiten (ed.), *Natural Theories of Mind*, Oxford: Basil Blackwell, 109–26.

Schmidt, Charles F., Sridharan, N. S., and Goodson, J. L. (1978) 'The Plan Recognition Problem: An Intersection of Psychology and Artificial Intelligence,' *Artificial Intelligence*, vol. 11, no. 1–2: 45–83.

Simonov, Pavel Vasilievich (1962) *Metod K. S. Stanislavskogo i fiziologiia emotsii*, Moscow: Akademiia nauk.

Stanislavski, K. (2008) *An Actor's Work: A Student's Diary*, Jean Benedetti (trans), London: Routledge.

Stanislavskii, K. S. (1986) *Iz zapisnykh knizhek*, 2 vols., Moscow: VTO.

——— (1988–1999) *Sobranie sochinenii*, 9 vols., Moscow: Iskusstvo.

Vinogradskaia, I. (ed.) (2000) *Stanislavskii repetiruet: Zapisi i stenogrammy repetitsii*, Moscow: Moskovskii khudozhestvennyi teatr.

——— (ed.) (2003) *Zhizn' i tvorchestvo K. S. Stanislavskogo: Letopis'*, 4 vols., Moscow: Moskovskii khudozhestvennyi teatr.

Whyman, Rose (2008) *The Stanislavsky System of Acting: Legacy and Influence in Modern Performance*, Cambridge: Cambridge University Press.

Wimmer, H. and Perner, J. (1983) 'Beliefs about Beliefs: Representation and Constraining Function of Wrong Beliefs in Young Children's Understanding of Deception,' *Cognition*, vol. 13: 103–28.

2

THE IMPROVISER'S LAZY BRAIN

Improvisation and cognition

Gunter Lösel

Good actors are not always good improvisers and good improvisers are not always good actors. Although these skills overlap, there seem to be specific abilities and cognitive functions required for improvisational theatre. Improvisers have more freedom; they deal with countless options and constantly make choices while playing with other improvisers. So improvisation might be described as a process of decision-making. How do improvisers make these choices? How do they arrive at, and select, new options? How can they do this without hesitation, 'thinking' or the audience noticing? Is improvisation simply a process of quick decision-making? How does an improvising actor's mind work? Can cognitive science help us understand improvisation? In this section, I will look at three approaches that might help explain how improvisers' minds work.

1. The Specific Language of Improvisation

   There are numerous 'How to improvise' books, and a specific language of improvisation has evolved within modern improvisational theatre over the last 40 years. Might the key concepts in this literature tell us something about the way improvisers think and communicate? This section reconstructs the specific language of improvisational actors, deducing certain higher-level patterns of thinking.
2. Improvisation as Problem Solving

   In classical cognitive theories, improvisation can be seen as a process of solving the 'problem of the scene,' that is, proposing options and selecting 'actions' as solutions. This approach is more focused on the single actor than on the group. Viewing improvisation as problem solving should allow cognitive computing to simulate improvisation in digital agents. This section will outline a cognitive model and previous attempts to build improvising robots.
3. Beyond Problem Solving: Embodiment and Social Cognition

   Psychoneurological studies, communication studies and theories of embodiment understand improvisation differently, focusing on group interaction more than on the single actor. In this section, I will examine the current state of research in embodied cognition, social cognition and its relevance for improvisation.

The specific language of improvisation

For over 40 years, the producers of modern improvisational theatre have been experimenting with concepts to describe what they are doing, to communicate with each other and to teach improvisation to students. The concepts in numerous 'how to improvise' books can mostly be traced back to two schools of improvisation: The British-Canadian school, represented by Keith Johnstone, and the Chicago school, based on Viola Spolin, Paul Sills, David Shephard and Del Close's work. Keith Sawyer (Sawyer 2003) was the first to analyse improvisation's evolving language and construct a 'grass-root theory of discursive action,' as well as an ethno-theory of improvisational theatre. His empirical base was the Chicago school's language, an approach I followed in 2013, adding British-Canadian concepts and differentiating between *rules* and *attitudes*.

Rules

The grass-root level: turn after turn

On a very basic level, the process of improvising can be described as a sequence of turns, usually sequentially taken by the actors. Each turn consists of an offer and a response. The most important rule is that each prior offer is accepted. Great emphasis is placed on this rule by both schools (Salinsky and Frances-White 2008, 57). An offer can be accepted in many ways—either in the obvious content or in the subtext—which makes offer/response much more complex than it first appears. The Chicago school coined the term the 'yes-and principle,' highlighting that accepting an offer and adding new information, and thus creating a new offer, are always connected and depend on each other.

Example:

A: 'Do you remember aunt Martha's birthday?'
B: 'Yes! And how she could not open the present!'
A: 'Yes! And how you brought her the big knife!'
B: 'Yes! And how her husband suddenly came in and thought you wanted to kill her?'

This principle guarantees that each player's contribution is integrated into a chain of communication that builds up a fictional reality: 'The whole point of the Yes And game is to build a chain of ideas, each linked to the previous one' (Salinsky and Frances-White 2008, 59). 'Yes' is not always literal. If, for example, a murderer pointed a gun at a character stating that he wants to shoot her, it would not be adequate to say 'Yes.' On the contrary, it would be considered blocking because the offer is an invitation to play a game, not to end it. It also is inadequate to soberly accept an offer; it should be accepted emotionally, enthusiastically and meaningfully (Halpern, Close, and Johnson 1994, 45–56; Johnstone 1999, 101–29; Salinsky and Frances-White 2008, 94). The offer itself does not hold any meaning; the meaning emerges through the manner of acceptance. Highlighting this point, Johnstone coined the term 'to overaccept' (Johnstone 1999, 110).

Working together, turn after turn, prevents a single player from contributing too little or too much to a scene, which is important to balance the collaboration, as I will show presently. In theory, there are no 'bad offers'; in practice, however, improvisers look for a certain quality in offers. Johnstone emphasises the importance of 'blind offers' (Johnstone 1987, 172), which are open to interpretation and provide a starting point for associations. Improvising actors discern four types of offers, depending on how much space they provide for interpretation (Table 2.1).

Table 2.1 Four kinds of offers

Blind offers	*Open offers*	*Closed offers*	*Controlling offers*
Open up the maximum possible interpretations	Open up a broad spectrum of possible interpretations	Open up a small spectrum of possible interpretations	Open up the minimum of possible interpretations
Example: Improviser putting forward his or her hand	Example: Improviser putting forward his or her hand and saying: 'I made this for you!'	Example: Improviser putting forward his or her hand and saying: 'I made this wooden puppet for you!'	Example: Improviser putting forward his or her hand and saying: 'I made this wooden puppet for you! You said you would give me 20 dollars for it.'

Most improvisers agree that controlling offers generally have a negative effect on improvisation: They do not free the stage partner's creativity and take control of the situation. Blind offers, on the other hand, are generally valued, but can lead to an option-overload for the stage partner, if they are too vague. Long-form improvisation relies on the slow construction of a fictional reality, so blind and open offers are preferred. In short-form improvisation, the fictional reality builds quickly, so closed and controlling offers are acceptable. More terms have evolved around the concept of offers like 'advancing,' 'extending,' 'heightening' and 'raising the stakes,' to refer to different forms (Sawyer 2003, 94–5), indicating the importance of the concept.

Making offers and accepting them builds a collaborative relationship and responsiveness between the stage partners—but this process is highly unstable. Improvisers have therefore sought to identify forms of behaviour that hinder the collaborative process and have coined terms to describe them. Johnstone introduced the term 'Blocking,' which involves hindering the offer of your stage partner from affecting the scene, thus neutralising it. For Johnstone, the player not only blocks the story, but also the partner's creativity, and possibly his or her own (Johnstone 1987, 127 ff; Johnstone 1999, 101–29). The Chicago School prefers the term 'denial' (Salinsky and Frances-White 2008, 57). A well-known example is found in 'Truth in Comedy': a couple is discussing their divorce. He says: 'But honey, what about the children? She replies, 'We don't have any children!' (Halpern, Close and Johnson 1994, 48). Different terms have evolved around the concept of blocking or denial, like cancelling, shelving and ignoring (Sawyer 2003, 96), indicating the importance of the concept.

Beyond the grass-roots level

There is common ground between the two schools of improvisational theatre at the grass-roots level, but differences become apparent at higher levels of performance. While the British-Canadian school emphasises the importance of storytelling, the Chicago school focuses on the concept of the 'game of the scene,' described later.

Rules for storytelling

Johnstone claims that humans have a natural ability for narration that can be freed through training. When improvisers learn not to block, the story evolves, taking on a quasi-natural shape, often a circle or spiral. According to Johnstone, stories take shape by exploiting their beginning, so it is crucial to start with a routine, a 'platform,' as Johnstone calls it, that consists of all information shared among the players and with the audience; a good 'platform' contains a lot of specific information that the actors can reintroduce later.

A general rule is to 'Start positive!'; it is considered a mistake to start a scene with interpersonal conflict. Improvisers talk of 'Instant trouble' (Johnstone 1999, 128), a very common mistake for beginners (and actors trained in scripted theatre), who tend to think that conflict is interesting and have to be trained to 'just be average' (Johnstone 1999, 64).

When moving through the story, prior elements get reincorporated and new moves are generated by following the 'what comes next' rule; every next move should inspire the players and audience. While following this, a story will emerge.

The game of the scene

The Chicago school places less emphasis on storytelling; its tradition rests on the concept of the 'game of the scene.' This is derived from Spolin, but has grown into a more sophisticated concept. In Spolin's work, games have a clear setting and well-defined rules, while in the concurrent Chicago school, games emerge during the scene without a clear setting or rules; actors have no time to explicitly agree upon rules. Instead, they are taught to 'Find the game' (or 'Listen to the game'), focusing on the ability to detect patterns and participate in an ongoing game. A game here means a sequence of interactions, in which the participants have interdependent objectives, usually in an antagonistic way; the closer A gets to his or her objective, the more distanced B will be towards his or her objective, and vice versa. For example, when two characters are both trying to gain a high status, they will inevitably have interdependent objectives and start a game. Any interaction can be turned into a game. It can take the shape of a conflict, but improvisers are taught not to rely on conflict too much in order to be open to any type of game that might emerge. Usually a game appears within the first five stage actions. It should not be constructed or forced upon the fellow players, but should emerge 'by itself.' Once found, the game can be heightened until a maximum is reached (usually three rounds, with every round taking things to an extreme). The scene ends when the game ends.

Attitudes

While at the grass-root level there are a lot of rules — only a few of which can be described here — there is a consensus among improvisers that attending to impro-rules will not necessarily generate a good scene. Mick Napier (2004) has made it his mission to remind the community of this fact because 'How-to-improvise' books regularly give a wrong impression. Advanced improvisers don't cling to rules, but, like sportsmen or musicians, follow their embodied knowledge, which consists mostly of attitudes that foster good improvisation. As attitudes have to sink deep into the mind and body of an improviser, they often take the form of mantras. Though the wording varies, I list some of the most common attitudes in the following section.

'Don't tell jokes'

Improvisational theatre is often misunderstood as a form of comedy, because it tends to be funny, but all pioneers of this theatre agree that improvisers should not try to be funny or original (Halpern, Close, and Johnson 1994, 23–8; Johnstone 1999, 30–3, 125). Jokes usually destroy the most precious resource of improvisation, the players' commitment to the team and the scene's fictional reality. A joke always lets the teller shine, but destroys the story, the game and the stage partners' motivation. An improviser who seeks a laugh loses his or her freedom and power to play, so improvisers are taught not to try to be funny and to ignore laughter.

'Show, don't tell!'

This rule dates back to Viola Spolin (Sawyer 2003, 109) but is a shared attitude in both schools. Physicalisation and emotion are valued more than just talking (Spolin 1977, 237–52). Emotions or ideas should be acted out spontaneously rather than be expressed verbally. Too much talking is considered a mistake and called 'hedging' (alternative: talking heads) (Johnstone 1999, 123). A similar error is 'waffling' (Sawyer 2003, 109), talking about things to do instead of doing them.

'No playwriting'

Spolin used the term 'playwriting' for the mental activity of planning the scene ahead (Spolin 1977, 388). Johnstone would agree that planning ahead is not useful for improvisation. Usually when involved in pre-planning, improvisers will lose their attentiveness; they will cut off the ongoing exchange of clues between the players and miss out on important information (Sawyer 2003, 99). They will not be able to really listen to the stage partners or the game (Halpern, Close, and Johnson 1994, 70).

'Don't do your best'

This attitude is one of Johnstone's most important lessons, which he seems to never tire of teaching to his students (Johnstone 1999, 64). If you try to do your best, you will inevitably fail in improvisation. Instead, you should feel safe and confident on stage, not expect too much and not punish yourself for being average. The Chicago school emphasises a similar concept of self-forgetting, flow and freedom from fear. I will return to this important attitude in Section 'Beyond Problem Solving: Embodiment and Social Cognition,' because it is potentially key to understanding underlying brain-functions.

'Do not fear mistakes – there are none' (Miles Davis)

This attitude originated in jazz; the concept was imported into improvisational theatre by the Chicago school: *'There is no such thing as a mistake'* (Halpern, Close, and Johnson 1994, 79). Not only will an improviser who tries to avoid mistakes get fearful and block his or her creativity, but improvisational theatre in general has flourished in objection to, and rejection of, most principles of theatre aesthetics; it had to reject scripted theatre's claims of perfection and artistic virtuosity, in order to develop its own criteria and a 'poiesis of imperfection' ('Poiesis des Imperfekten' (Bormann, Brandstetter, and Matzke 2010, 193)). In part, this references oral cultures that embraced improvisation. Jonathan Fox, founder of the playback theatre movement, writes:

> Failure is inadmissible in a literary-technological society, but in oral culture – in the world of improvisation, of putting together in the crucible of the moment – the tangible possibility of failure is what makes success possible. Failure is not the poison but the spice of oral composition.
>
> *(Fox 1994, 96)*

The poiesis of imperfection, rooted in oral cultures, is also a very modern trait, with connections to early twentieth-century *avant-garde* movements in Europe (Lösel 2013, 53 ff).

'Give up your self'

Both schools describe the ideal of self-forgetting, or at least reduced self-consciousness, and the idea of conscious authorship is criticised as not useful for improvisation. For Spolin, one of the main goals was to free the actor from the tendency to 'perform':

> A moment of grandeur comes to everyone when they act out their humanness without need for acceptance, exhibitionism, or applause.
>
> *(Spolin 1977, 44)*

Spolin describes an ideal state of self-forgetting in spontaneity, which can be reached through attention to a 'Point of Concentration' and complete commitment to the game. In her later work, she further emphasises this in the rule 'Follow the Follower.' In this, the improvisers are taught never to make an active offer, but to follow the previous offer. This leads to emerging group-creativity and a trance-like state, in which the improvisers will not conceive of themselves as authors of the work. Similarly, Johnstone saw attempts at originality as a problem of improvisation (Johnstone 1999, 124). He discusses this problem at length, summarising his belief that people who try to be original always come up with the same old boring answers. Improvisers should act as if they are controlled by an outer force: 'Great improvisers "go with the flow" accepting that they're in the hands of God, or the Great Moose' (Johnstone 1999, 341).

Other evidence for the ideal of self-forgetting is the existence of 'Footing-Games' (Sawyer 2003, 29). Sawyer introduced this term to describe a category of games that cover up agency. Two players merge into one person, speaking with one voice, dubbing each other, miming the other's hands and so forth. Neither the actors nor the audience can detect who is the author of this scene or responsible for its content.

'Take risks'

Improvisers must be able to make unpredictable moves, produce crazy ideas and surprise their fellow players and even themselves. Johnstone emphasises the use of material that might seem obscene and even psychotic. One of his important ideas is the concept of tilts – surprising actions or lines that reframe the fiction of the scene (like a player saying: 'I died last year in a car accident.'). In the Chicago school, Close devised the rule 'Always take the active choice' (Johnson 2008, 53), meaning that an improviser should always choose the option that maximises change on stage. These attitudes are in contradiction to the instruction to 'be obvious' and stay inside the 'circle of expectations.'

Conclusion

While attitudes and their wording vary between countries and schools, they have one thing in common: They take on very simple, repeatable forms, so they can become 'second nature' for the players. Like sportsmen, improvisers have no time to consciously think about options and rules while they are performing, but must rely on embodied knowledge available often after years of training. The ten Impro Commandments, probably dating to the 1980s in Australia, are an example of this sacred simplicity:

1. Thou shalt not block.
2. Thou shalt always retain focus.
3. Thou shalt not shine above thy team-mates.
4. To gag is to commit a sin that will be paid for.
5. Thou shalt always be changed by what is said to you.
6. Thou shalt not waffle.
7. When in doubt, break the routine.
8. To wimp is to show thy true self.
9. S/he that tries to be clever, is not; while s/he that is clever, doesn't try.
10. When thy faith is low, thy spirit weak, thy good fortune strained, and thy team losing, be comforted and smile, because it just doesn't matter.

(Charles 2003, 247)

Game building and game breaking

As noted before, some rules of improvisation contradict others. This is not just a side phenomenon but pertains to some of the basic instructions. To understand this, I suggest distinguishing between two phases of improvisation: a phase of game building and a phase of game breaking. In game building, players will act in collaboration, attend to the rules (especially the 'yes-and principle') and slowly build up a fictional reality. In game breaking, an individual improviser will break the rules, act in unpredictable, anarchistic ways and destroy the emerging patterns of the improvisation. The criteria for a good choice thus depend on the improvisation phase. In a game-building phase, an option is good when it is obvious and adds to shared mental models of the scenic reality. In a game-breaking phase, an option is good when it is divergent, unexpected and leads to much stage action. The improviser should develop a sharp sense of when game building or game breaking is needed (Lösel 2013, 311).

Embodiment of rules and attitudes

The role of domain knowledge in improvisation has long been discussed, beginning with Jeff Pressing's early work 'Cognitive processes in improvisation' (Pressing 1984). As described earlier, the rules and attitudes have to become 'second nature' through a training that resembles training for team sports. While there is a specific language to analyse problems *post performance*, the improviser on stage draws from embodied knowledge *peri-performance*. As Berkowitz points out:

> Once installed in memory, however, the elements of the knowledge base must be organized and refined – 'enriched' in Pressing's words.
>
> *(Berkowitz 2010, 40)*

It takes years for an improviser to learn to become relaxed and skilful at the same time, to not give his or her best and to let his or her brain work without interfering too much.

Improvisation as problem solving

Classical cognitive theories conceive of cognition as problem solving, and it is certainly worth looking at improvisation through this conceptual lens. The specific language and concepts of improvisation can be thought of as a toolbox for problems that arise.

Brian Magerko et al. (2009) conducted a study of improvising actors by conceptualising improvisation as a process of problem solving. Using video-cued-recall, they explored what improvisers conceive as the 'problem of a scene,' and what they do to solve it: Experienced improvisers were video-recorded while improvising in a laboratory situation. After the performances, the participants were shown their scenes and asked to comment on their thoughts at specific points in the performance. This method has limitations; it can only uncover what the improvisers consciously experience and what they can verbalise. Nevertheless, it is probably the best current method of 'looking inside improvisers' heads.' So, Magerko et al. gathered verbal reports on cognition in improvised scenes. One result shows that improvisers do use their specific language to describe and explain the problems of improvisation.

Modelling improvisation

In a second step, Magerko et al. suggested that improvisation can be modelled using a well-established cognition model – the decision circle from Newell's Unified Theory of Cognition. This model is connected to a computational model called SOAR (State, Operator and Result), and can therefore be directly codified (Figure 2.1).

SOAR is an input/output concept, with perception as input and action as output. Between input and output, working memory, recognition memory and production-match work together closely to generate two learning circles, of which one is designated to

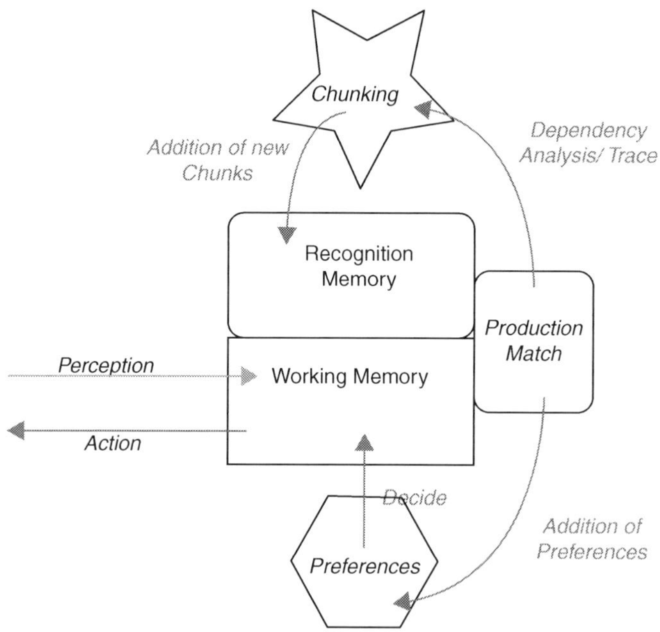

Figure 2.1 SOAR model.

proposing options. If none of the stored options fit, the problem and strategies have to be broken down into smaller pieces—subgoals—until new options can be generated. The other feedback loop serves as *decision-maker*, using past experiences from an episodic memory continually supplying preferences and updated with every new experience. SOAR uses production rules to generate states that gradually bring the system closer to the goal state. The main link to the outside world is the Working Memory, which controls the input (perception) and the output (action), completing and evaluating the input and selecting the most promising action. SOAR follows Allan Newell's decision circle, which states that, whenever reasonable, cognitive acts can be separated into five steps:

1 receiving input
2 the elaboration of new knowledge based on other knowledge or inputs
3 the proposal of new operators/actions/goals to pursue
4 the selection of one of the proposed courses of action
5 the execution of the selected action (Figure 2.2)

While Magerko et al. assume that this decision circle applies to improvisation in the same way it does to other cognitive actions, there are major differences: (1) improvisers are trained *not* to evaluate the input, but to greet every impulse with joy and acceptance; they skip the process of input evaluation. (2) The production of options is maximised. Improvisers follow a brainstorming technique, generating many of the options as quickly as possible; they produce 'quick-and-dirty' options. (3) The decision about which option should be selected also works without evaluation. Improvisers are trained to take any choice without worrying. So, to a certain extent, it appears as 'selection by chance.' The decision cycle in improvisation might be visualised as shown in the figure (Figure 2.3):

Compared to everyday life decision-making, improvisation decision-making seems to follow a reduced and simplified form of the decision cycle, omitting the evaluation of both input and output. The 'normal' circle of decisions is not the rule in improvisation, and it seems possible to propose an improvisational mode of cognition, which applies when normal problem-solving strategies have failed and/or when there is time-pressure and/or no real problem at all, as in a playing situation.

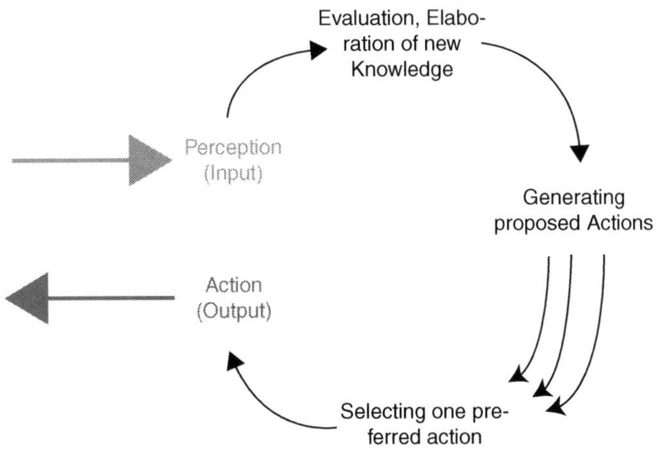

Figure 2.2 Newell's unified theory of cognition.

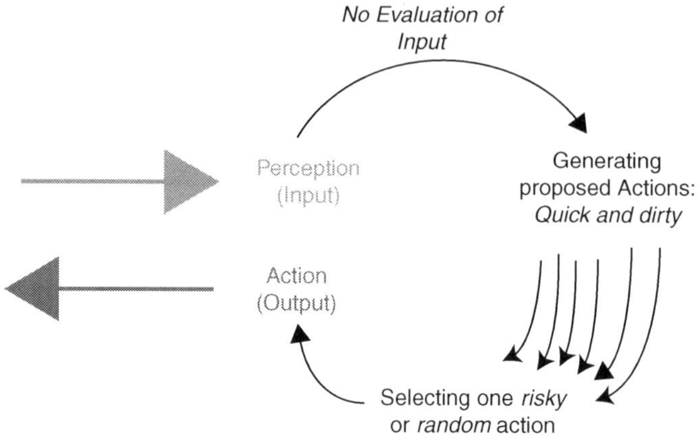

Figure 2.3 Newell's model adapted to a mode of improvisation.

Improvisation, anticipation, prediction

Yoshiro Endo (Endo 2008) worked out a computational framework for improvising robots, assuming that improvisation is a kind of emergency mode for the system in order to be able to react when anticipation fails. He constructed a system that does not rely on episodic memory but on a transcoded form of memory derived from Antonio Damasio's concept of 'somatic markers' (Damasio 1996): Instead of searching for specific memories, the system will search for more abstract markers like emotions, vectors and intuitions. It also includes sensory data from the robot's 'body.' This model is in accord with the more recent approach of predictive processing (Clark 2016), suggesting that the brain works as an automatic prediction machine on multiple levels. Only when prediction fails on one level will the next (higher) level spring into action, recoding the input in an attempt to make better predictions.

What does this mean for improvisation? At the beginning of a scene, when every option is possible and predictability is very low, the improviser's brain has to deal with high prediction error. Andy Clark describes a similar situation to illustrate the start of predictive processing under very special conditions:

> Imagine you are kidnapped, blindfold, and taken to some unknown location. As the blindfolds are removed, your brain's first attempts at predicting the scene will surely fail. But rapidly processed, low special frequency cues soon get the predictive brain into the right general ballpark. Framed by these early emerging gist elements [...] subsequent processing can be guided by specific mismatches with early attempts to fill in the details of the scene. These allow the system to progressively tune its top-down predictions, until it settles on a coherent overall interpretation pinning down detail at many scales of space and time.
>
> *(Clark 2016, 42)*

Quite similarly, the improviser's brain will start generating quick and uncertain guesses about the environment, starting from gist elements, on a very general level, and slowly filling in the details. There is one crucial difference to a real situation though: no 'true' information exists to validate the guesses. Instead, every suggestion will be accepted by fellow players and become part of the fictional reality of the scene. So there is no resistance by the outside

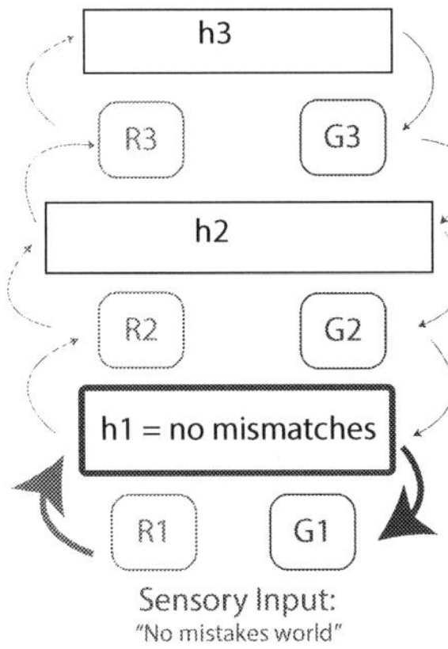

Figure 2.4 Predictive processing in improvisation.

world to the proposals of the inside world, or, as I call it, improvisation creates a 'no mistakes world.' Improvisation here exceeds a simple input/output model of cognition, instead generating a very special environment that triggers automatic prediction processes, but avoids mismatches. Presumably this will lead to a processing mode that does not involve higher levels of processing, because those would only spring into action when a mismatch occurs (Figure 2.4).

Having no 'resistance' of real-world data, predictive processing will presumably lead to a fast mode of processing that excludes higher levels of processing, involving a high degree of automatisation, which may account for the trancelike state, improvisers often report (Scheiffele 1995; Lösel 2013). This model suggests the minor importance of domain knowledge (like the rules and attitudes) in the process of improvising as they would presumably be stored in higher levels of the cognitive system. Instead, the interaction with the very specific environment of improvisation would be processed mainly on a very basic level, with high speed and an increasing confidence in the 'no mistakes world.' This is what I refer to as a 'lazy brain.'

Chance, selection and emergence

Johnson-Laird, a cognitive scientist and jazz musician, was one of the first to apply cognitive concepts to improvisation (Johnson-Laird 1989). According to him, only three types of algorithms truly facilitate improvisation: neo-Darwinian, neo-Lamarckian and a hybrid of the two. Both Jean-Baptiste Lamarck and Charles Darwin developed a theory of evolution in the nineteenth century. While Darwin proposed mutation and selection as the only mechanisms for evolution, Lamarck doubted that species evolved in such a trial and error way. Instead, he suggested the inheritance of acquired characteristics as a motor of evolution. A neo-Darwinian algorithm generates a new piece by randomly blending different pieces together, generating a large number of options and leaving selection to the environment. In the case of collective improvisation, the environment consists of the other improvisers and their choices. If they pick up an idea it will evolve; otherwise it disappears. In a neo-Lamarckian algorithm, a new piece is derived from relevant domain knowledge, leading to more elaborated improvised outcomes that hardly require selection. Johnson-Laird notes that even though the neo-Darwinian approach, because of the randomness of the production process, is fated to produce a substantial amount of unwanted pieces, it is the only way to improvise when no expert knowledge is available. On the other hand, the neo-Lamarckian approach might be doomed to stay inside the frame of the already known, since it heavily draws on expert knowledge. Both approaches seem to be compatible though: Shimon, an improvising robot musician, incorporates both neo-Lamarckian and neo-Darwinian aspects (Hoffman and Weinberg 2011).

Conclusion

Starting from classical models of cognitive science, there are interesting conclusions to be drawn: (1) rather than embedding improvisation into 'normal' processes of cognition, improvisation seems to call for a separate mode of cognition. (2) The improvisational mode seems to be called for when anticipation fails or is suppressed. (3) The improvisational mode of cognition either produces numerous options (neo-Darwinian) or one specific option (neo-Lamarckian). (4) The improvisational mode must enable proactive behaviour, even though there is insufficient information to build anticipations, so it will presumably make extensive use of predictive processing. (5) The improvisational mode is a 'lazy' way of processing, since it omits exhaustive procedures of comparing the current situation to experience and evaluating options.

The improvisational mode probably uses less neural capacity than everyday problem solving, with a faster, but less precise outcome. Clayton Drinko explored cognition and improvisation in 2013, and he suggests that the improvisational mode might rely on the same structures as Kahnemann's system 1 (Drinko 2013). Kahnemann, exploring economic decision-making, identifies two systems: System 1 makes fast choices, relying on heuristics. It can complete a sentence someone else has started, know when someone is angry before they speak, understand simple phrases, drive, read simple words or react when something is disgusting. System 2, on the other hand, relies on a slow mode of cognition. It is responsible for the continuous monitoring of one's behaviour, building an internal monologue, and is the conscious problem solver (Kahnemann 2011). If the improvisational mode of cognition is identical with Kahnemann's system 1, this explains why improvisers put so much emphasis on staying in a state of effortlessness and ease while improvising. When they fail to reach this state, System 2 springs into action and a completely different way of processing starts. This accords with Johnstone's mantra to never give your best and with the general emphasis on reducing stress and fear of failure. The ideal of effortlessness thus appears to have a neurological basis; to improvise, the brain has to be in a laid-back state. It has to be lazy.

Beyond problem solving: embodiment and social cognition

Psychoneurological research hints towards an understanding beyond problem solving. Charles Limb, a jazz-musician and researcher, conducted several functional Magnetic Resonator Imaging (fMRI) studies to detect areas of the brain that increase or decrease activity during improvisation (Limb and Braun 2008, Donnay et al. 2014). He constructed devices to make improvisation and musical interplay possible under fMRI's restricted conditions. His findings indicate: (1) a decrease in activity in the lateral prefrontal cortex, an area associated with self-monitoring and self-control, (2) an increase of activity in the medial prefrontal cortex, an area ascribed to creativity and connected to the limbic system, which processes emotions and (3) a close connection to Broca's area, which is crucial in producing language and gesture. These results fit the theories of improvisation. Actors are trained to minimise self-control, activate areas of spontaneity and creativity and to be in a close dialogue with their fellow improvisers. Drinko emphasises links between language and gesture as key concepts for acting, arguing that all forms of communication—music, language or acting—might share the same neurological basis, referencing the 'shared syntactic integration resource hypothesis' (Patel 2008). These findings suggest that improvisation should be more aptly framed as communication instead of being reduced to problem solving.

Bodies in sync and an altered state of mind (ASM)

The idea of synchronised movement and a hidden link between the bodies (and minds) of improvisers is shared among improvisational actors and strongly emphasised by all pioneers of modern improvisational theatre, from Jacob Moreno to Viola Spolin and Keith Johnstone. The Chicago school refers to the Group Mind, which is a similar concept. Synchronisation or entrainment becomes obvious in mirror games—where improvisers strive for synchronised movement – and in 'follow the follower' games, as well as in Johnstone's concept of status. Bodies are trained to be responsive to each other, without higher cognitive functions interfering, often resulting in a trancelike state of mind. Synchronisation of movements also links to experimental approaches in cognitive psychology and behavioural sciences. Using an advanced method, Walton et al. studied the synchronisation of movement in the bodies of improvising musicians (Walton et al. 2015), identifying certain parts of the body that tend to sync more closely than others, especially heads and hips. Research is also being undertaken on brain synchronisation during improvisation. Early results indicate that brains probably synchronise at lower frequencies (e.g., delta and theta waves) (Müller, Sänger, and Lindenberger 2013), but the results are not consistent. Synchronisation appears to foster a positive relationship, while de-synchronisation tends to destroy it. Thus, mirroring helps create a collaborative mode.

An altered state of mind (ASM)

In improvisation, body, mind and environment are connected in specific ways that differ from everyday consciousness. There is some evidence that the improvisational mode of cognition is a form of trance, or altered state of mind. Eberhard Scheiffele, a psychologist, explored improvisation as an altered state of mind, finding that most criteria of hypnosis and trance fit to mind-states in improvised acting (Scheiffele 2001). Clayton Drinko explored the closeness of improvisation to trance and to alcohol intoxication, suggesting that similar neurological structures are involved (Drinko 2013). There are other works linking improvisation with the concept of flow (*cf.* Kießling 2011). Johnstone also establishes trance as a key concept of improvisational acting, especially in his mask work. Trance phenomena have specific effects on cognitive functions like attention and memory, as Drinko points out, and while there is not yet a brain study to prove ASM during improvisation, improvisers' subjective accounts support this hypothesis. Improvisers often do not remember exactly what they did on stage, just as trance media usually have no memory of what they did during trance. In another similarity to trance, improvisation calls for a paradoxical form of attention. On one side it needs to be extremely focused. Spolin used the term 'point of concentration (POC),' reminding her students to 'keep your eyes on the ball,' creating narrowed attention (Spolin 1977, 21, 28). Johnstone creates a similar effect using masks through which the actor can see only a very small part of the environment. So, just as a medium in trance is asked to focus on a monotonous object like a clock, the improviser's attention is also narrowed. Paradoxically, openness to environmental information is simultaneously extremely high, so a double-attention is generated that is typical for trance (Figures 2.5 and 2.6).

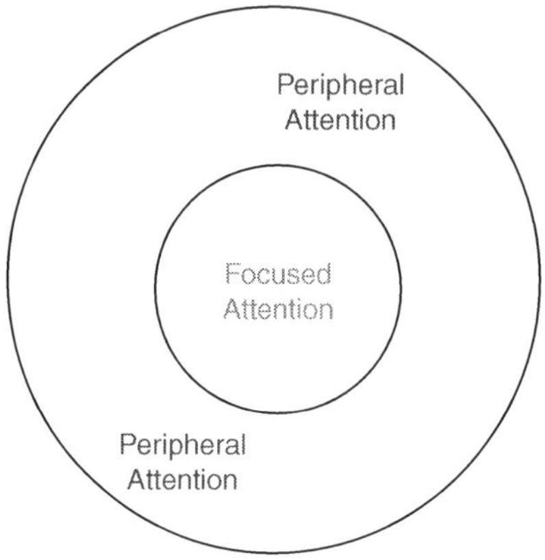

Figure 2.5 Attention in everyday cognition.

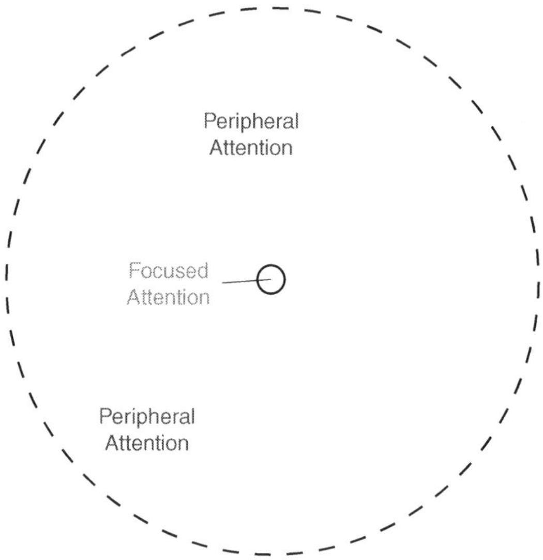

Figure 2.6 Attention in improvisation or trance.

Improvisation as communication and collaboration

Improvisers enter a specific state of connectedness with their environment. In what follows I describe two concepts of improvisation as interaction or communication: Keith Sawyer's concept of the interactional frame and my own concept, in the tradition of Niklas Luhmann's systems theory.

Social emergence

Sawyer's (2003) work on improvisation applies the concepts of communication science, in the tradition of George Bateson and Irving Goffmann, to improvised dialogues, building a theory of social emergence in improvisation. His model integrates bottom-up and top-down approaches, introducing two levels (Figure 2.7):

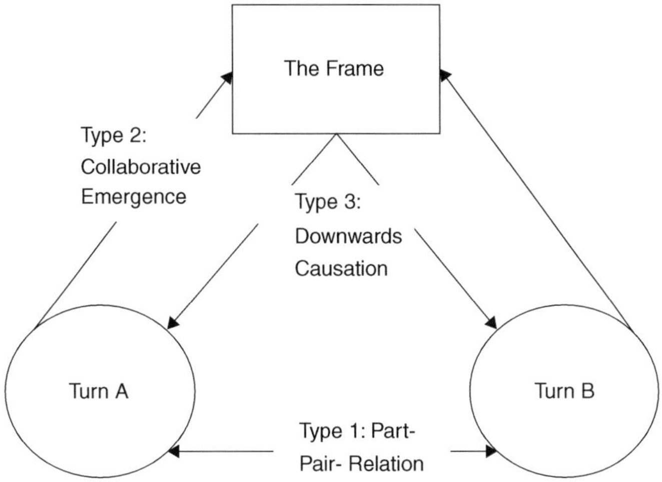

Figure 2.7 Sawyer's model of social emergence in improvisation.

At the lower level, actors exchange pieces of dialogue (Type 1 relation: Part-to-Pair Relation), which lead to the emergence of an interactional frame on the higher level of the system (Type 2 relation: Collaborative Emergence). The interactional frame will then be the frame for new interactions on the lower level, facilitating certain moves and making others improbable (Type 3 relation: Downward Causation). Sawyer suggests understanding improvised interactions as complex systems, meaning these causal relationships are not deterministic. The pair-to-pair interaction cannot predict the emergence of the interactional frame, the frame does not determine the following interactions and so on. Unpredictability is introduced into the model, while still allowing causal relationships to be assumed. Sawyer underpins his model with elaborated analyses of improvised dialogues to explore the interactional frame's emergence. Table 2.2 shows one example.

The dialogue is segmented into separate turns and the interactional frame becomes increasingly specific; every turn leads to a changed interactional frame. It is important to note that a turn (or offer) not only provides information, but also retains information to keep the space of possibilities open. Improvisers do not give the information directly, but are trained not to deliver too much information and wait for the frame to emerge 'by itself' through interaction. Sawyer provides much evidence for social emergence as a key

Table 2.2 The beginning of a two-minute scene at the Improv Institute

1	<Andrew steps to stage center, pulls up a chair and sits down, miming the action of driving by holding an imaginary steering wheel>
2	<Ben steps to stage center, stands next to Andrew, fishes in pocket for something>
3 Andrew	'On or off?'
4 Ben	'I'm getting on, Sir!' <continues fishing in his pocket>
5 Andrew	'In or out?'
6 Ben	'I'm getting in! I'm getting in!'
7 Andrew	'Did I see you tryin' to get in the back door a couple of stops back?'
8 Ben	'Uh...'

concept for improvisation, claiming that the specific language of improvisational acting sets up perfect conditions for social emergence. Unpredictability emerges through interaction, not through the improvisers' wittiness, an assertion in sync with the concepts of both schools mentioned earlier.

Niclas Luhmann and the concept of connectability

While Sawyer provides a model for improvised dialogues in general, he does not explain the difference between everyday interaction versus stage improvisation. A model derived from Niklas Luhmann's operative system theory might fill this gap. The term 'operative' emphasises that the focus of this concept is not so much on the elements of the system, but on the operations within the system. A key concept is connectability: an operation only adds to the system when it can connect to a prior operation and when it provides possibilities for future operations to connect. Thus, Luhmann's systems comprise chains of operations and cease to exist when there are no more operations that follow. This chimes with the improvisers' view that they are dealing with operations (= offers) and their connectability (= yes and) all the time. Luhmann's theory can also integrate Newmann's circle of decision by modelling operations before one of them is selected. The following diagram visualises a chain of operations, which I call the 'fir branch model of improvisation' (Figure 2.8).

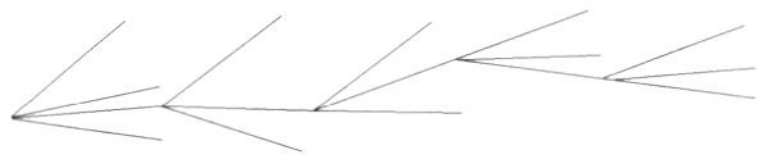

Figure 2.8 The fir-branch model in everyday interaction.

Each operation provides a couple of options to build upon. The following contributor selects one, adds an operation and provides more options to build on. For the system to be sustained, it is crucial to generate both operations that possess connectability and a code that allows operations to connect to each other. The code enables the selection of certain operations. In improvisation, connectability is maximised by preferring open offers and by applying the yes-and principle (Figure 2.9):

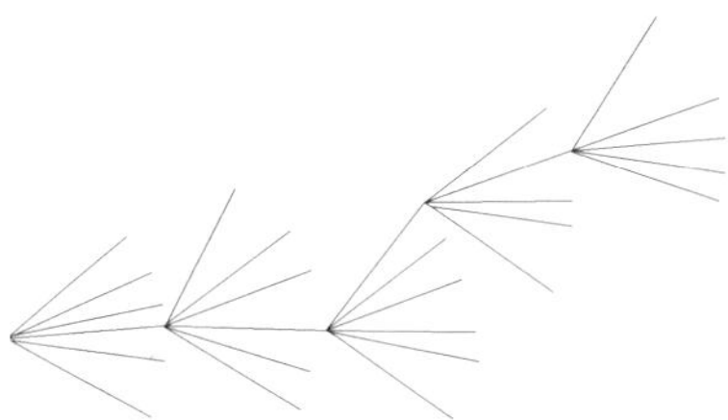

Figure 2.9 Maximised connectability in improvisation.

At each turn of the chain of interactions, the improvisation can be described as a process of decision-making in response to a spectrum of options opened up by the prior operation. This spectrum is much wider than everyday communication, leading to more choice. The rules of improvisation guarantee connectability; every offer has to be accepted. Where improbable and inadequate options would be rejected in the everyday system, in improvisation they are included and the probability of their selection rises. In this way, improvised interaction becomes more robust than interaction in everyday life because someone will always find a connection. This is why improvisation can assimilate a lot of 'crazy' choices without stopping; it allows for spontaneity, which might kill a normal conversation. Such a system is likely to become complex and produce emergent phenomena.

Improvisers strive to build larger components, like a fictional reality, dramatic interaction and/or a story. Since there is no plan, the shape and structure of the improvisation emerges through the application of a code that is implicitly agreed upon, usually after 5–10 turns. After that, the selection of options will follow this code. So, while the improvisation is progressing, the usage of prior knowledge gets more important, and the improvisation changes from a neo-Darwinian to a neo-Lamarckian mode (Figure 2.10).

The code, which in Sawyer's terms would correspond to the interactional frame, specifies what types of operations are selected and thus connected to a following operation, while the others die off. In this way, a selection is possible that will shape the system/improvisation. The code could be Del Close's instruction to 'Always take the active choice,' for example. Every selection by every player follows this code, and the improvisation becomes increasingly risky. The code could also be 'Always take the most emotional option' or 'Always take the most insulting option.' In other words, the code is the emerging rule of the game and performance will form into self-similar shapes (in the visualisation, they take the shape of a spiral) (Figure 2.11).

The code stays valid until a dramatic peak is reached; usually three repetitions please both audience and players. After this, the system, or game, ends (not necessarily the improvisation because it might comprise several or even many games). In the course of the improvisation, improvisers must listen carefully to the new code and break the game twice; after the 'platform' is established, and after the game plays out. The timing of these two points of game-breaking seems crucial for successful improvisation. In the field of dance, João Fiadeiro (Fiadeiro 2016) developed similar models within the research project Blackbox Cognition & Art (Fernandes 2014), indicating that some features of improvisation might be considered universal. The importance of these rather abstract models is that they break down the process of improvisation into single turns that can be described as individual decisions, while still allowing the improvisation to be viewed as a collaborative enterprise involving the improvisers' bodies and minds. Cognitive science can help develop and test abstract models of improvisation.

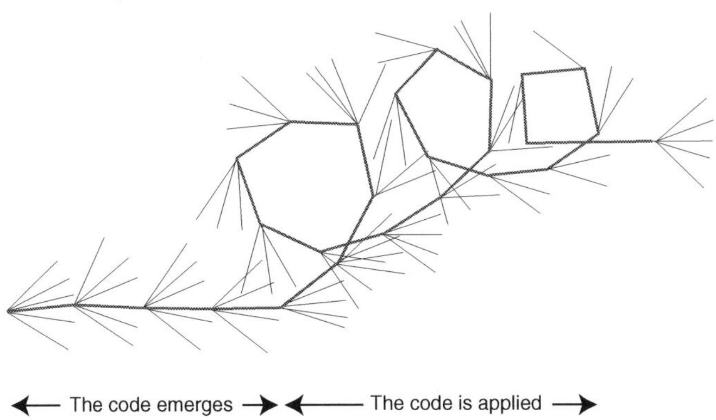

Figure 2.10 The emergent code.

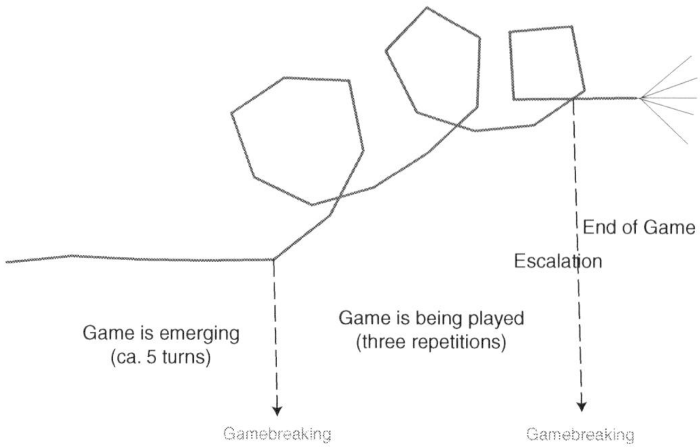

Figure 2.11 The emerging game and game breaking.

Conclusion

The specific language of improvised acting consists of rules and attitudes that help improvisers communicate and shape their cognitions while improvising. When analysing scenes and teaching improvisation, they refer to their explicit concepts, describing improvisation as a form of problem solving. There is some evidence, however, that while improvising on stage, they have only minimal access to self-reflective parts of the brain and to explicit domain knowledge. Rules and attitudes therefore have to become embodied knowledge to allow for responsiveness 'without thinking,' that is without reflective thought. Predictive processing seems to be a promising concept to describe and explain the way the individual actor is connecting with his or her fellow player by predicting their actions on a very basic level. Presumably their training enables them to deal with a very high degree of prediction error without interrupting the interaction. This mode is quite different from that used for everyday problem solving and could possibly be described as an altered state of mind. While the individual improviser learns to enter this altered state of mind, improvisation is also a collaborative enterprise, building mutual understanding and anticipation. Borders of the self are no longer maintained and, in this state, social emergence can occur spontaneously. The rules and attitudes of improvisation not only generate a specific mindset for the single improviser, but also a collaborative environment that allows the improviser to switch on his or her 'lazy brain' without danger.

References

Berkowitz, Aaron L. 2010. *The Improvising Mind: Cognition and Creativity in the Musical Moment.* New York: Oxford University Press.

Bormann, Hans-Friedrich, Gabriele Brandstetter, and Annemarie Matzke, eds. 2010. *Improvisieren – Paradoxien Des Unvorhersehbaren.* Bielefeld: Transcript Verlag.

Charles, David. 2003. 'The Novelty of Improvisation: Towards a Genre of Embodied Spontaneity.' Louisiana State University.

Clark, Andy. 2016. *Surfing Uncertainty: Prediction, Action, and the Embodied Mind.* New York: Oxford University Press.

Damasio, Antonio R. 1996. 'The Somatic Marker Hypothesis and the Possible Functions of the Prefrontal Cortex.' *Philosophical Transactions of the Royal Society of London* 351 (1346): 1413–20.

Donnay, Gabriel F., Summer K. Rankin, Monica Lopez-Gonzalez, Patpong Jiradejvong, and Charles J. Limb. 2014. 'Neural Substrates of Interactive Musical Improvisation: An fMRI Study of 'Trading Fours' in Jazz.' *PLoS ONE* 9 (2). doi:10.1371/journal.pone.0088665.

Drinko, Clayton. 2013. *Theatrical Improvisation, Consciousness, and Cognition*. New York: Palgrave Macmillan.

Endo, Yoichiro. 2008. 'Countering Murphy's Law: The Use of Anticipation and Improvisation via an Episodic Memory in Support of Intelligent Robot Behavior.' ProQuest Dissertations and Theses.

Fernandes, Carla. 2014. 'Blackbox Cognition & Art.' http://blackbox.fcsh.unl.pt.

Fiadeiro, Joao. 2016. *Graphic Models Developed by Joao Fiadeiro. BlackBox Art & Cognition*. https://vimeo.com/149281774.

Fox, Jonathan. 1994. *Acts of Service. Spontaneity, Commitment, Tradition in the Nonscripted Theatre*. New Paltz, NY: Tusitala pub.

Halpern, Charna, Del Close, and Kim Johnson. 1994. *Truth in Comedy – the Manual for Improvisation*. Colorado Springs, CO: Meriwether Pub Ltd.

Hoffman, Guy, and Gil Weinberg. 2011. 'Interactive Improvisation with a Robotic Marimba Player.' *Autonomous Robots* 31 (2–3): 133–53. doi:10.1007/s10514-011-9237-0.

Johnson, Kim. 2008. The Funniest One in the Room – the Lives and Legends of Del Close. Chicago, IL: Chicago Review Press.

Johnson-Laird, Philip Nicholas. 1989. *The Computer and the Mind : An Introduction to Cognitive Science. Fontana Masterguides*. London: Fontana Press.

Johnstone, Keith. 1987. *Impro – Improvisation and the Theatre*. New York: Routledge.

Johnstone, Keith. 1999. *Impro for Storytellers*. London: Faber & Faber.

Kahnemann, Daniel. 2011. *Thinking, Fast and Slow*. New York: Farrar, Straus and Giroux.

Kießling, Nele. 2011. 'Flow und Improvisations theater.' Universität Hildesheim.

Limb, Charles J., and Allen R. Braun. 2008. 'Neural Substrates of Spontaneous Musical Performance: An fMRI Study of Jazz Improvisation.' *PLoS ONE* 3 (2). doi:10.1371/journal.pone.0001679.

Lösel, Gunter. 2013. *Das Spiel Mit Dem Chaos – Zur Performativität Des Improvisationstheaters*. Bielefeld: Transcript Verlag.

Madsen, Patricia Ryan. 2005. *Improv Wisdom – Don't Prepare, Just Show up*. New York: Random House.

Magerko, Brian, Waleed Manzoul, Mark Riedl, Allan Baumer, Daniel Fuller, Kurt Luther, and Celia Pearce. 2009. 'An Empirical Study of Cognition and Theatrical Improvisation.' *Seventh ACM Conference on Creativity and Cognition*, no. c: 117–26. doi:10.1145/1640233.1640253.

Müller, Viktor, Johanna Sänger, and Ulman Lindenberger. 2013. 'Intra- and Inter-Brain Synchronization during Musical Improvisation on the Guitar.' *PLoS ONE* 8 (9). doi:10.1371/journal.pone.0073852.

Napier, Mick. 2004. *Improvise: Scene from the Inside Out*. Portsmouth: Heinemann Drama.

Patel, Aniruddh. 2008. *Music, Language, and the Brain*. Oxford: Oxford University Press.

Pressing, Jeff. 1984. 'Cognitive Processes in Improvisation.' In W. Ray Crozier & Anthony J. Chapman (Eds.), *Cognitive Processes in the Perception of Art*. Amsterdam, North Holland. Elsevier Science Publishers BV.

Salinsky, Tom, and Deborah Frances-White. 2008. *The Improv Handbook*. New-York and London: Continuum Publ.

Sawyer, Keith. 2003. *Improvised Dialogues – Emergence and Creativity in Conversation*. London: Ablex Publ.

Sawyer, Robert Keith. 2003. *Improvised Dialogues: Emergence and Creativity in Conversation*. Westport, CT: Ablex Publishing.

Scheiffele, Eberhard. 1995. *The Truth in Theatre*. Berkeley, CA.

Scheiffele, Eberhard. 2001. 'Acting – An Altered State of Mind.' *Research in Drama Education* 6 (2) 179–191.

Spolin, Viola. 1977. *Improvisation for the Theater*. London: Pitman Publishing.

Walton, Ashley E., Michael J. Richardson, Peter Langland-Hassan, and Anthony Chemero. 2015. 'Improvisation and the Self-Organization of Multiple Musical Bodies.' *Frontiers in Psychology* 6 (MAR): 1–9. doi:10.3389/fpsyg.2015.00313.

3

DEVISING – EMBODIED CREATIVITY IN DISTRIBUTED SYSTEMS

Rick Kemp

How can cognitive science assist makers of devised theatre? While this form has become a familiar feature of Western performance, little has been written about principles that underlie its many disparate expressions. Indeed, publications on the topic describe widely differing approaches to making theatre (e.g. Williams 1988, 1999; Oddey 1994; Callery 2001; Bicât and Baldwin 2002; Govan, Nicholson and Normington 2007; Graham and Hoggett 2009; Mermikides and Smart 2010; Swale 2012; Syssoyeva and Proudfit 2013, 2016; Robinson 2015). A prolific diversity of processes and styles is evidence of the exuberant health of the form, but students, practitioners and researchers could be assisted by concepts and terms that identify both commonalities and differences among and between different practitioners. Theatre people have long had components (evolving mainly from Aristotle and Stanislavski) with which to analyse, describe and prepare scripted theatre, but in written drama these derive from the linear temporal structure of a fixed narrative – the script. Devised theatre, however, is authored through action, with meaning arising dynamically from the simultaneous interaction of multiple people, objects and phenomena. Consequently, we need different concepts and terms to describe both the processes and products of devised theatre. In this chapter I'm going to describe some ideas from cognitive science that identify underlying principles that may be applied across varied devising processes. In doing so, I hope to provide practitioners, researchers and students with some concepts that can assist in describing features of devising that might otherwise be difficult to define or even recognise. I also hope that researchers from disciplines other than theatre may find these conceptualisations useful in examining how devised theatre can be a useful forum for the exploration of intersubjective creativity. This possibility arises through the application of concepts and terms from cognitive science to performance processes, thus creating a degree of 'consilience' (see Bruce McConachie's Chapter 26 in this book).

Also known as collaborative creation or ensemble theatre, devised theatre has moved from an esoteric fringe status in the second half of the twentieth century to a presence at multiple levels of activity – community theatre, regional theatre, training programmes and even the most commercial of mainstream projects. Despite the prevalence of devised theatre, it is difficult to define. It is a genre in which dramatic material is created through a variety of processes in a multiplicity of forms and contexts, and for many different purposes. For many people, the default description of devised performance has been what it is not; the

Western tradition in which a playwright creates a script which is then interpreted and staged by director, designers and actors. Indeed, one of the most significant works on devising in theatrical scholarship acknowledges that the common understanding of devising is that it is 'a mode of work in which no script – neither play-text nor performance score – exists prior to the work's creation by the company' (Heddon and Milling 2006, 3). While this statement is generally applicable as a description, it has some limitations as a definition of devised theatre. Firstly, as the description can apply to improvisatory performances, it does not distinguish these from devised productions. While devising processes incorporate improvisation, a devised piece that is presented to an audience generally has a 'performance score' that enables it to be repeated. Improvisatory performances are, by their nature, productions of meaning that are simultaneously generated and expressed.

Secondly, techniques of devising are now applied to existing scripts, playwrights participate in devising processes, and collaborative improvisations can result in published plays that are then performed by other practitioners. These phenomena have only increased in scope in the decade since Heddon and Milling published the first edition of their book. For example, one of the best-known devising companies, Complicite, uses the same practices on both script-based and non-script based shows. In describing the company's first approach to working with an existing text (Friedrich Dürrenmatt's *The Visit*), Complicite member Mick Barnfather said about the rehearsal process: 'Our approach was the same [...] It still felt just as exploratory. Yes, there was a text, but we were experimenting and doing exercises in a similar way to find the style' ([Fry, 2013] in Saunders 2015, 177).

Recognising the limitations of the common understanding of devising, Heddon and Milling suggest that devising might 'be best understood as a set of strategies' (2006, 2) that generate, shape and edit material into an original performance. These strategies include improvisation, researching, designing, writing, choreographing, discussion and debate, editing and rehearsing, among others. Again, while this description is broadly applicable, it has limitations if we seek to use it as a definition. It is both possible and likely that a playwright might also employ some or all of these strategies in writing a script. Since similar strategies are taking place in the creation of both devised and scripted theatre, the strategies themselves are not definitive of one form or another, but are illustrative of tendencies. It is probably useful to think of these strategies forming a spectrum of theatrical creation. In one area of this spectrum, a solo playwright might be mentally improvising, choreographing and editing dramatic material that will be recorded in the form of a written script. In another area of the spectrum, a group of practitioners would also be improvising, choreographing and editing a performance – but with a crucial difference – *they incorporate the embodiment of meaning in its creation*. The intersubjective embodiment of a solo playwright's script does not occur until it is rehearsed and performed. (I use the term 'intersubjective embodiment' to distinguish this phenomenon from the way that language is embodied in the solo playwright's body-mind as he or she writes). As I described in *Embodied Acting* (2012), our brains process written language differently from speech. As the linguistic component of a devised piece arises in the form of speech, its meaning is expressed simultaneously with that of the non-verbal communication that accompanies it. The embodied nature of the process emphasises the extent to which non-verbal communication carries meaning and thus encourages multi-modal expression, rather than prioritising language as a written script does. Additionally, as devising is generally (but not always) conducted in groups, it facilitates polyvocal expression. This disrupts the implicit hierarchy of creativity in traditional theatre in which director and actors 'interpret' an individual playwright's propositional meaning. Devising practitioners also recognize that the social and physical environments within which they work provoke, affect and contextualise meaning. Thus, people (performers, designers, writers, musicians, technicians, etc.) and physical features (space, location, objects, light, sound, temperature, textures and so on) are all recruited in the creation of a piece. So, as the first step towards identifying distinctive tendencies of devising, one can say that it is the embodied creation of multi-modal (rather than predominantly linguistic) meaning.

Situated cognition theory claims that what people perceive, how they conceive and what they do all develop interactively and are tied to their environment. In shows that are created entirely from improvisation, pre-known knowledge can be said to reside in the declarative and procedural memories and imaginations of the improvisers. The interactions of the improvisers create new and often unexpected knowledge/meaning, especially since devising practices deliberately incorporate imagined circumstances both in preparation and performance, a feature that links embodied and material phenomena with fictional worlds. I propose that this, among other factors, prompts the stimulation of ideational combination, identified by neuropsychologist Arne Dietrich as a central process of creativity (2016). Dietrich rejects theorisations of creativity that make it a specialised activity with its own neural signature. Instead, he insists that as a brain mechanism, creativity must be embedded and distributed. Drawing on evolutionary biology, he posits a generate-and-test process of ideational formation among neural networks and identifies a feature known as a Predictive Goal Representation, which establishes merit parameters for the fitness of hypothetical functions. This predictive goal state links explicit and implicit systems, which takes account of both conscious deliberation and intuitive insight in the creative process. As ideas are combined within the merit parameters, their degree of fitness determines whether they will move into fringe working memory and from thence to core working memory for conscious consideration. I suggest that the multi-modal activity that is prevalent in devising at the behavioural level is a significant stimulator of mental cross-modality, encouraging variety in the process that Dietrich describes. An example of how this process might be expressed in behaviour can be found later in this chapter, in the description of studio work in embodying the qualities of paper. In the journey of transposing this abstracted movement into a human context, the movement qualities of the practitioners behaving like crumpled-up or torn paper can be thought of as a set of merit parameters for which the body-mind is seeking a match that is congruent with 'human behaviour.' Multiple ideas can arise about what would provoke a person to 'crumple,' to 'tear' or be 'torn.' For Toby Jones, the actor describing this activity, the hypothetical 'fitness' expresses itself as the emotion provoked by being 'torn apart' from a loved one. Jones associates this with the world of Chekhov's plays. Other practitioners could well find different 'matches,' such as a personal memory of parting, or an imagined scene of a family being separated. These and other possible 'matches' demonstrate the fluidity of potential meaning in a devising process, which leads to my next proposal.

A common theme among devising practitioners is the importance of process – prioritising this over 'product.' So I propose that, because the meaning of a devised performance is developed over time through process, devising is fundamentally improvisatory in nature, whether or not practitioners use the term or follow established modes of verbal or physical improvisation. What I mean by improvisatory is that at the start of the process, the full content or meaning of the eventual performance is unknown. Evidently there is a known goal to create a show, and that goal can be focused by various criteria – for example, by theme, by intent, by process, or by source material, among many other possibilities. These criteria form sets of parameters (or constraints) for Predictive Goal Representations in the minds of the members of the devising group. As they work together to generate material, the meaning of each component is contingent on its interaction with others during the process of devising. Consequently, the overall meaning of the final performance is both emergent and distributed. This way of understanding improvisation follows studies of verbal improvisation conducted by cognitive psychologist Keith Sawyer (e.g. 2002, 2009). In these studies, Sawyer identified contingent meaning in a chronologically sequential structure of verbal turn-taking in improvisation; the overall meaning that arises is emergent because each participant's contribution is dependent on preceding contributions and not apparent until the whole temporal process is completed: 'Emergence is commonly observed in complex dynamical systems – systems with many elements, organised into multiple levels of subcomponents, with multiple interactions among elements and subcomponents' (Sawyer and DeZutter 2009, 83). This concept of emergent meaning within a dynamic system plays an important part in identifying the creativity of devising as 'distributed.'

Cognitive neuroscientist Edwin Hutchins is one of the most significant researchers in the field of distributed cognition, and has developed the concept of the cultural-cognitive ecosystem to describe the ways in which all human cognition is 'distributed.' As Bruce McConachie states, '[...] we could say that all performance events are parts of one or several cultural-cognitive ecosystems' (McConachie 2015, 93). McConachie demonstrates how Hutchins' theory can be applied to theatrical performance by using it to analyse the emergence of musical comedy in London and New York in the early twentieth century. While I do not have space here to apply Hutchins' theory to devising, it serves to demonstrate the way in which cognitive scientists acknowledge the influence of multiple factors, including evolutionary biology and culture, in human cognition and behaviour.

Sawyer and DeZutter's application of the concept of 'distributed cognition' to creativity challenged the traditional concept of it as a mental process that takes place in an individual's head. By applying interaction analysis to some theatrical performances developed through improvisation, they described a non-individualistic creative process that they referred to as 'distributed creativity' (Sawyer and DeZutter 2009, 81). While the focus of this study was predominantly on verbal turn-taking (although non-verbal communication was acknowledged), devising can also be thought of as distributed creativity. In contrast to the sequential turn-taking of verbal improvisation, the interaction of components of meaning can be both simultaneous and reciprocal because of the tendency of devising to be collaborative and multi-modal, de-emphasising the primacy of language. While speech may be involved, practitioners are also sensitive to factors such as space, image, sound, rhythm and physicality. This approach allows for non-linguistic meaning to play its part in constructing the meaning of a performance.

Examples of this multi-modal approach to devising can be found in the teaching of the school that has produced more devising companies than any other single institution worldwide. This is the Ecole Jacques Lecoq in Paris. I describe aspects of Lecoq's pedagogy and its relation to embodied cognition in Chapter 16, and that chapter can fruitfully be read in conjunction with this one. As we described in our General Introduction, our section aims are teleological and there will inevitably be some overlap of topics between sections. Here, the connection between the two chapters is intentional. Using Lecoq's work as a route to identifying principles of devising is not intended to prioritise his approach over others; as I mentioned earlier, there are a multiplicity of devising methods. Given Lecoq's accessible articulation of theatrical creativity, I hope that some common principles can emerge through an examination of his work, and that of devising practitioners who use his methods. I'm going to focus on some aspects of Lecoq's pedagogy that, given a cognitive analysis, can be seen to promote creativity in devising practice. These are his neutral mask 'identifications' work; transposition – the situating of embodied phenomena in imagined environments; and the explicit and conscious use of constraints, which creates a flexible and fluid conceptualisation of dramatic structure that is consistent with Dynamic Systems Theory (DST), a framework for modelling interactive behaviour that I will describe in more detail later.

Lecoq's connection with the growth of devised theatre in the twentieth century was stimulated by his explicit identification of improvisation, analysis of movement and the exploration of collaborative creativity as the three pillars of his pedagogy. Within this overall framework, three pedagogic features that he developed have contributed to the remarkable amount of devised theatre companies that originated at his school or that use his techniques. Firstly, '*Autocours*' (self-taught courses) were sessions in which groups of students worked independently to create short performances on given themes that they presented to the rest of the school. Secondly, the '*enquête*' ('investigation') required students to participate in the activities of a working environment like a hospital, fire station or factory for four weeks and then create a short performance that 'replayed' their experiences in a compressed time span. Thirdly, the '*commande*' was a commission for a final self-created work presented at the end of the second year. Each of these practices required students to discover how they could autonomously generate original dramatic material. As theatre practitioner Nir Paldi's comments later in this chapter demonstrate, Lecoq graduates consciously acknowledge the connection between these activities and their professional devising work.

As I mention elsewhere, Lecoq believed that the starting point for theatre is not the text of a play, but an actor's playful engagement with sensorimotor experiences of his or her physical and social environments. This focus on the actional domain of experience correlates with one of the foundational concepts of embodied cognition – that sensorial and motor experiences form the neural foundations for mental concepts. Lecoq's pedagogy explores the experiential domain of action through a framework that he calls 'dynamics' – combinations of rhythm, force and space. This framework is significant as he is encouraging his students to initially engage with experience at its sensorimotor level, rather than beginning with language as one does in the tradition of scripted theatre. These experiences are initially framed by training with the Neutral Mask, a full-face mask with a neutral expression. In the Neutral Mask 'identifications' work (described in Chapter 16 of this book), actors consciously embody the rhythms of movement found in natural and social environments. In *Embodied Acting* (2012), I demonstrated how the non-habitual movement patterns of Neutral Mask training can stimulate an altered sense of self through proprioceptive awareness. Importantly, with regard to the concept of ideational combination, Neutral Mask identifications work encourages performers to experience themselves metaphorically – for example as dough, wind, water, cardboard, liquids or, in the example described here by British actor Toby Jones, as paper:

> when we were studying paper and studying the ripping apart of different densities of paper and then studying embraces and watching paper un-crumple on its own or being crumpled back up … at the end of the lesson just seeing all of this paper strewn all over the classroom and a class that had really been about embracing and tearing and the tearing apart and coming together and the sheer emotion in the room, it just looked like an autumnal scene from a Chekhov play of people parting or coming together … it's very concrete and real.
>
> (*Jones in Evans and Kemp 2016, 205*)

The embodied experience of the Neutral Mask identification provokes Jones's imagination to provide a context – this is in itself an example of ideational combination – seeing shreds of paper as leaves in autumn, and the actions of crumpling and tearing as embracing and pulling apart. Arne Dietrich talks about speed of processing being a factor in ideational combination. A pre-calibrated and weighted neural network creates processing speed differentials for different kinds of information. Some types of information are processed more expediently than others, moving them more rapidly from 'fringe working memory' to conscious awareness. For Jones as a trained actor who frequently inhabits the fictional worlds of playwrights, the association that most rapidly contextualises his experience is the emotional world of a Chekhov play. This particular leap of creativity has happened spontaneously and is an example of the situating of embodied phenomena in imagined circumstances.

This process is consciously encouraged by Lecoq through activities that he terms 'Transpositions,' where the patterns of movement that are experienced in the Neutral Mask identifications are transposed into human behaviour and social contexts – as Lecoq explains:

> The work done on identifications has to be reinserted into the dramatic dimension. For this purpose I use the transference method which consists in basing oneself on natural dynamics, on action gestures, on animals, on materials using them for expressive purposes in order to achieve a better playing of human nature. The objective is to achieve a level of theatrical transposition, going beyond realistic performance.
>
> *(Lecoq 2001, 44)*

Lecoq proposes two approaches – firstly to humanise an element or an animal giving it a behaviour or a voice and relating it to other elements or animals. The second is to reverse the process by beginning with a human character and gradually allowing the elements or animals in which it is grounded to show through.

Beyond the stylistic results that support performers in creating non-realist theatrical forms, I propose that Lecoq's transposition has another effect that stimulates creativity. Geoffrey Rush, another Lecoq-trained actor, refers to the process of the school as 'synaesthetic discovery' (Rush in Evans and Kemp 2016, 405). In the clinical sense, synaesthesia is a neurological phenomenon in which stimulation of one sensory or cognitive pathway leads to automatic, involuntary experiences in a second sensory or cognitive pathway. Himself a synaesthete, Rush is using the term more loosely than in its clinical sense to describe instances where embodied experience is shifted from one mode or context to another – Lecoq's transposition. For some students, this process can be extremely vivid. Theatre practitioner and teacher Jenny Gilrain writes of her experience of working with Lecoq on the dynamics of music in 1991. After listening to Bartok's *Andante Tranquillo*, students are asked to describe it verbally using the vocabulary they developed while working with the Neutral Mask, a vocabulary of natural phenomena and man-made materials.

> *LECOQ: What color is it? Where is it in the space? In what direction does it move? How does it move? What path does it take? What is the dynamic? The space? Shape? Element?*
>
> The answers come pouring out of students: Circular. Green. Brown. Purple. Twisting. Creeping. Up. Diagonals. Undulations. Snake. Vines. Disc. Falling. Spiraling. Hole. Down. Abyss.
>
> *(Gilrain in Evans and Kemp 2016, 128)*

Lecoq then plays the music again and asks the students to stand and watch it:

> Twenty-four years later, I can still see the music in my mind as I saw it then: One thin creeping vine enters through the crack around the door, then another and another. The

> vines grow thicker and climb the walls, sprouting leaves, covering the ceiling. Then the whole room begins to spin. A hole opens up in the middle of the floor, sucking everything down. Spiraling down, down, down.
>
> *(Ibid)*

Gilrain reports that her experience of seeing the music in this way was not unique – all the students in the class pictured the music in varying ways. Rather than considering this phenomenon as induced synaesthesia, which is, after all, an involuntary condition, this could be thought of as consciously stimulated metaphorical thinking resulting in imaginative projection. In neuroscience, the question of the relationship between metaphorical thinking and synesthesia is very much a live one. However, there are studies on linguistic synesthetic metaphors (cross-modal metaphors) that show that their mapping is consistent with Lakoff and Johnson's conceptual metaphor theory – in other words, that they are embodied metaphors (Yu 2012). Glenberg and Gallese's Theory of Action-Based Language (2011) suggests that language acquisition, comprehension and some aspects of production, including gesture, can be related to the sensorimotor neural networks involved in physical actions. If we accept that language is based in action, as Glenberg and Gallese propose, then it is possible that the same neuronal phenomenon involved in linguistic synaesthetic metaphors could apply to the physical cross-modality of Lecoq's work. This hypothesis is supported by researchers examining the effect of action on thought. In this field, several studies suggest that (to quote the title of a particularly relevant article) 'Doing Gesture Promotes Learning a Mental Transformation Task Better Than Seeing Gesture' (Goldin-Meadow et al. 2012).

In the exercises that I've described, Lecoq is consciously using a variety of constraints. In the Bartok class, the music is a constraint, both for the attention and subsequent physical actions of the performers. It is not proscriptive – each performer is free to respond in his or her own way, but the music creates a set of parameters for his or her behaviour. This behaviour has already been categorised through the Neutral Mask identifications, but in those the parameters are more defined – for example, the movement dynamics of dough are different from those of fire. Lecoq refers to spatial, temporal and numerical constraints, but he also considers dramatic styles to be constraints. Conceiving of theatre as sets of interacting constraints creates parameters of attentional focus that facilitates cross-modality. In the Bartok example, the parameter of the neutral mask vocabulary depends on the modes related to sensorimotor and visual perception. The conscious awareness that students have developed of their own proprioceptive and interoceptive patterns allows them to apply these modes to the aural mode in a targeted way that structures interaction and provokes the imagination. This way of minding means that performers who train with Lecoq come to conceptualise features of performance as sets of parameters that can intersect with one another, can contain one another and can interact with one another. This conceptualisation is apparent in this extract from an interview with drag artist Nir Paldi (N.P. in the following extract), a 2005 Lecoq graduate who formed the London-based devising company Ad Infinitum:

> N.P. In the auto-cours at Lecoq –this is the part of the training where you make a new piece of theatre every week – you learn to work within a constraint: We work with constraints too, and as a collective, but unlike auto-cours in which there is no leader, there is always a person who is leading our process.
>
> V.A. What do you mean by constraint?
>
> N.P. So for *Light*, essentially, the constraint is complete darkness and actors lighting themselves. For *Ballad*, we are talking about war and the Israeli-Palestinian conflict. Or

> the Israeli psyche through cabaret and drag. In *Translunar*, it's without words, and we are traveling from youth to old age using hand-held, full-face masks.
>
> V.A. So when you start the rehearsal process, you don't know where it will lead, all you have is the constraint?
>
> N.P. That's right. We just went through this process again with *Light*. This is how we practice what we took from Lecoq: we challenge ourselves to completely reinvent the form every time we approach a new theme and a new story. You start with the constraint, then you develop the constraint, then develop the form within the constraint; then you ease the constraint to let the story breathe.
>
> *(Ackermann in Evans and Kemp 2016, 331)*

Paldi identifies several different features as constraints: the use of darkness, the topic of the Israel–Palestine conflict and the topic of ageing. He also acknowledges that a performance arises through the process – in the terms of cognitive science, that it is emergent meaning. The constraints can be understood as sets of parameters for Predictive Goal Representations as described earlier. An example of the ideational combination that arises can be seen in the show *Ballad of the Burning Star* that Paldi refers to, which uses drag and cabaret to address the Israeli–Palestinian conflict. These features were considered highly original by many critics, with the show winning several awards.

The devising practice of using varying forms of constraints can be understood with DST. A dynamic system is a flexible, complex, time-dependent and emergent system in which simultaneous contingencies stimulate the reciprocal interaction of components.

DST provides a model that describes the flow of relationships among the components of a whole phenomenon. As this model acknowledges the real-time 'circular causality' of elements within a system, it is well-suited to describe the ways in which people, environment and actions simultaneously affect one another in the context of devising a dramatic performance, as well as presenting that performance. Originating in mathematical theory, DST has been applied by neuroscientist J.A. Scott Kelso to model properties of human cognition and social behaviour in a field known as coordination dynamics. Kelso's overall goal is to understand how human beings and human brains coordinate activity. In theatre scholarship, John Lutterbie has applied DST to performance in *Towards a General Theory of Acting* (2011). Lutterbie uses a description of UK director Katie Mitchell's process of adapting Virgina Woolf's *The Waves* at the UK's Royal National Theatre as an analogy for the neural activity in the brain – how patterns of neural networks are established and reconfigured in response to different stimuli. (For more on DST, see Chapter 17 by Gabriele Sofia in this book.)

Lutterbie also points out that, in certain ways, neuronal activity is patterned like everyday activity. In this, he follows the principal claim of coordination dynamics, which is that the coordination of neurons in the brain and the coordinated actions of people and animals share a common mathematical or dynamical structure. It is this claim that leads me to suggest that the collaborative creativity of devising processes could be a useful forum for understanding creativity at both the social (distributed creativity) and the neuronal levels (ideational combination). As Sawyer and DeZutter state: 'When cognitive processes are distributed across groups, they become visible, and scientists can observe them by analyzing the verbal and gestural interactions among the participants' (2009, 81). The examples of Lecoq's practice that I've given serve, I hope, as specific examples of some of the tendencies of devised theatre. These highlight how its social and behavioural organisation stimulates both interactive and neural processes of creativity.

Summary

In this chapter, I've proposed that we can identify devised work as having the following tendencies:

1 The embodied creation of multi-modal (rather than predominantly linguistic) meaning, which arises in a context formed of multiple agents of meaning; not only people (performers, designers, writers, musicians, technicians, etc.) but also the physical environment: space, location, objects, light, sound, temperature, textures and so on.
2 The interaction of these agents of meaning links embodied and material phenomena with imagined circumstances, both in preparation and performance. This, among other factors, facilitates the stimulation of ideational combination, a central process of creativity.
3 As meaning emerges from the exploratory devising process, practitioners select aspects to create a form that is (to a greater or lesser extent) 'fixed' to create a performance score that facilitates repeated presentations. This factor distinguishes devised theatre from improvisatory performances.
4 All of these features interact with one another in a manner that can be modelled as a Dynamic System in which process is intentionally structured to promote emergent meaning.

References

Bicât, Tina and Chris Baldwin, eds. 2002. *Devised and Collaborative Theatre: A Practical Guide*. Marlborough: Crowood Press.
Callery, Dympna. 2001. *Through the Body: A Practical Guide to Physical Theatre*. London: Nick Hern Books.
Dietrich, Arne. 2016. *How Creativity Happens in the Brain*. New York: Palgrave Macmillan.
Evans, Mark and Rick Kemp. 2016. *The Routledge Companion to Jacques Lecoq*. Abingdon and New York: Routledge.
Graham, Scott and Steven Hoggett. 2009. *The Frantic Assembly Book of Devising Theatre*. Abingdon: Routledge.
Glenberg, Arthur M. and Vittorio Gallese. 2011. 'Action-Based Language: A Theory of Language Acquisition, Comprehension, and Production.' *Cortex*, Vol. 48, No. 7, pp. 1–18.
Goldin-Meadow, Susan, Susan C. Levine, Elena Zinchenko, Terina KuangYi Yip, Naureen Hemani and Laiah Factor. 2012. 'Doing Gesture Promotes Learning a Mental Transformation Task Better than Seeing Gesture.' *Developmental Science*, Vol. 15, No. 6, pp. 876–84, November 2012.
Govan, Nicholson and Normington. 2007. *Making a Performance: Devising Histories and Contemporary Practices*. Abingdon: Routledge.
Heddon, Deirdre and Jane Milling. 2006. *Devising Performance: A Critical History*. Basingstoke: Palgrave Macmillan.
Kemp, Rick. 2012. *Embodied Acting: What Neuroscience Tells Us about Performance*. Abingdon and New York: Routledge.
Lecoq, Jacques. 2001. *The Moving Body (Le Corps Poétique), Teaching Creative Theatre*, translated by David Bradby. London: Methuen.
Lutterbie, John. 2011. *Towards a General Theory of Acting*. New York: Palgrave Macmillan.
McConachie, Bruce. 2015. *Evolution, Cognition and Performance*. Cambridge: Cambridge University Press.
Mermikides, Alex and Jackie Smart. 2010. *Devising in Process*. Basingstoke and New York: Palgrave Macmillan.
Oddey, Alison. 1994. *Devising Theatre: A Practical and Theoretical Handbook*. Abingdon: Routledge.
Robinson, Davis. 2015. *A Practical Guide to Ensemble Devising*. New York: Palgrave Macmillan.
Saunders, Graham. 2015. *British Theatre Companies: 1980–1994*. London: Bloomsbury Publishing. Kindle Edition.

Sawyer, R. K. 2002. 'Improvisation and Narrative.' *Narrative Inquiry*, Vol. 12, No. 2, pp. 319–49.

Sawyer, R. K. and Stacy De Sutter. 2009. 'Distributed Creativity: How Collective Creations Emerge from Collaboration.' *Psychology of Aesthetics, Creativity, and the Arts*, Vol. 3, No. 2, 81–92.

Swale, Jessica. 2012. *Drama Games for Devising*. London: Nick Hern Books.

Syssoyeva, Kathryn Mederos and Scott Proudfit. 2013. *Collective Creation in Contemporary Performance*. New York: Palgrave Macmillan.

Syssoyeva, Kathryn Mederos and Scott Proudfit. 2016. *Women, Collective Creation, and Devised Performance: The Rise of Women Theatre Artists in the Twentieth and Twenty-First Centuries*. New York: Palgrave Macmillan.

Williams, David. 1988. *Peter Brook: A Theatrical Case Book*. London: Methuen.

Williams, David. 1999. *Collaborative Theatre: The Theatre du Soleil Sourcebook*. Abingdon: Routledge.

Yu, Xiu. 2012. 'On the Study of Synesthesia and Synesthetic Metaphor.' *Journal of Language Teaching and Research*, Vol. 3, No. 6, pp. 1284–9, November 2012.

4 EMBODIED COGNITION AND SHAKESPEAREAN PERFORMANCE

Darren Tunstall

The theory of embodied cognition has underpinned a broad and expanding field of inquiry within cognitive studies since the mid-1970s (see Lakoff 2012: 773–5). It has also received attention from those theatre scholars who, since the early 1990s, have been exploring the topics of embodiment and cognition in relation to Shakespearean production, actor training and methods of rehearsal. A by no means exhaustive list of such pioneering scholars would include Mary Thomas Crane (2001), Bruce McConachie (with F. Elizabeth Hart 2006; see also 2008, 2013, 2015), Rhonda Blair (2008), Amy Cook (2010), Rick Kemp (2012, 2017), Evelyn Tribble (2011), John Lutterbie (2011) and the contributors to *Embodied Cognition and Shakespeare's Theatre* (Johnson, Sutton and Tribble 2014). In this chapter, I mean to outline what embodied cognition is, single out some aspects of it that pertain to performance theory and practice and relate these aspects to different strategies for the Shakespearean text that I have modelled upon some influential theatre practitioners. In doing so, I'm not suggesting that all of these practitioners were precursors or proponents of embodied cognition; rather, the intention is to see what light an understanding of embodied cognition might shed upon rehearsal methods. Thus, I aim to move from the general to the particular, from the theoretical to the practical.

What is embodied cognition? It can be understood as the proposition that an individual's cognitive and bodily processes should be seen as interdependent. As George Lakoff and Mark Johnson (1999) demonstrated, this conception marks a radical break between 'first generation' cognitive science, which tended to consider thinking as disembodied – a computational activity somehow separate from the material reality of the human brain – and 'second generation' cognitive science, which sees thinking as grounded in both neuroanatomy and bodily (sensorimotor) experience. There are variations of emphasis within the field, from the somewhat guarded claims of Alvin Goldman (2013) that an individual's cognition is embodied to a significant degree, to the 'extended mind' hypothesis of Andy Clark (2008), who argues that the idea that cognition is embodied may be applied beyond the human frame into features of the environment itself, such that (to use Clark's most famous illustration) a person's notebook can be seen as playing the role of a biological memory. Differences aside, it is the focus upon brain physiology combined with bodily experience that gives embodied cognition its radical uniqueness. Many experiments have demonstrated that people's engagement with the environment through their senses, and even the behaviour of their internal organs, affects what and how they think. For example, it has been shown that our thought patterns are influenced by tiredness, by hunger, by carrying weight, by cold or heat, by the shape, by relative hardness and roughness of an object we are holding, by the posture we happen to be in and so on (see e.g. Williams and Bargh 2008; Zhong and Leonardelli 2008; Jostmann, Lakens and Schubert 2009). Evolutionary processes have resulted in human brains that are adapted to seek out and to respond actively to external cues (called *affordances* by the perceptual psychologist James J. Gibson), or to internal physiological events, in terms of their relevance to our needs and their functional usefulness. Humans actively perceive the environment in terms of what it can do for them (or against them) in every moment. Indeed, more than any other animal, we imagine and construct objects and environments that furnish us with affordances.

We might see such an engagement as a kind of adaptive priming effect. Some psychologists have considered this priming effect to be an indication that humans suffer from implicit bias in their judgements. For example, we might feel that on a pleasantly sunny day we are more disposed to be generous towards someone who asks us for help. The next day, if the weather is downright miserable, we might not be in the giving vein. On the surface, this behaviour seems to be a mistake; after all, the person who asks for our help is the same whatever the weather, and the person may think the worse of us because our behaviour appears arbitrary. However, we shouldn't necessarily think that embodied cognition causes us to be deluded in our decisions. For one thing, it seems unlikely that millions of years of evolution would lead to such maladaptive behaviour. Perhaps there is a hidden logic to the workings of embodied cognition even if information from the environment leads us into counter-factual thinking. On a pleasant day, when it is relatively easy to stay warm and dry, you may feel that you have enough energy to spare to help someone out; on a cold, rainy day, you may feel that many activities are more demanding in terms of time and effort since you have the weather to deal with, and thus you feel the need to devote more energy to yourself.

Nearly everything that goes under the name of embodied cognition happens under the radar of conscious awareness. Our thoughts pop into our head, and it hardly seems relevant to wonder how they got there. This is also a trick of evolution, designed to keep us focussed on what matters to us. After all, if you are facing danger and you get an impulse to run for your life, it probably won't benefit you to stop and reflect upon what mental process led to that impulse; you're more likely to survive if you just run. Conscious or not, embodied cognition makes sense when it is set against a backdrop of evolutionary constraints and motives. Some research into the evolutionary drives that undergird it has unearthed intriguing results. An important paper by Klein, Cosmides, Tooby and Chance (2002), for instance, demonstrates how priming helps us understand other people's behaviour by introducing boundary conditions into our judgements. Since behaviour can change according to the situation (e.g. the weather), it follows that personality traits cannot be set in stone. So, in order to understand what someone is like, we draw upon general information from semantic memory ('this person is usually helpful and friendly') and specific instances from episodic memory ('this person wasn't helpful and friendly when it was raining yesterday'). In any human interaction, it's usually necessary to make quick inferences on the basis of what are unlabelled signals, because most behaviour doesn't arrive with a signpost declaring its meaning. It is because cognition is embodied that we are able to determine not just the unchanging facts about a person but those aspects of their behaviour that can serve as more or less reliable signals, and in this way embodied cognition feeds forward into decision-making.

There is a useful connection to be made here to Lisa Zunshine's concept of 'embodied transparency' (Zunshine 2012). Zunshine draws upon a range of examples from poetry, films, television comedies, paintings and novels to build an intriguing argument: that cultural products trade on our eagerness to read people's minds on their bodies. At the same time, though, since people tend to agree that 'the body speaks,' we attempt to hide the potentially compromising or threatening signals of our own inner lives by adopting behavioural masks. For Zunshine, the power of fiction lies to a considerable extent in its capacity to provide representations that offer privileged moments of access to the thoughts and feelings of characters: 'We get to see fictional characters at the exact moment when their body language betrays their real feelings. This is in contrast to real life, in which there is always a possibility that we will misinterpret seemingly transparent body language, particularly in a complex social situation, or that people will perform transparent body language to influence our perception of their mental states' (Zunshine 2012, xix).

This strikes me as a particularly successful application of embodied cognitive science that helps to clarify what is often vaguely labelled as 'truthfulness' or 'authenticity' in acting.

Zunshine was among the first wave of path-breaking scholars mentioned earlier. Two recent examples give a flavour of the ways in which newer scholars are building upon their explorations. A 2016 collection of essays edited by Rhonda Blair and Amy Cook, *Theatre, Performance and Cognition: Languages, Bodies and Ecologies*, contains pieces by Neil Utterback and Christopher Jackman on actor training, by Edward Warburton on dance and by Laura Seymour on *Julius Caesar*. Seymour draws upon the work of George Lakoff and Mark Johnson in her discussion of the significance of gestures such as kneeling for relations of political power in the play. And an edition of the journal *Connection Science* devoted to 'Embodied Cognition, Acting and Performance' (2017) seeks, amongst other things, to broaden the discussion into the situation of conservatoire training (for example, in Ysabel Clare's essay on Stanislavsky's system as 'enactive cognitive embodiment'). On that subject, it may seem as though there's nothing especially new about the principle of embodiment in itself. For example, in the field of actor training, during the 1990s, many conservatoires in the UK latched onto the continental traditions of movement-based practices championed by pedagogues like Jacques Lecoq – although conservatoires tend to be selective in their delivery of Lecoq's curriculum. There was also a rethink of Stanislavsky's model of realism in acting in the wake of interventions by scholars like Sharon Carnicke (1998/2009). And along with that has come a promotion of the work of those disciples who took issue with their master Stanislavsky – most notably Vsevolod Meyerhold and Michael Chekhov – as well as practices adopted from other disciplines such as martial arts. Thus, much conservatoire training – at least, in the UK – has led to a renewed focus on the 'psychophysical' resources of the actor. However, it is one thing to declare that actors must find ways to embody their characters, and quite another to adhere to the more radical principle that all cognition is inherently embodied. It is not uncommon for actor trainers to insist upon embodiment while clinging stubbornly to the kinds of mystical ideas about a nonmaterial mind, self or spirit that were a feature of the world-views of canonical ancestors like Stanislavsky and Laban. Here too, Zunshine's argument applies, since such still-prevalent mind-body dualism speaks of the common human desire to be accorded privileged status as an expert in mind reading.

Of course, artists have always wanted to get out of the body, and the yearning to do so has resulted in some extraordinarily beautiful art. But, to repeat, embodied cognition theory stresses that every idea that comes to us is grounded in our bodily anatomy and experiences – including the very idea that we would yearn to get out of the body. We simply cannot think, we cannot imagine, we cannot yearn, outside the body. According to Lakoff and Nunez (2000), even the most rarefied abstract thinking, such as higher mathematical reasoning, has its origin in basic physical facts such as our bilateral symmetry and our sense of spatial orientation.

It's time to see how we might apply such an idea to the performance of a play by Shakespeare. Since earlier I was talking about how the weather affects people's thoughts, why not *King Lear*, a play well known for attempting to put a thunderstorm on stage:

KENT: Who's there, besides foul weather?
GENT: One minded like the weather, most unquietly

(3.1.1–2)

How do we rehearse these opening lines to the scene? Here are some, by no means exhaustive, possibilities for two actors playing Kent and the Gentleman, who I shall refer to as 'you':

1. Sit around a table with the script and talk through what it means. You might try speaking the lines a few different ways, using intonation patterns to communicate different meanings. Then stand up and say the lines whilst moving about in the rehearsal space away from the table. Perhaps another person reacts to what you do, gives instructions, offers direction, ideas and encouragement, and so on.

On the surface, this 'table talk, then on your feet' approach doesn't *seem* like the most embodied way of starting work on the script. However, embodied cognition theory reminds us that all experiences are processed through the body, and that includes sitting around a table talking about the play. So as soon as you sit down at a table, your thought processes will be affected; the issue is then whether or not the kind of embodied experience you're having will help you in performing the text (and perhaps it might). The relative hardness of the chair you sit in will impact what you imagine and feel about the text, as will the shape of the table, the distance between you and the other people around the table, where you sit at the table (e.g. at its corner, in the middle), along with any objects placed on the table…along with what you were already feeling when you sat down (for example, you were hungry). If this table and these chairs, and this physical relationship to the other people, are not actually relevant to the staging of the scene, you could make inappropriate choices in your intonations and gestures. There's a possibility that the scene will fall into a kind of passive state. The moment an actor sits in a soft chair, for instance, the effort to resist gravity is reduced and the result is often a loss of projected energy in the body and voice as the actor begins to sink into the furniture.

Arthur Glenberg and colleagues have developed a strategy to help children read, called 'Moved by Reading,' which draws upon a simulation theory of language comprehension. In Glenberg's words, 'we understand language much like we understand situations: in terms of the actions the situation, or described situation, affords' (Glenberg 2011: 7). The situation described in the text is simulated using our neural systems of action, perception and emotion. In my experience, actors who are inexperienced or not confident with Shakespearean text can be helped further at the stage of 'table talk' using an approach based on Glenberg's strategy. The actors are presented with toys to play with, representing the characters and the environment. In the first reading, the actors act out the situation by physically manipulating the toys on the table (as if they were puppeteers, or directors creating an action plan for the scene using a model box). In the second reading, the toys are removed and, while speaking the text, the actors imagine moving them in the same way. Of course, there is nothing new about asking an actor to create mental imagery; the difference is that by using toys at the reading stage, the actor – like the child learning to read – is, in Glenberg's words, 'likely to engender a significant motor component in addition to visual imagery' and this motor component 'increases the range of information encoded' (ibid: 12).

2. Learn the lines in advance of rehearsal. Stand in the rehearsal space and, without discussing what the lines mean, move about in the space saying the lines to your scene partner.

Here, you've put the script down so your hands are free for gesturing; you're working out what the scene is about on your feet rather than being potentially stuck in a physical posture like sitting. Perhaps you move with what you sense as changes of thought or intention. The voice coach Cecily Berry used to teach actors a preparation exercise for that: you walk and talk, and change direction every time you hit a punctuation mark in the script. It might begin to seem as if your feet are doing the thinking. The notion that different parts of your body can do the thinking, which I have borrowed from Michael Chekhov, has been explored by Tom Cornford (2012). The strategy tends to require, though, that you have either done your homework or you are experienced at handling Shakespeare's language.

3. On your feet, improvise verbally around the content of the scene before moving onto the text.

This exercise bears a resemblance to what Stanislavsky called 'active analysis.' With each pass, you move closer to using the actual words written in the script. Meanwhile, the physical actions of the scene are being worked on in ever greater detail.

One variation of this strategy is to render the text into your own words first. You might do this around a table (with or without toys as you see fit). Then speak your contemporary prose version with a commitment to an intonation pattern that suggests the purpose behind the speech. At the same time, you are invited to make gestures that illuminate the purpose still further:

KENT: (*holding up a hand as if to stop the GENTLEMAN*) Who's out *there* in this awful *weather*?
GENT: (*holding up his hands in submissive greeting*) I'm *feeling* like the weather – very *troubled*.

Having worked out a pattern of intonation and movement which makes sense of your response to your partner, return to the script and perform as far as possible with the same intonation and the same movements:

KENT: (*holding up a hand as if to stop the GENTLEMAN*) Who's *there*, besides foul *weather*?
GENT: (*holding up his hands in submissive greeting*) One *minded* like the weather, most *unquietly*.

Having performed this to your satisfaction, the next stage is to eliminate the gesture, but perform as though you are making the gesture in your imagination. The gesture takes on something of the quality of what Michael Chekhov called Psychological Gesture (Chekhov 2002), an internalised representation of something essential about the character; it also bears a resemblance to the second part of Glenberg's reading strategy. The results of this exercise can be very encouraging if you stick to the game of performing the script with the same purposeful intonation and gesture patterns as your personalised version of it, and if you commit to the stage of performing the gesture in the imagination while speaking. It's useful to have a director in the room to insist that you don't alter the intonation as you move towards internalising. The psycholinguist and gesture scholar David McNeill, in an interview I conducted with him by email, wrote about my adoption of this exercise:

> Gesture and prosody are so alike, and the gesture would orchestrate it most directly. Then, at last, they move into a mode of secure performance where the gesture seems to have dropped out (though I suspect it is present internally).
>
> *(Tunstall 2016: 158)*

Gestures embody the speaker's mental imagery, and furnish evidence for the proposition of Lakoff and Johnson (1999) that thinking happens through the body. It follows from this that, as McNeill contends, a composed written structure like *King Lear* contains within its language hidden gestures that are in principle recoverable; in this exercise, the actor is attempting to unearth them and make them visible. In a similar vein, Rick Kemp shows how gesture analysis can support 'the common practice in rehearsing Shakespeare of identifying the "active" word in a line, and consciously executing a gesture to accompany it. This gesture may or may not be carried through to performance, but serves to enrich the actor's physical experience of the text, thus integrating written language with speech and gesture' (Kemp 2012: 36–7).

> 4. You have a bamboo cane about four feet long. You hold one end of it, and your scene partner holds the other, with one finger or in the palm of your hand. You move together, keeping the stick between you, pushing and being pushed around the space through the stick, and you explore this relationship for a while. Then you play the same game whilst speaking the dialogue. After a while, remove the stick but continue playing with an idea of physical influence, of pushing and pulling with your partner, at the same distance. The possibilities of the game with respect to timing, synchronicity, mimicry, force, scale and so on can be developed as you play with the text.

Here, a game derived from Jacques Lecoq is used to bring you, the actor, into a conscious awareness of how your body moves in relation to the affective dynamics of the scene. The patterns of movement can be as large, expansive and stylised as you please. Other games from the traditions of 'movement-based' performance can be used in a similar manner, such as the viewpoints approach of Anne Bogart (Bogart and Landau 2005), or the choreographic language of Rudolf Laban (see Bradley 2009). As with rehearsal exercise 3, the movements can be reduced, disguised or even dropped altogether while you continue the game in your imagination. The style of performance may be wedded to a behavioural verisimilitude, but there would still be an intensified awareness of your body in its relation to your partner's body, coupled with mental imagery.

These are just four approaches to rehearsal drawn from various practitioners that suggest ways in which attention may be placed upon the body's primary role in the creation of meaning. Of interest to me, following the conception of embodied transparency coined by Lisa Zunshine, is how Shakespeare's dialogue offers Kent (and the audience) immediate access to the inner life of the Gentleman, a character we know nothing about. In doing so, he accomplishes many other things: he reinforces any offstage sound effects or onstage body language attempting to conjure up the weather for the audience; he puts us in Kent's shoes as the character who is doing the mindreading; he creates the potential for a trusting relationship between the two men; and, in suggesting how the weather influences people's thoughts and feelings, he reinforces the cosmic metaphorical significance of the storm.

But here I want to stress that the agenda is not to make Shakespeare's writing appear more embodied than it is; on the contrary, it's hard to imagine a verbal structure more intensely engaged with embodiment than his dramatic language. My crude paraphrasing of the lines of Kent and the Gentleman reveal this – I have failed, for instance, to find an acceptable substitute for the almost offhand manner in which Shakespeare embodies the foul weather. It becomes a person with an unquiet mind, a personification that the scene amplifies as it continues:

KENT: I know you. Where's the king?
GENT: Contending with the fretful elements:
Bids the wind blow the earth into the sea
Or swell the curled waters 'bove the main,
That things might change or cease.

(3.1.3–7)

In the ensuing scene, the weather is amplified still further into an apocalyptic assembly:

LEAR: Blow winds and crack your cheeks! Rage, blow,
You cataracts and hurricanoes, spout
Till you have drenched our steeples, drown the cocks!

(3.2.1–3)

The weather gods behave like vengeful, destructive people, and as Zunshine would say, they get inside your head. Lear cannot think of the weather in a disembodied way, because no thinking is disembodied. As one works through any play by Shakespeare, one finds this over and again: the words are full to the brim of 'body-ness.'

I have discussed embodied cognition in terms of individuals responding subconsciously, and yet actively, to the environment. This response is not the product of a brain that is a blank slate. As the work of James Gibson and Ulric Neisser (2014) revealed, we do not look at the world with unprepared eyes. To return to an earlier theme, we perceive the world with sensory equipment that has been primed by evolution to discover what the environment may offer us by way of affordances. Further, we try to construct a world that will furnish us with those affordances we, largely unconsciously, seek. With that in mind, there is a fifth rehearsal strategy to be added, one which owes something to the utilitarian methods of Meyerhold (see Paavolainen 2012: 53–92) and which finds an interesting parallel in Rhonda Blair's description of working on Anton Chekhov's *Three Sisters* by inviting her actors to make sculptures out of human bodies (Blair 2008: 100–3):

5. Learn your lines before you enter the rehearsal room. Then, using objects, sound and light, construct an environment in the room which affords you opportunities to move, to manipulate the objects, and/or to relate to your scene partner in different ways. This world and the things in it may not necessarily bear any obvious relation to the world described by the playwright. They need only provide stimulating challenges to your senses. Now – play.

References

Blair, Rhonda (2008) *The Actor, Image, and Action: Acting and Cognitive Neuroscience*. Abingdon and New York: Routledge.

Blair, Rhonda and Amy Cook, eds. (2016) *Theatre, Performance and Cognition: Languages, Bodies and Ecologies*. London and New York: Bloomsbury.

Bogart, Anne and Tina Landau (2005) *The Viewpoints Book: A Practical Guide to Viewpoints and Composition*. New York: Theatre Communications Group.

Bradley, Karen K. (2009) *Rudolf Laban*. Abingdon and New York: Routledge.

Carnicke, Sharon Marie (1998/2009) *Stanislavsky in Focus*. Abingdon and New York: Routledge.

Chekhov, Michael (2002) *To the Actor*. Abingdon and New York: Routledge.

Clare, Ysabel (2017) 'Stanislavsky's system as an enactive guide to embodied cognition?', *Connection Science* 29 (1): 43–63.

Clark, Andy (2008) *Supersizing the Mind: Embodiment, Action and Cognitive Extension*. Oxford: Oxford University Press.

Cook, Amy (2010) *Shakespearean Neuroplay: Reinvigorating the Study of Dramatic Texts and Performance through Cognitive Science*. Basingstoke and New York: Palgrave Macmillan.

Cornford, Tom (2012) 'The importance of how: Directing Shakespeare with Michael Chekhov's technique', *Shakespeare Bulletin* 30 (4): 485–504.

Crane, Mary Thomas (2001) *Shakespeare's Brain: Reading with Cognitive Theory*. Princeton, NJ and Oxford: Princeton University Press.

Glenberg, Arthur M. (2011) 'How reading comprehension is embodied and why that matters', *International Electronic Journal of Elementary Education* 4 (1): 5–18.

Goldman, Alvin I. (2013) *Joint Ventures: Mindreading, Mirroring, and Embodied Cognition*. Oxford: Oxford University Press.

Johnson, Laurie, John Sutton and Evelyn Tribble, eds. (2014) *Embodied Cognition and Shakespeare's Theatre: The Early Modern Body-Mind*. Abingdon and New York: Routledge.

Jostmann, Nils B., Daniël Lakens and Thomas W. Schubert (2009) 'Weight as an embodiment of importance', *Psychological science* 20 (9): 1169–74.

Kemp, Rick (2012) *Embodied Acting: What Neuroscience Tells Us about Performance*. Abingdon and New York: Routledge.

Kemp, Rick (2017) 'The embodied performance pedagogy of Jacques Lecoq', *Connection Science* 29 (1): 94–105.

Klein, Stanley B., Leda Cosmides, John Tooby and Sarah Chance (2002) 'Decisions and the evolution of memory: Multiple systems, multiple functions', *Psychological Review* 109 (2): 306–29.

Lakoff, George (2012) 'Explaining embodied cognition results', *Topics in Cognitive Science* 4: 773–85.

Lakoff, George and Mark Johnson (1999) *Philosophy in the Flesh: The Embodied Mind and Its Challenge to Western Thought*. New York: Basic Books.

Lakoff, George and Rafael E. Núñez (2000) *Where Mathematics Comes from: How the Embodied Mind brings Mathematics into Being*. New York: Basic Books.

Lutterbie, John (2011) *Toward a General Theory of Acting*. Basingstoke and New York: Palgrave Macmillan.

McConachie, Bruce (2008) *Engaging Audiences: A Cognitive Approach to Spectating in the Theatre*. Basingstoke and New York: Palgrave Macmillan.

McConachie, Bruce (2013) *Theatre & Mind*. Basingstoke and New York: Palgrave Macmillan.

McConachie, Bruce (2015) *Evolution, Cognition and Performance*. Cambridge: Cambridge University Press.

McConachie, Bruce and F. Elizabeth Hart, eds. (2006) *Performance and Cognition: Theatre Studies and the Cognitive Turn*. Abingdon and New York: Routledge.

Neisser, Ulric (2014) *Cognitive Psychology*. Classic Edition. London and New York: Psychology Press.

Paavolainen, Teemu (2012) *Theatre/Ecology/Cognition: Theorizing Performer-Object Interaction in Grotowski, Kantor and Meyerhold*. Basingstoke and New York: Palgrave Macmillan.

Tribble, Evelyn (2011) *Cognition in the Globe: Attention and Memory in Shakespeare's Theatre*. Basingstoke and New York: Palgrave Macmillan.

Tunstall, Darren (2016) *Shakespeare and Gesture in Practice*. London and New York: Palgrave Macmillan.

Williams, Lawrence E. and John A. Bargh (2008) 'Experiencing physical warmth promotes interpersonal warmth', *Science* 322 (5901): 606–7.

Zhong, Chen-Bo and Geoffrey J. Leonardelli (2008) 'Cold and lonely: Does social exclusion literally feel cold?', *Psychological Science* 19 (9): 838–42.

Zunshine, Lisa (2012) *Getting Inside Your Head: What Cognitive Science Can Tell Us about Popular Culture*. Baltimore, MD: Johns Hopkins University Press.

5

THE REMAINS OF ANCIENT ACTION

Understanding affect and empathy in Greek drama

Peter Meineck

In this chapter, I will explore the application of cognitive theory to ancient drama and how this approach can help us understand more about how these plays functioned in antiquity. In doing so I also hope to dispel a few theatrical myths that still surround Greek drama, and inform the way in which these ancient plays are interpreted by modern practitioners. The focus of most studies on ancient Greek drama has tended to be either textual – that is the recovery, interpretation and translation of what has come down to us via the manuscript tradition, or, to a lesser extent, material – the archaeological remains of the performance spaces and the iconographic evidence found on ancient vase paintings and relief sculpture. My work involves using this evidence alongside cognitive theory to try and gain a better understanding of Greek drama from an experiential perspective. My basic question is: what was the live experience of watching a play in Athens in the fifth century BCE like? How did the extra-textual elements of the performance (movement, music, masks, props, environment, surprise, non-linguistic verbalisations, etc.) contribute to the entire theatrical event? The small amount of evidence we do possess for the reception of these plays in antiquity all allude to how the experience was a highly emotional one, and how drama could 'move the soul' (Plato, *Minos* 231a, Isocrates, *Evagoras* 2.49 and 2.10, and Aristotle, *Poetics* 1450b.16–21). Therefore, an approach that applies research from the affective sciences, cognitive theory and social psychology can be useful in better understanding how Greek theatre functioned in antiquity. Here I will briefly explore three areas based on my recent research on cognitive approaches to the experiential elements of ancient drama: environment, masks, and movement (Meineck, 2017).

Re-thinking the *seeing* place

An ongoing archaeological survey of the remains of the Theatre of Dionysos in Athens, together with new interpretations of relevant inscriptions, indicates that it was far smaller than was once thought, not arranged around a circular orchestra, and primarily built of wood (Csapo, 2007; Meineck, 2012; Papastamati-von Moock, 2015). Thus, the theatre of Aeschylus, Sophocles, Euripides and Aristophanes had 5000 to 6000 seat spaces with a wooden *theatron* (seating area), rectilinear *orchestra* and a simple *skene* – a wooden hut with a roof, single door (later three) and quite possibly a low wooden stage. What is clear from these new findings is that instead of thinking about the Greek theatre space as architecture, we should reconsider the *theatron* ('seeing place') in terms of its location and relationship to the surrounding environment. In other words, what the Greeks saw when they gathered on the southeast slope of the Acropolis was not only the masked actors and chorus members, but the Sanctuary of Dionysos Eleuthereus below, the old southern city with its ancient shrines and natural features, the Attic hills, the countryside (Rehm, 2002, 14–15) and, perhaps most importantly, what was the largest element in the visual field of the audience – the sky. What did that mean for the *theates* (spectators) who gathered there and how did it affect their experience of the plays?

Barthes (1985) wrote of the Greek theatre, 'the open-air cannot have the same image repertoire as the dark theater: the latter is one of evasion, the former of participation.' He meant that the open, natural environment was ever changing, full of sound and movement and possessed a fragility that could never be repeated twice, rather like a live performance. In such a space, the audience are also fully aware of each other as co-participants in the event. But there is a deeper level of participation that is activated by sky-space that a cognitive approach can help explain. Our relationship to space is an embodied and enactive one, in that we create our perception of the space around us depending on where our body is placed within that space and how we navigate within it. Space is therefore not a fixed immutable element that we enter and experience passively, but the dynamic creation of our brains and bodies as we move through it. In effect, we create the space we happen to inhabit; it is a construction of our own distributed systems of perception.

One highly successful model of this kind of distributed theory of space was advanced by Previc and known as the four realms of three-dimensional space (Previc, 1998). Previc was working with military pilots and trying to solve the problem of discombobulation when they lost visual sight of the horizon and crashed. He inverted the way in which people thought about space, and conceived of it in four embodied realms. These are peripersonal space – which is closest to us and where we can reach; extrapersonal space – where we look to navigate towards; focal extrapersonal space – where we focus our foveal vision; and ambient extrapersonal space – the distal areas such as faraway countryside and the sky. In flight simulator testing, Previc found that when pilots became mentally confused between the peripersonal space of their instrument panels and the extrapersonal distal sky space, they became disorientated and crashed. But Previc also noticed something else – the pilots reported feeling 'spiritual' at these times, or being 'outside of their bodies'; this was something unexpected and strange (Previc, 2006).

Further research indicated what was going on. Previc found that most people when engaged in thoughts about abstract concepts, such as remembering the last time they had a favourite meal, tended to look up even when inside – you can try this on someone and watch their gaze direction (Previc, et al., 2005). He found that the act of upwards vision into even imagined distal space activated the dopamine receptors in the dorsolateral prefrontal cortex, situated next to the frontal eye fields in the brain. Subsequent studies have shown that looking up into LAN (Looking at Nothing) space creates a sense of perceptual disembodiment and releases dopamine (Previc, et al., 2005).

While there are several neural networks connected to the processing of external space and complex electrochemical interactions that contribute to the operation of these functions, dopamine is one neurotransmitter essential for the retrieval of memories and needed for motor planning, movement, prediction formulation and abstract thought (Collins and Frank, 2016).

The reason why Previc's pilots were reporting feelings of spirituality was that their engagement with ambient extrapersonal space was flushing these spatial cognitive systems with dopamine, excessive amounts of which have been shown in people who describe themselves as 'very religious' or 'highly spiritual,' as well as in people suffering from schizophrenia. It is also notable that many cultures tend to linguistically arrange spiritual and religious concepts using vertical spatial arrangements; thus, gods tend to be 'on high,' and mountaintops, clouds and the sky are regarded as places of divinity and spirituality. In terms of distributed cognition, extrapersonal space is physically out of reach and therefore the most abstract, the opposite of peripersonal space where we can hold and grasp objects near to us. In their flight simulators, Previc's pilots were cognitively conflicted by rapidly shifting between perceiving their peripersonal cockpit instruments and gazing out into the expanse of the extrapersonal sky. Thus, contemplating extrapersonal space seems to involve 'the tendency of projecting the self into mental dimensions that transcend sensorimotor contingencies' (Crescentini, et al., 2014).

Plato uses such an embodied concept of the sky as a form of cognitive enlightenment in his famous allegory of the cave in the *Republic* (514a–420a) where a prisoner escapes the flickering shadows and climbs up into the light to gaze upon the true sight of the forms in the sky. Plato's *theoric* traveler must return to the cave to impart the knowledge he has gained, a dangerous task that I think reflects what the ancient theatre was also striving to do – expose its audience to different perspectives, create empathy with others and explore new possibilities. Unlike most modern theatres in the West where the sensory experiences of the audience are controlled by darkness, artificial and focused lighting, sound design and visual settings, Athenian audience members were encouraged to look up and engage with distal sky space.[1]

When Aristophanes created a comedy about escaping from the pressures of the Peloponnesian War and Athenian political life, he invited his audience to imagine that the theatre could become a fantasy city in the sky. Since then, 'Cloudcuckooland' has become a byword for any incredulous fantastical scheme. But the serious side of *Birds* is the idea that the Athenian theatre space was an environment that encouraged abstract thought and the contemplation of alternate ideas. On a neurochemical basis, we can even say that the Theatre of Dionysus was a dopaminergic environment, and by experiencing the dramas staged under the expansive open sky, people were more cognitively primed to abstract thought and the contemplation of different perspectives. The Greek theatre was a place that promoted a sense of disorientation, possibility and spirituality. It was the environmental starting point for mimetic journeys of alterity, dissociation, cognitive absorption and even empathy. This should be considered when we translate such an expansive art form to inhabit interior and artificially lit theatre spaces, where the outside world and the sky are usually occluded.

The enactive mask

In such an expansive space, how did Greek drama manage to create the kind of intimacy or emotional intensity necessary for the successful presentation of complex plots and character interactions? I think this was due in large part to the qualities of the Greek dramatic mask that made it a superb material anchor for spectator projection. Yet, it remains one of the most misunderstood elements of the ancient stage, partly because varied mask traditions and techniques have tended to be conflated and held in similar regard despite their differences. For example, the movement techniques of Commedia are not the same as the kata system of Noh. In antiquity, there was also quite a big difference in form and function between Greek Hellenistic masks, the Roman versions they influenced and the masks used in fifth-century

Athens. If we focus here on classical Greek tragic masks, the limited evidence we possess suggests that what was in use was a full-face mask no larger than a human head with a soft skull cap with realistic hair attached. It was probably made of linen or wood and painted with eyes (the actor looked out of the small pupil holes). Despite the pervasive myth of the megaphone mouth, for which there is absolutely no evidence, Greek masks had small mouth apertures. Furthermore, the evidence found mostly on contemporary vase paintings and some sculptures also show that these masks were not neutral but had ambiguous facial expressions, and like Noh masks were deliberately rendered with a thick bottom lip and high forehead, which, when the mask was tilted at different angles, helped it seem to change its expression (Meineck, 2011).

Neuroscientific research on Japanese Noh masks can be applicable to classical Greek masks as they share similar features. These have shown how human facial recognition systems are provoked by the tilt of the mask and the ambiguous expressions to project certain affective states (Lyons, et al., 2000). Thus, the mask was not fixed in its aspect but was able to seem to change its expression. Neuroscience can also indicate how the mask may have been perceived differently by the ancient Greek audience. For example, Dehaene's fMRI studies on illiterate and semi-literate people have shown how reading re-purposes the existing neural systems involved in facial processing situated in the Visual Word Form Area located in the left fusiform gyrus of the brain (Dehaene, 2013). The left and right fusiform gyri are both associated with facial processing as part of the visual ventral stream. Dehaene found that there was a marked difference between illiterate and literate people when they looked at faces in that the acquisition of reading pushes activation from the left gyrus to the right. It has also been suggested that the left fusiform gyrus is more holistic and the right more analytical and that illiterate and semi-literate people may process faces more emotionally than contextually favouring the left gyrus (Hirshorn, et al., 2016). Considering the low level of literacy in fifth-century Athens, this kind of research can indicate that the mask may have appeared even more emotionally variable than most of us are able to perceive it today.

Recent theoretical work in cognitive archaeology can also help us to better understand how the mask functioned in antiquity. For example, Malafouris' Material Engagement Theory (MET) from the field of cognitive archaeology places things such as ancient artefacts, objects and material sign systems in cognitive equilibrium with brains, bodies and environments across permeable mental boundaries (Malafouris, 2013). If we consider the mask as such an enactive object and not just a representation of a character or a means of disguise, we can start to understand how it was such a superb anchor for material projection. Its schematised features and ambiguous expression promoted abduction and abstraction in much the same way that most people tend to be able to recognise a caricature quicker than photographs of a real famous subject (Benson and Perrett, 1994). It was also enlivened by the actor's own movements, the play of shadows on its three-dimensional features and an uncanny ability to capture the viewers foveal (focused) vision, thereby creating attention and the physical means for denoting a performance.

The ancient Greek mask also conflated several elements of the performance – the presence of the god Dionysos, the means by which a dramatic performance was signalled, an actor's device for playing one or more roles, a character (from the Greek to 'carve') in a play and a physical anchor for the projection of changing affective states. It was far more than the mere symbolic representation it has become today, and hence its name in Greek, *prósōpon*, meant both 'mask' and 'face' and is derived from *pro* and *opsis* – 'before the gaze.'

As well as its enactive multiplicity, the mask possessed an ambiguity of agency – is the actor somehow 'possessed' by the mask, is the mask enlivened by the actor or is the mask a surface for spectator projection? The answer is probably that it was a blend of each. An element of this uncanny aspect of the mask can be seen on some vase paintings such as the Athenian Pronomos vase of 420 BCE.[2] On it, the actors' faces seem to have 'melded' to resemble the features of the masks they are holding. This phenomenon is something I have observed in rehearsals of productions that have used masks; after a few days the mask starts to look like the actor wearing it. This may be a product of whole-body processing in that we recognise people more effectively from their movements and by our own kinesthetic responses to them, something I will explore later in the chapter. Because the mask seems at once to confuse and to magnify this process, this may be another reason why the act of wearing it creates an instant frame for performance and demands spectator attention.

In terms of distributed cognition, Kirsh (2009) has shown how enactive projection acts as a foundation for sense-making and how projection is different from perception in that projection affords us the ability to see beyond what is physically manifest to imagine what could be. Projection also goes beyond working memory and builds on distributed mental scaffolding attached to an existing material anchor, in this case, the mask. Thus, we project possibilities on these anchors as we attempt to make sense of them and then further develop what we first perceived. This in effect, externalises the mental process and allows for even deeper projection, which, in turn, frees up working memory and increases attention and absorption This helps make the mask visually compelling, highly enactive and emotionally mutable. To the Athenian audience, seated in the environment of the ancient theatre, the dramatic mask may well have been far more effective and personally affecting than the human face. They did not perceive the emotional expressions chosen by the actor, but instead projected their own conceptualisations onto the surface of the mask, provoked by gesture, movement, speech, song and music. This helped create a kind of personal intimacy that was essential if narrative drama was to be successful in such a large open-air environment.

The Greeks recognised this centrality of the mask to the development of drama by crediting their first mythic actor, Thespis, with the act of masking his face and stepping out of the chorus (West, 1989). So much of the affective intimacy of the Greek theatre that we find lacking in our texts or difficult to comprehend in the vast ancient theatre spaces was stimulated by the mask. I am not advocating that modern productions of ancient drama should necessarily be masked, rather that this understanding of the enactivism and affective mutability of the mask frees the actor from approaching Greek drama as if it is some grand exercise in emotional restraint and psychological detachment. For the ancient audience, the mask enabled a deeply personal experience, and modern actors and audience might likewise embrace the emotional power that lies at the heart of Greek drama.

The active inference of movement

Another prevailing myth that persists about Greek drama is that it was a static, stately and magisterial art form. This idea has more to do with the elite Western universities of the late nineteenth and twentieth centuries than anything the ancient Athenians would have experienced.[3] Athenian drama was a theatre of almost constant movement, both individual and collective, and a cognitive approach to what we know of it can help us understand just how much dance, gestures and movement contributed to the emotionality and narrative action of these works.

The chorus was central to the performance of drama in the fifth century and was rooted in a deep culture of chorality and procession, the most pervasive performance forms in antiquity.

At the culminating point of such processions, people would gather to watch a sacrifice or other performed ritual. The Theatre of Dionysos developed from this kind of use, its wooden *theatron* set up over the Sanctuary of Dionysos, the ending point for the processional worship of the god. We should view the chorus with the procession in mind. Whereas most have tended to view the chorus as interludes to actor scenes, we should reverse this notion and envision the actor scenes as 'erupting' out of dynamic choral songs and dances.

We know very little of the music, dance and the non-verbal vocalisations (shouts, cries and other emotive utterances) that were woven through the plays, but we can make some useful deductions. For example, we can see frozen images of dance steps and gestures on Greek iconography, and it is striking how the hands and feet and distinctive gestures tend to be emphasised in many of them. Thus, like Noh, Kathakali or Topeng, Greek drama seems to have also placed great importance on the precise communication of gestures and movement. The chorus members also wore masks, the use of which in such an expansive performance environment would be an advantage when combined with body movements. Research on whole-body perception has indicated that the body is far better at communicating affective states over long distances than the face, which is only effective when in close proximity to the observer (Baylor, 2009).

When we do focus on the emotional communicative properties of a face, we tend to infer the mental/emotional state of the individual under view, whereas when we observe the whole body we infer action and this is magnified when observing a group, especially one moving in unison. Some studies have indicated that when shown a disembodied face displaying surprise, many people are confused as to how to categorise the emotional state, yet when the same facial expression is observed as part of a whole-body schema, emotional recognition tends to be far more accurate (de Gelder, et al., 2010). As described earlier, the area of the brain most closely associated with facial processing is the fusiform face area; however, whole-body perception involves a broader neural network that also includes the cortical and sub-cortical areas. Thus, viewing the whole body in motion is more cognitively engaging than just looking at a face, and the action of the chorus would have been a powerfully dynamic emotional communicator (Peelen and Downing, 2005), especially considering that due to the large numbers of male citizens required to dance at the Dionysia, the audience members would have been expert dancers themselves. Furthermore, fMRI studies have indicated that areas of the brain associated with movement will activate when a dancer watches the dance form they have trained in (Calvo-Merino, et al., 2005).

This kind of embodied affective engagement has been called 'kinesthetic empathy' – the way in which emotions are communicated by the observation of the facial and body movements of others such as dance, gestures, postures and facial expressions (Fogtmann, 2007). The result of this kind of kinesthetic empathy has been called 'emotional contagion,' which is a cognitive process where the affective states of others under observation are mirrored subliminally by the micro-expressions of the observer (Hatfield, et al., 1994). 'Empathy' is a difficult term, sometimes described as meaning 'feeling with' as opposed to 'sympathy' – 'feeling for.' I take the position that it is impossible to feel the exact same emotions as another person, as we can never fully know what has influenced them or the full psychological makeup of the person under view (Goldie, 2011; Slaby, 2014). We can, however, make inferences about what we observe and be stimulated by the actions and affective state of another to feel something ourselves. Greek drama aimed to provoke its audience to feel for others, often marginalised, powerless and foreign characters, or mythological personages in the depths of despair, and it is in this context that Aristotle famously noted that from *eleos* and *phobos* (pity/empathy and fear) and other emotions, one can be made to experience *catharsis* (*Poetics* 1449b27–28).

There are, of course, personal and cultural complexities to how emotional contagion operates and how receptive people are to it. However, we all perceive the world around us by making constant active inferences about the sensory information we receive – this includes the movements of others and the way they can make us feel.

Active inference is the cognitive process central to Friston's 'free energy principle' – the theory that the brain needs to generate constant predictions to minimise 'free energy' or face total entropy. Andy Clark calls this 'predictive processing' and declares human cognition as a kind of 'bootstrap heaven' where bottom-up incoming information is rapidly compared to top-down mental models to 'predict' the world around us (Ondobaka, et al., 2015). Kinesthetic empathy and the emotional contagion it helps generate are essential to the mechanisms of the active inference system. In kinesthetic terms, inferences can be generated by proprioception – the awareness of one's own body movements; by exteroceptive means – eyesight and sensation; and by interoceptive – feelings generated by human internal autonomic systems such as the stomach, nerves and blood vessels.

The ancient audience members were able to perceive, predict and project the emotions proprioceptively and seemingly instantaneously. Additionally, the gestural system of the actors generated exteroceptive signs and induced activated representations and thus powerful communicators of generalised concepts that further enhanced the transference of knowledge from actor to audience (Novack, 2016). The spectator's interoceptive systems could be stimulated by micro-mirroring such movements, which could, in turn, produce autonomic responses such as a gulp, a pit in the stomach, a slight shudder or a tightening of the throat. Such embodied predictive generation stimulated by the movements and gestures of the masked chorus members and actors of Greek drama facilitated the kind of active inferences that induced emotional contagion and empathy. To us, the chorus is perhaps the most challenging aspect of ancient drama, but for the ancient spectator, it was probably among the most viscerally affecting, dynamic and thrilling.

We tend to look backwards at Greek drama and perceive it as the genesis of our own theatrical traditions. As a result, artists can be over-reverent to this material and approach it with trepidation that can lead to a disconnection between their experience and training and the material they are trying to make sense of. Although Athenian drama had a profound influence on Hellenistic and Roman theatre, our relationship to it is one of re-discovery, re-interpretation and reception and dependent on scholarship rather than performance. An approach anchored in cognitive theory can bring us closer to the experience of the audience in antiquity, help demonstrate how compelling, absorbing and affecting Greek drama must have been and inspire modern artists to free the emotional and empathetic spirit of these remarkable ancient plays.

Notes

1 There are textual examples of this. For example, Euripides' *Bacchae* 1264–70; Aristophanes' *Clouds* 228–34, *Peace* 82–153, and *Birds* 175–80. See also Plato, *Timaeus* 90a.
2 See Meineck (2017: 81).
3 For references, see Tarvin (1990).

References

Barthes, R. 1985. *The responsibility of forms*. Howard, R. (Tr.), Berkeley, CA: UC Press.

Baylor, A.L. 2009. 'Promoting motivation with virtual agents and avatars: Role of visual presence and appearance' in *Philosophical Transactions of the Royal Society of London B: Biological Sciences* 364.1535: 3559–65.

Benson, P.J. and Perrett, D.I. 1994. 'Visual processing of facial distinctiveness' in *Perception* 23.1: 75–93.

Calvo-Merino, B., Glaser, D.E., Grezes, J., Passingham, R.E. and Haggard, P. 2005. 'Action observation and acquired motor skills: An FMRI study with expert dancers' in *Cerebral Cortex* 15.8: 1243–9.

Clark, A. 2015. *Surfing uncertainty: Prediction, action, and the embodied mind*. Oxford: Oxford University Press.

Collins, A.G.E. and Frank, M.J. 2016. 'Surprise! Dopamine signals mix action, value and error' in *Nature Neuroscience* 19.1: 3–5.

Crescentini, C., Aglioti, S.M., Fabbro, F. and Urgesi, C. 2014. 'Virtual lesions of the inferior parietal cortex induce fast changes of implicit religiousness/spirituality' in *Cortex* 54: 1–15.

Csapo, E. 2007. 'The men who built the theaters: Theatropolai, Theatronai, and Arkhitektones' in Wilson, P. (Ed.), *The Greek theatre and festivals: Documentary studies*. Oxford: Oxford University Press: 97–121.

Gelder, B. de. 2009. 'Why bodies? Twelve reasons for including bodily expressions in affective neuroscience' in *Philosophical Transactions of the Royal Society B: Biological Sciences* 364.1535: 3475–84.

Dehaene, S. 2013. 'Inside the letterbox: How literacy transforms the human brain' in *Cerebrum* 7: 1–16.

Fogtmann, M.H. 2007. 'Kinesthetic empathy interaction – Exploring the possibilities of psychomotor abilities in interaction design' in Ramduny-Ellis, D., Dix, A. and Gill, S. (Eds.), *Second International Workshop on Physicality in proceedings of the 21st British HCI Group Annual Conference on People and Computers: HCI... but not as we know it*. London: British Computer Society: 217–18.

Goldie, P. 2011. 'Anti-empathy' in Coplan, A. and Goldie, P. (Eds.), *Empathy: Philosophical and psychological perspectives*. Oxford: Oxford University Press: 301–17.

Hatfield, E., Cacioppo, J.T. and Rapson, R.L. 1994. *Emotional contagion*. Cambridge: Cambridge University Press.

Hirshorn, E.A., Wrencher, A., Durisko, C., Moore, M.W. and Fiez, J.A., 2016. 'Fusiform gyrus laterality in writing systems with different mapping principles: An artificial orthography training study' in *Journal of Cognitive Neuroscience* 28.6: 882–94.

Kirsh, D. 2009. 'Projections, problem space, and anchoring' in Taatgen, N.A. and van Rijn, H. (Eds.), *Proceedings of the 31st annual conference of the Cognitive Science Society*. Austin, TX: Cognitive Science Academy: 2310–15.

Lyons, M.J., Campbell, R., Plante, A., Coleman, M., Kamachi, M. and Akamatsu, S. 2000. 'The Noh mask effect: Vertical viewpoint dependence of facial expression perception' in *Proceedings of the Royal Society of London B: Biological Sciences* 267.1459: 2239–45.

Malafouris, L. 2013. *How things shape the mind*. Cambridge, MA: MIT Press.

Meineck, P. 2011. 'The neuroscience of the tragic mask' in *Arion: A Journal of Humanities and the Classics* 19.1: 113–58.

Meineck, P. 2012. 'The embodied space: Performance and visual cognition at the fifth century Athenian theatre' in *New England Classical Journal* 39: 3–46.

Meineck, P., 2017. *Theatrocracy: Greek drama, cognition and the imperative for theatre*. Abingdon and New York: Routledge.

Novack, M.A. and Goldin-Meadow, S. 2016. 'Gesture as representational action: A paper about function' in *Psychonomic Bulletin & Review*. doi:10.3758/s13423-016-1145-z.

Ondobaka, S., Kilner, J. and Friston, K. 2015. 'The role of interoceptive inference in theory of mind' in *Brain and Cognition* 112: 64–8.

Papastamati-von Moock, C. 2015. 'The wooden theatre of Dionysos eleuthereus in Athens: Old issues, new research' in Frederiksen, R., Gebhard, E.R. and Sokolicek, A. (Eds.), *The architecture of the ancient Greek theatre, acts of an international conference at the Danish Institute at Athens 27–30 January 2012*. Aarhus: Aarhus University Press: 39–79.

Peelen, M.V. and Downing, P.E. 2005. 'Selectivity for the human body in the fusiform gyrus' in *Journal of Neurophysiology* 93.1: 603–8.

Previc, F. 1998. 'The neuropsychology of 3-D space' in *Psychological Bulletin* 124.2: 123–64.

Previc, F. 2006. 'The role of the extrapersonal brain systems in religious activity' in *Consciousness and Cognition* 15: 500–39.
Previc, F., Declerck, C. and de Brabander, B. 2005. 'Why your 'head is in the clouds' during thinking: The relationship between cognition and upper space' in *Acta Psychologica* 118: 7–24.
Rehm, R. 2002. *The play of space: Spatial transformation in Greek tragedy.* Princeton, NJ: Princeton University Press.
Slaby, J. 2014. 'Empathy's blind spot' in *Medicine, Health Care and Philosophy* 17.2: 249–58.
Tarvin, W.L. 1990. 'Tragic closure and "tragic calm"' in *Modern Language Quarterly* 51.1: 5–24.
West, M.L. 1989. 'The early chronology of Attic tragedy' in *The Classical Quarterly* (New Series) 39.1: 251–4.

6
MINDING IMPLICIT CONSTRAINTS IN DANCE IMPROVISATION

Pil Hansen

Introduction

Improvisation in dance ranges from the widespread practice of contact improvisation, which involves interpersonal exploration of physical weight sharing and balance, through forms of structured and task-based improvisation, to the more controlled dance-generating systems in which source materials, tasks and rules of interaction are predefined. In dance for the stage, improvisation is often used as a method to create new movement material, which then is refined and set in repeatable choreography. However, an improvisation score or system can also be presented as performance, either by itself or in combination with set movement sections. Dance improvisation is, furthermore, commonly used as a form of somatic therapy and as a teaching tool.

Approaches to dance improvisation have historically pursued a form of intensified, bodily performance presence in which the dancer is responding spontaneously to stimuli in the present moment instead of relying on past learning or anticipation of the future. However, contemporary practitioners and researchers of dance improvisation are challenging this concept of bodily presence in the improvisation discourse with the aim of further understanding the layers of kinaesthetic and cultural memory that the practice involves. At the same time, cognitive scientists (mostly from neurobiology, experimental psychology and biomechanics) have collaborated with artists to empirically examine implicit (i.e., not conscious) cognitive dynamics through dance improvisation. These cognitive dynamics include procedural memory and entrainment. 'Procedural memory' refers to a series of both cognitive and motor tasks that are automated and thus can be performed without awareness, and 'entrainment' is a generic term for the human tendency to coordinate movement and language rhythmically to one another. Although dance improvisation at first glance may seem spontaneous, discernible patterns of interaction do tend to emerge and have been analysed as a form of self-organization. The implicit cognitive dynamics studied by cognitive scientists contribute to constraining dancers' improvisation options in ways that attract patterns of interaction and self-organisation. In dance-generating systems, explicitly designed constraints (i.e., tasks, rules, cues) are added to affect the self-organisation of dance improvisation. Such tasks typically direct the dancer's conscious attention towards the otherwise implicit constraints mentioned earlier, and thus enable the dancer to draw on, inhibit or manipulate both implicit and explicit constraints. This agency reaches beyond the spontaneous present of established improvisation discourse and involves attentive engagement with memory and planning, albeit of a radically different kind than the performance of set choreography or the planning of movement sequences.

I aim to qualify these introductory observations through a review of relevant studies in the cognitive sciences, and then position the concept of presence within the established discourse of dance improvisation. I then give examples of contemporary improvisers who directly challenge this concept. Shifting perspective, I draw upon interdisciplinary studies of cognitive science and dance improvisation to elaborate on the role of implicit and explicit constraints. Putting these different viewpoints in conversation with one another provides the context within which I discuss how the addition of explicit constraints may give dancers the ability to actively work levels of memory and planning that otherwise affect the improviser implicitly.

Dance improvisation

Genealogy

The studies discussed in this contribution address improvisation from contemporary dance practices in Europe, the United States, Canada and Australia that build directly or indirectly on ideas from the American avant-garde of the 1970s[1]. At that time, Anna Halprin drew on the child's playfulness to develop improvisation approaches meant for psychological therapy, educational purposes and community building on the west coast (Carter 2000, 184–5; Ross 2003, 45). Although Halprin's work introduced dance improvisation to these contexts, it is contact improvisation, developed by Steve Paxton at Judson Dance Theatre in New York, that has been most widely implemented (Novack 1986; Paxton 2003; Biasutti 2013). Another member of Judson Dance, Trisha Brown, explored norms of movement and perspective through improvised, pedestrian engagement with physical materials and urban structures (Berger 2002, 17–28). Brown has inspired environmental improvisation approaches used in site-specific dance and the usage of impossible tasks like 'inhabit two places at once' as improvisation instigators (Meg Stuart qtd. in Reiter 2010, sixth paragraph). To date, Brown's and Paxton's collaborator, Deborah Hay, continues to develop and teach complex improvisation systems comprising specific perceptual orientations and rules, dual tasks and poetic movement scores (Hansen and House 2015, 67–8), which, in turn, have influenced the development of dance-generating systems by choreographers such as William Forsythe in Germany and Ame Henderson in Canada (Hansen 2015, 126–8).

Performance presence

Regardless of whether the aforementioned approaches aim to develop an embodied sense of self and community, connect with another through a shared point of physical balance, explore our environment and everyday norms or manipulate the body as a perceptual instrument, they all seek spontaneity in the present moment. This spontaneous presence is often purchased by inhibiting planning of the future and conscious repetition of movement material from the past while responding to present stimuli that derive from the environment and its inhabitants or from the improviser's proprioception and associative imagination (Hansen and Oxoby 2017, 77–80). An ideal of being in the present, detached from past learning and anticipation of the future, is often referred to as a form of freedom from formal or trained choreographic structures (Burrows 2010, 24; Goldman 2010, 94–111). This freedom is furthermore believed to liberate the dancer from structures of social identity and hierarchy embedded in set choreography (Novack 1986, 11; Little 2014).

This ideal concept of presence was and is troubled by the improviser's phenomenological experience of time during performance, which is not limited to the present. This experience has been described by influential scholars and founders of improvisation approaches from the Judson Dance Theatre generation, who otherwise prioritise the need to detach from past knowledge when improvising.

Choreographer and scholar Susan Leigh Foster described dance improvisation as a talking 'back and forth between the *known* and the *unknown*' (2003, 4). According to Foster, the known has been established in the past and comprises the dancer's trained tendency to move in patterns and predetermined structures that limit the improviser's choices. She argued that dancers need to 'extricate ourselves from that which is known' to arrive at the unpredictable (4). Foster furthermore emphasised that 'improvisation's bodily mindfulness summons up a kind of hyperawareness of the relation between immediate action and overall shape, between that which is about to take place and is taking place and that which has and will take place' (7). To Foster, the improviser thus pushes through and beyond learnt movement tendencies and structures of the past, focusing his or her conscious attention on the movement responses and choreographic opportunities from the near past and near future that give shape to the present.

Steve Paxton, the founder of contact improvisation, eventually sought to develop techniques of perceptual orientation, which could produce awareness of a form of procedural memory he referred to as 'reflex action.' Interestingly, this pursuit addresses an aspect of the dancer's implicit body knowledge that is not normally explicitly 'known' to the dancer, including learnt movement tendencies. Here, it is useful to consider that contact improvisers undergo an implicit learning process that gradually allows them to take greater risks of falling and trust their ability to connect with their partner in finding a shared point of balance. Instead of 'extricating' improvisers from this memory, Paxton stated that reflex action 'successfully preserves the integrity of the body' during improvisation, arguing that conscious awareness of such implicit memory is necessary to both draw on past learning and 'learn from the moment' (Paxton 2003, 177).

In cognitive psychology, it is recognised that human beings have two individual, but connected, learning systems (Stevens 2017, 56–7):

1 A slow and implicit system which, for example, is involved in procedural, motor knowledge and immersive language acquisition. This implicit system tends to extend rather than depart from prior learning.
2 A fast and explicit system through which episodic knowledge (situated interactions) and semantic knowledge (information) are learnt more systematically and synthesised by directing attention. This explicit system involves the possibility of choice and departure from prior knowledge.

With these systems in mind, Paxton's pursuit of awareness can also be understood as an attempt to support the implicit learning system with the explicit learning system as a way of extending past learning into future choices. Thinking beyond Paxton's writing, it is furthermore possible that this awareness speeds up the learning process and enhances the possibility of choice-making.

There are contemporary dance improvisers who see the discursive matrix of spontaneity, presence, internal self-expression and freedom from the 1970s as a limitation (Hansen, Kaeja, and Henderson 2014, 30; Midgelow 2015, 107–8, 112–13). They work from the hypothesis that explicit awareness of otherwise implicit factors that affect the improvisation, including movement tendencies, affords the improviser the ability to make reflective choices.

One such example is Malaika Sarco-Thomas' *touch + talk* practice. Dance improvisers are asked to verbally respond to questions about their motivation to improvise, strategies of perceptual recalibration and anticipated audience reception at regular intervals during public performances of contact improvisation (2014, 198). With this practice, Sarco-Thomas pushes beyond the proprioceptive and internal focus typically involved in the performance of presence in contact improvisation. Through explicit articulation of factors and reflection on choices, she directs attention towards pre-established motivation, anticipation of the future and the experience of a group that is external to the dancer's proprioceptive focus (194).

Vida Midgelow's improvisation practice and research involves a form of dramaturgical consciousness that reconciles the improviser with the many layers of memory that improvised responses draw on and express. While she, like Paxton, recognises the importance of implicit, procedural memory involved in movement skill, she has developed a series of improvisation and reflection tasks that help the improviser reveal trained and socially conditioned knowledge. When rendered explicit, this knowledge is used as source material and to make choices about emergent structural possibilities while improvising—choices that embrace or challenge the revealed knowledge (Midgelow 2012, 2015). Supported by this dramaturgical consciousness, the improviser's choices in any given moment can reach farther into the past and the future than the near past and future that Foster wrote about. An ideology of liberation through presence is replaced with agency.

The cognitive science of dance improvisation

Midgelow's and Sarco-Thomas' tools direct conscious attention towards factors that inform improvisational acts and responses as they are experienced subjectively by the improviser. In order to access knowledge about factors that are harder for the dancer to register consciously, observation from a third-person perspective can be useful. In such an effort, I now turn to arts-science studies of dynamical systems and cognition in dance improvisation.

The role of constraints in the self-organization of dynamical systems

The factors that I have discussed thus far, can be and have been considered as constraints of behaviour within the framework of Dynamical Systems Theory (DST). This cognitive framework is applied in order to understand how such factors contribute to the generation of self-organising patterns of interaction between improvisers (see Stevens and McKechnie 2005; Buckwalter 2010, 134; Hansen and House 2015).

DST looks at how complex systems, ranging from chemical processes to birds' flight patterns, tend to self-organise over time. In my DST studies of dance-generating systems, I typically begin by identifying the implicit variables (individual movement tendencies) and explicit parameters (tasks, rules) that constrain interaction possibilities. I also name preselected sources of energy (memories of movement, recycled movement scores, images, songs, etc.) that dancers use as sources for movement generation. By systematically processing video recordings of improvised performances, I then proceed to analyse patterns of self-organised interaction. Note that the emergent composition of improvisation comprises such patterns. Central to this task is identification of the particular variables or parameters that attract these patterns in shifting phases (Hansen 2015; Hansen and House 2015).

As explained by cognitive psychologists Thelen and Smith (1996, 45–70), a dynamical system depends on a continuous influx of external sources of energy in order to self-organise. On the one hand, if the system becomes too closed off from such sources, then interaction will become repetitive and cease to generate new forms of organisation. If the system, on the other hand, is pried fully open to new sources and its boundaries thus become less defined, self-organisation is less likely to occur and certainly harder to observe. There are two ways in which an improvisation system could become too closed: (1) if the dancer primarily relies on the implicit constraints of movement tendencies and thus repeats movement without the ability to generate original responses; and (2) if the improvisation context stops providing fresh energy because its sources have all been encountered and learned by the dancer.

The DST method makes such issues detectable by outside observers. An improvisation approach designed with awareness of these issues will include parameters such as tasks, principles and rules that explicitly constrain interaction possibilities and thus solve the problems caused by too much openness. Gradual addition of new source material, new improvisers or new environments can address the risk of repetitiveness caused by exhaustion of energy sources. However, the equally closing effect of implicit (i.e., non-conscious) repetition of movement tendencies warrants further examination.

Implicit constraints: procedural memory and kinaesthetic perception

The implicit memory that is referred to as body knowledge, movement tendencies, habit or reflex by the dance improvisers cited in this chapter is in cognitive terms known as procedural, motor memory. It retains the movement patterns that have proved functional in response to specific forms of interaction in the past and that have been learnt implicitly over multiple rounds of repetition. One rarely achieves awareness of such patterns, and the enactment of them is experienced phenomenologically as an original, present and perhaps even intuitive response and not as the habitual repetition of memory that we know it is.

The same can be true for how dancers direct their attention and thus their kinaesthetic perception. Contemporary dancers are trained to use peripheral vision and proprioceptive information, and to perceive spatial relationships and relationships to other bodies. Thus, experienced dancers are able to perceive their own body and movement in relation to their surroundings in ways that non-dancers rarely master, but this expertise is typically implicit and automated.

Kinaesthetic perception and procedural memory deliver the foundation for other areas of expertise, including memorisation of choreography (Bläsing 2010; Stevens 2017), imaging of movement (May et al. 2011) and joint action (also called unison). While imaging is usually drawn on consciously and memorisation is supported by sounding and marking techniques that render specific landmarks explicit (Vass-Rhee 2010; Kirsh 2011; Warburton et al. 2013), joint action depends on the cognitive processes of entrainment, which are implicit.

Combining implicit and explicit constraints: entrainment and attention

Entrainment is typically defined as the simultaneous rhythmical coordination of two individuals in time and space (Phillips-Silver and Keller 2012). It has been established experimentally that dancers entrain more effectively to music than non-dancers (Washburn et al. 2014), likely because of their enhanced kinaesthetic perception. Although entrainment to music is the most common area of study, entrainment to internal and external bodily rhythms in dance has also received significant attention. George Sofianidis et al. (2014) did, for example, find that swaying non-dancers entrain to expert dancers' sway rhythm when sharing touch, while expert dancers are able to inhibit such influences from their partners, and thus lead the entrainment.

To date, the most comprehensive and interdisciplinary case study of entrainment to non-musical rhythm in dance examines the rehearsal process of William Forsythe's dance work *Duo* (Waterhouse, Watts, and Bläsing 2014). This study reveals important relationships between implicit entrainment and strategies that afford dancers additional explicit control. These findings indicate that the condition, which enabled dancers to inhibit bodily entrainment in Sofianidis and colleagues' study, was the dancers' ability to become explicitly aware of competing rhythmical stimuli and direct their attention to the sway task. In other words, dancers are entrainment experts and, furthermore, have a superior ability to consciously manipulate this otherwise implicit dynamic.

Waterhouse, Watts and Bläsing advance the study of entrainment from simplified limb coordination, sway or improvisation tasks to far more complex forms of unison in dance (2014). *Duo* is partly set as memorised and repeatable choreography and partly open for improvised choices. The dancers synchronise and de-synchronise their movement in different proxemic relationship to one another, without touch or music. Instead of these stimuli, they employ visual cues and 'sounding' (i.e., audible cues like breath) to 'counterpoint' movement: oscillating between points of synchronisation and individual variation. With reference to Phillips-Silver and Keller's concept of 'coordination smoothers' in entrainment (2012), the authors suggest that simultaneous auditory and visual cues are used deliberately to modify behaviour in order to produce predictability (Waterhouse, Watts, and Bläsing 2014, 13). In terms more familiar to DST, the cues can be said to constrain the improvisation possibilities and thus increase the likelihood that unison will be arrived at. These explicit markers and points of multisensory attention are coupled with what the dancer, Watts, describes as an 'elastic temporal integrity' (as quoted in ibid.). This term refers to each performer's ability to make expressive choices regarding the compression or stretching of time and tempo, but it also encompasses their responsibility to arrive together at counterpoints in time. Although the authors do not discuss the possible relationship between this inter-subjectively perceived and acted on temporality and the forms of movement or speech rhythm that are typically associated with entrainment, it is plausible that variation in movement tempo is also experienced as variation in rhythm, and thus responded to through entrainment to rhythm. This interpretation of their data matches the authors' conclusion:

> We suggest that this example of expert entrainment is a skilful perceptual activity of rhythmic collaboration based on sensorimotor knowledge. The case study demonstrated that skill-based entrainment in dance requires the integration of implicit and explicit processes ….
>
> *(Ibid., 14)*

Attention to multisensory, rhythmical cues that deliberately constrain improvised movement variation is thus combined with implicit entrainment to movement rhythm that is based on trained and automated motor memory. Together, these two levels of skilful entrainment make joint action (unison) possible.

Although the case of *Duo* belongs on the more closed end of the improvisation scale, the complementary role of explicit attention to and conscious awareness of rhythmic cues as a skilful aspect of entrainment in dance can be applied strategically to the broader range of improvisation practices. Indeed, this role may contribute to what Midgelow refers to as 'dramaturgical consciousness' (2015) or to what the dance improviser Kent De Spain calls 'improvisational awareness':

> the intensity and density of improvisation causes practitioners to develop survival strategies and skills that help streamline choice-making and open doors to creative resources. One such skill is a heightened sensitivity to what is happening during the improvisation, a layered, synchronic process of embodied cognition that I refer to as "improvisational awareness."
>
> *(De Spain 2012, 26)*

Conclusion

The knowledge of the past, which Foster recognises as part of the dance improviser's path, involves implicit procedural memory that is expressed in movement tendencies, in enhanced entrainment to rhythmical stimuli and in skilful kinaesthetic perception. If the improvised present is to reach into the unknown and bring something new into the world, these implicit competencies must be accompanied by strategic points of kinaesthetic attention and conscious engagement with past influences that enable the dancer to inhibit, manipulate or make choices regarding parts of these processes. Adding explicit constraints, such as audio-visual cues and rules of engagement are effective examples of such strategies, as they demand continuous awareness and repeatedly pull the dancer back to a point of adjustment and choice-making. The reflection challenges proposed by contemporary improvisers are other examples of effective strategies. Together, these strategies both manipulate the self-organisation of improvisational interaction and ensure that dynamical improvisation systems in dance do not become too closed in repetition of procedural memory to generate new forms of self-organisation.

Note

1 Improvisation is also a central aspect of historical and cultural dance practices that precede this source, such as the Japanese Noh Drama, the Flamingo from Moorish Spain (Carter 2000, 183) and social jazz dance with its Caribbean, African and urban influences (ibid., 184).

References

Berger, Maurice. 2002. "Gravity's Rainbow." In *Trisha Brown: Dance and Art in Dialogue, 1961–2001,* edited by Hendel Teicher, 17–28. Andover, MA: Addison Gallery of American Art.

Biasutti, Michele. 2013. "Improvisation in dance education: Teacher views." *Research in Dance Education* 14 (2): 120–40. doi:10.1080/14647893.2012.761193.

Bläsing, Bettina. 2010. "The Dancer's Memory: Expertise and Cognitive Structures in Dance." In *The Neurocognition of Dance: Mind, Movement and Motor Skills*, edited by Bettina Bläsing, Martin Puttke-Voss, and Thomas Schack, 75–98. New York: Psychology Press.

Buckwalter, Melinda. 2010. *Composing While Dancing: An Improviser's Companion*, Madison: University of Wisconsin Press.

Burrows, Jonathan. 2010. *A Choreographer's Handbook.* New York: Routledge.

Carter, Curtis L. 2000. "Improvisation in dance." *The Journal of Aesthetics and Art Criticism* 58 (2): 181–90. doi:10.2307/432097.

De Spain, Kent. 2012. "Improvisation and intimate technologies." *Choreographic Practices (Online)* 2 (1): 25–42. doi:10.1386/chor.2.25_1.

Foster, Susan Leigh. 2003. "Taken by Surprise: Improvisation in Dance and Mind." In *Taken by Surprise: A Dance Improvisation Reader,* edited by Ann Cooper Albright and David Gere, 2–12. Middletown, CT: Wesleyan University Press.

Goldman, Danielle. 2010. *I Want to Be Ready: Improvised Dance as a Practice of Freedom.* Ann Arbor: University of Michigan Press.

Hansen, Pil. 2015. "The Dramaturgy of Performance Generating Systems." In *Dance Dramaturgy: Modes of Agency, Awareness and Engagement,* edited by Pil Hansen and Darcey Callison, 124–42. Basingstoke: Palgrave.

Hansen, Pil, and Christopher House. 2015. "Scoring the generating principles of performance systems." *Performance Research* 20 (6): 65–73. doi:full/10.1080/13528165.2015.1111054.

Hansen, Pil, and Robert J. Oxoby. 2017. "An earned presence: Studying the effect of multi-task improvisation systems on cognitive and learning capacity." *Connection Science* 29 (1): 77–93. doi:full/10.1080/09540091.2016.1277692.

Hansen, Pil, Karen Kaeja, and Ame Henderson. 2014. "Transference and transition in systems of dance generation." *Performance Research* 19 (5): 23–33. doi:abs/10.1080/13528165.2014.958350.

Kirsh, David. 2011. "How marking in dance constitutes thinking with the body." *Versus: Quaderni Di Studi Semiotici* 113–115: 179–210.

Little, Nita. 2014. "Restructuring the self-sensing: Attention training in contact improvisation." *Journal of Dance & Somatic Practices* 6 (2): 247–60. doi:10.1386/jdsp.6.2.247_1.

May, Jon, Beatriz Calvo-Merino, Scott deLahunta, Wayne McGregor, Rhodri Cusack, Adrian M. Owen, Michele Veldsman, Cristina Ramponi, and Philip Barnard. 2011. "Points in mental space: An interdisciplinary study of imagery in movement creation." *Dance Research* 29 (2): 404–432. doi:10.3366/drs.2011.0026.

Midgelow, Vida L. 2012. "Dear practice ...: The experience of improvising." *Choreographic Practices (Online)* 2 (1): 9–24. doi:10.1386/chor.2.9_1.

Midgelow, Vida L. 2015. "Improvisation Practices and Dramaturgical Consciousness: A Workshop." In *Dance Dramaturgy: Modes of Agency, Awareness and Engagement,* edited by Pil Hansen and Darcey Callison, 106–23. Basingstoke: Palgrave.

Novack, Cynthia Jean. 1986. *Sharing the Dance: An Ethnography of Contact Improvisation.* Ann Arbor, MI: University Microfilms International.

Paxton, Steve. 2003. "Drafting Interior Techniques." In *Taken by Surprise: A Dance Improvisation Reader,* edited by Ann Cooper Albright and David Gere, 174–84. Middletown, CT: Wesleyan University Press.

Phillips-Silver, Jessica, and Peter E. Keller. 2012. "Searching for roots of entrainment and joint action in early musical interactions." *Frontiers in Human Neuroscience* 6: 26. doi:10.3389/fnhum.2012.00026.

Reiter, Sonja. 2010. "Interview with Meg Stuart." *In Dance* (July/August). www.damagedgoods.be/en/about/interviews/2010/in-dance-one-on-one-2. Accessed July 2017.

Ross, Janice. 2003. "Anna Halprin and Improvisation as a Child's Play: A Search for Informed Innocence." In *Taken by Surprise: A Dance Improvisation Reader,* edited by Ann Cooper Albright and David Gere, 41–52. Middletown, CT: Wesleyan University Press.

Sarco-Thomas, Malaika. 2014. "Touch + talk: Ecologies of questioning in contact and improvisation." *Journal of Dance & Somatic Practices* 6 (2): 189–204. doi:10.1386/jdsp.6.2.189_1.

Sofianidis, George, Mark T. Elliott, Alan M. Wing, and Vassilia Hatzitaki. 2014. "Can dancers suppress the haptically mediated interpersonal entrainment during rhythmic sway?" *Acta Psychologica* 150: 106–13. doi:10.1016/j.actpsy.2014.05.002.

Stevens, Catherine J. 2017. "Memory and Dance: 'Bodies of Knowledge' in Contemporary Dance." In *Performing the Remembered Present: The Cognition of Memory in Dance, Theatre and Music,* edited by Pil Hansen and Bettina Bläsing, 39–68. London: Bloomsbury Methuen.

Stevens, Catherine J., and Shirley McKechnie. 2005. "Minds and Motion: Dynamical Systems in Choreography, Creativity, and Dance." In *Tanz im Kopf: Yearbook 15 of the German Dance Research Society 2004,* edited by J. Birringer and J. Fenger, 241–52. Münster: LIT Verlag.

Thelen, Esther, and Linda B. Smith. 1996. *A Dynamic Systems Approach to the Development of Cognition and Action.* 1st paperback ed. Cambridge, MA: MIT Press.

Vass-Rhee, Freya. 2010. "Auditory turn: William Forsythe's vocal choreography." *Dance Chronicle* 33 (3): 388–413. doi:10.1080/01472526.2010.517495.

Warburton, Edward C., Margaret Wilson, Molly Lynch, and Shannon Cuykendall. 2013. "The cognitive benefits of movement reduction: Evidence from dance marking." *Psychological Science* 24 (9): 1732–39. doi:10.1177/0956797613478824.

Washburn, Auriel, Mariana DeMarco, Simon de Vries, Kris Ariyabuddhiphongs, R. C. Schmidt, Michael J. Richardson, and Michael A. Riley. 2014. "Dancers entrain more effectively than non-dancers to another actor's movements." *Frontiers in Human Neuroscience* 8: 800. doi:10.3389/fnhum.2014.00800.

Waterhouse, Elizabeth, Riley Watts, and Bettina Bläsing. 2014. "Doing *Duo*: A case study of entrainment in William Forsythe's choreography 'Duo'." *Frontiers in Human Neuroscience* 8: 812. doi:10.3389/fnhum.2014.00812.

7

APPLYING DEVELOPMENTAL EPISTEMIC COGNITION TO THEATRE FOR YOUNG AUDIENCES

Jeanne Klein

Theatre for Young Audiences (TYA) provides artistic productions for young people, ranging in age anywhere from one through 18 years, with works involving one or more young protagonists that serve as metonymic representations of respective age groups. However, when staging works for elementary students, too many TYA companies operate from misguided assumptions and implicitly childist beliefs (Young-Bruehl 2012) about children's 'short or weak' attention spans and 'limited' comprehension abilities that disrespect and denigrate their actual cognitive competencies (Goldberg 2006, 184–89). Companies also cling to romanticised beliefs that children have vast imaginations through which they 'suspend their disbelief' and 'identify' with child characters, often performed by adult actors who exaggerate childish behaviours simply to make children laugh (Nolan 2007). Such tacitly derisive caricatures of childhood are especially deleterious because child spectators do not necessarily recognise prejudicial discriminations directed against them, and critics are often led to prejudge such productions for children as 'inferior' to theatre for adults.

In order to optimise child spectators' aesthetic experiences, I have long argued that TYA practitioners need to create artistically driven productions from children's cognitive-affective perspectives (Klein 2005). In my view, optimal experiences occur when spectators recognise, articulate and apply artists' intended meanings in performances to themselves and society. However, all too often, TYA practitioners presume that the emotional behaviours of children during performances (e.g., silence, restlessness, giggles and laughter) support their intended meanings, without actually knowing, asking, discussing or necessarily caring whether their intended meanings coincide with children's interpretations after performances. Whilst such behavioural evidence clearly demonstrates embodied minds in action during performances, children seldom have public opportunities to share their private meanings after performances unless teachers, parents and actors provide adequate time for post-performance conversations. Unlike published reception studies, enactive theories in cognitive science alone do not fully account for children's subsequent meanings constructed after performances (e.g., McConachie 2015). Furthermore, theory of mind evidence during early childhood presupposes second-order continuities through middle childhood, without considering how cognitive epistemologies change over lifespans, as Iordanou (2016) reviews. Likewise, cumulative evidence regarding the impact of theatre education in elementary schools (Winner, Goldstein and Vincent-Lancrin 2013) has yet to integrate the impact of screen media on young people from infancy through adolescence (e.g., Anderson and Kirkorian 2015) to help explain the cognitive mechanisms of dynamic spectating systems in theatre.

In this essay, I seek to counter childist practices and pervasive myths regarding young minds, specifically aged 6 to 12 years, by applying the theories and developmental evidence of epistemic cognition to TYA. As producing director Moses Goldberg advises, 'The more one understands the stages of a child's development, the more one can provide suitable stimuli, moments that reach out to that level of aesthetic understanding' (2006, 94). Given the ephemerality of one-time performances, artists need to seize their present opportunities to create favourable and long-lasting memories that may persuade adults to bring children back to venues, rather than subscribe to the myth of turning children into future audiences in our digitally mobile worlds.

Epistemic cognition

When TYA practitioners claim to know children by prejudging their minds, they tend to accentuate their cognitive deficiencies in comparison to adult norms. Instead, they need to consciously recognise how their own thinking habits and epistemic beliefs coincide with children's inferential ways of knowing theatre during all production phases. Epistemic cognition may be defined broadly as any thinking related to knowledge about particular domains, and specifically, as metacognitive knowledge, procedural strategies and criterial standards used to evaluate and justify claims and epistemic beliefs in action (Iordanou 2016, 108).

Adults share four recursive epistemological orientations with children that include thinking as imitative realists (birth to age four), dualistic absolutists (roughly ages five to twelve), multiplistic relativists (around ages 12 and up) and evaluative critics (around ages 25 and up) (e.g., Kuhn 2005, 30–3). When constructing meanings, realists rely on perceiving the phenomenal realities of performances to assert their preferences, whilst absolutists compare the so-called 'true or false facts' within performances against their own 'real' life experiences. Relativists express multiple opinions, relative to their self-selected theories and speculative beliefs regarding artists' intentions, whilst critics evaluate performances by coordinating dynamic theories and existing evidence using critical argumentation.

From first-stage orientations, artists and spectators alike implicitly expect to use theatre for personal gratification along a transactional continuum of diverting entertainment and instructive enlightenment, which then determines how much cognitive effort they are willing to invest to construct and interpret meanings. From artistic perspectives, entertainment (*divertissement*) means diverting or distracting minds *away* from problematic concerns by generating entertaining emotions so children may feel good about themselves, whilst enlightenment (*éclaircissement*) means directing minds *towards* dramatised conflicts by clarifying emotion-driven circumstances so children may recognise their lives on stage. All artistic decisions depend on the extent to which playwrights, directors, actors and designers consciously or unconsciously intend to entertain and enlighten spectators along this highly nuanced continuum. In general, imitative and absolutist thinkers prefer to lean towards entertainment, whilst relativists and critics shift towards enlightenment.

Within the first-stage domain of aesthetic judgements, toddlers gratify themselves by engaging in object substitutions and developing personal preferences. Valkenburg finds that when 6- to 18-month-olds view screen media, they prefer animated characters and voices, physical humour with moderate to rapid movements, bright colours, peculiar sounds, startling surprises, visual transformations, music, singing and laughter (2004, 17–27). As preschoolers engage in imitative pretend play at age three and blend distinctions between actors and characters, they enjoy clownish gestures and slapstick, but they prefer slow pacing and repetition to master meanings in stories. By age four, they take on others' cognitive perspectives, having mastered the concept of false beliefs in which they recognise that others hold divergent beliefs about observable realities and pretenses (e.g., Harris 2000).

Likewise, TYA practitioners who think as imitative realists seek to gratify mutual pleasures by making artistic decisions intended solely to entertain child spectators. To keep youngsters happily diverted, productions imitate animated media with physically humourous cartoon characters who speak with loud, high-pitched voices and move frequently and rapidly. When insecure actors hear children giggle, they tend to exaggerate their facial expressions and buffoonish gestures even further to raise their own self-esteem. Making children laugh as loudly as possible satisfies all too many egocentric desires to please children from realist orientations. In effect, when practitioners rely solely upon imitative caricatures of childhood, they tacitly position older audiences as preschoolers who recognise this patronising content as intended for 'babies,' otherwise known as 'schlock.'

When faced with adult actors portraying child characters, realist and absolutist thinkers respond to grown-ups as actual adults who do not know how to act their age when they try to imitate and exaggerate childish behaviors. Children's uproarious laughter suggests they are laughing at (not with) childish adults making silly fools of themselves in public. When asked to explain why they laughed, they may point out an actor's 'funny' behaviours but feel reluctant to label him or her a 'bad' actor for fear of hurting the actor's feelings.

Most TYA companies also operate from realist orientations when they reproduce dramatic and musical adaptations of literary and screen media, because adult ticket-purchasers desire to re-experience commercial titles associated with childhood. When staging adaptations, realist and absolutist spectators expect much the same imagery, as they compare and evaluate live renditions against the original sources they've already experienced. Any significant changes that depart from original models may be criticised for failing to reproduce initial mediations. For instance, preschoolers may wonder why Winnie-the-Pooh is 'naked' if he isn't wearing Disney's red shirt, and older children may question casting Mowgli as a girl.

As children enter formal schools and face social pressures to conform, they begin to think as dualist absolutists by using authoritarian either/or criteria to justify good/bad values and true/false claims. Likewise, practitioners who intend to entertain through instruction may make either/or decisions based on their childist beliefs. To satisfy teachers' expectations of curricular content, they may present educational concepts theatrically, often by breaking the fourth wall, thereby inviting eager six- to eight-year-olds to answer dialogic questions because they want to help protagonists win and conquer obstacles. However, 'preachy' dialogue, in which characters instruct and repeat educational messages unnecessarily, patronises older children who grow restless for many 'been there, done that' reasons. In addition, absolutists ignore or distort any evidence that contradicts or doesn't fit their existing beliefs (Iordanou 2016, 109). For example, one teacher criticised a production of *More of a Family* for casting an African American actor as a homeless man, the play's pivotal character, even though he also played the spouses of three white women in this double-cast ensemble.

From third-stage orientations, adolescents (and relativist artists) examine multiple criteria to assert their opinions, relative to their interpretations of self-selected theories and epistemic beliefs over what counts as contextual theatre evidence. Practitioners who intend to inform spectators through entertainment may make multiple decisions relative to their selective analyses of performative texts. Knowing they have the creative right to express whatever they think best, they rely upon their artistic instincts and justify their opinions with explanations that fit with their beliefs about theatre and childhood. When artistic opinions clash, collaborators may simply agree to disagree when they are unable to dismantle their colleagues' absolutist claims. As Iordanou points out, the 'danger' of individuals' interpretations may 'stall in a radical relativism with the evil of subjectivity seen as overpowering the quest for any knowledge beyond subjective opinion' (2016, 114).

TYA practitioners who think as evaluative critics from fourth-stage orientations intend to enlighten spectators by clarifying arguments made about childhood in dramatisations. Whilst artists cannot control each spectator's private meanings, metacognitive thinkers can control play choices and script analyses, casting and staging, and design and technical decisions. Playwrights craft compelling conflicts that raise provocative questions regarding the sociopolitical impact of adults' power over young lives, despite absolutists' rejection of such taboo topics as child poverty, violence, death, sexuality and even sadness (Kruckemeyer 2012). Directors and actors work critically to keep minds invested in solving the mysteries of unfolding actions. Designers clarify metaphoric concepts organically with theatrical styles that range from detailed realism to minimalist abstractions. Those artists who grant children agency by including them in developing original works often discover solutions to artistic problems they might not have otherwise considered (Newman 2006, 190–98, 408–12; Elnan 2012).

Above all, critical artists take full, ethical responsibility for their artistic choices, knowing that children's behavioural responses during performance are never 'wrong' or 'inappropriate.' When unintended responses arise, evaluative actors 'lean back' and pause momentarily, until a modicum of silence returns, and then pull children back into the story's mysterious secrets. When such strategies fail, directors re-evaluate whether such problematic moments stem from the script or directing and acting choices, and make necessary adjustments to further clarify their intentions (Goldberg 2006, 142–46). For instance, during initial performances of *The Red Badge of Courage*, wise-cracking adolescents heckled Henry whenever he took off his boots. To win over sceptical audiences, the actor revealed their thinking and subverted their joking by smelling his stinky boots (Church 2017).

Evidence of epistemic cognition in action

Contextualised evidence from theatre and media reception studies explains how realist and absolutist thinkers recall meanings after adult-mediated performances. As mnemonic studies reveal (e.g., Schneider 2015), the major cognitive differences between child and adult minds involve the ongoing maturation and synaptic efficiencies of young brains that are 90 percent developed by age six, their uses of metacognitive strategies, the capacities of their working memories and their speeds of processing explicit and implicit imagery in analogical meta-representations. Visual, verbal/aural and kinaesthetic images constitute the bases of human imagination.

In general, what children actually see and hear on stage is what they get from performances. What actors do on stage physically, vocally and emotionally matters far more than what they think internally, because young minds process moment-to-moment actions more quickly and efficiently visually than linguistically. When 6 to 12-year-olds are asked 'how they know' 'main ideas' in performances, most rely on explicit evidence in staged enactments by describing visualised actions and paraphrasing verbalised dialogue to interpret, connect and evaluate protagonists' goal-directed actions, antagonistic conflicts and thematic concepts. Given their mnemonic focus on explicit imagery, they tend to miss any offstage characters or major events discussed solely in dialogue. As egocentric tendencies weaken between ages seven and nine, they increasingly make inferences from implicit imagery to the same extent as adults, based on concepts they already know and those that affirm their experiences. Upon realising that theatre offers purposeful applications to their lives around age nine, they also decipher analogies between themselves and society from characters' objectives, often as moral prescriptions (e.g., people shouldn't lie) (e.g., Saldaña 1996, 81; Scollen 2012).

When evaluating performances, children prioritise their criterial values as comprehensibility, engaging enactions, humour, child characters, comparative realism within stylistic frameworks, and lastly, aesthetic qualities (Klein and Schonmann 2009). Dramatised content matters far more than artistic forms, because children are driven to comprehend meanings that, in turn, propel attentions in order to predict what may happen next. They search for causal connections between episodes in clearly connected, climactic plots (van den Broek, Bauer, and Bourg 1997) and prefer informative child characters with sufficient agency to resolve their own formidable conflicts, ideally without adult assistance. During performances, children let actors (and playwrights) know when stories stop progressing by growing restless, unlike adults who politely control their behaviours even when their attentions stray.

Rather than 'identify' with characters, as many practitioners presume, realists and absolutists dwell upon actors' appearances and actions by comparing and contrasting their physical, active, emotional, social and moral traits with themselves. Consequently, TYA practitioners are better advised to use proportionately sized puppets to portray young characters than costume actors in ludicrous anthropomorphic suits. When black-dressed puppeteers re-direct attentions to their characters by keeping their eyes focused solely on their puppets, children readily blend such couplings as shown in their drawings that often depict characters without puppeteers (e.g., Reason 2010, 66–84). However, double-casting and child-adult embodiments of the same character may confuse absolutists if they discern no plausible reasons. For example, at the end of a production of *Wolf Child*, when Joseph, the Wolf Child, exited behind a wall and the narrating Traveler re-entered on the other side, he revealed himself explicitly as the Wolf Child grown up. However, according to teachers' post-performance evaluations, some children missed this temporal transformation, perhaps because the cast included two adult actors of the same height and the playwright's stilted speech patterns spoken by the Traveler clouded understandings.

Unlike artists who take casting conventions for granted, children do not know and appreciate the historical and contemporary bases for cross-identity casting. Like any spectator who desires to see her self-identified features embodied on stage, child spectators prefer (and deserve) to see child bodies cast as child characters (Grady 1999). Age-appropriate casting not only enhances perspective-taking but also showcases child actors' competent abilities which can vary as much as adults' abilities (Magnasco 2015). Casting by gender also affects child spectators' attitudes towards young characters. Media producers have long known that most boys refuse to involve themselves with female characters and actors, unless they behave in masculine ways, whilst girls are willing to engage with both genders, likely because adventurous males dominate family films (Smith et al. 2010).

Whilst many six- to nine-year-olds find visual and aural humour hysterically funny (e.g., butt jokes and cartoonish voices), parody and slapstick involve more cognitively complex and deceptively antagonistic forms of 'adult' humour (Buijzen and Valkenburg 2004, 150–52). Full appreciation of parody, such as camp in drag, requires prior knowledge of satiric and ironic intentions, and ironic vocal intonations may confuse six- to nine-year-olds who misunderstand why anyone would knowingly say something false yet not intend to deceive a listener (Winner 1988, 133–59). Consequently, name-calling of any kind not only implies prejudicial harassments but also raises teachers' ire about bullying. Children also do not know the historical contexts of slapstick in *commedia dell-arte* wherein Harlequin used his slapstick against Pantalone to avenge his oppressions. Therefore, artists need to remember that rapid-fire rates of slapstick comedy diverts young children's attentions away from victims' tacit oppressions and also does not allow sufficient time for evaluating underlying motivations.

Asking spectators to explain artists' intentions reveals the effectiveness of intended meanings and suggests epistemic orientations (Omasta 2011). In my *Pipe Dream* production and study (1993), when asked why Magritte's paintings were projected on a screen, many nine- and eleven-year-olds inferred these images as explicit visualisations of René's mind, whilst realist viewers thought they simply showed his paintings. When asked why red lights flashed during a chaotic classroom scene, 11-year-olds more than 7-year-olds surmised that 'They were *thinking so much* and the *feeling* just wasn't enough,' so the lights made René's mind appear 'dizzy,' 'crazy' or 'driven nuts.' However, some realists were led astray when Magritte's Father told René that his Mother was 'lost,' rather than found dead, at the river. Regarding a virtual reality production of *Dinosaurus*, when asked why artists included both screen-projected and human dinosaurs, children surmised pragmatic purposes, such as making the screen dinosaurs bigger and more realistic because people don't look like dinosaurs but actors were needed to make the dinosaurs talk (Klein 2003, 44–5).

To avoid interpretative problems that diverge from artistic intentions, artists need to remember that six- to eight-year-olds need sufficient time and slower pacing to process, connect and integrate the consequences of characters' dialogic actions, as the following two examples illustrate. In my production and study of *Crying to Laugh* (Klein 1995), the character of Seluf directly explains to Mea, the protagonist, the explicit, metaphoric meaning of a black balloon that must be popped in order to release her pent-up sadness over the death of her dog (a puppet). Twelve minutes later, at the end of the play, Mea releases her frustrations and joyfully pops multi-coloured balloons that fall from above. However, spectators' responses to collective questions revealed that no one, including undergraduates, spontaneously reported that Mea's balloon-popping signified the physical release of her stress, perhaps because I did not ask them directly to explain this metaphor. In retrospect, I realised that we had created a magnificent spectacle that diverted attentions away from this metaphor's significance. Instead, we should have first dropped one black balloon to allow Mea and spectators sufficient time to recall its meaning. After remembering that she needed to pop this balloon to relieve her stomach ache, she could then make a more purposeful decision to pop it, and then gradually pop a second and third falling balloon one at a time before releasing an onslaught of remaining balloons.

Divergent interpretations about *Dinosaurus* also reflected mnemonic and epistemic differences when children were asked what Bunk, the oil worker, 'decided to do at the end of the play' and 'what he learned' as a consequence. Eleven-year-olds were more likely than younger children to remember causally connected episodes; that is, in return for a dinosaur having saved his life in an earlier episode, Bunk decided to save the dinosaurs' lives by exploding the cave's entrance (offstage) so that no one, including himself, could ever return. As a result, he learnt, for example, 'that money wasn't more important than somebody's life' and 'you should respect other people or animals and not just try to barge into their area and their lifestyle.' In contrast, most seven-year-olds relied primarily on the director's final tableau, in which Bunk returned to the stage (the cave) and stood beside the dinosaurs. Based on this last image, they concluded that he decided to stay with the dinosaurs because 'he learned they were nice' – the opposite meaning intended by the playwrights (Klein 2003, 45–7).

Whilst many TYA practitioners would argue that mismatched meanings do no harm, these examples highlight the interdependent epistemic and mnemonic complexities involved when artists seek to clarify intended meanings. To my knowledge, no published studies have engaged child critics in arguing and justifying their divergent interpretations and relative epistemic criteria about one performance in relation to their entertainment-enlightenment goals. Unfortunately, this dearth of reception studies constrains further evaluative integrations of existing evidence with dynamic cognitive theories.

Nevertheless, TYA artists have an ethical responsibility to optimise aesthetic experiences by clarifying their intended meanings about the multifarious complexities of childhood that may remain in spectators' memories for years to come. By consciously recognising and co-ordinating cognitive developmental and epistemic theories with even intuitively derived evidence, TYA artists may change their epistemic orientations and criterial values when judging performances. In these ways, TYA may shed its childist propensities for the common benefit of all multi-generational spectators.

References

Anderson, Daniel R., and Heather L. Kirkorian. 2015. "Media and Cognitive Development." In *Handbook of Child Psychology and Developmental Science. Vol. 2, Cognitive Processes*. 7th ed., edited by Lynn S. Liben and Ulrich Müller, 949–94. Hoboken, NJ: Wiley.

Broek, Paulus Willem van den, Patricia J. Bauer, and Tammy Bourg. 1997. *Developmental Spans in Event Comprehension and Representation: Bridging Fictional and Actual Events*. Mahwah, NJ: Erlbaum.

Buijzen, Monica, and Patti M. Valkenburg. 2004. "Developing a Typology of Humor in Audiovisual Media." *Media Psychology* 6 (2): 147–67.

Church, Jeff. E-mail message to author, May 29, 2017.

Elnan, Merete. 2012. "The Notion of Children: How Can the Idea of Childhood, or Children as Spectators and of Understanding Influence Theatre for Young Audiences?" In *TYA, Culture, Society*, edited by Manon van de Water, 165–78. Frankfurt am Main: Peter Lang.

Goldberg, Moses. 2006. *TYA: Essays on Theatre for Young Audiences*. Louisville, KY: Anchorage Press Plays.

Grady, Sharon. 1999. "Asking the Audience: Talking to Children about Representations in Children's Theatre." *Youth Theatre Journal* 13: 82–92.

Harris, Paul L. 2000. *The Work of the Imagination*. Malden, MA: Blackwell.

Iordanou, Kalypso. 2016. "From Theory of Mind to Epistemic Cognition: A Lifespan Perspective." *Frontline Learning Research* 4 (5): 106–19.

Klein, Jeanne. 1993. "Applying Research to Artistic Practices: *This Is Not a Pipe Dream*." *Youth Theatre Journal* 7 (3): 13–17.

Klein, Jeanne. 1995. "Performance Factors That Inhibit Empathy and Trigger Distancing: *Crying to Laugh*." *Youth Theatre Journal* 9: 53–67.

Klein, Jeanne. 2003. "Children's Interpretations of Computer-Animated Dinosaurs in Theatre." *Youth Theatre Journal* 17: 38–50.

Klein, Jeanne. 2005. "From Children's Perspectives: A Model of Aesthetic Processing in Theatre." *Journal of Aesthetic Education* 39 (4): 40–57.

Klein, Jeanne, and Shifra Schonmann. 2009. "Theorizing Aesthetic Transactions from Children's Criterial Values in Theatre for Young Audiences." *Youth Theatre Journal* 23 (1): 60–74.

Kruckemeyer, Finegan. 2012. "The Taboo of Sadness: Why Are We Scared to Let Children Be Scared?" In *TYA, Culture, and Society*, edited by Manon van de Water, 85–92. Frankfurt am Main: Peter Lang.

Kuhn, Deanna. 2005. *Education for Thinking*. Cambridge, MA: Harvard University Press.

Magnasco, Julia. 2015. "Age Appropriate Casting: Best Practices from First Stage." 29 January, http://tyatodayonline.org/2015/01/29/age-appropriate-casting-best-practices-from-first-stage/.

McConachie, Bruce. 2015. *Evolution, Cognition, and Performance*. New York: Cambridge University Press.

Newman, John D. 2006. "Spotlight on Process: New Play Development at the Bonderman Playwriting Symposium." PhD diss., New York University.

Nolan, Ernie. 2007. "What Adults Sometimes Forget When Playing Kids." *TYA Today* 21 (2): 12–16.

Omasta, Matt. 2011. "Artist Intention and Audience Reception in Theatre for Young Audiences." *Youth Theatre Journal* 25 (1): 32–50.

Reason, Matthew. 2010. *The Young Audience: Exploring and Enhancing Children's Experiences of Theatre.* Sterling, VA: Trentham.

Saldaña, Johnny. 1996. "'Significant Differences' in Child Audience Response: Assertions from the ASU Longitudinal Study." *Youth Theatre Journal* 10: 67–83.

Schneider, Wolfgang. 2015. *Memory Development from Early Childhood through Emerging Adulthood.* Dordrecht: Springer.

Scollen, Rebecca. 2012. "'Olivia Has Lost Her Voice!': An Audience Reception Study of Children's Responses to New Australian Play *Spirits in Bare Feet.*" *Youth Theatre Journal* 26 (2): 158–72.

Smith, Stacy L., Amy Granados, Marc Choueiti, Sarah Erickson, and Allison Noyes. 2010. "Changing the Status-quo: Industry Leaders' Perceptions of Gender in Family Films." Los Angeles: Geena Davis Institute on Gender in Media, https://seejane.org/research-informs-empowers/.

Valkenburg, Patti M. 2004. *Children's Responses to the Screen: A Media Psychological Approach.* Mahwah, NJ: Erlbaum.

Young-Bruehl, Elisabeth. 2012. *Childism: Confronting Prejudice against Children.* New Haven, CT: Yale University Press.

Winner, Ellen. 1988. *The Point of Words: Children's Understanding of Metaphor and Irony.* Cambridge, MA: Harvard University Press.

Winner, Ellen, Thalia R. Goldstein, and Stéphan Vincent-Lancrin. 2013. *Art for Art's Sake? The Impact of Arts Education. Educational Research and Innovation.* Paris: OECD Publishing.

8

4E COGNITION FOR DIRECTING

Thornton Wilder's *Our Town* and Caryl Churchill's *Light Shining in Buckinghamshire*

Rhonda Blair

Staging traditional text-based theatre can be described as moving from the page to the stage, doing things with words or making the word flesh. Theatre artists create worlds-within-the-world that are meaningful for and affect those who make them and see them. Using two case studies, Thornton Wilder's *Our Town* and Caryl Churchill's *Light Shining in Buckinghamshire,* this essay considers how to apply principles of 4E cognition to processes of making theatre, in negotiating the relationships among text, research and embodiment. Terms from cognitive science illuminate the inter-relationships of perception and meaning involved in performance, and also in human experience more broadly. 4E cognition is a basic feature of human existence – we operate by its principles all the time every day. This essay uses some of the research in 4E cognition to study rehearsal and performance processes in the theatre, in order to better understand both theatrical practice and aspects of human cognition more generally; rehearsal processes and performances provide discrete models of cognitive ecologies that are broadly reflective of how we operate in life. Through the case studies, I consider dramatic text, actor-as-individual, actor-as-company-member, physical material given (e.g., space, set, costumes, props – or 'properties,' things actors manipulate and use with their hands) and audience, and how these work together. Applying enactivist concepts can move actors from an intellectual grasp of historically complex materials into a fully embodied and collectively vital engagement. I begin with a brief reminder of the 4E terms: embodied, embedded, extended and enacted.

Embodied: Cognition isn't separable from our physical being, but rather occurs throughout our physical being. One of many proofs of the interconnectedness of these different aspects of cognition is the inseparability of language production, language comprehension and perception of intent in the brain. Some of the same brain areas are crucial to language production *and* to language comprehension *and* to perceiving the intention of physical actions such as grasping and manipulation (Fadiga et al. 2006): 'the sensorimotor cortices are crucial to semantic understanding' of action (Kaag 2014:186), that is, whether you pick up a box or someone tells you 'Pick up that box,' many of the same neurons are activated (186).

Embedded: Cognition depends heavily on off-loading cognitive work and taking advantage of potentials, or affordances, in the environment, for example, the handle on a cup of hot coffee allows us to pick it up, we stand on a chair to reach a high shelf; a fundamental aspect of cognition derives from the individual's interactions with the environment (see Robbins and Aydede 2009:10).

Extended: Cognition can be understood as extending beyond the boundaries of the individual to encompass aspects of our material environment as well as our social, interpersonal environment. The ecology in which we live and to which we react includes other people. This engages things such as neural and kinaesthetic mirroring systems and mind-reading, or 'Theory of Mind,' which posits how we respond based on our ability not just to perceive the actions of others, but to read their intentions, consciously and preconsciously. This is an obviously fundamental aspect of the actor's work.

Enacted: Cognition is inseparable from action, and is an outgrowth and even an attribute of action. A particularly valuable insight for actors is Alva Noë's that perception is 'something we do. ... *What we perceive* is determined by *what we do* (or what we know how to do); it is determined by what we are *ready* to do. ... we *enact* our perceptual experience; we act it out' (Noë 2004:1–2). In this view, to act is to have a mental image and vice versa (Noë 2004:130–1); no '"translation" or transfer [from perception into action] is necessary because it is already accomplished in the embodied perception itself' (Noë 2004:80).

4E principles reframe our understanding of what language is and how it works. Language is not an abstract or merely 'mental' representation of meaning, but arises out of bodies in specific situations as a component of both action and perception. Language is a tool that participates in constructing meanings of our experience; the state of the body is not only an input into language interpretation, it is also an output. Linguist George Lakoff and philosopher Mark Johnson draw heavily on three major findings of cognitive science and argue that our sense of our bodies, indeed, the fact of having a body, is the source for major metaphors of thought, meaning and values: first, the mind is inherently embodied; second, thought is largely unconscious; and, third, abstract concepts are largely metaphorical. They hold that fundamental metaphors of time (e.g., 'time is money'), space ('she felt distant'), events and causation ('I was walking on eggshells'), selfhood ('I just let myself go') and morality ('he had no backbone') that pervade our thinking and speech, and that are the principal tools by which we construct meaning, grow directly out of our sense of *physical* being: 'An embodied concept is a neural structure that is actually part of, or makes use of, the sensorimotor system in our brains. Much of conceptual inference is, therefore, sensorimotor inference' (Lakoff and Johnson 1999:20, ital. added). That is, a concept is a particular state in the brain – and, hence, *de facto* a physical state. A crucial implication of this is that the metaphors we live by (to use the title of their first book together [1980]) are not just conceptual or poetic, but are of our bodies in the most immediate way. Cognitive scientists and linguists Gilles Fauconnier and Mark Turner also give prominence to metaphor. Their network model of conceptual integration begins to account for the way we put together associations from widely divergent situations and experiences; they assert that, first, imagination is the central engine of meaning, and, second, (echoing Lakoff and Johnson), metaphor is central to cognition. Disparate 'inputs' are blended to create new knowledge, insight or experience that go beyond that contained in the initial inputs, that is, to create a blended conceptual space. One example they give is of using the image of a waiter carrying a tray as an aid in learning to ski downhill more effectively; intuitively, this association makes sense to us, though the two inputs in fact involve very different movement problems. Nonetheless, we are somehow able to blend the inputs kinaesthetically to accomplish the task of skiing. Conceptual blending is consciously cognitive on one level – we are able to describe or depict it – but it operates largely unconsciously; significant, unconscious acts of translation and, especially, transformation must occur for the things being blended, for example, tray-carrying being 'like' skiing, to fit together.[1] For theatre makers, viewing language as embodied action takes us beyond traditional close reading to engage a more nuanced assessment of language, images and structures of the text. As Amy Cook says, it could be that 'language is less a system of *communicating* experience than actually *being* experience; we do not translate words into perceptions, we perceive in order to understand' (Cook 2009:589). Body, breath, voice and language are inseparable – in life and in acting.

A global way of thinking of about 4E cognition is in terms of how the 4Es help us articulate different aspects of the cognitive ecologies in which we live. Cognitive ecology is a holistic term for engaging

> the multidimensional contexts in which we remember, feel, think, sense, communicate, imagine, and act, often collaboratively, on the fly, and in rich ongoing interaction with our environments. ... The idea is not that the isolated, unsullied individual first provides us with the gold standard for a cognitive agent, and that mind is then projected outward into the ecological system: but that from the start (historically and developmentally) remembering, attending, intending, and acting are distributed, co-constructed, system-level activities.
>
> *(Tribble and Sutton 2011:94)*

We constantly engage shifting and varied cognitive ecologies as we move from 'environment' to 'environment.' In theatre, one can think of any play as being its own cognitive ecology,[2] in which we engage embodied, embedded, extended and enacted aspects.[3]

What follows are two case studies of theatre productions which I directed as models of how we might apply principles of cognitive ecologies and 4E cognition, basic features of our existence, to the rehearsal and performance processes and practice.

Our Town, Southern Methodist University, fall 2010

In our work on Thornton Wilder's *Our Town,* we leveraged 4E concepts of embodiment by keeping preliminary discussions to a minimum and engaging the actors' minds explicitly through their bodies and senses; embeddedness by having actors work on their feet, figuring out spatial orientation and how to utilise a minimal number of set pieces; extendedness and enaction by engaging spatial relationships to discover and create connectedness among actors and with the audience. Elements such as group singing and shared responsibility for the live soundscape reinforced the feeling of communion, of extendedness, allowing the actors to be more sensitively attuned to the moment.

Our department casts our fall shows in late spring, so I met with the company once before leaving for the summer; this was the only time we sat around a table. Theatre rehearsals often begin with an initial phase called 'table work,' typically lasting anywhere from three or four days to a week, in which the company reads through and discusses the text, gets background information about the play and playwright, considers possible interpretations and shares thoughts and ideas. This compartmentalises and, I believe, limits an actor's ability to engage cognitive processes holistically. Our single spring meeting included my introductory comments about the play, Wilder, and the play's era; a presentation by our design team about sets, lights and costumes; and a read-through of the script. I asked the actors to come back in late August off-book (the theatre term for having lines memorised), and to familiarise themselves with the hymns in the play. Over summer I emailed them writings by Wilder, PowerPoints with photographs of people, towns and landscapes from early twentieth-century New England and other visual images to engage them with the characters and their world. I asked the actors not only to think of the photographs as 'information,' but also to imagine smells, sounds, textures and feelings conveyed or implied in these images. This worked directly with the actors' bodies – their embodied thoughts and imaginations – by engaging their senses with Wilder's world and the world of the play.

From the beginning, being in physical relationship to the space and each other was part of the 'thinking' and construction of meaning and interpretation, that is, consciously engaging the 4Es. While 4E operations are occurring in the particular cognitive ecology of even just sitting at a table with others, I engaged the usual processes from this perspective. Our theatre space sat 120 in two parallel seating units. Actors could enter around all sides of the seating units, as well as go up or down the center aisle of each unit. It was a single environment for the 19 actors and the audience, as some of the scenes and moments happened on and in the seating areas. The outline of our set had been taped out on the rehearsal room floor, and we had a full set of rehearsal furniture. Actors carried scripts only for prompting (i.e., to check for a line when memory momentarily failed) and to take notes and record blocking (the term for where to move and when). While human beings always operate by principles of 4E cognition and cognitive ecologies, from the first day of rehearsal we worked with an eye towards deploying and leveraging these concepts. Actors' bodies were immediately engaged more fully than they would have been by just sitting at a table (embodiment), using space and properties (embeddedness), in relationship to others (extendedness), engaged in action (enactedness). The actors were on their feet, in space and time with each other, dealing with text in terms of spatial relationships and movement. I was strongly informed by Noë's construction of perception as 'something we do,' and Gallagher's idea of spatiality of situation, as space being more about what's happening than about location. Actors always have to answer 'Who am I?' 'What am I doing?' and 'Where am I?' but the 4E terms moved us immediately into a more explicitly collective engagement of relationship to self, environment, objects and others as a cognitive ecology. (This has some resonance with techniques such as Stanislavsky's and Viewpoints, but differs from them by being framed in cognitive science terms.)

We relied as much as possible on the actors to create the material and imaginary aspects of the play's world, that is, the ecology of Wilder's Grovers Corners. During the few scene changes that take place, company members moved the limited set pieces (e.g., wooden chairs) from 'kitchens' to 'church.' We used no hand props, and action was sometimes almost in the laps of the people in the front rows or by the aisles. Except for recorded organ music for the wedding, the company made all sound cues live and were often visible to the audience while doing so; they used, among other things, a child's toy train whistle, a crate of milk bottles shaken during the milkman's deliveries, an empty metal pipe struck to serve as the church bell and a rolled up newspaper thwacked in the hand of an offstage actor as the newspaper boy tossed his imaginary papers on his route; actors visible on the periphery made animal noises – one crowing for the rooster at dawn, five clucking while Mrs. Gibbs fed her flock of chickens and one neighing and snorting as Bessie, the milkman's horse. The entire cast also worked extensively on the three hymns in the play. The actors, from the deeply religious to the atheist, came to experience a connectedness through singing together. This also drew the audience in when the actors sang two verses of 'Blessed Be the Tie' at the end of the curtain call.[4]

We wanted the audience not only to feel that we were telling a story both specific and 'timeless' (something about which Wilder writes), but also to be included in and embraced by the life of the town. In short, much of this work was about evoking connectedness and feeling through our awareness of the dynamic nature of 4E aspects of cognition.

Light Shining in Buckinghamshire, Southern Methodist University 2016

Caryl Churchill wrote *Light Shining in Buckinghamshire* in 1976, when the UK was struggling out of a three-year recession involving stagflation (a deadly combination of high unemployment and major inflation), decline in the GDP and labour strikes. Using the events of the English Civil War primarily from 1647 to 1649, she illuminates contemporary problems of power, access, property, class, gender, morality and faith. The play follows historical and fictive characters who rise up against the monarchy to demand freedom and basic rights such as food and pay. Religious forces of the time ranged from the establishment of the Church of England to the anti-Catholic Protestant ethos of Cromwell to the millennial fervour of the Ranters, who believed Christ would return any day to reign over the Earth. Churchill explodes aspects of traditional narrative and representation, and, like Brecht, uses elements of Verfremdung, historicisation, episodic structure, songs and scene titles. Characters range from Cromwell to landowners to vicars to the poorest of the poor, engaged in the project of reenvisioning and surviving in a turbulent England.

Historically and politically dense, the play has 23 scenes set in the Putney Debates hall, poor homes, dark roadsides, a vicar's comfortable rooms, a church, a tavern, among others. Originally done with two women and four men, the actors played over 30 characters, who sometimes were played by a different actor in a different scene; this mirrored the collaborative way in which the script was developed under Churchill's leadership and embodies the historical argument against property made by the Levelers and Diggers, who wanted to dismantle ownership of land and property. In our production, cast with six actors (two Masters of Fine Arts students, four Bachelor of Fine Arts students, two of them women, four men), a given role was always played by the same actor. (A 2015 National Theatre production had a cast of 18 plus 40 community members.)

While one rehearsal at the table sufficed for *Our Town*, *Light Shining* required more sitting, reading and explication: actors needed to know about historical facts, historical figures, issues and conditions of the English Civil War, and specifics of the play's language to have a foundational sense of the meaning and urgency of these things for the characters. This was fundamentally for clarifying meaning and argument, part of which was helping actors find images that connected them personally and affectively to the play, strengthening their sense of investment and urgency (or, in acting terms, a sense of the 'stakes' of the scene or play). This included relating the play to fall 2016, with the political upheaval of the presidential election, factions, religious fundamentalism, economic inequalities and their manifestation in things such as food, shelter, sex, gender and religion. This discussion and analysis engaged the actors' minds through bodily, sensory images, looking for associations and analogues with their particular experiences. This focus on embodiment – engaging and waking up the actors' psycho*physical* being – laid the ground for supporting relationships (extendedness) and action (enaction).

Psychologist James Gibson's concept of affordances is useful here to understand how we engaged particular embedded and extended aspects of 4E cognition, once the actors got on their feet in the space. Briefly, an affordance is what the environment provides for a person to accomplish certain things, for example, a knife can be used for cutting, a stick for scratching a hard place to reach on one's back and a rock for sitting or climbing on. Affordances are fundamentally connected to intentionality and action; an individual's sense of an affordance is dependent on what she is trying to accomplish or needs to do. As Gibson states, 'Needs control the perception of affordances (selective attention) and also initiate acts' (Gibson 1982:411).

When the actors got onto their feet and had different stage areas and levels, set pieces, props and costume pieces, we played with what these afforded in terms of character, action and emphasis.

Gallagher's 'space is about situation' was again a touchstone focusing actors on characters *doing* something in the space in relationship to the others, and responding to and building on what they were receiving from the others; this would eventually include the audience. As with *Our Town*, elements such as the cast singing the four songs in the play and the shared responsibility for scene changes reflected principles of extended cognition, in the anticipating and coordinating of actions, related to the actors being more attuned to each other and engaging the audience in the life of the play and its situations. Extended cognition is one prism through which to look at this kind of coordination and 'fellow-feeling'; this can also be discussed in other cognitive science terms outside the scope of this paper, for example, mirroring, Theory of Mind, and empathy. Another touchstone was Noë's perception-action dyad, in which sensorimotor skills enable perceptual access to the world, that is, 'an understanding of the way in which sensory stimuli change as a result of movement,' in which to see something is 'to interact dynamically with it in a manner that one understands … To see an object is to stand in a relation to a thing that is characterized by the exercise of a range of characteristic sensorimotor skills' (Noë 2008: 661). Cognition and perception through action was fundamental throughout: doing – leading to conscious and articulated understanding, meaning and feeling – being 'first among equals,' rather than privileging 'interpreting' or talking.

Embodied metaphor was fundamental to designing the space. We were inspired by Isaiah 24, xvii–xx, which begins the play and is sung by the cast. It describes the pit of hell, the key image informing our scenic design:

> Fear, and the pit, and the snare are upon thee, O inhabitant of the earth.
>
> And it shall come to pass that he who fleeth from the noise of the fear shall fall into the pit; and he that cometh out of the midst of the pit shall be taken in the snare; for the windows from on high are open, and the foundations of the earth do shake.
>
> The earth is utterly broken down, the earth is clean dissolved, the earth is moved exceedingly.
>
> The earth shall reel to and fro like a drunkard, and shall be removed like a cottage; and the transgression thereof shall be heavy upon it; and it shall fall and not rise again.
>
> *(Churchill 1976:191)*

'Fear,' 'the pit of hell,' 'the earth utterly broken down' and 'clean dissolved' capture the feeling of the world of the play and the hellishness that is part of many of the scenes. A central playing area – the 'pit' – was surrounded by four seating areas on risers (84 seats total), a main entrance, three performance areas with platforms of different heights and configurations, all of which were deconstructed lumber and 'rock,' painted to look as though charred and ruined from fire and decay. Through its intimacy and structure, this space intertwined audience and performer. The proximity of actor and audience was crucial to the cognitive 'field.' The songs and certain speeches were delivered directly to the audience, and, in the Putney debates scene, the audience was addressed as New Model Army members to be persuaded by the different factions. Actors were as close as two feet to audience members and never farther than ten.

Each actor had a base costume with contemporary elements to facilitate quick additions or removals of select period or period-inflected pieces for each character. These pastiches grounded the actors and kept visible for the audience the situatedness of the story as both contemporary and period and the actor's identity as she or he transformed from role to role, sometimes in front of the audience. An example: in view of the audience, at the end of a scene an actor playing a pregnant woman removed her pregnancy pad, tossed it aside and tucked her skirt to become a man being recruited at a pub; later, she donned a rich jacket, a cavalier's hat and gauntlets to play Cromwell in the Putney Debates.

Props operate literally and metaphorically in regard to the 4Es. A simple example: late in the play, after the king's defeat and Cromwell's success have changed nothing for the poor, a butcher confronts the audience with their privilege, stuffing themselves with meat while children starve. All 4Es can be seen at work here: the actor donned a bloody apron and carried a large cleaver, extending the physical reach of his arm, the blade threatening, associations being triggered in the audience's body-minds. The bloody apron and cleaver played on audience members' bodily experiences of and associations with blood and large, sharp things; the impact was not just abstractly intellectual, but experiential, that is, embodied. The actor used the cleaver to actually expand his reach and his power (embedded), expanding the size of his gestural 'circle' and the intentional nature of his action, which carried more threat and force than if his hand had been empty (extended and enactive). A more complex example: in an early scene, 'The Vicar Talks to His Servant,' a smug, well-off royalist vicar sits at his table, drinking wine and eating oranges from a big silver bowl, and desultorily asks the servant waiting on him about his sick baby (who is in fact dying); at the scene's end, he gives the servant a single orange for the baby. The actors needed to understand the significance and rarity of oranges as a product grown far away and imported at great expense into England, an unimaginable beneficence for the servant. Late in the play, in 'The Vicar Welcomes the New Landlord,' the vicarage has been confiscated by Parliament and bought by one of Cromwell's men. The tables have turned. There are no stage directions, so we had Star, an officer for Cromwell and the new owner, assume the vicar's place at the table from the earlier scene and eat oranges taken from his jacket, the bowl on the table now empty, while the vicar stood where his servant had before. Near the scene's end, Star hands an orange to the displaced and frightened vicar. No mention of oranges in this scene, but it was an explicit, embodied repetition of aspects of situation and action.

The 20-minute long Putney Debates scene at the end of the first act can be seen as a case study for addressing language and embodiment. Composed of verbatim extracts from transcripts of the first three days of the New Model Army's debates about how to structure Britain's new constitution, it provided a rich testing ground for how to apply 4E cognition tools and terms to embodying the text. The scene has theatrical pitfalls: there are dense arguments about issues such as franchise and property, in the conceptual frameworks and syntax of 1647. It's a long scene of political debate at the end of a long act. It could seem abstract and impersonal. However, for the characters, the stakes (an acting term for what is to be gained or lost, which in this case is the fate of England) could not be greater. The actors had to grasp this in their bodies and grab the audience so that they not only understood, but felt the urgency of the arguments. This involved engaging aspects of extended and enactive cognition, as the actors collaborated to find the 'dance,' the discoveries and the argument of the scene, and 'weaponize' their arguments to affect the audience. Lakoff and Johnson's embodied metaphor and Fauconnier and Turner's conceptual blending underpinned our approach, keeping physical imagery, physical stakes and contemporary associations and cultural-temporal 'translation' (from the latter 1640s England to 2016 Dallas) foremost and central. The scene provided a rich ground for understanding the profound inseparability of language from physical and affective feeling, as a response to environment.

Franchise, property and their relation to each other are central to the scene's argument. Commissary General Henry Ireton, a Cromwellian, argues for franchise being linked to property, while Colonel Thomas Rainborough, a Leveller, holds that any male resident should have a vote, regardless of owning land. In rehearsal we linked this to 2016 election campaign, drawing parallels between the Debates' arguments and those of October 2016 about money-as-speech, wealth inequality, property, voting rights, among other things, to spark actors' engagement with speech as action, as performative and agential, as a tool to get something important done. (In discussions I had with many who attended the performances, there was indeed a clear grasp of *plus ça change, plus c'est la même chose.*)

4E and enactivist perspectives provide tools and categories to address key directorial, acting and design tasks, allowing us to engage the work of any psychophysically based practitioner more effectively because we now know more of the science behind it. These provide a more specifically articulated focus on questions such as 'How can I make this physical?' 'How can I describe this in terms of relationship or environment?' 'How can I describe this in terms of experiences the artist has had (metaphor, analogy, metonymy)?' 'How can the actor or designer connect this to something sensory?' 'How can I use personal relationships and our actual physical environment to bring the material to life?' These can connect us more consciously to the problem of embodying a text – or making a piece of any kind. This has not to do with psychology or motivation as defined in traditional psychologically based approaches to acting. Because they are dynamic and relational, 4E approaches are *de facto* social and political, as well as personal. The default mode is 'embodied, embedded, extended, and enactive': How can we use mind, language, bodies, environment and our connection to each other to engage material, whatever it is, in an enlivened, even urgent way that will make us smarter and more attuned to it – and the world?

Notes

1 This discussion of language and cognition is a version of my discussion in *The Actor, Image, and Action*, pp. 16–18.

2 This has resonance with Elinor Fuchs's foundational essay on text analysis, 'EF's Visit to a Small Planet: Some Questions to Ask a Play,' which are derived from notes at her course on dramaturgy at the Yale School of Drama. It is an invaluable resource for how to read a play.

3 This echoes Stanislavsky's concept of characters in given circumstances – bodies in shared environments, trying to accomplish actions; I have written about this at length in *Actors, Image, and Action: Acting and Cognitive Neuroscience,* which also discusses 4E cognition's resonances with other Stanislavskian terms such as circles of attention and communion.

4 Some of this may sound familiar to those who saw David Cromer's production of the play in Chicago or New York; I was in fact deeply inspired by it.

References

Blair, Rhonda (2008). *The Actor, Image and Action: Acting and Cognitive Neuroscience*. London: Routledge.

Churchill, Caryl (1976). *Light Shining in Buckinghamshire*. In *Churchill: Plays One* (1985). London: Methuen, 181–242.

Cook, Amy (2007). 'Interplay: The Method and Potential of a Cognitive Scientific Approach to Theatre.' *Theatre Journal* 59:4, 579–94.

Fadiga, Luciano, Laila Craighero, Maddalena Fabbri Destro, Livio Finos, Nathalie Cotillon-Williams, Andrew T. Smith and Umberto Castiello (2006). 'Language in Shadow.' *Social Neuroscience* 1:2, 77–89.

Fauconnier, Gilles and Mark Turner (2002). *The Way We Think: Conceptual Blending and the Mind's Hidden Complexities*. New York: Basic Books.

Fuchs, Elinor (2004). 'EF's Visit to a Small Planet: Some Questions to Ask a Play.' *Theater* 34:2, 4–9.

Gibson, James J. (1982). *Reasons for Realism: Selected Essays of James J. Gibson. Resources for Ecological Psychology*. Hillsdale, NJ: L. Erlbaum.

Kaag, John (2014). *Thinking Through the Imagination: Aesthetics in Human Cognition*. New York: Fordham UP.

Lakoff, George and Mark Johnson (1999). *Philosophy in the Flesh: The Embodied Mind and Its Challenge to Western Thought*. New York: Basic Books.

Noë, Alva (2004). *Action in Perception*. Cambridge, MA: MIT Press.

Noë, Alva (2008). 'Précis of Action in Perception: Philosohy and Phenmenological Research.' *Philosophy and Phenomenological Research* 76:3 (May): 660–665.

Robbins, Philip and Murat Aydede (2009). 'A Short Primer on Situated Cognition.' *The Cambridge Handbook of Situation Cognition*. Cambridge: Cambridge UP.

Tribble, Evelyn and John Sutton (2011). 'Cognitive Ecology as a Framework for Shakespearean Studies.' *Shakespeare Studies* 39, 94–103.

9
ACTING AND EMOTION

Vladimir Mirodan

A few questions?

A friend, six months into an actor-training course, is working on a monologue requiring him to become 'enraged' and 'let rip.' His acting teacher is not satisfied. During one rehearsal, he gets hold of my young friend's arms and requests that he deliver his speech while fighting against the pressure. Muscles strain, blood rushes to the head and much shouting ensues. 'That's it, now you've got it!,' the director beams. Our young actor, instantly becalmed, replies: 'Ah, I see… That's what you wanted. OK.' To which the director: 'Now I'm not so sure. How can you be so calm, so soon after? *Did you really feel it*?' The young actor is perturbed: what is he meant to 'feel'? Why should he not keep his cool? And what can the director possibly know about his intimate feelings?

Two hundred miles to the east, a well-known Hollywood star is filming in a Paris suburb. The script requires her to make a transatlantic telephone call, during which she hears that her relationship with her boyfriend is at an end, and then break into tears. The movie is set in the 80s, before the advent of mobile phones, so an old-fashioned public telephone had to be found. But the location scouts did not notice that the street where this rare period cabin is situated was under the flight path of planes landing at Charles de Gaulle airport. Eleven takes are needed before a decent sound track can be recorded. Our highly skilled actress manages to break down in tears 11 times in succession, always on the same exact word in the script. In the hotel bar that evening, the crew cannot help but ask themselves: how can someone appear so genuinely, visibly affected and yet control her emotions to such an extent?

I meet an acquaintance, a much-liked British actress in her 50s, and I tell her that I am writing something about emotion. 'Ah,' she says, 'you must write that actors possess a special ability to feel the emotions of others. I always know when I meet a true actor,' she adds, 'because of their special sensitivity.'

Such stories give rise to a number of posers, common to all actors: 'How do I achieve intense emotions? How do I repeat these night after night, take after take, on cue? Do *I* have to feel them for others to feel? Indeed, how is it that others respond emotionally to what I do? Is it because I have some unusual ability? Can anyone apart from me know what happens emotionally when I am acting?'

Contemporary cognitive neuroscience offers a set of competing explanations regarding the arousal, perception and transmission of emotion. I will outline the principal theories first, then look at the ways in which certain widely used acting approaches map themselves onto them, as well as at how such explanations – while still partial and uncertain – might begin to influence the development of acting practices.

What's an emotion?

Prevalent theories almost always begin by examining the relationship between the physical signs of emotion and their penetration into, and interpretation by, consciousness. The so-called 'read-out' theories (Damasio 1994, 1999, 2010; LeDoux 2002; Rolls 2005) use the term 'emotion' to describe physiological changes. Our bodies strive to maintain a balance between key internal functions and the environment – homeostasis. Physiologically considered, emotions are departures from homeostasis: under the impact of fear, rage or joy, our skin changes its electrical charge, our palms sweat, hearts race or are frozen in terror, the pituitary and adrenal glands adjust their outputs, certain peptide modulators are released into the bloodstream and the immune system undergoes rapid modifications. Some changes can also be observed with the naked eye: we blush or turn pale, our breathing rate increases, our facial muscles rearrange themselves in configurations typical of anger or disgust, our body postures indicate aggression or submission. Emotion thus refers to chemical and neural patterns, triggered automatically by innate brain mechanisms, without the intervention of consciousness. 'Read-out' perspectives on emotion would explain the experience of my young friend, the trainee-actor, in terms of his awareness of the rapid onset and equally rapid diminution of physiological arousal. Emotions are intense ('salient' in psychological terminology) and of relatively short duration. They are not to be confused with long-lasting 'moods,' such as exuberance or serenity. States of arousal with high salience focus the attention, dislodge other thoughts and take you over. As the psychologist Jerome Kagan (2007, 22) explains in a telling metaphor: 'There is always some form of weather, but we award special status to the infrequent, distinct arrangements of humidity, temperature, and wind velocity called hurricanes, blizzards, and thunderstorms.'

Neuroscientist Antonio Damasio's notion of 'background emotions' is also useful:

> When we sense that a person is 'tense' or 'edgy', 'discouraged' or 'enthusiastic', 'down' or 'cheerful', without a single word having been spoken to translate any of those possible states, we are detecting background emotions. We detect background emotions by subtle details of body posture, speed and contour of movements, minimal changes in the amount and speed of eye movements, and in the degree of contraction of facial muscles.
> *(1999, 52)*

'Feeling,' on the other hand, is the 'private, mental experience of an emotion' (Damasio 1999, 42). Feelings occur when physiological changes penetrate into consciousness and are experienced as a recognition, a 'realisation' of the link between a stimulus and the body alterations this has provoked. Feelings involve evaluation: the bodymind has an inbuilt set of values enabling it to judge whether a stimulus is good or bad (Damasio 1999, 30). Such judgements depend in the first place on how familiar or strange the new event is, and then on how it measures against our sensory experiences – is it going to be pleasant or unpleasant? In psychological terminology, feelings have 'valence.' There are background feelings just as there are background emotions: a background feeling 'corresponds… to the body state prevailing *between* emotions. When we feel happiness, anger, or another emotion, the background feeling has been superseded by an emotional feeling' (Damasio 1994, 150–1, italics original).

Crucially, from an acting perspective, the evaluative function of feeling is closely linked to action. Feelings rise in consciousness *after* we have reacted physically to a stimulus: 'The feeling of fear came after you jumped and after your heart was already pumping – the feeling itself did not cause the jumping or pumping,' writes the neuroscientist Joseph LeDoux (2002, 206). The scheme he proposes is therefore as follows: physiological change in the body > registered by the brain (emotion) > consequent action (fight or flight) > connections with memory of past, similar occurrences > feeling (registering the change in consciousness).

Our conscious consideration of the world, not least of other people, also generates images. The mind/brain develops structures enabling it to respond to these images automatically, non-consciously. Damasio calls these brain structures the 'somatic markers': 'somatic' because they are about the body (*soma* in Greek), 'markers' because they 'mark,' fixing into brain circuitry images arising, however fleetingly, in the mind. Somatic markers are 'automated alarm signals' (Damasio 1994, 173), important weapons in our defensive armoury which, by linking present events to the memory of past dangers, enable us to make life-saving choices in milliseconds, without the need for a laborious examination of possible courses of action. Emotions can therefore be triggered in the brain not only by an actual object or event, but also by 'conjuring up from memory' (Damasio 1999, 56) an image of that object or event.

Feeling is the construct that brings emotion to the mind; when feelings arise, we tell ourselves: 'I am afraid' or 'I am excited.' For 'read-out' theorists, the ability to think in words is therefore essential to the generation of feelings (Rolls 2005). The words we choose are influenced by the immediate context of the event, by our other thoughts and actions at the time and by our culture (Kagan 2007, 42). Are we watching *The Blair Witch Project* or are we actually lost in a dark wood? – the same state of arousal can be described as excitement, terror or impatience. Emotions are therefore context-dependent:

> consider a woman who has just learned that her ninety-year-old mother, who had been suffering from a painful cancer for three years, died. If asked to describe her state she is likely to say 'sad' or 'relieved', depending on the most salient feeling, when her actual stage is a blend of both emotions.
>
> *(Kagan 2007, 8)*

Two other contextual elements are important in shaping the emotion. First, the nature of the target towards which we direct it – is Romeo in love with Juliet or with Rosaline? Second, while reading or hearing about an event, the words activate perceptual representations rooted in our individual experiences: direct experience of parenthood will affect the emotions aroused by words such as 'son,' 'daughter' or 'children.' In the recall of emotion-laden events, our thoughts direct themselves towards the context in which these arose and in particular towards their presumed causes. We always ask 'what led to this?' These thoughts colour the unconscious sensations into becoming a specific feeling.

A competing, albeit minority, view on the nature of emotion (Izard 1991; Panksepp and Biven 2012) minimises the role played by consciousness and emphasises instead the operation of those strata of the brain which lie deep below the 'thinking cap' of the neocortex. Experiments have demonstrated that one can elicit avoidance or approach reactions in animals through the direct stimulation of their brains. Applying electrical or chemical stimuli to the areas associated with fear, for example, will make even young animals, who had never been exposed to danger, cower, and if the stimulus is strong enough, scurry away in terror (Panksepp and Biven 2012, 20). Such reactions amount to coherent emotional responses and contradict those theories positing that systems in the neo-cortex 'read out' physiologically induced signals in order to transform emotion into feeling.

Neuroscientist Jaak Panksepp and his allies thus propose a view of consciousness that differs from the one outlined by Damasio, LeDoux and Rolls, yet chimes with certain common artistic discourses on imagination and creativity:

> neither cognitive ability nor the ability to think in words is a necessary condition for affective consciousness. Felt experience can be *anoetic* – an unreflective, unthinking primary-process kind of consciousness that precedes our cognitive understanding of the world, or our so-called *noetic* (learning, knowledge-based) secondary-process consciousness... As we feel our affective states, we do not need *to know* what we are feeling.
>
> *(Panksepp and Biven 2012, 14, italics original)*

Emotion is thus seen as a primary, sub-cortical response which – if it needs to be generated at will, as in acting – is best engendered by physical means that echo biological triggers for fear, anger, lust and so forth, encountered 'in the wild.' From *this* perspective, my young friend's drama school exercise would be explained as an attempt, however unsophisticated, to access this innate mechanism.

I can now hear the theatre-orientated reader who, having ploughed through the preceding pages, grumbles: 'This is all rather technical and distant, detached from what *it actually feels like* to be on stage or to watch a performance.'

The philosopher of cognitive science Giovanna Colombetti (2014) also vents her dissatisfaction with these scientific accounts of the emotions. In particular, she complains, 'affective neuroscience has so far neglected emotion *experience*' (2014, 143, my italics). Her analysis makes extensive use of data derived from the so-called 'third-party' methods – the technologies and methods scientists use to measure physiological changes related to emotion, including their sophisticated ways of looking 'inside' the brain. But – argue Colombetti and the phenomenological school within which her critique inscribes itself – in relying heavily on physical measurements, affective neuroscience has neglected first-hand, subjective reports of emotion. A different framework is therefore proposed: the so-called 'enactive' approach, integrating first- and third-party accounts and forcing scientists to 'focus not only on *what* [the subjects] experience but also on *how* they experience it' (Colombetti 2014, 147, italics original).

'Enactivism' is a synthesis of ideas drawn from cognitive science, biology and phenomenology (Varela et al. 1991). One of its central tenets is *embodiment*: the brain alone is not considered a sufficient physical basis for the mind. Rather, the mind is generated or 'enacted' by the entire living organism: not only its sensorimotor functions, but also the viscera and the circulatory, immune and endocrine systems. Another core assumption is that the mind operates on organisational principles common to all living organisms. Organisms, it is asserted, are defined by their drive 'to make sense' of the world and the concept 'mind' encompasses this drive. 'Enactivity' is the organism's capacity to 'make sense' of the world and of necessity adopt an attitude or perspective on it, and not necessarily in the form of articulated thought. As the philosopher biologist Evan Thompson (2007, 128) puts it: 'Mind is life-like and life is mind-like.'

By adopting this broad definition, enactivism readily incorporates conclusions on human emotion derived from animal observation (the 'sub-cortical' or 'anoetic' model described by Panksepp). Conversely, the 'read-out' model, with its several phases (see LeDoux's scheme earlier), is considered uneconomical, over-complicated and ineffective. This is because this model assumes that the specific kind of emotion experienced depends on a series of cognitive appraisals of an event in terms of its significance for the organism's survival and well-being. From a biological perspective, emotion generated by the entire organism, without the intervention of a complex system of appraisal checks, feels more in keeping with the way in which nature organises itself. The 'evaluating activity... is realised entirely immanently, in virtue of the organism's autonomous organisation,' Colombetti writes (2014, 101). As an extension of this position, no meaningful distinction between cognition and emotion can be drawn: 'The mind, as embodied, is intrinsically or constitutively affective; you cannot take affectivity away from it and still have a mind' (Colombetti 2014, 1).

The generation and perception of emotions cannot therefore be 'parsed' into separate cortical and sub-cortical processes. The brain integrates 'vertically' and uses all its resources to generate cognitive/emotional responses. This perspective rejects the sequence first posited by William James and essentially adopted by Damasio, LeDoux and many others, who consider that a sensory stimulus is the trigger for the generation of emotion in the body/mind. If the body acts 'as a whole,' then the brain is already primed to act: 'There is', Colombetti writes, 'no first cause in this process... Emotion is not a distinct step in a perception-action sequence...emotion is rather an inescapable, pervasive dimension of brain activity on which sensory information impinges and from which action progresses' (2014, 63).

This holistic, all-encompassing perspective supports acting methodologies which seek to do away with any artificial separation between analytical ('table-work' – text analysis or 'actioning,' for example) and intuitive (e.g., 'neutral/larval' mask work or contact improvisation) approaches. I also find talk of an 'immanent' relationship between the body/mind and the environment seductive, humanising and capturing the sense we have of our lived experience. Nonetheless, this remains a mostly philosophical construct. Some (as yet modest) support is emerging from studies showing high levels of integration in brain functions, suggesting that emotion and cognition operate together in distributed networks (Gu et al. 2013; Hardcastle 1999, 242–3; Kassam et al. 2013, 1; Pessoa 2008). Colombetti's preferred model for conceptualising emotions is therefore as 'dynamical patterns' – self-organising patterns of the organism, best described with the tools of dynamical systems theory (DST). DST maps mathematically the mutual interdependence of variables within a system or 'state.' When organisms react, their reactions are not 'directed' or 'determined' by a primary cause, but are shaped by the mutual interaction between all the factors involved. A telling example of such mutual interdependence is what occurs when two grandfather clocks are placed next to one another on the same wall. After a while, their pendulums synchronise, with the wall acting as a conduit, through vibrations. Neither clock can be said to 'lead' the other – they are both engaged in continuous, mutual adaptations.

As the descriptions detailed earlier will have made clear, the emotions – seen as aspects of cognitive activity – are being extensively examined from several perspectives: physiological–biological, experiential–psychological and expressive–social. However, there is as yet little consensus regarding their neural structures or the processes that give rise to them. Nowhere is this disagreement more in evidence than on the question of whether emotions may be classified in accordance to the ways in which they are expressed and communicated (through the face in particular) and whether inferences regarding their underlying neural mechanisms may be drawn from these classifications. Two main approaches sum up this debate: the 'basic' versus 'modal' classifications.

Following widely quoted research by Paul Ekman (1982) and others, six 'basic' or 'primary' emotions are sometimes isolated: happiness, sadness, fear, anger, surprise and disgust. These are so defined because their facial expressions are universal, cutting across cultures and continents. The classical theory is that each of these is generated by an 'affect programme' (Tomkins 1962, 244), a brain structure developed through evolution and transmitted genetically. It has been very difficult, however, to observe in real life the distinct facial patterns Ekman assigns to each of the basic emotions. As a result, adherents of the affect programme conception of emotion tend of late to explain variability in real-life expressions as 'blends' of the basic emotions. Ekman has even suggested viewing the very notion of 'affect program' as a metaphor. Nonetheless, while acknowledging that affect programmes can be altered by culture or experience, Ekman (2003, 67) continues to postulate that 'there must be different [brain] circuits for the different responses that characterise each emotion'; in other words, that emotions occur when the brain 'executes a pre-written program' in response to a stimulus.

This position, though still influential, is less secure than it was considered a couple of decades ago. An alternative theory – the componential emotion model – argues that emotions cannot be defined according to set configurations of the face but that, when we experience an emotion, face muscles move and rearrange themselves fluidly. The expression of emotion is therefore seen as a *process,* 'during which different elements of the expression surface and combine at different points in time' (Scherer and Ellgring 2007, 116). The inference is that emotions are not discrete entities, recognisable through set response patterns, but rather 'fuzzy sets' of cognitive activity, motor expression, physiological arousal, action tendencies and subjective feeling states. These components of emotion (hence 'compositional') are linked, or even synchronised, and their expressions *emerge* over time in a variety of psychophysical configurations.

This view of emotion gives some scientific support to the enactive theory. From the same perspective, the fraught question of which comes first – arousal, consciousness or action – is being replaced by an emphasis on the interactions between these different phases: while it is acknowledged that emotions are generated more or less automatically by a central mechanism, it is considered that this can be triggered by any one of these elements and, once triggered, reinforced and modified by the other two.

As far as performance is concerned, I expect there will be little dissent if from the descriptions earlier I pick out the intrinsic link between emotion and action. Emotional states that do not result in perceptible expression, however diminutive, are generally speaking of little use in the communication of meaning.

Emotion and acting

Three conclusions of significance to acting theory can be drawn from these descriptions:

- Cartesian distinctions between reason and feeling on the one hand, and between control and abandonment on the other, are comprehensively rejected
- Assumed hostilities between the conscious and the unconscious, in their various post-Freudian incarnations, are replaced by an understanding of the interdependence of neo-cortical (conscious) and sub-cortical (non-conscious) processes
- Great emphasis is placed upon the indivisibility and mutuality of emotional arousal and physical action.

Some widespread assumptions of Western schools of acting are, however, still rooted in the old distinctions. One entrenched belief is that convincing acting involves the creation of 'genuine' emotions, contrasted with the 'technical' imitation of their outward signs (Strasberg 1965). If the emotion is to be believable and affect the spectator, it is stated, the actor must experience it at least in some degree. This formulation needs to be rephrased in the light of scientific findings: neither spectators nor performers *can help* being 'moved' (changed) by emotion-generating actions. The amount of conscious control actors can exercise over the generation as well as the transmission of emotions, and the equivalent control spectators have over *their* emotional reactions, are tightly circumscribed by our biology. Emotional expression, like laughter, is outside our voluntary control and so is its detection.

But what is the most effective route to arousing it? Two principal approaches define Stanislavsky's legacy in this area: one takes as its starting point a mental trigger, the other a physical trigger.

The first, perhaps best known among non-specialists, revolves around summoning 'emotional memories.' The theory underpinning it assumes that our experiences leave traces in the nervous system, and that memories get fixed in our brains by association with some of the sensations which accompanied their creation: 'a polka-dot tie, an ivy leaf on a stucco wall, the smell or sound of sizzling bacon, a grease spot on the upholstery... [an] apparently insignificant object had been unconsciously perceived and associated with the original emotional experience' is how Uta Hagen (1973, 48) describes it. These memories are not available to the conscious mind, but are triggered when we recollect details associated with the original experience. One of the early adopters of Stanislavski in the United States describes the procedure thus:

> The theory is that if, quietly relaxed, you think back to a certain incident in your life which moved you strongly at the time, and if you can remember and recreate in your mind the physical circumstances of that moment (where you were, who was there, what happened, the time of day, the place, surroundings) and start reliving it... it is possible that a feeling similar to what you felt at that time will recur. If it was a very strong emotion and you can bring it back successfully three times in a row, it is quite possible you have something that will work for you for a long time.
>
> *(Lewis 1958, 35–6)*

This technique implies a high degree of self-knowledge, and actors are often encouraged towards soul-searching. Introspection is sometimes mitigated by balancing emphases on turning reflection into action and on openness to the emotions of others, as in the teachings of Uta Hagen and Sanford Meisner, respectively. Yet the technique remains essentially static, enabling access to emotion by means of a mental process, with a minimum of physical activity. As has by now been comprehensively established, Stanislavsky's notion of emotional memory was based on a partial and in places erroneous reading of the ideas of the French psychologist Théodule Ribot on the nature of memory (Benedetti 1988, 31–2; Carnicke 2009, 154ff.; Kemp 2012, 157). In American Method schools, this reading was then amplified by the introduction of significant Freudian practices – 'quietly relaxed' often meant lying down, as if on an analyst's couch. But, as theatre scholars attentive to the messages of cognitive science point out, the assumptions underpinning emotional memory exercises are both confirmed and challenged by much that the former has to say about the nature of emotion.

Records of objects and events with which we have interacted are indeed stored ('co-registered') in memory in association with the motor activities (reaching, looking, tasting, etc.) and emotional reactions which formed part of our encounter with them. And while motor, affective and sensory storage occurs in different systems in the brain, retrieval is always holistic: we recall a landscape in association with the memory of the effort needed to reach it and the feeling of awe or surprise when we first saw it. 'As a consequence, when we recall an object... we recall not just sensory characteristics of an actual object but the past reactions of the organism to that object' (Damasio 1999, 161). Memory is indeed 'affective.'

It is, however, erroneous to speak simply of 'retrieving the image' of an object.

Cognitive approaches to memory distinguish between inactive 'long-term memory' and 'basic working' memory – the ability of the brain to hold on to images for periods from tenths of a second to a number of consecutive seconds (Damasio 1994, 197). Reactions to emotional stimuli are a dialogue between long-term and working memories, directed by purposeful attention. Associations are not only imbued with the sensations and emotions of the past, but are also shaped by the context in which they arise in the present; they are, Joseph LeDoux (2002, 203) explains, 'constructions assembled at the time of retrieval.' Thus, each 'construction' will be slightly different from past remembrances of the same event. An actor seeking to recall a personal emotion for the highly specialised purpose of transferring it to a fictional character in fictional circumstances does not therefore engage in the like-for-like retrieval of a past event, but in its imaginative reconstruction. This is also affected by what psychologists call 'mood dependence' – our mood today will influence the past we bring up (Connolly and Ralley 2007, 254).

There is a fair bet that the way in which the Hollywood actress I described earlier managed to burst into tears 'on cue' was in part through the use of emotional memory techniques. Yet neuronal and conscious memories are not identical. When our actress broke down in tears, her technique involved just as much being able to access 'triggers' from her past as being aware of the context and requirements of the present. What she did was an act of emotional *imagination*. Whatever the memories she summoned, they were no more 'real' than Michael Chekhov's famous improvisation when he reenacted his father's funeral to great effect, only to disclose at the end that his parents were still hale and hearty. This got him expelled (temporarily) from Stanislavski's classes, yet forced the master to reassess his demand for 'absolute truth' in the recreation of emotionally laden biographical events. From a cognitive perspective, emotional recall can only be a recreation, affected just as much by traces left in the brain by past events as by the emotional context in which the retrieval takes place and, above all, by its *current purpose* (see also Blair in McConachie and Hart 2006, 174).

What then of emotions generated as a result of physical exertion or the reproduction of facial arrangements and/or gestures commonly associated with certain affects, the so-called 'outside-in' process? It has frequently been observed that actors can become angry by repeatedly banging their fists on a table or distressed by imitating the outer actions of crying (Carnicke 2009, 187; Hagen and Frankel 1973, 50). Such observations are supported by scientific findings: the psychologist Ernst Gellhorn (1964) thought that emotions could be aroused and becalmed at will by the action of the muscles. Research indeed shows that emotions can be triggered when physical movement is accompanied by somatic sensory-feedback to the brain, consequent changes in neo-cortical circuits and autonomic nervous system (ANS) and hormonal changes (Panksepp and Biven 2012, 171). However, it also appears that a 'tipping point' needs to be reached in the size of the outer expression before this becomes effective. In a number of studies, participants reported that a film was funnier or the electrical shocks they received more painful when they exaggerated their amusement or pain than when they kept their faces still (cited in Hatfield et al. 1994, 54). As we saw, Ekman demonstrated that the facial expressions corresponding to the six 'basic emotions' could be accurately and universally read. However, a study comparing Ekman's static, laboratory-based experiments with reactions to facial expressions taken from several realistic films only found limited recognition of the emotions portrayed (Carroll and Russell 1997). It appears that, within certain limits (Wallbott and Scherer 1986, 697), exaggerating the facial expression corresponding to an emotion increases its intensity as well as the accuracy with which it is perceived.

Overall, in light of the discoveries of cognitive science, ancient quarrels between actors who work from the 'inside-out' and those who start with physical actions appear increasingly sterile. After all, Stanislavsky, Meyerhold, Laban and Copeau all emphasised the unbreakable circuit of psychology and physicality. Cognitive science confirms the idea of a continuous loop between stimulus, physiological reactions, non-conscious changes in the brain and consciousness. Where in this loop one starts is irrelevant, as emotion/feeling, once initiated, will surge and intensify involuntarily.

How is it, then, that actors are to a large extent able to control the physiological effects of emotional arousal? Here a further distinction needs to be drawn between the generation and transmission of emotions in real life and in a performance context. Whether as actors or spectators, we come to performance in a peculiar frame of mind. Actors direct their attention to elements in their material which provoke both conscious and non-conscious inferences regarding character, relationships and actions. Most, if not all, of these are emotionally charged. One might call this process 'reading for emotion' rather than 'reading with emotion.' The actor reads the text purposefully, with the goal of emotional engagement leading to physical representation constantly in mind. Working on a role involves a to-and-fro between the circumstances described by the script and the personal attitudes of the actors. This iterative process involves a controlled arousal of emotion: reading-with-a-purpose in preparation for acting, which is then refined and modified in the crucible of the rehearsal room.

Moreover, actors 'read for emotion' with trained eyes. The literary critic R.A. Zwaan (1993) has argued that the ways in which readers construct meanings reflect their *purpose* in engaging in reading and that, with repeated exposure to such purposeful reading, they develop special capacities. Reading for emotion will similarly include specialised forms of attention and concentration. In particular, I would argue, actors acquire highly developed abilities to disguise and modify emotional behaviours in accordance with the so-called social 'display rules': 'conventions, norms and habits that… specify who can show what emotion to whom, and when' (Ekman 1980, 87). Psychologist Nina Bull (1951/1968, 44–7) once carried out a series of experiments in which the subjects were asked to react while under hypnosis to emotionally laden words triggering fear, disgust, triumph, etc. Their physical responses amounted to preparation for action, but never to the action itself. In anger, for example, the 'subjects would clutch their hands in readiness to strike, but they never actually struck at anything.' Moreover, when questioned (while still under hypnosis), the subjects described being aware of the feeling of the emotion (e.g., the feeling of being angry), as well as, separately, of the physical sensations they were experiencing. The two were perceived as closely related, but the subjects 'seemed always aware of a difference' (Bull 1951/1968, 47). It appears that the mechanisms of control lie deep in our minds and that acting training exploits and enhances this universal trait.

Equally, repeated exposure to certain types of performance – whether classical verse plays, contemporary dance or action movies – conditions audiences' aptitudes for specific emotional responses. Jerome Kagan cites an interesting experiment in this context: when monkeys are given a sweet drink, specific neurones connected with feelings react in their brains. When the animals have had enough of the glucose solution, these circuits fall silent, but show up again when they are offered a different food. The implication is 'that an individual, like the hungry monkey, must be psychologically or biologically prepared for an emotional state' (Kagan 2007, 145).

Two further questions need addressing:

- How is emotion transmitted from actor to actor and from actors to audience in the course of performance?
- Do actors possess a special 'gift' which enables them to access and transmit emotion more than others?

The philosopher Robert Gordon wrote: 'You can catch an emotion, just as you can catch a cold, without knowing who you caught it from' (cited in McConachie 2008, 67). Psychologists Elaine Hatfield and John Cacioppo (1994) promoted the notion that our social interactions are characterised by pervasive forms of transmission and reception of emotion they describe as 'primitive emotional contagion.' It has been long known that mimicry of posture, gesture and in particular of facial expressions is a salient feature in our interactions. We react to the pitch, rhythm, speed and length of pauses in others' speech and are attuned to the ways they stand as well as to their gait and hand gestures. Such observations are so routine, so much part of our daily interactions, that they become automated. That is, they take place with great speed and with minimal demands on consciousness (Hatfield et al. 1994, 12).

Automatic contagion through mimicry frees our brains from having to assess consciously the emotions of our interlocutors. We are thus able to focus on what they are saying or on our own tasks. Nonetheless, even when in awareness we pay attention to concrete facts and actions, subliminally we are still affected, moment by moment, by the feedback from changes in our muscles and ANS induced by mimicry (Hatfield et al. 1994, 48).

When we watch a performance, complex conditioned responses to its codified language come into play – certain gestures, facial expressions or vocal qualities affect spectators, who have deliberately placed themselves in a state of receptivity. But this codified language is itself based on everyday facial expressions, sounds, and so forth, which register non-consciously, as part of our primitive contagion mechanism. The discovery of mirror neurones in macaque monkeys (and their presumed existence in humans) has led to an understanding of imitation not as a process of conceptual reasoning, but as direct simulation, the actual experience of others' emotions and actions (Gallese et al. 2004). Strong evidence is also emerging of emotion and sensation mirroring through bodily functions other than those of the brain (Uithol and Gallese 2015 for a review). This has led the neuroscientist Vittorio Gallese (2003, 2007; Gallese and Sinigaglia 2011) to introduce the concepts of 'embodied simulation' and 'intercorporeality' to account for interactions between subjects which cannot be assigned to the classical 'mind-reading' cognitive model. 'By this,' Uithol and Gallese write, 'it is implied that mindreading is preceded, both from a phylogenetic and ontogenetic point of view, by non-propositional forms of social understanding' (2015, 11). 'We know through the body – kinetic consciousness,' writes philosopher Maxine Sheets-Johnstone (2009, 383). Intuited through the experience of acting, this type of direct experience was memorably described by the Stanislavsky-influenced literary critic Francis Fergusson as the 'histrionic sensibility,' a state in which the actor 'responds mimetically with his whole being' (1955, 24).

Are our reactions in direct proportion to the emotions observed in others? Psychologist Kenneth Craig (1968) designed a classic series of experiments to test this question. His conclusions and other experimental evidence suggest that we feel others' distress, but at lower levels than our own, directly experienced pain (Hygge and Ohman 1976).

Further qualifications pertain when applying discoveries based on daily life experiments to the specialised context of performance. It is true that, as a matter of biological efficiency (the brain does not like to waste resources), emotional responses follow the same neuronal pathways in reaction to real life as to fictional stimuli. What the necessary control mechanism might be has been the subject of much speculation. I find of particular interest the view that, in fictional contexts, *beliefs* play a role far greater than in daily life in modifying emotion-contingent actions. In responding to fictional stimuli, the brain receives two complementary sets of messages: one (probably sub-cortical) from the 'reality' experienced, the other (cortical) from our pre-formed understanding of its fictional nature. As a consequence, emotional reactions are attenuated: 'one's belief that certain events are only being played by actors has a very modest impact upon one's feelings, but this belief is quite enough to keep one seated during a convincingly depicted murder' (Schroeder and Matheson in Nichols 2006, 33).

If spectators (receivers) engage emotionally, albeit at attenuated levels, are there people (transmitters) who are more successful than others at generating emotional contagion? And are such people better at it because they are themselves capable of higher levels of arousal or, on the contrary, because they keep a cool head while signalling arousal?

Elaine Hatfield (1994, 128) recognises that there are both powerful 'spreaders' of the emotional virus (the 'Typhoid Marys') and people who are particularly vulnerable to contagion, the 'Marcel Prousts' of this world. In terms of both physical and psychological characteristics, some people display greater natural abilities to express emotions; in psychological terminology, there are 'dispositional differences' between stronger and weaker senders. There thus appears to exist a sub-set of the population with particular abilities in this area. So-called 'externalisers' show emotions in their faces, but have low-level ANS responses. A different sub-set, 'internalisers,' have strong ANS responses, but few, if any, outer manifestations (Hatfield et al. 1994, 133). One can reasonably conjecture that performers, not least actors, are generally drawn from within the first group. Few studies have been carried out to test this hypothesis, but one study (Stern and Lewis 1968) concluded that actors practised in the systematic use of emotional memory were able to increase their galvanic skin responses more than control groups.

The experimental director and critic Charles Marowitz once wrote that an actor's utter absorption when repeating words, moves and gestures during rehearsals leads to 'a mild form of self-hypnosis' (1978, 100), thus reducing interfering conscious thoughts and opening up the ability to react spontaneously. A recent study (Panero et al. 2015) looked at whether acting students were more hypnotisable than control groups of musicians and non-artists. The conclusion of the study was that they were, as a result of their greater openness to imaginative suggestibility, proneness to fantasy and acquired skill in focusing their attention in order to become absorbed. In general, people who are inclined towards (and practiced at) 'reading' both others' and their own emotional responses, who display a natural propensity towards mimicry and who are generally 'emotionally reactive,' are probably more vulnerable to emotional contagion (Hatfield et al. 1994, 148–9). Again, one recognises here some commonly encountered descriptions of the kind of person who enters and thrives in the acting profession.

Acting on emotion

Thus far, much of the writing on acting and cognitive science – Blair (2008), Lutterbie (2011) and Kemp (2012) – has been framed by debates around certain core practices of the American conservatoire. Strasberg is the main target at which all the writers take aim, concerned that his approach downplays the roles of the imagination and of physicality as well as setting up confusing notions of what 'real' emotions are. Blair, for example, draws on LeDoux to reconsider how affective memory works and to call for a reassessment of the way in which this is taught and used (Blair 2008, 81). Overall, a return to original Stanislavskian sources, as opposed to American developments, is favoured. Blair declares original Stanislavsky writings 'more valid' because their descriptions of acting processes are being confirmed by cognitive research. Kemp also finds in physical action 'a certain prescience of the current understanding of the emotions...' (Kemp 2012, 163). Indeed, the term 'prescient' is a *leit-motif* in Kemp's book and is applied equally to Diderot, Stanislavski, Meyerhold, Grotowski and Lecoq, with the latter brought strongly into focus.

To correct what they see as the dominance of an introvert, psychologising tradition, these writers adduce Stanislavskian physical action, Grotowski's extensions of Stanislavski's ideas, as well as explicitly body-based approaches such as Laban's and Lecoq's, in support of their calls for psychophysical integration. Acting approaches which start with the reproduction of physical outcomes of emotion and rely on these to generate emotions proprioceptively are extolled, in opposition to passive emotional memory. Kemp, for example, cites approvingly Grotowski's use of songs and sonic vibrations as an approach to arousing emotions by physical means. Blair finds in Damasio's scheme strong support for classical 'outside-in' approaches – Meyerhold and physical action – as points of departure for the creation of emotionally charged 'images,' rooted in the body. A methodology is thereafter proposed whereby reactions to texts and given circumstances lead to the creation of a stream of 'images' and thence to a 'detailed kinesthetic score that supports the body-mapping of those images' (Blair 2008, 81). In this context, most writers also cite Ekman's studies showing that one can trigger a basic emotion by arranging one's face muscles. At the same time, any simplistic notion that emotions may be engendered mechanically, by physical means alone, is rejected. Reshaping of the body, such as by means of animal, element or object work, can be an important contributor, but has to be done in dialogue with the internalised awareness of the psychological attributes of the emotion being sought and with a conscious, purposeful intent to generate it. Kemp (2012, 188ff.) describes a series of interesting exercises based on the eyes: students are taught to isolate what are normally involuntary movements and recognise how certain directions and frequencies link to different basic emotions. At the same time, Kemp recognises that, while feelings 'arise without conscious bidding,' they do so 'through a *combination* of physical actions and empathetic responses to a fiction' (Kemp 2012, 149, my emphasis). In other words, firing our histrionic sensibility necessitates both a physical approach and the deliberate, systematic mental engagement with how we *imagine* the emotion. This holistic approach accords with cognitive science principles, and its purpose in training is to sensitise the actor to the interaction between the physiological aspects of emotion and the primary metaphors with which we access them mentally (see also Kemp 2016).

While it is recognised that the science remains fluid and undetermined, a few acting methodologies have sought to adapt experimental findings directly. These approaches distinguish themselves from traditional acting methodologies by explicitly and deliberately using as their starting points *quantifiable* physiological elements, as opposed to traditional, psychoanalytical, qualitative methods (see also Blair 2008, 47).

In the 1970s, Susannah Bloch (1993; Bloch et al. 1987) developed 'ALBA Emoting,' an early process informed by scientific principles. Bloch's students were taught to control their breathing rhythms as well as arrange their facial features and postures into patterns corresponding to certain emotions frequently encountered in performance. This enabled them to generate emotions with relative accuracy, sustain them over longer periods and rapidly reduce their effects when the context no longer required them. The approach was a pioneering endeavour, started at a time when cognitive science was still in its infancy. Its teaching continues sporadically and has recently been incorporated into the curriculum of a UK conservatoire (Jackson 2017, 82). However, the ALBA technique has been criticised on the grounds of a certain lack of rigour in its experimental methodologies and resultant data and because some of its assumptions do not reflect current scientific findings (Kemp 2012, 186–187).

A lesser-known technique is that developed by the German director Stephan Perdekamp. Known as the Perdekamp Emotional Method (PEM), this combines certain cognitive scientific principles with less secure psychological concepts, such as 'bio-energy.' Like ALBA Emoting, PEM strays from the scientific cannon in, for example, extending the list of recognised 'basic' emotions to include grief and happiness. While one can see how these may be useful in acting exercises, from a scientific perspective their inclusion appears somewhat permissive. Both approaches are also open to the criticism that their basic design assumes that one can generate specific emotions by replicating their physiological signs in isolation, regardless of the context in which these emotions occur. These include character as well as actor personality traits, the circumstances imposed by the narrative or the immediate context of the rehearsal room/studio.

Overall, methodologies based on the rigorous application of cognitive principles are still to establish themselves widely in actor training, but the field is ripe for development.

References

Benedetti, J., 1988, *Stanislavski: A Biography*, London: Methuen.

Blair, R., 2008, *The Actor, Image, and Action (Acting and Cognitive Neuroscience)*, London and New York: Routledge.

Bull, N., 1951/1968, *The Attitude Theory of Emotion*, New York and London, Johnson Reprint Corporation, 1968 (original published 1951 as *Nervous and Mental Diseases Monographs No 81*, Coolidge Foundation, publishers).

Bloch, S., 1993, 'ALBA Emoting: A Psychophysiological Technique to Help Actors Create and Control Real Emotions', *Theatre Topics* 3 (2), 121–38.

Bloch, S., Orthous, P., & Santibanez-H, G., 1987, 'Effector Patterns of Basic Emotions: A Psychophysiological Method for Training Actors', *Journal of Social Biological Structure* 10, 1–19.

Carnicke, S. M., 2009 (1998), *Stanislavsky in Focus: An Acting Master for the Twenty-First Century*, 2nd edition, Abingdon: Routledge.

Carroll, J. M., & J. A. Russell, 1997, 'Facial Expressions in Hollywood's Portrayal of Emotion', *Journal of Psychology and Social Cognition* 72, 164–176.

Chekhov, M., 1953, *To the Actor: On the Technique of Acting*, New York: Harper and Brothers.

Chekhov, M., 1991, *On the Technique of Acting*, edited by Mel Gordon, New York: Harper Collins.

Colombetti, G., 2014, *The Feeling Body, Affective Science Meets the Enactive Mind*, Cambridge, MA and London: The MIT Press.

Connolly, R., & R. Ralley, 2007, 'The Laws of Normal Organic Life or Stanislavski Explained: Towards a Scientific Account of the Subconscious in Stanislavski's System', *Studies in Theatre and Performance* 27 (3), 237–59.

Craig, K. D., 1968, 'Physiological Arousal as a Function of Imagined, Vicarious and Direct Stress Experiences', *Journal of Abnormal Psychology* 73, 513–20.

Damasio, A., 1994, *Descartes' Error: Emotion, Reason, and the Human Brain*, New York: Putnam.

Damasio, A., 1999, *The Feeling of What Happens: Body and Emotion in the Making of Consciousness*, London: William Heinemann.

Damasio, A., 2010, *Self Comes to Mind: Constructing the Conscious Brain*, New York: Random House.

Ekman, P., 1980, 'Biological and Cultural Contributions to Body and Facial Movement in the Expression of Emotions', in Rorty, A. O., ed., *Explaining Emotions*, Berkeley: University of California Press, 73–102.

Ekman, P., ed., 1982 (reprinted 2013), *Emotion in the Human Face, Second Edition*, Los Altos, CA: Malor Books Reprint.

Ekman, P., 2003, *Emotions Revealed*, New York: Times Books.

Fergusson, F., 1955 (1949), *The Idea of a Theatre, A Study of Ten Plays, The Art of Drama in a Changing Perspective*, Garden City, NY: Doubleday & Co.

Gallese, V., 2003, 'The Roots of Empathy: The Shared Manifold Hypothesis and the Neural Basis of Intersubjectivity', *Psychopathology* 36 (4), 171–80.

Gallese, V., 2007, 'Before and Below 'Theory of Mind': Embodied Simulation and the Neural Correlates of Social Cognition', *Philosophical Transactions of the Royal Society B-Biological Sciences* 362 (1480): 659–69.

Gallese, V., & Sinigaglia, C., 2011, 'What Is So Special about Embodied Simulation?', *Trends in Cognitive Sciences* 15 (11): 512–19.

Gallese, V., Keysers, C., & Rizzolatti, G., 2004, 'A Unifying View of the Basis of Social Cognition', *Trends in Cognitive Sciences* 8 (9): 396–403.

Gellhorn, E., 1964, 'Motion and Emotion: The Role of Proprioception in the Physiology and Pathology of Emotions', *Psychological Review* 71, 457–572.

Gu, X., P. R. Hof, K. J. Friston, & J. Fan, 2013, 'Anterior Insular Cortex and Emotional Awareness', *The Journal of Comparative Neurology* 521 (15), 3371–88. doi:10.1002/cne.23368.

Hagen, U., & H. Frankel, 1973, *Respect for Acting*, New York and London: Collier Macmillan.

Hardcastle, V. G., 1999, 'It's OK to Be Complicated: The Case of Emotion', *Journal of Consciousness Studies* 6, 237–49.

Hatfield, E., J. T. Cacioppo, et al., 1994, *Emotional Contagion*, New York: Cambridge University Press.

Hygge, S., & A. Ohman, 1976, 'Conditioning of Electrodermal Responses through Vicarious Instigation and through Perceived Threat to a Performer', *Scandinavian Journal of Psychology* 17 (1), 65–72.

Izard, C. E., 1991, *The Psychology of Emotions*, New York: Plenum.

Jackson, D., 2017, 'Stanislavski, Emotion and the Future of the UK Conservatoire', *Stanislavski Studies* 5 (1), 75–83, doi:10.1080/20567790.2017.1298195.

Kagan, J., 2007, *What Is Emotion?: History, Measures and Meanings*, New Haven, CT: Yale University Press.

Kassam, K. S., A. R. Markey, V. L. Cherkassky, G. Loewenstein, & M.A. Just, 2013, 'Identifying Emotions on the Basis of Neural Activation', *PLoS ONE* 8 (6): e66032. doi:10.1371/journal.pone.0066032.

Kemp, R., 2012, *Embodied Acting, What Neuroscience Tells us about Performance*, London: Routledge.

Kemp, R., 2016, 'Lecoq, Emotion and Embodied Cognition', in Evans, M. and R. Kemp, eds., *The Routledge Companion to Jacques Lecoq*, London: Routledge, 199–207.

LeDoux, J., 2002, *Synaptic Self: How Our Brains Become Who We Are*, New York: Penguin.

Lewis, R., 1958, *Method or Madness*, with an Introduction by Harold Clurman, London: Heinemann.

Lutterbie, J., 2011, *Toward a General Theory of Acting: Cognitive Science and Performance*, New York: Palgrave.

McConachie, B., 2008, *Engaging Audiences: A Cognitive Approach to Spectating in the Theatre*, New York: Palgrave Macmillan.

McConachie, B., & F. E. Hart, 2006, *Performance and Cognition: Theatre Studies and the Cognitive Turn*, London: Routledge.

Nichols, S., ed., 2006, *The Architecture of the Imagination: New Essays on Pretence, Possibility and Fiction*, Oxford: Clarendon Press.

Panero, M. E., T. R. Goldstein, R. Rosenberg, H. Hughes, & E. Winter, 2016, 'Do Actors Possess Traits Associated with High Hypnotizability?' *Psychology of Aesthetics, Creativity, and the Arts* 10 (2), 233–9.

Panksepp, J., & L. Biven, 2012, *The Archaeology of the Mind, Neuroevolutionary Origins of Human Emotion*, New York and London: W.W. Norton & Company.

Pessoa, L., 2008, 'On the Relationship between Emotion and Cognition', *Nature Reviews Neuroscience* 9, 148–158.

Rolls, E., 2005, *Emotion Explained*, Oxford: Oxford University Press.

Russell, J. A., 2003, 'Core Affect and the Psychological Construction of Emotion', *Psychological Review* 110, 145–172.

Scherer, K., & H. Ellgring, 2007, 'Are Facial Expressions of Emotion Produced by Categorical Affect or Dynamically Driven by Appraisal?', *Emotion* 7, 113–130.

Sheets-Johnstone, M., 2009, 'Animation: The Fundamental, Essential, and Properly Descriptive Concept ', *Continental Philosophy Review* 42, 375–400.

Stern, R. M., & N. L. Lewis, 1968, 'Ability of Actors to Control Their GSRs and Express Emotions', *Psychophysiology* 4, 294–9.

Strasberg, L., 1965, *At the Actors Studio, Tape-Recorded Sessions*, edited by Robert H. Hethmon, New York: Theatre Communications Group.

Thompson, E. 2007, *Mind in Life: Biology, Phenomenology, and the Sciences of Life*. Cambridge, MA: Harvard University Press.

Tomkins, S. S., 1962, *Affect Imagery Consciousness: Volume I. The Positive Affects*, London: Tavistock.

Uithol, S., & V. Gallese, 2015, 'The Role of the Body in Social Cognition', www.academia.edu/13946451.

Varela, F. J., Thompson, E., & Rosch, E., 1991, *The Embodied Mind: Cognitive Science and Human Experience*. Cambridge, MA: MIT Press.

Wallbott, H. G., & K. R. Scherer, 1986, 'Cues and Channels in Emotion Recognition', *Journal of Psychology and Social Cognition* 51, 690–9.

Zwaan, R. A., 1993, *Aspects of Literary Comprehension*, Amsterdam and Philadelphia, PA: John Benjamins.

PART II: LEARNING

Introduction

Bruce McConachie

Although we commonly tie human learning to specific environments and tasks – school and reading, job training in the workplace, actors learning lines at a rehearsal, and so forth – learning happens everywhere, involves mundane actions and perceptions as well as specific tasks, and occurs throughout a person's lifetime. Children learn interpersonal skills and language, teens learn emotional control and group norms, young adults typically learn parenting and intercultural strategies and older people must learn to accept the limits of their bodies and their social authority. Genetic evolution has primed our body-minds to learn, and all cultures depend upon learning to survive and flourish. Cognitive scientists seek to understand learning in all of these contexts. As the chapters in this part explore, learning often involves **embodied** accounts of **empathy**, **memory**, **consciousness**, **play**, **imagination** and the **emotions**. Within **enactivism**, learning occurs all the time; it is a normal part of the perception-action cycle.

The chapters in Part II begin with an essay focused on hominins (our direct ancestors on the evolutionary family tree) who improvised communication before the invention of language. They proceed to the kind of learning that occurs during religious rituals, the public education of a community of citizens who have chosen to perform a commemorative dance together and students in a university classroom who learn how to design a 'living newspaper' that explores a significant social issue in their lives. The last three chapters focus on the education of actors. They include an overview of the developmental psychology that supports the ability to perform a fictitious role before an audience, a teacher's use of metacognitive strategies to enhance student learning in an acting class and Jacques Lecoq's approach to the teaching of acting technique.

I wrote the first chapter in this part – 'Improvising Communication in Pleistocene Performances.' While studying linguistic and anthropological approaches to the gradual evolution of language during the Pleistocene epoch (2.5 million years ago to 11,000 BCE), I was struck by the scholarly consensus concerning the growth of cooperation and interpersonal trust among our hunter-gatherer ancestors with regard to their hunting, cooking and childcare activities. Scholars had speculated that early, proto-language communication among these hominin bands involved mimetic kinds of sounds and gestures, but none had applied any knowledge about theatrical improvisation today to the situations and abilities of these hominins, who clearly needed to communicate with each other about common practical problems roughly 2 million years ago. Perhaps, I suggest, the same kind of interpersonal trust that supported many of their other activities could provide the foundation for improvising proto-languaging among our Pleistocene ancestors.

In his 'Ritual Transformation and Transmission,' David Mason begins by questioning sociologist Emile Durkheim's assumption that religious believers understand the ritual actions they perform as a straightforward expression of their religious beliefs. Instead, Mason sides with anthropologist Marcel Mauss and those who agree with him that the embodied actions of the performers, not their professed beliefs, must be the primary key to understanding what the performers mean by doing their ritual. Mason notes that how ritual actions shape beliefs and self-concepts is now the dominant approach in ritual studies. He builds on this performative understanding of ritual through the cognitively inflected ideas of anthropologist Roy Rappaport and the phenomenology of Maurice Merleau-Ponty. Mason applies this understanding of the embodied **emergence** of ritual meaning to a temple ritual performed among the Mormon faithful in 1846.

A contemporary dance performance by 100 citizen-volunteers is not the same as a nineteenth-century religious ritual, of course, but both deployed movement to create meaning for themselves and others. Ariel Nereson, in her 'Communities of Gesture: Empathy and Embodiment in Bill T. Jones/Arnie Zane Dance Company's 100 Migrations,' analyses a community-based work in Charlottesville, Virginia, performed in 2009, to commemorate the 200th birthday of Abraham Lincoln. Bill T. Jones choreographed the piece with amateur dancers, he said, to find out what 'democracy moving' might look like. Dressed in costumes of blue and grey to signify the antagonists in the U.S. Civil War and dancing across a space that featured Lincoln's deathbed in the centre, the dancers performed gestures they had improvised and supported each other physically in the midst of often dangerous moves. Nereson draws on the discourse of public history among historians and the insights of philosopher Mark Johnson on embodied cognition to probe the likely meanings of '100 Migrations' for the dancers and their audience.

Our part chapters move from cultural and historical contexts for learning to educational classrooms with Nancy Kindelan's essay, 'Creative Storytelling, Crossing Boundaries, High-Impact Learning, and Social Engagement.' Kindelan designed and taught an Honours seminar for non-Theatre majors that culminated in each student presenting a 'Treatment' for the presentation of a Living Newspaper play, the kind of documentary theatre produced initially by the Federal Theatre Project (FTP) as a part of President Roosevelt's New Deal. In terms of their learning, the seminar challenged students to deploy metacognitive skills – skills that required them to reframe and synthesise what they knew in a general way about journalistic evidence, narrative structure and performative effectiveness – in order to enable them to take a position on a matter of significant social concern and then revise and shape it for a specific audience. The students studied several varieties of Living Newspaper performance, from the FTP productions in the 1930s, to the documentary work of Anna Deavere Smith, and performances by the Rimini Protokoll (a contemporary German-based coalition of artists), and then selectively applied what they learned to improve their final projects.

Our final group of essays on learning acting begins with Thalia R. Goldstein's chapter, 'From Banana Phones to the Bard: The Developmental Psychology of Acting,' which surveys the cognitive and social-emotional capabilities that children normally develop before they are able to perform a role in a play. In terms of cognitive skills, children must be able to quarantine the real world from fictitious dramatic action, and they must learn how to lie persuasively, usually the first step on the way to role-playing. Although engaging in pretend play is a necessary prerequisite for acting, this normal toddler activity is not as demanding and complex as performing a role. A child's ability to recognise the self as a separate entity is usually the first socialisation step that facilitates acting, but later development must also include emotional regulation, which helps the growing child shift her or his attention back and forth between the world of the play and audience response during performance. In short, Goldstein provides copious empirical evidence that problem-solving, a capacious memory, interpersonal sensitivity and executive control are some of the more important primary skills that growing children must learn before they can become good actors.

Little has been written about how student actors can learn effectively while watching their acting teacher coach others during an acting class. Claire Syler addresses this problem with a chapter enticingly titled, '"I'm giving everybody notes using his body": Framing actors' observation of performance.' As she notes, the acting teacher who made that statement wanted to be sure that students used his comments and their observation of other student actors to advance their classroom learning. To declare that he was using one actor's body as a means of coaching all of his students was a 'metacomment,' defined by Syler as a deliberate act of framing that structured his students' attention and interactions. As in Kindelan's essay on teaching and learning, Syler celebrates metacognition as a cognitively rich teaching practice that can help students to take charge of their own learning in the theatre classroom.

Part II concludes with Rick Kemp's essay, 'Acting Technique, Jacques Lecoq, and Embodied Meaning.' Noting that Lecoq's and Stanislavski's teaching of acting training are complementary, Kemp asserts that Lecoq's pedagogy can open up many techniques and ways of thinking about acting that would remain mostly unconscious in a class that emphasised Stanislavski's approach to script-based performance. Kemp focuses on two major principles of **embodied cognition** that undergird Lecoq's pedagogy – (1) physical experiences are the foundation for meaning-making, and (2) the embodied simulation of meaning is dynamic. He ends his chapter with a case study that links learning and meaning to culture and underlines the specificity of gestures in performance. This chapter nicely sums up the major links between embodiment and learning (in classrooms, rehearsals, and cultures) and also points us towards Part III: scholarship, which continues to discuss the importance of embodied cognition but also calls attention to **enactive** and **distributed** cognition for understanding theatre and performance.

10

IMPROVISING COMMUNICATION IN PLEISTOCENE PERFORMANCES

Bruce McConachie

Invited to deliver a keynote address at the next conference of the Association for Cognitive Semiotics early in 2016, I embarked on a crash course in the evolution of human communication to prepare for my talk in mid-June. I had recently completed *Evolution, Cognition, and Performance* (2015), so had some notion of the vast territory I was entering. But my survey of the evolutionary dynamics of performance for that book had, I knew, left many questions unanswered, even unasked. What I soon discovered was a rich interdisciplinary conversation among anthropologists, biologists, linguists, psychologists, philosophers, sociologists and many others that, despite its many winding paths, was beginning to converge into a general consensus about the likely origins and evolution of human communication and language during the Pleistocene epoch, roughly 2.5 million years ago to about 11,000 BCE. Mostly missing from this general narrative, however, was what I found to be the importance of performance – specifically improvisation – as a necessary initiator of gestural and vocal communication among our *Homo erectus* ancestors around the middle years of the Pleistocene. While there was general interdisciplinary agreement about the significance of ritual for the evolution of communication among later hunter-gatherer bands, these scholars had overlooked the evidence for the kind of interpersonal trust that, still today, provides the necessary seedbed for improvisation among actors and musicians. The talk I presented in June at the conference about improv as a step toward language, plus the encouraging feedback I received there from my colleagues, is the basis for this short chapter.

The need for a new narrative to explain the evolution of language came about because of the gradual collapse of the old one. Following the success of *Language and Mind* in 1968, many linguists came to accept Noam Chomsky's claim that language was an internal, gene-centred human ability that did not depend upon social and cultural dynamics for its evolution. This consensus began to fall apart in the mid-1980s as a result of two developments. First, evolutionary biologists began to recognise that major changes can and have occurred in cognition and behaviour without alterations in genetic make-up; social innovations might actually precede and help to determine genetic changes. Second, evidence began piling up that our *Homo erectus* ancestors must have needed some kind of proto-languaging skills to accomplish what they did in terms of the use of tools, fire for cooking and other cooperative developments during the Pleistocene. Perhaps, then, the evolution of language might be traced over the course of a million years or so, rather than understanding it as the result of genetic change that came all of a sudden with the emergence of *Homo sapiens* around 300,000 years ago. One of the first scholars to persuasively put together the new evidence from evolutionary biology and physical anthropology was Merlin Donald, whose 1991 *Origins of the Modern Mind: Three Stages in the Evolution of Culture and Cognition* argued that the second of his three stages, which he termed 'mimetic culture,' necessarily involved everyday performances of gestures, mime and imitation. Donald updated and expanded his understanding of this predecessor to language-based communication in 2001 with *A Mind So Rare: The Evolution of Human Consciousness.*

Since then, several scholars have adopted a version of Donald's notion of mimesis to explain how and why hominin performances during the Pleistocene departed from the rudimentary communication skills that all primates exhibit. Although Donald did not elaborate on the details of mimetic communication, subsequent scholars have made it clear that the ability to imitate must have been an important skill in the emergence and evolution of early proto-human communication. But before a gesture or a sound would have been widely copied, the group of hominin hunter-gatherers that invented that particular visual-aural sign had to provisionally accept the signal as a carrier of meaning. I will argue that selected bands of *Homo erectus* improvised their way towards the sharing and understanding of communicative intentions and meanings that eventuated in performances of proto-languaging.

This mode of communication would not have occurred among our ancestors without the co-evolution of human intersubjectivity and the beginnings of social morality to support it. To set the stage for my remarks on improvisation, I will summarise some of the main points of linguist Jordan Zlatev's synthesis of the relevant scholarship on these two necessary predecessors of symbolic language. Zlatev builds on Robert Boyd and Peter J. Richerson's 2009 essay, 'Culture and the Evolution of Human Cooperation,' which extends their 2005 understanding of the co-evolution of human genetics and culture. Countering traditional notions of evolution, Richerson and Boyd discovered mechanisms and sketched suggestive narratives that demonstrated how evolution could operate on group as well as individual levels. Dependent on the interplay of genes and culture over time, co-evolution favoured groups of hominins that practiced social cooperation by rewarding them initially with survival and epigenetic changes that further enhanced their pro-social proclivities. Finally, note the two anthropologists, 'Moral systems enforced by systems of sanctions and rewards increased the reproductive success of individuals who functioned well in such environments, and this in turn led to the evolution of other-regarding motives like empathy and social emotions like shame' (Boyd and Richerson 2009, 3281–2). (See my Chapter 26 in Part IV of this book for a longer discussion of Boyd and Richerson's understanding of co-evolution and its consequences.)

Working within the implications of Boyd and Richerson's framework, Zlatev examined some of the leading scholarship on the evolution of human intersubjectivity, morality and language to find key points of convergence that would allow him to sketch the probable causal relations among these evolutionary elements. Specifically, Zlatev found significant commonalities regarding the evolution of human sociality in the work of anthropologist Robin Dunbar, evolutionary psychologist Michael Tomasello and sociobiologist Sarah Hrdy. These included insights about vocal grooming and gossip from Dunbar, shared intentionality and pro-social motivations from Tomasello and cooperative child rearing (i.e., alloparenting) from Hrdy. Zlatev discovered that all of these elements depend, implicitly or explicitly, upon Boyd and Richerson's evolutionary models for multi-level selection. 'The co-evolutionary scenario of intersubjectivity, morality, and language that we are led to,' concludes Zlatev, 'is, in brief, that intersubjectivity (in an alloparenting context) spearheaded the way, followed by morality and language which evolved co-temporally, in spirals of increasing complexity' (Zlatev 2014, 265).

Because it likely provided the initial contextual basis for improvising proto-languages, hominin alloparenting will provide a necessary preface to my discussion of improvisation. In her *Mothers and Others* (2009), Hrdy uses neuroscience, primatology, archaeology and social and developmental psychology to conclude that 'there emerged in Africa a line of apes that began to be interested in the mental and subjective lives – the thoughts and feelings – of others, interested in understanding them. These apes were markedly different from the common ancestors they shared with chimpanzees, and in this respect they were already emotionally human' (Hrdy 2009, 31). Why the capabilities for shared intentionality, sympathy and empathy emerged among one line of apes and not others had to do with alloparenting, group co-operation among the hominin adults and older children of the same band – primarily women and older girls – for the care and provisioning of infants and the young. As Hrdy explains, 'Both before birth and especially afterward, the mother needed help from others; and, even more importantly, her infant would need to monitor and assess the intentions of both his mother and these others and to attract their attentions and elicit their assistance in ways no ape had ever needed to do before' (31). Indeed, other ape mothers hold fast to their infants, entrusting them to no one else out of fear for their lives. Among Pleistocene hominins, however, 'it took a village' – as it does today for our species – to raise the children.

Hrdy draws on a variety of research to demonstrate the evolutionary and psychological advantages of alloparenting. Studies from several countries, for example, show that 'infants nurtured by multiple caretakers grow up not only feeling secure but with better developed and more enhanced capacities to view the world from multiple perspectives' (Hrdy 2009, 132). In terms of hormonal rewards, it is now clear that female alloparents derive many of the same benefits, including dopamine and oxytocin, that biological mothers gain from cuddling and caring for infants. Evidently, co-evolutionary dynamics gradually favoured hominin cultures engaged in alloparenting. New archaeological evidence about hominin residence patterns during the Pleistocene suggests, as well, that *Homo erectus* bands continued to reward grandmothers for their useful skills, probably for child care and food gathering. Compared to hypothetical babies raised by mothers alone, alloparented ape babies, says Hrdy, 'would [have been] more aware of distinctions between self and others, better able to read the mental states of conspecifics, and capable of integrating information about their own intentions and those of others' (139).

These are some of the same intersubjective attributes that facilitate good improvisation. While caring for infants and toddlers, the older children and adults of the hominin band would have encountered many situations in which they had to think quickly while gauging the intentions of the child under their care. If more than two alloparents were working together, both would have found that each needed to validate and support the choices of the other caregiver or risk upsetting the child. These and other situations would have tested and extended each alloparents' ability to empathise with others in the band. Several of Hrdy's explanations and examples in *Mothers and Others* have to do with empathy. Often defined as the ability to adopt the perspective of another person, empathy encourages good improvisors to set aside their own immediate desires, perceive what others in their improv group are thinking and feeling and then act on those perceptions in concert with them.

Several cognitive scientists acknowledge that our mirror neuron systems (MNS) are the gateways to empathy. Shaun Gallagher, for instance, understands activation of the MNS as the first step towards the kinds of everyday social perceptions and interactions that constitute empathy. Gallagher disagrees with Vittorio Gallese, however, that mirror neuron interactions among two or more people are, by themselves, sufficient for empathy. Evan Thompson, too, notes that the sensorimotor coupling that occurs when this system mirrors the intended actions of another body in motion puts the first body in tune with that other person and can lead to the second stage of empathy, which Thompson calls 'imaginary transposition' (Thompson 2007, 395). As the name of this second empathetic process suggests, imaginary transposition allows the empathiser to attempt to imaginatively place herself or himself into the mind and body of the other person in order to take that person's perspective. Most toddlers can accomplish the rudiments of imaginary transposition by the time they are nine months old. The ability to empathise with others is now recognised as an important and (mostly) hard-wired part of human social development.

It is no coincidence that Viola Spolin, the great American teacher of improvisational techniques, used what is conventionally called 'the mirror exercise' to warm up and sensitise beginning improvisers to the embodied feelings of other students in her classes. Still practiced today by many acting teachers who know nothing about the MNS, the exercise remains an important pedagogical element in most improv classes. Acting partners stand face-to-face about two arms' length from each other, and one initiates whole-body movements that the other – moving only the eyes and the relevant body parts – must follow. This physical extension of the MNS, of course, involves mimesis, the attempt to copy (in reverse) the movements of another, and this direct embodiment of Merlin Donald's insight can be extended to improvising with sounds and music as well. Doing these exercises encourages performers to trust the impulses sent by their mirror neurons and to let go of self-censoring fears that might prevent them from fully responding to the other in an improvisation. Some confirmation for the playful and freeing effects of improvisation comes from Clayton Drinko, a theatre historian and advocate of improvisation in psychotherapy. In his *Theatrical Improvisation, Consciousness, and Cognition* (2013), Drinko cites a scientific study focused on improvising jazz musicians. Using fMRI techniques, researchers at Johns Hopkins University scanned the brains of several musicians while they were playing and found that an area of the prefrontal cortex linked to planning and self-censoring actually slowed down during their improvisation. At the same time, another area of the cortex linked to self-expression and creativity 'increased in activity,' according to Drinko (Drinko 2013, 8).

There can be little doubt that evolution had already outfitted our Pleistocene ancestors with some type of MNS; mirror neurons were initially discovered in macaque monkeys and have been found in many other primates. Hrdy's copious evidence for alloparenting also implies that *Homo erectus* bands could also practice an early form of 'imaginary transposition,' the next stage in Thompson's four levels of empathy. This suggests that proto-languaging in the middle Pleistocene probably began among alloparents in bands large enough to require the coordination of infant care and child rearing among several people. Mothers would not have left their children in the care of others without knowing that they would be safe. Face-to-face communication through grunts and pointing could be effective for what we might call short-term babysitting, but if mom was off in the jungle or wandering the savannah in search of food, hominin alloparents had to know what to do, how to seek help and how to call mom back in the event of an emergency. Before the elaboration of symbolic language, iconic signs would have been useful for planning the future and communicating at a distance. Even before leaving their infants in another's care, however, mothers also had to know whom they could trust. Finding out from others in the *Homo erectus* band who did what with your kids while you were away would certainly have been a strong incentive for chatting up other women. So, in addition to making plans and calling to others at a distance, finding out about the past would also have been a motive for proto-languaging in the context of alloparenting.

How might improvisation have helped to accomplish these tasks? First, we must assume that alloparenting had at least begun to spread among many of the women and perhaps some of the men in a given band to establish the kind of trust that necessarily supports improvisation. Unless people believe that others will cooperate with them in establishing their own small world of meaningful sounds and gestures, improvisation cannot get off the ground. Next, what has been called the 'foundational rule' of improvisation must be in effect. In *Truth in*

Comedy, one of the best guides for improvising comic sketches, authors Charna Halpern, Del Close and Kim Johnson assert that ' "Yes, &..." is the most important rule in improvisation' (Halpern, Close and Johnson 1994, 46). According to the authors, Yes, &... 'means that whenever two actors are on stage, they agree with each other to the Nth degree' (46). In order to build their improvised situation together, performers make 'offers' about the reality of their given circumstances and the others in the group nearly always accept that offer, no matter how ridiculous or contradictory it might be. They say, in effect, 'Yes, and [here's something else to add to our reality].' The 'Yes, &...' rule sustains the necessary web of trust and mutual support among the improvisers. Partners in an improv nearly always agree with the assertions of the other(s) as they construct a world together of specific characters in given situations.

It is not difficult to imagine the 'Yes, &...' rule helping to structure a Pleistocene proto-conversation. A mother has somehow asked a caregiver how her baby acted while she was away and the grandmother alloparent has to invent a sound and perhaps a gesture that might be translated as 'the brat cried all day long.' So she points to the child and gives a long, baby-like cry with a gesture of baby-like rage that tries to communicate what she means. Let's represent this sound-gesture as 'WAAAA^*!' The mother has enough empathetic skill to put herself in grandma's place and figure out what she's trying to say. But perhaps mama wants to know why her kid never stopped crying, so she repeats 'WAAAA^*!' and adds a gesture to express her question. In effect, the mother has used the rule of 'Yes, &...' to affirm the meaning of 'WAAAA^*!' and both participants in the improvisation can now continue to use that sign to carry on their proto-conversation. Maybe that is the only time the sign is used and it quickly fades from the memory of both adults. Or perhaps 'WAAAA^*!' catches on among other women in the band to indicate babies that cry often. Over time, it gets modified, shortened and passed down to the next generation; it is now a sign that indicates cranky babies among most, perhaps all, alloparents in the band. Maybe that particular pantomime with sound carries that meaning for a thousand years. By that point, it is no longer an improvised invention; it has become an accepted, normative part of the band's proto-language. Long term, the rule of 'Yes, &...' can lead to semantic conventionalisation.

The process of establishing semantic conventions is important, of course, because it moves the subjunctive proposition, 'What if we allow "WAAAA^*!" to stand in for all cranky babies' from a 'what if' possibility into a declarative reality. In terms of co-evolution, what began as a playful improvisation could end up as a cultural fact. Beginning in a game of 'Let's pretend,' 'Yes &...' can institutionalise norms of behaviour among groups of people over time if they allow the continuing improv to do so. In this way, as several anthropologists have affirmed, initially playful religious beliefs and informal economic trading can grow to become the Ten Commandments and the World Bank. I agree with anthropologists Emily Wyman and Jerome Lewis, who emphasise the importance of hominin play in the gradual emergence of language and other institutionalised forms of symbolic culture from hominin signals.

The extent to which hominin adults might have been able to extend childhood play into adulthood during the Pleistocene cannot be known with any certainty, of course. Today, we are the only species that allows for and even encourages 'let's pretend' activities in adulthood, as evidenced by the proliferation of porno films, football games and Internet cat videos that crowd into what we sometimes call adult entertainment. Simply on the basis of our present fascination with subjunctive performances in multi-modal formats, it is likely that there would have been some evolutionary pressure on Pleistocene hominins to continue engaging in play activities after childhood. Also, as we know from social behaviour today, most adults enjoy playing with children, and *Homo erectus* alloparents would have had many opportunities and some incentives to do so. Playful contests among adults today, such as sports and games, deliver dopamine rewards to losers as well as winners, a good reason to continue to play well past puberty. Did this genetic change begin during the Pleistocene? Probably. Improvising signals for communication is a type of play, of course, and if that activity proved enjoyable, our hominin adult ancestors may have had several incentives to continue it.

I have sketched a possible scenario for the invention and elaboration of proto-languaging based on hominin improvisation, but how credible is it? Conventional ideas about innovation usually emphasise the importance of pinpointing a need, gathering as much relevant information as possible, sifting through it for possible ways of fulfilling the need in question and then trying out various options, first as thought experiments and then as marketing ploys. In contrast, as Gunter Lösel emphasised in his chapter in Part I, improvisation is quick, messy and playful. It is probable that most of these hominin signs, invented on-the-fly, were not remembered past their first use. And for those that were, any band that continued to communicate with a few hundred gesture-sounds for more than five generations would likely have replaced many of them. These signs, however, were improvised neither for long-term use nor for proto-linguistic coherence. They were invented in a moment of communicative need, provided a roughly iconic equivalent of the object or action imitated and served their purpose well enough to effect some necessary communication in that situation. Put another way, it is difficult to conceive that the creation of these gesture-sign sounds could have been any more deliberative than hominin improvisation allowed. Assuming their level of communication was not much more than that of gorillas and chimps in the wild today, our Pleistocene ancestors had no tools for careful deliberation. But they had learnt alloparenting trust and empathy well enough to craft and share some gestural proto-languaging necessary for their caregiving, an evolutionary innovation that would eventually lead our ancestors to fully symbolic spoken language.

References

Boyd, Robert, and Peter J. Richerson, 2009. 'Culture and the Evolution of Human Cooperation.' *Philosophical Transactions of the Royal Society B* 364: 3281–8.

Chomsky, Noam. 1968. *Language and Mind*. New York: Harcourt, Brace.

Donald, Merlin. 1991. *Origins of the Modern Mind: Three Stages in the Evolution of Culture and Cognition*. Cambridge, MA: Harvard University Press.

Donald, Merlin. 2001. *A Mind So Rare*. New York: Norton.

Drinko, Clayton D. 2013. *Theatrical Improvisation, Consciousness, and Cognition*. New York: Palgrave Macmillan.

Dunbar, Robin I.M. 1996. *Grooming, Gossip, and the Evolution of Language*. London: Farber and Farber.

Halpern, Charna, Del Close, and Kim Johnson. 1994. *Truth in Comedy: The Manuel of Improvisation*. Colorado Springs, CO: Meriwether Publications.

Hrdy, Sarah B. 2009. *Mothers and Others: The Evolutionary Origins of Mutual Understanding*. Cambridge, MA: Belknap Press.

Richerson, Peter J. and Robert Boyd. 2005. *Not by Genes Alone: How Culture Transformed Human Evolution*. Chicago, IL: University of Chicago Press.

Spolin, Viola. 1963. *Improvisation for the Theatre: A Handbook of Teaching and Directing Techniques*. Evanston, NJ: Northwestern University Press.

Thompson, Evan. 2007. *Mind in Life: Biology, Phenomenology, and the Sciences of Life*. Cambridge, MA: Harvard University Press.

Tomasello, Michael. 2008. *The Origins of Human Communication*. Cambridge, MA: MIT Press.

Wyman, Emily. 2014. 'Language and collective fiction: From children's pretense to social institutions.' In *The Social Origins of Language*, Daniel Dor, Chris Knight, and Jerome Lewis, eds. Oxford: Oxford University Press, 171–83.

Zlatev, Jordan. 2014. 'The co-evolution of human intersubjectivity, morality, and language.' In *The Social Origins of Language*, Daniel Dor, Chris Knight, and Jerome Lewis, eds. Oxford: Oxford University Press, 249–66.

11
RITUAL TRANSFORMATION AND TRANSMISSION

David Mason

Conventional wisdom has held that rituals communicate meaning. Emile Durkheim launched sociology as a discipline based on the ideas that religion consists of rituals and beliefs, or 'states of opinion,' and that the sanctity of ritual objects comes from the objects' representation of these 'states of opinion.' Beliefs, in this thinking, provide a premise for ritual, the acts of which *mean* the content of beliefs (Durkheim 2001, 36). The study of ritual that followed from Durkheim operated on structuralist principles, casting ritual as a sort of literature to be read as symbolically expressive of beliefs, so that, under analysis, the symbolic actions that comprise a ritual would reveal psychological pretexts shared across a culture. Descending through Mircea Eliade and Claude Levi-Strauss, this intellectual tradition exemplified the structuralist divorce of phenomenon from value: the thing itself that is here, in the moment, for our experience, does not matter so much as the way that the thing that is here points to something that is absent. One thing this approach did was excuse scholars of their own absence from the rituals about which they wrote. As long as the value of a ritual resides in the meaning it signifies, a scholar can assess the ritual's value through someone else's written description. Durkheim could *literally* read a ritual and tease out the meanings in Aboriginal totemism without visiting Australia. Similarly, Eliade could interpret shamanism without interacting with any shamans in Siberia, Belize or Arizona.

There were alternatives. Some early attention to ritual minded the value of bodies, *per se*. Talal Asad has pointed out that, in the 1930s, Durkheim's nephew Marcel Mauss set concern for beliefs aside in favour of attention to 'the mode in which the living human body, as a thing, exists, acts, and is acted upon' (2012, 42). In what he called *technique*, Mauss identified actions that do not merely signify, but that are 'simultaneously real, physically effective gestures' (2009, 76). Mauss's interest in action anticipated attention to what ritual *does* and the trend of the later twentieth century to set belief and meaning to the side, so as 'to let the activities under scrutiny have ontological and analytic priority' (Bell 1998, 211).

The interest in ritual action has made ritual study much more interdisciplinary, with an emphasis on anthropological field work and ethnographic labour. The interest in bodies and action has also drawn ritual studies alongside disciplines like cognitive studies and performance studies, effectively reversing Durkheim's premise that ritual serves beliefs. Often, it seems, ritual action informs the 'states of opinion' that constitute beliefs, not symbolically indicating a *meaning* that is apart from the action, but creating *value* that is indissoluble from the action. The way in which action—especially the structured and stylised action of ritual—shapes embodied identities can form beliefs, worldviews and self-concepts. Rituals can be seen to transmit and transform, working to conserve tradition in the individual identities that it refashions for its purposes. The mechanisms through which rituals thus function involve the intervention of intentional action in the circulation of perception and self-consciousness through a particular individual's body. A performative theory of ritual does not concern how to understand *reality*, as such, so much as it concerns how to understand processes by which the body in ritual shapes itself, and, consequently, the manner in which the body-brain can manipulate and transform the portions of reality that a body *can* experience as reality. The following essay summarises performance and ritual theory on which this performative theory of ritual rests, particularly the characterisation, via Roy Rappaport, of ritual action as creative, *per se*, and Merleau-Ponty's theory of perception. I then offer a mid-nineteenth-century Mormon temple ritual as a case study in which some creative features of ritual are evident.

The late twentieth-century notions of *performance* that emerged from J.L. Austin's theory of effectiveness in a *speech act*, and which are now crucial to performance studies, characterised some action as 'performative,' suggesting that they actually change things, as opposed to symbolically indicate things.[1] Building on Austin's *performativity*, ritual theory now wields two understandings of *performance* that are not necessarily mutually exclusive, but that are certainly different. On the one hand is the kind of *performance* that Caroline Humphrey and James Laidlaw identify in *performance-centred* ritual, involving dramatisation—mimetic and, often, narrative—in which belief in what is being dramatised is important. Humphrey and Laidlaw describe this kind of activity as 'weakly ritualized' (2004, 8).[2] We could also call this kind of ritual activity *weakly performed*.

On the other hand, we might call Humphrey and Laidlaw's *liturgy-centred* ritual—the kind of ritual for which correct form is paramount—*strongly performed*. Relying on Austin, anthropologist Roy Rappaport argued in the 1970s that the customary invariability that we find in *liturgy-centred* ritual not only creates the category of sacredness that Durkheim placed prior to ritual, but it also transforms everything involved in it—including the human agents. Perhaps ritual *especially* transforms human agents, since 'the use of the body defines the self of the performer for himself and for others' (1996, 436). Rappaport then describes, anecdotally, a ritual act with a *strongly performed* quality:

> In kneeling, for instance, [the ritual agent] is not merely sending a message to the effect that he submits in ephemeral words that flutter away from his mouth. He identifies his inseparable, indispensable, and enduring body with his subordination. The subordinated self is neither a creature of insubstantial words from which he may separate himself without loss of blood, nor some insubstantial essence or soul that cannot be located in space or confined in time. It is his visible, present, living substance that he 'puts on the line,' that 'stands up (or kneels down) to be counted.' As 'saying' may be 'doing,' 'doing' may also be an especially powerful—or substantial—way of 'saying.'
>
> *(1996, 436)*

The mimetic quality of the action—the *weakly performed* aspect of the rite—merely reiterates or signifies concepts of deference and authority. The *weakly performed* aspect of this ritual act is that it *means* subordination. The *strongly performed* aspect of the ritual resides in the blending of the agent's visible, present, living substance with tradition in the doing of the act. Among other things, the *strongly performed* quality of ritual kneeling transforms the kneeling agent into something much more than a *meaning*. In kneeling, the kneeler becomes a device of the ritual, personally, bodily transmitting in each instance of the act the cultural structure of authority and subordination that the *weakly performed* qualities of the rite only *mean*. Strongly performed, ritual is an extraordinarily powerful transmitter of culture.

Rappaport insinuates that doing ritual kneeling operates, somehow, on the identity, the self, of the kneeler (and, indeed, on the selves of all present), but without a material theory of ritual action, we remain in the realm of symbolic meaning or social function. Austin's *performative utterances* and *illocutionary acts*, as evocative as they are, rely on symbolic formulae whereby some actions have effective power on account of the conventions and traditions they cite, indicatively. By the 1970s, Maurice Merleau-Ponty's phenomenology of perception, reflections on habit and skill, and concept of personal identity as necessarily wrapped up in consciousness had nudged the concern for *speech acts* in ritual toward a consideration of ritual, *per se*, as a matter of embodiment, aesthetics, reflexivity, etc., providing a way of understanding how the kneeling itself, in Rappaport's example, affects the self and the reality of which the self is a part.[3] (For more discussion of phenomenology and cognitive science, see Chapter 18)

Merleau-Ponty (2004, 139) characterised perception as 'communion.' A person cannot come to know an external object *as* an external object, but only as a thing that physical sensation has made available, more or less *internally*. Rather than an object, *per se*, which necessarily exceeds the competence of perception, physical sense offers an individual a selection of an object's qualities, a *reduction* of an object, to a phenomenon that is part of the individual's experienced sensations. Every sensation, then, incorporates (i.e., integrates as a physically constituent part of the human body) some object, drawing the object from an external reality into a person's experience of self that is indistinguishable from the internal consciousness that grounds personal identity. One consequence of the limited interface between a person's self and reality—as senses provide it—is that every perceptual encounter results in a modification of the body (Merleau-Ponty 2004, 144).[4] Because the body provides for selfhood, such modifications have implications for personal identity. We compose ourselves of what we sense, in a way that is not merely analogous to 'you are what you eat.'

Art, for example, shows us that people can manipulate the qualities of an object that sense can make available to a person, and, thus, manipulate the ever-developing sense of self held by a person who internalises the object via perception. Manipulation of this sort can be active in time and space, in the form of dance, for instance, in which one object that is available to sense is another person's equally sensible body. In Merleau-Ponty's view, a person's *perception* of a dancer, in the action of dancing, constitutes a communion whereby the observer internalises the qualities of the dancer that are available to sense –including such things as rhythm, and even things that are not clearly material, such as the dancer's *intention*, to the extent that intention appears in the narrative arc of a leap, and might also manifest in grace, style, verve, and so forth. The perceiving person, here, incorporates the dancer, becoming a new self of which the dancer in motion is a constitutive part.

Ritual, similarly, manipulates perception, reducing, organising, and channeling reality into specific perceptual phenomena. The participant in a ritual confronts a field of objects selected out of reality, arranged and modulated. In perceiving that field of objects, the participant internalises and incorporates them and the organised field of which they are constituent parts. In distinction from the conventional mode of watching dance, the ritual participant's mode of *participating* in the ritual amplifies the effects of perception. Following Merleau-Ponty, ethnologist Edward Schieffelin noted in the 1980s that a person incorporates an action to the extent that the person's embodiment of an action achieves a level of skill, of *ease*. Habit is partly identified by how little a person must attend to it to do it. In this case, less distinction between self and act remains. A person's Heideggerian *being* can be constituted by acting, and can be conflated with the world that the act makes possible 'in the mode of participation' (Schieffelin 1985, 619). But ritual tends to attend to everything, or, at least, to modulate and manipulate all its mundane action so as to demand attention to it. The ritual participant perceives the ritual's field of objects, the other participants who act in the ritual and the ritual's peculiar action. The ritual participant also *interacts*, physically and sensibly, with those objects and with those other participants, and that interaction is often so alienated by the ritual that the participants must attend to their own action and sense their own bodies as objects in the ritual, which results in a reincorporation *of* themselves as ritual agents *in* themselves. Ritual systems, thus, involve circulations of perception and action from which a participant is not distinct, and, therefore, by which a participant changes.

The way in which perception and ritual participation construct and reconstruct the identity of an individual can have a radically ontological result, such as what anthropologist Jon P. Mitchell has characterised as a 'transformation of the existential grounds of selfhood' (2009, 56).[5] In the case of Rappaport's hypothetical kneeler, ritual brings into material fact an individual's subordination. The stylised (ritualised) action draws the kneeler's attention to the sensation of his body as a kneeling object, alienating the kneeler's body, and reincorporating it through reflective perception in a way that reconstructs the self that kneels as a kneeling thing. Ritual *does* what it *is*.

Insofar as ritual tends to be a collective, communal enterprise, the performative transformations can also be communal. Merleau-Ponty's philosophical rationalisation of perception has a neurobiological correlate in the human brain's 'mirror neurons' (2015, 39), which facilitate 'motor attunement' in higher mammals (40). The mechanism, as Bruce McConachie summarises it, operates, thus:

> If a spectator watches an aerialist take a step on a high wire or sees a hockey player flick the puck toward the stick of a teammate, for example, the same group of neurons in the empathizer's brain is activated as in the player's brain; neurologically, it is almost as if the observer had taken the step or flicked the puck himself. By working through our perceptions, bodies, and minds, our networks of mirror neurons unconsciously attune us to each other.
>
> *(2015, 121)*

The way that mirror neurons might provide a foundation for empathy, as McConachie suggests, is beyond the scope of this essay.[6] Most important, here, is that mirror neurons seem to respond exclusively to *action*, and, for that matter, they seem to respond exclusively to action that reveals some intentional agency. In whatever ways that mirroring combines with other structures and processes, and to whatever complex psychological ends, the fundamental operation of the mirror neuron systems produces in a subject's body the activity in which the perceived other is engaged, along with the physical sense of the perceived other's self-defining agency.[7] The consequence can be understood as a genuinely material intersubjectivity. Which is to say that evolution has given people, it seems, a neurological mechanism by which perception – in this case, primarily, sight – facilitates the Merleau-Pontian incorporation of other people in a way that is not simply analogical or metaphorical. The subject embodies both the other's action and the other's intent, such that the subject inhabits a new ontological status, a new self, combining both subject and other. As neuroscientist Vittorio Gallese would have it, mirror neurons facilitate a 'literal' embodiment of what can be perceived in others (2009, 524). Where ritual is concerned, people engage in deliberately stylised, often repetitive, often mimetic and often collective action, which is fraught with agency, such that participants seem necessarily bound to influence and change each other at a sensorimotor level. There seems no reason not to describe this neurological process as an always ongoing transformation of the self, if not *into* another's body, then, at least, as a consequence of contact *with and under the influence of* another's body.

It also appears that mirror neurons provide a neurological ground for the kind of distributed agency that William Sax identifies in ritual (2009). Given the ways in which mirror neuron systems join the acts and intents of people, ritual can have a teleological purport that does not issue from a particular individual, but coalesces in a collective (Sax 2008, 478). We can hypothesise that the ritual action effects in participants, a mirroring attitude that approximates a sharing of intention and action through a group. In Rappaport's hypothetical case, those who observe the action of the kneeling agent in-corporate the kneeler's action and partake of the kneeling experience. How participants understand and express the kneeling that mirrors in them will be contingent on what is made possible for each participant in the particular cultural complex that the kneeling, inevitably, affirms.

Ritual, then, persists as an especially robust means of transmitting culture. Ritual participants often do not so much *learn* culture in the form of beliefs as much as they *become* culture and its transmission to the next generation. Pierre Bourdieu insisted that the performative consequence of ritual not only does not depend on beliefs, but can shape and even create the beliefs on which ideologies are founded.[8] Indeed, in its direct operation on physical bodies, ritual conserves culture across generations in a way that can be so effective that it can be oppressive. Bourdieu argued that rituals such as initiations—which he preferred to call *rites d'institution*—preserve authority in a society by embedding the function of that authority in the very self-experienced identities of initiates.[9] By in-corporating others' action and intention, the ritual participant comes to be, personally, the physical function of authority and the vehicle by which that authority asserts its exclusivity and reproduces itself, through time.

Case study: Mormon temple ritual in 1846

Ritual's capacity to transmit and transform can be seen in the temple rituals of the mid-nineteenth-century Mormon community, based in Nauvoo, Illinois. These ceremonies were distinct in some important ways. First, for some years, these rites were open to only a handful of people, though scores of members of the community laboured on the temple building in Nauvoo, Illinois. Consequently, through the early 1840s, the Nauvoo rites accrued a certain mystique within the Mormon population. Second, the rites not only opened up to the larger community late in 1845, but became suddenly obligatory for most of the population of thousands. Third, the attempt to accommodate the new interest of thousands in satisfying a ritual obligation ran just ahead of what the community perceived as a necessary mass evacuation of the city they'd built around the temple, and an emigration, altogether, from the United States. In this context, the Mormons' Nauvoo Temple rituals operated as the process by which members of the community performed their bona fides. The rites expressed individuals' commitment to a community that faced dire calamity, and the rites also worked to inscribe a new, communal identity on individuals. Through cognitive mechanisms, the Nauvoo ritual fashioned new identities for its participants that were composed of, and which affirmed, the unique qualities of the community. The effectiveness of the Nauvoo rites in transforming Mormon identity played a not insignificant role in the event of more than 10,000 people crossing the Mississippi to accept homelessness in the dead of winter. One of every ten of them would die during the following year.

In February 1846, Increase Van Deusen went into the Mormon temple in Nauvoo, Illinois, to receive, he was told, a 'reward' for faithful service in helping to build the unusual building (1847, 3). Van Deusen was among the last of the several thousand Illinois Mormons to pass through this temple ritual between December 1845, and February 1846, when the evacuation of Nauvoo began. Van Deusen participated in what had developed from a decade of Mormon experiments with temple rituals, involving washing and anointing ceremonies cribbed from the first books of the Bible, to which had been added a pageant that dramatised the Bible narrative of creation, Eden and The Fall. To this complex, ritual gestures and oaths borrowed liberally from Masonic rites were attached. Van Deusen was among the smaller portion of the community who did not follow Brigham Young to the Rocky Mountains, and who became quickly estranged from Mormonism as they had known it under Young's murdered predecessor, Joseph Smith. Since one premise of the rites was secrecy, historians have relied on those who ultimately rejected the community under Young's leadership for full descriptions of the ritual. Among the few of these accounts from the period, Van Deusen's seems to be the most at pains to describe the components of the rite, accurately (though not shying from opportunities to excoriate Young). In 13 pages, tightly packed with nineteenth-century type, Van Deusen describes 'The Drama,' as he calls it, that gave roles in its narrative to all initiates, so that they found themselves not viewing a dramatisation of Bible stories, but personally acting out the stories in the roles of Adam and Eve.

Van Deusen and his wife presented themselves at the temple early in the morning, on account of little more than an invitation, and with no more indication of what to expect than the charge to bring along their 'night clothes' (4). Directed to the attic storey of the building, husband and wife went into separate rooms. Increase Van Deusen then recounts being laid out horizontally in a tub, washed, anointed liberally with scented oil and dressed in simple but distinctly ceremonial attire, so that he proceeded through the remainder of the event in a tight, cotton undergarment, the markings of which, he was told, signified devotion and protection. During these initial rites, he was also ordained to be a king (15). The officiators then led him to another room where he followed directions to lie down on the floor. From his place on the floor, he heard a 'rumbling noise' and a voice declaring, 'Let the light be divided from the darkness.' Van Deusen himself understood the purport of this recitation not merely as a reading of scripture, but as a representational performance, '*as if* the Almighty himself is first in the act of creation' (7).[10] The dramatic recitation proceeded through the remaining stages of creation, as more or less dictated by Genesis, and then coalesced around Van Deusen himself. An individual in the role of God entered the room and mimed the formation of Adam from the earth, by patting the floor, and then Van Deusen's body, in turn. After having been awoken as Adam, and then returned to sleep, Van Deusen was reunited with his wife, who entered the ceremony as Eve. Van Deusen was fed the lines he was to speak in the drama, including, on the appearance of his wife as Eve, 'This is bone of my bone and flesh of my flesh' (8). As Adam and Eve, the two went to yet another room made up, to remarkable effect, as Eden (8).

The couple then acted the roles of Adam and Eve, interacting with others who played the roles of God and the devil. The ceremony called on the pair to eat the forbidden fruit (in the form of raisins tied to a tree), to play the discovery of their nakedness and to speak their parts of the biblical script (i.e., 'She gave me of the tree and I did eat'). In accord with genesis, the officiator playing the devil returned to the ceremony to be cursed, and then departed by crawling out of the room, 'on his belly, dragging himself slowly along' (9). The rite then expelled Adam and Eve from Eden, which sent the Van Deusens into a fifth room, representing 'the world,' where they encountered the devil again, bombastically encouraging them to be falsely religious. Eventually, God again appeared here to dispatch His nemesis, after which Increase and Maria Van Deusen were guided through a series of oaths accompanied by Masonic gestures and hand-clasps, at which point additional ceremonial items were added to their costumes (including a robe and a cap). The sixth room into which they were led represented, they were told, 'the Millenial [sic] Morning.' Here, as the culmination of the rite, the Van Deusens kneeled at an altar to take ritual oaths to become forever enemies of the United States and not to reveal any element of the temple ritual through which they had just passed (12–13). Van Deusen was then led through a ritual conversation with an officiator behind a screen. By repeating the appropriate responses to questions and by demonstrating that he had learnt a series of ritual gestures, Van Deusen found himself admitted to the seventh and final room, as though entering 'the kingdom' (15). There was no formal activity here. Instead, the room was crowded with individuals all dressed identically in the white robes of the ceremony, who were then instructed that the 'the laws of the land are no more binding on us' (15). Van Deusen left the temple still wearing the cotton underclothing that he had assumed as part of the ritual, along with the charge 'always to wear this garment under [his] clothes, while [he was] in the world' (6).

Although the source of this account directly separated himself from the community – and, thus, seems not to have been affected in the way the rite intended – others who would not report on the rite in deference to their secrecy vows reveal some experience of the ceremony as transformative. Norton Jacob described his passage with his wife through the rite in December 1845, as 'the most interesting scene of all my life and one that afforded the most peace and joy that we had ever experienced since we were married, which has been over fifteen years' (Buerger 2002, 79). Samuel Richards reports feeling 'lost to myself' in the Nauvoo temple, and seeing 'the earth reel to and fro, and [it] was moved out of its place' (Anderson 2011, 5). Wilford Woodruff, who, in his capacity as the president of the LDS Church decades later, would bring an end to LDS-Mormon polygamy, felt a 'connection' to the gestures and signs of the Nauvoo ritual 'as to the blessings of God' (Anderson 2011, 8). Jacob and Woodruff were both among the first group of 100 Mormons to enter the Salt Lake Valley in 1847. During the emigration to Utah, Richards was evangelising for the faith in Great Britain.

Where the rite effected such transformations, we can see physical interactivity operating on the participants' experience of self, shaping and transforming it into something collective, and also as constituted by elements of the rite. As witnesses to each other's participation in the ritual, the collected participants perceptively incorporated each other. Through basic processes of neurological mirroring, the acts and intents of Masonic gestures and postures of worship shared an agency throughout the performing and perceiving bodies. For one thing, the ritual was rather dense with points of direct physical contact between participants, which provided for particularly Merleau-Pontian intersubjectivity. As Van Deusen reports it, for example, the Nauvoo ritual involved a number of Masonic hand-clasps, or what Van Deusen calls 'grips,' performed interactively with a ritual officiator (Van Deusen 1847, 10). Van Deusen makes no attempt to interpret the meanings of these grips. In the ritual context, the meaning is not so important as the action, which, in this case, imposed direct perception of an other, sensed as a deliberate agent pursuing the intention of the ritual, in concert with the reflective perception of the self, acting similarly as an agent of the ritual. A participant such as Van Deusen could not escape perceiving the officiator's hand in the grip, his own hand in the grip, himself as an agent that has deliberately cooperated to form the grip, the officiator completing the grip and perceiving the participant as a feature of the grip, and so on. What I have called above the circulation of perception and action involves Van Deusen's consciousness of his hand, the actual position and shape of his actual hand, in contact with another hand, in an actual space and in a moment of time in which his body in that space and time was embedded, and in which it was available to his own conscious perception.[11] This circulation evokes Merleau-Ponty's doctrine of communion via perception: in addition to perceiving and in-corporating the other with whom he shared the ritual grip, the Nauvoo initiate necessarily perceived himself as a participant without whom the action of the ritual could not occur. The initiate experiences ritual action as expressive of his own identity, reciprocally comprising other participants and the action itself.

I would add that role-playing, as seen explicitly in the Nauvoo rite, also contributes significantly to ritual's substantive transformation of bodies and selves. Role-playing often enters into ritual's perceptual loop as a participant's experience of the ritual comes to include the reflective perception of the participant's own embodiment of a role from sacred lore and its narrative trajectory. Nauvoo ritual participants not only in-corporated and embodied each other and the action of the rite but also in-corporated Judeo-Christian roles. The Adam 'role,' in this case, existed prior to the ritual in a common religio-literary tradition, bearing both a cultural significance and an already-accomplished narrative. The Adam role, here, and all that Adam's story entails, became, by perceptual in-corporation, a part of the participant's own identity, lending the self a teleology according with the goals of the group.[12]

A final point. What the body recognises as a self, as Bruce Wilshire argues in *Role Playing and Identity*, emerges from the way in which the body's action reiterates the action of others and appropriates that action for itself through reflection on its own reiteration of the ways in which others perceive it.[13] This complex is always in operation, always shifting, ever renewing, so that the self – as Kierkegaard and others have intuited – is never being but only becoming. Acting the role of Adam – reciting Adam's lines and eating Adam's raisins, as the Nauvoo initiates did—brought an initiate into a field in which his action deliberately reiterated the action of others. The initiate's body perceived its own action as the self-referential activity of the temple and also as a reiteration of that activity. As evident in the self-conscious description he provides, Van Deusen's ritual role-play approached a reflective consciousness of his body as the mimetic mechanism that the rite's officiators and his predecessors in the rite expected. Van Deusen's ritual performance can be read, thus, as a strong move towards a new *habitus*, the lived, embodied character of this community, which was, at this moment, so desperate to separate itself from the United States.

Consider Rappaport's kneeling subject, again, in comparison with the Nauvoo ritual. Where 'subordination' might be a *meaning*, signified by the body's ritual act, the body can also be understood *to perform* subordination. The body's act creates the subordination that it signifies, and in a manner that participants experience sensibly, as a facet of reality. In the place of subordination, the Nauvoo rite employed bodies in acts of *insubordination*, not merely signifying, but *creating* treason, in reality, in the act of kneeling and swearing enmity against the state. The kneeling in this case also performed submission to the community, of course. In coordination with other such curated acts as ordination to kingship, kneeling, here, embedded an antipathy to the United States and a commitment to the triumph of the clan in an identity shared by thousands. For a genuinely embattled group, the rite created privilege, experienced as privilege, in affirmation of a divine will for the group's survival.

Although the man who documented the 1846 Nauvoo ritual did not leave the United States and disavowed the movement shortly after its mass emigration, thousands of Van Deusen's peers did throw themselves into a miserable, deadly Iowa winter, a plague of disease and malnutrition over the succeeding year and then a 1,000-mile trek to an unknown and unidentified homeland. The Mormons continued to perform parts of the Nauvoo rite in their temporary Iowa settlements, and even in the migration on the Midwest plains. The Nauvoo ritual continually regenerated a common teleological purport, an agency, and transmitted that agency throughout the community in the Mormons' own bodies. Intuiting that the Nauvoo temple rites were 'the cord which has bound this people together' (Buerger 2002, 75), one of the first things Brigham Young did upon arrival in the Great Salt Lake Valley in 1847 was designate a spot for the construction of a new temple, the model of which, lying abandoned in Illinois, not only lived large in the Mormons' imagination, but remained embedded in their bodies (Mason 2014, 72). Such effectively transformative rituals—those that are strongly-performed—do not succeed on account of meaning, even if meaning is important. Transformative rites work self-referential action as a tool on the very selfhood of participants, transforming them into intersubjective agents of the ritual itself, and, consequently, of the collective identity of the performing community. The ritual initiate becomes, then, the living transmission of that collective identity, that *culture*, to the next generation.

Notes

1 Austin's *speech act* does not come out of a vacuum. The ways in which Hans Vaihinger and Alexius Meinong wrestled with the paradox of fiction in the late nineteenth century, as well as ideas developed by Husserl, Brentano, Wittgenstein and Kenneth Burke earlier in the twentieth century, anticipated Austin.

2 Humphrey and Laidlaw attribute the term *performance-centred* ritual to Jane Atkinson.

3 In addition to Rappaport, important figures in the development of 'performative' theories of ritual include Stanley Tambiah, Bruce Kapferer, Jonathan Z. Smith and Edward Schieffelin.

4 For a summary of approaches to the 'paradox of human subjectivity,' denoting the problem of embodied self-perception, see Dorothée Legrand (2014).
5 Mitchell concurs that the 'heightened reflection' in ritual 'reconstitutes selfhood' through an incorporation of other ritual agents.
6 McConachie cites Evan Thompson's argument that empathy emerges from a multi-stage process, at the beginning of which is the activation of mirror neurons (2007, 121).
7 What I have called, here, 'self-defining agency' can be correlated with "intention" as understood by Caroline Humphrey and James Laidlaw (2004, 92–93). Their analysis of Jain ritual leads them to conclude that, whereas an agent's intention to extend a greeting informs and constitutes the act of waving, what distinguishes ritual is that the intention of ritual action is primarily the action itself, as itself, without additional signification.
8 'The habitus is the universalizing mediation which causes an individual agent's practices, without either explicit reason or signifying intent, to be none the less "sensible" and "reasonable"' (Bourdieu 2005, 79).
9 Bourdieu understands *belief* as 'socially fashioned dispositions' to recognise validity – an inclination that social practice imposes on the body *to defer in this situation in the manner that feels appropriate*, rather than propositionally stated ideas such as *this person deserves my loyalty*. The ritual inclination that the body inherits (re)produces a field in which the propositional belief works (Bourdieu 1991, 125).
10 Emphasis mine.
11 This circulation-perception loop echoes what performance theorist Erika Fischer-Lichte calls an 'autopoietic feedback loop' that arises in the interaction of performers and audiences (2008, 55). Fischer-Lichte borrows her term from Humbert R. Maturana and Francisco J. Varela (1980), where autopoiesis is the distinct feature whereby living systems, as opposed to non-living systems, change in response to environment. In recent work, Bruce McConachie acknowledges autopoiesis as a workable metaphor for the circulation of sense, intentions and understanding (2015, 141).
12 I am applying to performance, here, the 'role-theory of religious experience' developed by psychologist Hjalmar Sundén et al. See Källsted (1987).
13 A conscious self comes into being 'when it can experience objects or persons which are in fact other than itself—but with whom it is mimetically involved—and then can reproduce them *as* other in their absence.... The body must be able, to some extent, to appropriate *as* its own *its* mimetic reproduction of *them*' (Wilshire 1991, 152).

References

Anderson, Devery S., ed. 2011. *The Development of LDS Temple Worship 1846–2000*. Salt Lake City, UT: Signature Books.

Asad, Talal. 2012. "Thinking about Religion, Belief, and Politics." In *The Cambridge Companion to Religious Studies*, edited by Robert A. Orsi, 35–57. Cambridge: Cambridge University Press.

Bell, Catherine. 1998. "Performance." In *Critical Terms for Religious Studies*, edited by Mark C. Taylor, 205–24. Chicago, IL: University of Chicago Press.

Bourdieu, Pierre. 2005. *Outline of a Theory of Practice*. Cambridge: Cambridge University Press.

Bourdieu, Pierre. 1991. *Language and Symbolic Power*. Edited by John B. Thompson. Translated by Gino Raymond and Matthew Adamson. Malden, MA: Polity Press.

Buerger, David John. 2002. *The Mysteries of Godliness: A History of Mormon Temple Worship*. San Francisco, CA: Smith Research Associates.

Durkheim, Émile. 2001. *The Elementary Forms of Religious Life*. Translated by Carol Cosman. New York: Oxford University Press.

Fischer-Lichte, Erika. 2008. *The Transformative Power of Performance: A New Aesthetics*. Translated by Sashay Iris Jain. New York: Routledge.

Gallese, Vittorio. 2009. "Mirror Neurons, Embodied Simulation, and the Neural Basis of Social Identification." *Psychoanalytic Dialogues* 19: 519–36.

Gharavi, Lance. 2012. *Religion, Theatre, and Performance: Acts of Faith*. New York: Routledge.

Källsted, Thorvald, et al. 1987. "Symposium on Hjalmar Sundén's Role-Theory of Religion." *Journal for the Scientific Study of Religion* 26 (3): 366–412.

Humphrey, Caroline, and James Laidlaw. 2004. *The Archetypal Actions of Ritual: A Theory of Ritual Illustrated by the Jain Rite of Worship*. Oxford: Clarendon Press.

Legrand, Dorothée. 2014. "Phenomenological Dimensions of Bodily Self-Consciousness." In *The Oxford Handbook of the Self*, edited by Shaun Gallagher, 204–27. Oxford: Oxford University Press.

Mason, David. 2014. *Brigham Young: Sovereign in America*. New York: Routledge.

Maturana, Humbert R., and Francisco J. Varela. 1980. *Autopoiesis and Cognition: The Realization of the Living*. London: D. Reidel.

Mauss, Marcel. 2009. *Techniques, Technology and Civilization*. New York: Berghahn.

McConachie, Bruce. 2015. *Evolution, Cognition, and Performance*. Cambridge: Cambridge University Press.

McConachie, Bruce. 2007. "Falsifiable Theories for Theatre and Performance Studies." *Theatre Journal* 59 (4): 553–77.

Merleau-Ponty, Maurice. 2004. *Basic Writings*. Edited by Thomas Baldwin. New York: Routledge.

Mitchell, Jon P. 2009. "Ritual Transformation and the Existential Grounds of Selfhood." *Journal of Ritual Studies* 23 (2): 53–66.

Rappaport, Roy A. 1996. "The Obvious Aspects of Ritual." In *Readings in Ritual Studies*, edited by Ronald Grimes. Upper Saddle River, NJ: Prentice Hall.

Sax, William. 2009. *God of Justice: Ritual Healing and Social Justice in the Central Himalayas*. Oxford: Oxford University Press.

Sax, William. 2008. "Agency." In *Theorizing Rituals: Classical Topics, Theoretical Approaches, Analytical Concepts*, edited by Jens Kreinath, Jan Snoek, and Michael Stausberg, 473–81. Boston, MA: Brill.

Schieffelin, Edward. 1985. "Performance and the Cultural Construction of Reality." *American Ethnologist* 12: 707–24.

Van Deusen, Increase, and Maria Van Deusen. 1847. *A Dialogue between Adam and Eve, the Lord and the Devil; Called the Endowment*. Albany, NY: C. Killmer.

Wilshire, Bruce. 1991. *Role Playing and Identity: The Limits of Theatre as Metaphor*. Bloomington: Indiana University Press.

12

COMMUNITIES OF GESTURE

Empathy and embodiment in Bill T. Jones/Arnie Zane Dance Company's *100 Migrations*

Ariel Nereson

November 2008 was a historic month in the U.S.A., seeing the election of the nation's first African-American President, Barack Obama. At the University of Virginia, on a campus built by enslaved people belonging to another American president, Thomas Jefferson, work began on a new dance meant to respond to this legacy and comment on the approaching 2009 bicentennial of the birth of yet another president, Abraham Lincoln. In the heady post-election days filled with a hope promised by Obama in his acceptance speech (Obama 2009, 224), the choreographer Bill T. Jones and his company developed a community-based work, *100 Migrations*, that shared this campaign rhetoric: 'It's about a climate of hope wherein we conjure up the means to save ourselves' (Jones 2008a, n.p.). Jones places responsibility for change on the performers and spectators involved with *100 Migrations*, eschewing a top-down process of reform and renewal. *100 Migrations*, involving 90 community members and 10 company dancers, models one such horizontal process, addressing the legacy of a cultural icon whose very name invokes notions of community and its particular formation in democracy. Democratic ideals of freedom and equality (those frequently associated with Lincoln) often feel utopian given the daily lived realities of many Americans, and Jones's scepticism of their achievement via Obama's election leads him to focus on performance, not policy, as the base of possible grass-roots activism. Explaining his desire to work with local community, Jones stated, 'I thought that the question of Lincoln was very much a question of being a part of a society and that everybody who considers themselves an American must have the DNA of that man who we call the greatest president who ever lived. So I wanted to know what that looks like, democracy moving' (*100 Migrations*). Jones's concept of 'democracy moving' asks what it is to move and be moved, engaging the double meaning, kinaesthetic and affective, of 'to move,' a double meaning grounded in the body.

This essay assumes that embodiment is knowledge – namely, it is how we know the world, how we learn our various social roles and how we explore alternatives. I proceed from the theory of embodied cognition articulated by philosopher Mark Johnson (2007), which builds upon the insights of Francisco J. Varela, Evan Thompson and Eleanor Rosch (1991), Johnson's collaborations with George Lakoff (1980, 1999) and Vittorio Gallese's body of research. *100 Migrations* asks how might we move people emotionally towards social change by moving them physically around the stage? Jones develops communities of gesture by asking participants to learn each other's self-generated choreography and to be responsible for fellow performers' physical safety. These strategies, I argue, can trigger empathic responses that connect moving towards social change with moving around a stage. I position *100 Migrations* as a public historical project, one that utilises the performance to expand the range of subjects towards whom we demonstrate empathy and to embody a democratic ideal of inclusiveness through choreographing the lived, embodied experiences of diverse communities.[1]

Johnson's concept of embodied cognition helps us understand the causes and consequences of bodies in motion in experiences both pedestrian (walking down a street) and aestheticised (performing dance choreography). Johnson concludes that far from being a physical limitation that the mind transcends, 'our bodies are the very condition of our meaning-making and creativity' (Johnson 2007, 15). Embodied cognition further encompasses the embodied nature of emotions. Antonio Damasio claims that emotions result in physicalised response 'in the form of actions and behavior,' leading to his development of the somatic-marker hypothesis, wherein rational decision-making is impossible without emotional response (Damasio 1999, 47). Damasio's somatic-marker hypothesis and Johnson's embodiment hypothesis have real ramifications for anyone investigating practices of sense-making, history being one such practice and art-making another that intersect in *100 Migrations*. One such consequence is the realisation that activities usually associated with aesthetics are involved directly in cognition, often foregrounding modes of meaning-making outside of, or in complement to, the linguistic. Dance, in particular, makes explicit the implicit connection between movement and emotion, a word whose definition includes 'to cause to move.'[2] Drawing upon empirical theories of embodiment does not, however, inevitably lead to positivism or simple determinism, a common critique. While human development of, and reliance upon, sense-making tools, like narrative, are neurologically hardwired, these processes take place within specific material and social environments that profoundly influence meaning-making. For Johnson, material and social worlds are co-constitutive of experience, alongside the biological realities of the body-mind.

The body/mind division that is superseded by the concept of embodied cognition is not the only divide that shapes this conversation. The flexible division between art and politics frames Jones's career. He is often associated with the culture wars, commenting in 1997 that his company 'attract[s] a number of people who see my persona as being about social change' (Jones 1998, 132–33). By the time of *100 Migrations* in 2008, the company's core interplay between aesthetics and politics was clear: 'Our mantra: What is at stake here?... What's at stake now has to do with what I call the discourse – the ongoing way in which the society attempts to know itself' (Jones 2008b, 104). In the Lincoln trilogy (*100 Migrations* is the middle work), the company focuses on historical discourse – the ways in which American communities attempt to know themselves through endowing the past with particular meanings. They adopt a public historical approach, defined by historian Michael Frisch as a commitment to 'shared authority' wherein the goal is not simply a top-down 'distribution of knowledge' but rather 'a profound sharing of knowledges, an implicit and sometimes explicit dialogue from very different vantages about the shape, meaning, and implications of history' (Frisch 1990, xxii). *100 Migrations* considers the role of the past in the present and future – the evergreen question 'have we learned from the past?' – and how embodiment and emotional response engage with the challenges of intergroup relationships in an increasingly pluralistic democracy. Thus, the work is not simply a choreographic reflection of the universal human realities of embodied cognition; rather, it argues for a particular understanding of embodied emotion, manifested choreographically in touch and gesture, as the potential for actionable empathy, for *moving* people towards social change by moving them around the stage.

Empathy is much contested across the humanities, arts and social sciences but can briefly be defined as an often other-oriented human capacity with the potential to both reinforce and transgress group boundaries. I look to recent cognitive neuroscience for a working definition, understanding that our comprehension of what exactly empathy is and how it works for humans is far from settled both within and without cognitive studies. Neuroscientists Grit Hein and Tania Singer distinguish between Theory of Mind (ToM) – 'our ability to understand other people's beliefs and thoughts' – and empathy – 'our ability to share other people's feelings' (Hein and Singer 2010, 110). Vittorio Gallese and Hannah Wojciehowski propose 'Feeling of Body,' a result of embodied simulation, rather than ToM, as the foundation of empathy: 'empathy may be conceived as the outcome of our natural tendency to experience interpersonal relations first and foremost at the implicit level of intercorporeity' (Gallese and Wojciehowski 2011, n.p.). Importantly, empathy is not itself an emotion but rather a capacity of responding to other people's emotional states. Empathy does not automatically involve other-oriented, or prosocial, action: we can share another's affective state and know that our affective state was prompted by theirs; however, this response need not motivate us to do anything about their affective state, particularly if it is negative. Group membership influences the likelihood that empathic response becomes actionable, or transforms into what Hein and Singer term *empathic concern*: 'an other-oriented response congruent with the perceived distress experienced by another person' (116). Hein and Singer's experiments suggest that there are neurological, functional differences 'between judging the mental states of similar and dissimilar others' (112). Empathic concern is more likely when we perceive someone as belonging to our group. In *100 Migrations*, Jones and company create groups whose social relationships are not shaped by racial, gender or ability identification but rather are based in a shared task.

Alongside the work's demands on performers, Jones articulates a similar site of growth for the audience as the development of 'a genius public' (Bromley 2008, n.p.) Elaborating on the vision of *100 Migrations* as a vehicle for social change, Jones states, 'I am hoping that there is somebody out there who can see it, and more importantly, as an artist speaking, who can *feel* it – that's what I believe in, a kind of emotional intelligence' (Jones 2008a, n.p.). Rather than a genius public composed of members who can recite facts in the proper order, Jones advocates instead for a public with the capacity to move and be moved. Frisch offers 'historical intelligence' as the public historian's target, defined not as 'a commodity whose supply they [public historians] seek to replenish,' but rather a search for 'the sources and consequences' of our selective remembering and forgetting of the past (Frisch 1990, 27). Empathic response is a potential connective tissue between, in Frisch and Jones's terms, historical and emotional intelligences. Catherine J. Stevens offers yet another formulation of intelligence as bodily, linking perception-action-cognition to embodiment in an endless feedback loop: 'perception and action are coupled. The body moves intelligently. Intelligence is bodily. Actions constrain or shape thought. Thought guides action' (Stevens 2016, 123). For *100 Migrations*, the relationship between perception and action is a reciprocal process of feeling and moving towards democracy in which both performers and spectators can participate.

The very ability of performance to offer paths into future action is concentrated when its subject is historical because of our tendency to associate past, present and future as a narrative. Freddie Rokem writes, 'The theatre "performing history" seeks to overcome both the separation and exclusion from the past, striving to create a community where the events from this past will matter again' (Rokem 2000, xii). Making the past matter is, as public historians Ray Rosenzweig and David Thelen (1990) found in a survey of Americans' attitudes towards history, a matter of making the past usable in shaping the future, a task that begins with understanding how the past has been used to make our present: '[respondents] assemble their experiences into patterns, narratives that allow them to make sense of the past, set priorities, project what might happen next, and try to shape the future' (12). In Louis Filler's concept of history as the usable past, the past and future were not necessarily isolated points on a progressive continuum, but rather necessary partners in terms of how people conceptualise history as an active, useful endeavour: 'the problem of the "usable" past…is one of determining where we have been and where we intend to go' (Filler 1947, 344). Evidence from cognitive sciences supports the conclusion that humans are not bound to a specific vision of the future, but rather the body-mind's very changeability gives us great agency in the present to causally affect future experiences to multiple ends, not all of them necessarily progressive. From the point of view of embodied cognition, the arc of the moral universe does not inherently bend towards justice, social or otherwise. Understanding and, in many cases, developing new narratives of the past become critical activities in forging community identity.

Dancing through division

The past appears in *100 Migrations* conceptually, in the images, feelings, thoughts and narratives that participants already have about Lincoln, as well as in the use of Lincoln's speech archive as part of the soundscape and as fodder for the choreographic process. The past also appears materially: Lincoln's deathbed, flanked by two podiums, is the focal set piece. The bed is centre stage and performers dance various patterns in groups around it throughout the performance. The performers are dressed in blues and greys, referencing the Union and Confederate uniforms. The choreography throughout suggests that commemorating Lincoln might be a way through past and current divisions towards a future with less conflict. The work includes duets and group sections that choreograph human social behaviours of encountering, appraising and opposing or welcoming the perceived 'other,' behaviours that are expressed and negotiated through movement. It concludes with the entire cast circling and touching Lincoln's deathbed, then dissolving the divisions between the two armies as performers melt away towards the exit.[3] Civil War historian Eric Foner notes an urgency around diversifying our knowledge of this particular historical moment, writing 'In public history…a large void still exists when it comes to slavery' (Foner 2002, xiii).[4] *100 Migrations*, taking place mere days after Obama's election in a centre of the Old South built by enslaved people, invites participants and spectators to reimagine what Lincoln means in a time yearning to be post-racial but deeply tethered to a fraught past. The work's choreographic focus on touch foregrounds embodied emotional response as meaning-maker, connecting past, present and future through practices of moving and feeling together. Donald A. Ritchie defines public history as actions that 'shape public consciousness through the presentation of the past in public places' (Ritchie 2001, 93). In *100 Migrations*, the past, in this case the archival past of the Gettysburg Address text and the object of Lincoln's deathbed, is put in motion through the teaching and learning of gesture, advancing a double understanding of public history as history presented *in* public and history presented *by* the public.

While the primary organising conceit for the 100 bodies onstage is that of two armies representing North and South, many other groupings composed the work. Cognitive narratologist Patrick Hogan writes that when we encounter others in literary narratives, as in daily life, 'the first crucial division…is between us and them, in-group and out-group' (Hogan 2011, 37). This division is based in our experiences of the social world and need not be a hostile division, but is at its root a simple recognition of the myriad differences that liken people to others (or not). The consequences of in-group and out-group identification can be costly, however, to the social fabric, as Hogan notes, 'Our empathy is inhibited with respect to members of out-groups' (ibid.). Recent studies have shown that in-group/out-group identification is a core trait for humans, 'unique among living primates in the extent of their preference for similar others' (Haun and Over 2013, 84–5). *100 Migrations* uses embodied emotional response to encourage empathy amongst new group formations that circumvent the potentially exclusionary boundaries of traditional group identities.

Jones and company disrupted habitual patterns of group formation from the beginning of the creative process. One of the first steps was the division of the 100 performers into groups of ten. Prior to breaking off into these small groups, performers were randomly assigned a word or phrase from the Gettysburg Address. Groups were organised solely by where members' text fell within the original speech, crossing lines of age, race, gender and ability. Jones's mission was to adopt an unbiased method for generating choreography: 'trying to be as inclusive as possible, it says loads about what the promise of Lincoln was' (*100 Migrations*). And indeed the work's most atypical inclusivity is across the boundaries of age and ability; it features several performers who make modifications to choreography based on what their bodies can achieve. Performers were instructed to 'make a shape with the sense of that word,' a gesture that could be easily repeated not only by a particular performer but also by the entire cast (*A Good Man* 2011). These gestures became 'The Hundreds,' with each group responsible for mastering its set of gestures. The gestures themselves tend towards simple abstractions of the participants' assigned words or phrases – 'battlefield,' 'unfinished,' 'devotion' and so on. Many involve a change in level, such as lunging towards the floor, accompanied by an arm gesture, such as the action of wrapping one's arms around a large basket. In teaching their fellow performers their gesture, participants specified the body parts, initiation, direction, and so on of the movement, and also its qualitative aspects, such as whether or not a given gesture should reach, flick or embrace. 'The Hundreds' demonstrate the foundational role of embodiment in making meaning, by filtering 'sense' – 'make a shape with the sense of that word' – through the body (Figure 12.1).

For local participant Lyn-Dell Wood, the group identification through gesture became a way to find belonging in the piece, signing her correspondence with the company 'Lyn-Dell 10 in group 1 w Tonio' (Lyn-Dell Wood, personal communication to Bill T. Jones/Arnie Zane Dance Company). Wood identifies herself by her group number but also her gesture number – 10 – and with the group's leader – company member Antonio Brown – suggesting that embodiment, emotion and social relationships intertwine within Wood's sense of participation, making a seemingly arbitrary numerical designation meaningful. In her signature,

Figure 12.1 The cast of *100 Migrations* performs 'The Hundreds' gesture sequence. Image courtesy of Bill T. Jones/Arnie Zane Dance Company.

Wood offers one possible response to a fundamental challenge of public history: 'where is interpretive authority located? How are we to understand interpretations that are, essentially, collaboratively produced…whether the relationship is one of cooperation or tension?' (Frisch 1990, xx). Her blended signature parses the multifaceted and atypical group membership fostered by this process, locating Wood as an individual making meaning in collaboration.

Touching the past, valuing the other

The other half of Hein and Singer's formulation for empathic concern, valuing the other's welfare, is enacted in *100 Migrations* through choreographies of risk that rely heavily on touch. During a recitation of Lincoln's 1858 'House Divided' address, groups perform phrases that include a member of the group being supported and lifted into the air. In some groups, the individual takes a flying leap into the clump, is caught and then lifted. In others, the individual simply falls backwards into the group, who catches and supports them. In a third variation, the individual, walking through space with the group, is touched by others and yields their weight, allowing them to be lifted continuously as the group moves. Group members become responsible for the physical welfare of one of their own, performing actions to secure that person's safety that enact empathic concern. Significantly, these actions are accomplished through contact, through an embodied connection between group members that requires them to literally support each other through touch. The House Divided choreography centres on sequences of prosocial behaviour within the company groups, groups which embody public historian Richard Ned Lebow's claim that 'individuals are likely to belong to multiple memory communities, making contact and cooperation across these communities more feasible' (Lebow 2008, 38). Contact across communities is literal in this section, as performers touch and share their weight regardless of their distinctions across age, ability, race and gender. Their choreographies counter the narration of Lincoln's House Divided speech: the divisions structuring the groups are not based on any salient social rubric, and are porous. As performers run through the space, they cross paths with other groups, momentarily join their ranks and then return to their home group. This choreography proposes that maintaining division is a choice – the opposite choice, to seek union, is equally possible and does not require ignoring real distinctions in the human experience (Figure 12.2).

Figure 12.2 Cast members of *100 Migrations* perform a supported lift. Image courtesy of Bill T. Jones/Arnie Zane Dance Company.

Generating 'The Hundreds' involved participants' memories of their encounters with the Gettysburg Address, with Lincoln broadly, and with the legacy of slavery in the U.S.A.. The performance foregrounds these memories when community members make their private histories public at the podiums flanking Lincoln's deathbed. Including these individual stories is both an aesthetic choice and a strategy of public history, designed to 'generat[e] from within...the authority to explore and interpret their own experience, experience traditionally invisible in formal history because of predictable assumptions about who and what matters' (Frisch 1990, xxi). One participant, Doris, shares the story of her great-grandfather Paul, who served as a freed black soldier in the Civil War. Her narrative is paired with Jim Respess's familial history, which centres on the Baltimore Plot, an assassination conspiracy headed by white supremacist Cipriano Ferrandini, later revealed to be Jim's great-great-grandfather. Doris and Jim's histories represent opposing sides of the conflict, and it is fitting that their stories take place within the context of the House Divided speech. During the House Divided recitation, company dancers Paul Matteson and Antonio Brown perform a duet, their racial differences mirroring the historical context for Lincoln's speech. Collectively these choices respond to, though do not resolve, a common difficulty in public history, that of the public's resistance to the historian's reality that 'there often exists more than one legitimate way of recounting past events' (Foner 2002, xvii).

Our faculty of proprioception, of sensing where our bodies are in space and how to move them, is neurologically intertwined with our capacity for empathic response. Learning others' gestures is a tangible way to empathise with their emotional states. A study by Antonio Damasio of persons with severe brain trauma found that 'patients with damage to body-sensing regions of the cerebral cortex would not be capable of [empathy],' and further, that 'In the absence of this region [right somatosensory cortices], it is not possible for the brain to simulate other body states effectively' (Damasio 2003, 116). Because of the embodied nature of emotion, body simulation (both purely neurological but also physicalised simulation, as in the repetition of gesture) is a primary conduit through which we sense others' feelings, and the foundation of *100 Migrations*' communities of gesture. These communities encourage empathic response for performers and spectators alike through their formation based in an aesthetic task rather than on neurotypical in-group, out-group distinctions. The choreographic process rejects traditional in-group, out-group distinctions as they often stymy attempts at engaging our capacity for empathy across lines of social identity. These activities occur within the event of performance, and may function as an intense rehearsal for the empathic exchanges that happen in the 'real' world.

The *100 Migrations* residency was one of many initiatives at the University of Virginia (UVa) that would revise and literally excavate its own history and the role of slavery therein.[5] Directly preceding the residency, UCARE (University and Community Action for Racial Equity) was formed with the express goal 'to further understand the legacy of slavery and segregation in the university's history as well as in the Charlottesville community...to help the university and adjacent communities come together to identify actions that could improve their relationship and lead to reconciliation' (Faulkner 2013, 5). Revising UVa's history and repairing its relationship with the Charlottesville community were articulated as parallel goals wherein acknowledging past wrongs, specifically racial offenses, was a key strategy – and a necessary one, as several town-gown hall meetings revealed that many in the Charlottesville community referred to UVa not as the lofty 'Academic Village,' but rather as 'The Plantation' (ibid., 6). *100 Migrations* became part of this initiative, designed to help bridge 'the gap between Charlottesville and UVa' (Turner 2009, n.p.). Involving community members in generating, rather than responding to, formations of inclusive community found its expression in making dance. *100 Migrations* choreographs the task of forming a more perfect union, and offered a democratic vision of community wherein embodied emotional response is routed towards inclusion and concern for differential others.

Notes

1 Johnson's 2014 book *Morality for Humans: Ethical Understanding from the Perspective of Cognitive Science* defends an interdisciplinary approach to explanatory frameworks, arguing that 'our best strategy of inquiry is to look for converging evidence generated by different methods' (22). See Bruce McConachie's "Reenacting Events to Narrate Theatre History" for the argument that historiographical approaches of reenactment and simulation are empathic experiences, and that empathy is stock-in-trade for the historian.

2 The notion of emotions as a causal force, inciting us to literal action, is explained by the neuroscience of emotions summarised by Johnson in chapter 3 of *Meaning of the Body*. Johnson states that emotions function to appraise specific situations an organism finds itself in, 'often initiating actions geared to our fluid functioning within our environment. It is in this sense that emotional responses can be said to move us to action' (61).

3 The work was initially to be performed on the famous UVa lawn; on the day of the performance, it was rained out and relocated to the basketball stadium.

4 Critical studies of the relationship between slavery and UVa are few: an unpublished 2003 paper by Charlottesville, VA historian Gayle Schulman and a senior thesis from 2006 by UVa student Catherine S. Neal. See Faulkner, *Slavery at the University of Virginia: A Catalogue of Current and Past Initiatives* for these references.

5 Initiatives included archaeological excavations of Pavilion 4 (one of the University's original structures used as slave quarters) and of 67 grave markers near the University Cemetery, likely those of slaves. See Faulkner (2013) for details on these projects.

References

100 Migrations by Bill T. Jones/Arnie Zane Dance Company, directed by Bill T. Jones and Janet Wong, University of Virginia, Charlottesville, November 15, 2008.

A Good Man. 2011. Directed by Bill Hercules. Performed by Bill T. Jones/Arnie Zane Dance Company. USA: Kartemquin Films. 2011. Film.

Bromley, Anne. 2008. "Art Isn't Just the Pursuit of Beauty, According to Bill T. Jones." *The Carter G. Woodson Institute for African-American and African Studies Newsletter.* Nonpaginated. http://news.clas.virginia.edu/woodson/x14574.xml. Accessed September 10, 2013.

Damasio, Antonio. 2003. *Looking for Spinoza: Joy, Sorrow, and the Feeling Brain.* Orlando, FL: Harcourt, Inc.

———. 1999. *The Feeling of What Happens: Body and Emotion in the Making of Consciousness.* New York: Harcourt Brace.

"emotion, v.". OED Online. March 2017. Oxford University Press. www.oed.com.gate.lib.buffalo.edu/view/Entry/281674?rskey=SqVBF4&result=2&isAdvanced=false. Accessed May 31, 2017.

Faulkner, Meghan Saunders. 2013. *Slavery at the University of Virginia: A Catalogue of Current and Past Initiatives.* Produced internally for the University of Virginia. Charlottesville: University of Virginia IDEA Fund.

Filler, Louis. 1947. "America and the 'Usable Past.'" *The Antioch Review.* 7.3: 336–44.

Foner, Eric. 2002. *Who Owns History? : Rethinking the Past in a Changing World.* New York: Hill and Wang.

Frisch, Michael. 1990. *A Shared Authority: Essays on the Craft and Meaning of Oral and Public History.* Albany: State University of New York Press.

Gallese, Vittorio and Hannah Wojciehowski. 'How Stories Make Us Feel: Toward an Embodied Narratology.' *California Italian Studies.* 2.1 (2011): n.p. http://escholarship.org/uc/item/3jg726c2%3Fimage.view%3DgenerateImage%3BimgWidth%3D600%3BpageNum%3D1. Accessed August 10, 2017.

Haun, Daniel B. M. and Harriet Over. 'Like Me: A Homophily-Based Account of Human Culture.' In *Cultural Evolution: Society, Technology, Language, and Religion*, edited by Peter J. Richerson and Morten H. Christiansen. 75–85. Cambridge, MA: MIT Press.

Hein, Grit and Tania Singer. 2010. "Neuroscience Meets Social Psychology: An Integrative Approach to Human Empathy and Pro-Social Behavior." In *Prosocial Motives, Emotions, and Behavior: The Better Angels of Our Nature,* edited by Mario Mikulincer and Phillip R. Shaver. 109–25. Washington, DC: American Psychological Association.

Hogan, Patrick. 2011. *Affective Narratology: The Emotional Structure of Stories*. Lincoln: University of Nebraska Press.

Johnson, Mark. 2014. *Morality for Humans: Ethical Understanding from the Perspective of Cognitive Science*. Chicago, IL: University of Chicago Press.

———. 2007. *The Meaning of the Body: Aesthetics of Human Understanding*. Chicago, IL: University of Chicago Press.

Jones, Bill T. 2008a. *UVa Today Podcast*. By Deborah McDowell. November 11, 2008.

———. 2008b."What's at Stake?" In *Further Steps 2: Fourteen Choreographers on What's the RAGE in Dance?*, edited by Constance Kreemer. 96–106. London: Routledge.

———. 1998. "Bill T. Jones in Conversation with Thelma Golden." By Thelma.

Golden October 27, 1997. Rpt. in *Art Performs Life: Merce Cunningham, Meredith Monk, Bill T. Jones*. 126–35. Minneapolis: Walker Art Center.

Lakoff, George and Mark Johnson. 1999. *Philosophy in the Flesh: the Embodied Mind and Its Challenge to Western Thought*. New York: Basic Books.

———. 1980. *Metaphors We Live By*. Rev. ed. 2003. Chicago: University of Chicago Press.

Lebow, Richard Ned. 2008. "The Future of Memory." *Annals of the American Academy of Political and Social Science*. 617.1: 25–41.

Lincoln, Abraham. 1905. *Lincoln's Speeches and Letters*. Farmington, ME: D.H. Knowlton & Co.

McConachie, Bruce. 2010. "Reenacting Events to Narrate Theatre History." In *Representing the Past: Essays in Performance Historiography*, edited by Charlotte Canning and Thomas Postlewait. 378–403. Iowa City: University of Iowa Press.

Obama, Barack. 2009. "Yes, We Can! The Victory Speech." In *Barack Obama: Speeches on the Road to the White House*, edited by Maureen Harrison and Steve Gilbert. 224–8. Carlsbad, CA: Excellent Books.

Ritchie, Donald A. 2001. "When History Goes Public: Recent Experiences in the United States." *Oral History*. 29.1: 92–7.

Rokem, Freddie. 2000. *Performing History: Theatrical Representations of the Past in Contemporary Theatre*. Iowa City: University of Iowa Press.

Rosenzweig, Roy and David Thelen. 1998. *The Presence of the Past: Popular Uses of History in American Life*. New York: Columbia University Press.

Stevens, Catherine J. 2016. "The Body in Mind." In *Theatre, Performance and Cognition: Languages, Bodies and Ecologies*, edited by Rhonda Blair and Amy Cook. 122–7. London: Bloomsbury Methuen Drama.

Turner, Lindsey. 2009. "UVa and the Arts: '100 Migrations' Summary and Proposal for Future Guest Projects." Non-paginated. Charlottesville: University of Virginia.

Varela, Francisco J., Evan Thompson and Eleanor Rosch. 1991. *The Embodied Mind: Cognitive Science and Human Experience*. Rev. ed. 2016. Cambridge, MA: MIT Press.

13

CREATIVE STORYTELLING, CROSSING BOUNDARIES, HIGH-IMPACT LEARNING AND SOCIAL ENGAGEMENT

Nancy Kindelan

The pedagogy of theatre studies is by its very nature multidimensional and cross-disciplinary. Its immersive teaching and learning strategies can help students address 'big questions' through cultivating inquiry-driven learners whose problem-solving skills enrich their cultural understanding and promote civic responsibility. This essay explores how a theatre class was designed to help all undergraduates approach life's unscripted problems through high-impact educational practices and metacognitive strategies in order to understand and manage some of the significant issues associated with twenty-first-century learning, such as complexity, diversity, flexibility and social change.

Many courses in dramatic literature, history and theory as well as theatre's experiential activities adhere to the 'Principles,' 'Essential Learning Outcomes' and 'High-Impact' learning strategies recommended by the Association of American Colleges and Universities initiative, Liberal Education and America's Promise (LEAP). Launched in 2005, LEAP has guided educators who are committed to developing powerful, high-impact educational practices that help all college students become intentional in their learning in order to prepare to become knowledgeable and responsible citizens and valuable contributors to a global economy (See Association of American Colleges and Universities 2008, 2015).

LEAP supports high-quality learning through achieving specific twenty-first-century 'Essential Learning Outcomes' such as 'knowledge of human cultures and the physical and natural world,' 'intellectual and practical skills,' 'personal and social responsibility,' 'integrative and applied learning.' Theatre pedagogy champions many of LEAP's outcomes through using plays to explore human nature as well as to develop new ways to think about human dignity, freedom and diverse cultures. Students are provided with opportunities to develop intellectual and practical skills, to confront complex real-world problems and social challenges through an in-depth or capstone project. Described by the 'LEAP Challenge' as 'Signature Work,' these problem-based projects demonstrate how students acquired skills, insights and knowledge to help them develop the capacity to explore and share their findings with others (AAC&U 2015).

As defined by LEAP, high-impact practices, known as HIPs, include a variety of learning experiences such as first-year seminars, capstone courses and projects, thematic learning communities, writing-intensive classes, collaborative projects, intercultural programs, service learning/community projects and internships. 'High-Impact Educational Practices' encourage students to explore 'big questions.' HIPs strategies involve students in challenging, engaging, evidence-based activities where learning becomes intentional and students become informed, responsible and empowered learners (Kuh 2008, 9–11).

Following the lead of George D. Kuh's findings (2008) about the success of HIPs in higher education, this essay demonstrates how HIPs played a significant role in the development of a collaborative, intra- /interdisciplinary, immersive, project-based honours class that culminated in the creation of a 'Treatment' for a contemporary Living Newspaper play. This documentary form of theatre utilises historical facts, authentic interviews, videos, original dramatic scenes and innovative theatrical practices to promote social consciousness and responsible social action. As such, HIPs, seen through the lens of theatre pedagogy, are analogous to metacognitive experiences, which required my non-theatre majors to develop metacognitive skills in order to complete the assigned cognitive task.

Generally, metacognition refers to self-awareness about our cognitive processes and our ability to regulate cognitive skills for more affective learning (Flavell 1979). In 'Metacognition and Cognitive Monitoring,' Flavell proposed that metacognitive knowledge is beneficial in helping 'to select, evaluate, revise, and abandon cognitive tasks, goals, and strategies' in light of the project at hand. Metacognitive knowledge can 'lead to' metacognitive experiences that involve 'self, tasks, goals, and strategies' that help in interpreting 'the meaning and behavioural implications of these metacognitive experiences' (908). Flavell also suggests that the interplay between metacognitive knowledge and metacognitive experiences might cause students to assess their feelings that their goals are difficult to achieve whereupon they may employ the metacognitive strategy of asking questions of more knowledgeable people (mentors) to help them achieve the desired outcome.

The authentic experience of writing a Treatment for a Living Newspaper included 'situated modeling, coaching, and fading' (See Brown, Collins, and Duguid) and cognitive 'apprenticeship' opportunities (Rogoff 1990). Viewed as a group of 'active learners' (Rogoff, 39, 198), the students' work on their subsequent creative projects was continuously challenged, supported and guided by numerous mentors, visiting guest professors and myself, until they were empowered to complete the task. The social interaction between mentors and students allowed the novice artists 'to participate in skills beyond those that [they are] independently capable of handling.' Rogoff maintains that working closely with an expert who has intellectual tools follows Vygotsky's theory of social influence as a contributing factor to cognitive development and as such resembles apprenticeship (Rogoff, 140–1). Vygotsky (1978) argues that learning occurs by interacting with others in active learning environments that include guided mentorship.

The Living Newspaper project allowed students to participate intentionally, metacognitively and behaviourally in a process that involved self-regulated learning: identifying a topic, setting reasonable goals, careful planning, questioning, monitoring and evaluating. For example, after selecting their topics students began the metacognitive reflective practice of how to incorporate facts and journalistic interviews and where revision was needed. Individual metacognitive experiences often called for additional cognitive strategies (more research), which then led to deeper learning tactics (additional discussions and guided mentorship) that eventually improved the play's narrative or dramatic techniques. Throughout the cognitive task at hand, metacognitive experiences provided opportunities for reflection and monitoring the quality of their projects. Asking questions about the social conflicts in their projects and listening to the targeted comments of their instructors, mentors and peers as they evaluated each of the successive iterations of their creative projects became a very comfortable standard that led to further thinking about the social purpose of their work.

Reflective ePortfolio journals helped my novice artists take control of their learning and diminish their fears that they were incapable of writing a Treatment for a Living Newspaper. For example, students received frequent feedback from other members in the learning community: we reviewed their numerous online iterations and helped them assess what was working and not working. For example, later in this essay I describe how a student found it necessary to reconsider her initial response to a social issue, reframe her original questions or arguments and find ways to move beyond a surface understanding of the play's topic. Because work in the course was presented as a journey (process over product) and the written 'Treatment' for a Living Newspaper was not a fully developed play script, non-theatre students became more comfortable creating an artistic project that involved subjects that were controversial, methods of learning that were unfamiliar and documentary forms of theatre that encouraged civic responsibility. A supportive environment of learners helped these students move beyond their comfort zones and discover how to learn in social contexts.

High-impact educational practices and metacognition strategies in the development of a Living Newspaper play

HIPs strategies that encourage critical, reflective, integrative and collaborative educational practices helped students think about and develop authentic stage-worthy dramatic techniques and images capable of communicating real-world problems. The creative activity of researching and writing a Treatment for a twenty-first-century Living Newspaper play advanced 'intentional, informed, responsible, and empowered' learning through the supportive course-based community of learners.

My course began by having students write a series of analytical and reflective papers about the development of various forms of socially responsible theatre. These examples provided various models for their understanding of the cultural traditions and methods of Living Newspaper plays. The student's version of a contemporary Living Newspaper involved transferring what they had learnt from one context to another.

Learning became even more 'intentional' for the students when they were asked to write and present a preliminary statement about what specific social issue or question they would use to frame their play. Their Purpose Proposal was informed by rigorous undergraduate research strategies, which included inquiry-based information gathering substantiated by cross-departmental connections with faculty mentors, interviews with experts in the field and the critical analysis and synthesis of all pertinent fact-finding information. The process of transferring their knowledge about the journalistic tools employed in prior Living Newspaper plays and their research into a thought-provoking and engaging Living Newspaper 'artefact' (a Treatment) involved multiple iterations. Through participating in group creative explorations, collaborative self-assessment assignments and interdisciplinary discussions, students considered various ways to clarify their play's dramatic structure and explore the hidden psychological world of their characters. Because they were involved in creating a Treatment for a *contemporary* Living Newspaper play within the culture and community of learners, students individually and collectively pondered and discussed in class and online whether their selected presentational form of theatre (its dramatic structure and performance strategies) had the potential to engage the audience in civic action.

Historical/theoretical foundations and contemporary reverberations

Since the early twentieth century, there have been numerous forms of documentary theatre. These range from the Epic style of Erwin Piscator and Bertolt Brecht to the Living Newspaper plays of the 1930s, to contemporary performance groups, such as Pop-Up Magazines (a 2009 U.S. form of 'performed journalism' and a similar version involving *Guardian* journalists and the British Royal Court Theatre makers), to the Tectonic Theater Project and the Living Newspaper festivals sponsored by a Chicago-based company, the Jackalope Theatre.

My recent Honours Inquiry course, 'Creative Social Engagement: The Rebirth of the Living Newspaper,' concentrated on three forms of documentary theatre: the 1930s Living Newspapers of the Federal Theatre Project, Anna Deavere Smith's journalistic performance art work, and the artists' collaborative, Rimini Protokoll, which uses theatrical tools (e.g., site-specific theatre and living installations) to expand the audience's notion of life experiences through a new form of documentary theatre. These models provided the dramatic and theatrical contexts for understanding and envisioning possible forms for my students' Living Newspaper projects.

Through deep readings of two plays and writing critical analyses that focused on the dramatic form and theatrical images found in the Federal Theatre Project's *One-Third of a Nation* (squalid housing conditions) and the 2000 film version of Anna Deavere Smith's *Twilight: Los Angeles, 1992* (L.A. race riots), students identified, examined and appraised how the staging of these plays revealed relevant social issues. In addition, Rimini Protokoll's *Situation Rooms*, *Remote Houston,* and *Remote L.A.* were also discussed to contextualise how these experimental twenty-first-century documentary forms of theatre serve to illustrate, by comparison, the evolution of documentary theatre.

The course was divided into three reflective stages. Each stage emphasised some aspect of the learning journey, including being led by metacognitive knowledge that triggered additional metacognitive experiences such as participating in guided mentorship, engaging in group activities, and ePortfolio reflective activities to see if what was learnt was successful in creating a sense of the whole. In Stage One of this course, students described, examined, differentiated and evaluated numerous examples of documentary theatre, which provided models for their thinking about their contemporary Living Newspaper projects.

Stage One

Yesterday's social theatre: the Living Newspaper

Through analytical papers, students demonstrated their knowledge about why the original Living Newspaper, its purpose and theatrical techniques, continues to inspire today's theatre practitioners interested in innovative ways to dramatise social ideas. Their reflections influenced their Treatments, especially the ways they would use journalistic tools to achieve the experiential goals of the class.

What was the Living Newspaper? This documentary form of theatre with unique production values was part of the Federal Theatre Project (FTP) programme sponsored by Franklin Delano Roosevelt's Works Progress Administration. The FTP originated during the Great Depression as a way to accommodate thousands of unemployed journalists and theatre artists. As director of the FTP, Hallie Flanagan supported a visionary, sometimes controversial, form of journalistic theatre that offered economic relief and provided the audience with a deeper understanding of how national, economic and human rights struggles had affected a diverse population. These projects employed an editorial staff of journalists who assiduously researched, documented and distilled current events and topics of ongoing interest. The Living Newspaper projects used facts to bring current events to life in an effort to educate and to empower their audiences to become socially responsible. For example, *Triple-A Plowed Under* (1936) covered, in 26 scenes, the personal, economic and political plight of drought-stricken farmers during the 1920s and early 1930s. *One-Third of a Nation* (1938) pursued, in approximately 20 scenes, the social, economic and political history of crowded, unsafe, unsanitary housing conditions in U.S. metropolitan slums.

The Living Newspaper's dramatic structure and topics made it possible to hire hundreds of unemployed actors and journalists across the U.S.A. And the Living Newspaper's version of modern morality plays discovered a way to dramatise the news, facts and human struggles in strikingly theatrical ways. The FTP found contemporary European presentational staging techniques best suited the Living Newspaper's documentary form and numerous scenes. Episodic staccato scenes mimicked the rhythm of the 1930s rapid-fire newsreels. Interlocking platforms facilitated multiple scene changes and simultaneous action; metaphorical staging conveyed social messages; scenes that ended with a punch line emphasised the play's ideas; directional light, shadows and blackouts stressed thought over emotion; music underscored specific concepts; symbolic characters created the essence of status over the nuance of psychological development and the voice of a narrator provided commentary. Finally, the use of carefully placed informational slides and documentary films made sure that the audience's attention remained focused on the play's message.

Untold stories: the documentary theatre of Anna Deavere Smith

Students continued their analytical work by demonstrating through group discussions and papers why Anna Deavere Smith's innovative documentary theatre has created a new form of theatre that encourages civic discourse. Smith's artistic vision and her virtuoso performances are evident in works such as *Fires in the Mirror: Crown Heights, Brooklyn, and Other Identities* (1992), *Twilight: Los Angeles, 1992* (1993–1994), *Let Me Down Easy* (2008–2010) and *Notes from the Field: Doing Time in Education* (2015–2016).

In Smith's memoir, *Talk to Me: Listening Between the Lines* (2000), she provides insights into the purpose behind her new form of theatre and the strategies she uses to develop her one-woman shows. Her goal is to seek ways to strengthen the role the arts play in understanding the essence of the 'American character' through listening carefully to the words of *all* its citizens. Smith acknowledges that we live in a 'communications revolution' and that we have 'a need to communicate'; however, she is concerned about our ability to communicate, to discuss difference, to leave our familiar houses of personal identity, to take the time to hear deeply, to analyse and to trust the unique voices of others. Smith's journey to create a socially responsible theatre involves leaving what is familiar to her, such as family, race, class, nation and professional sphere of expertise. She explains:

> When you leave your safe house, you will end up standing someplace in the road. I would call these places that are without houses crossroads of ambiguity. On the one hand, they are not comfortable places. On the other hand, in them one acquires the freedom to move.

In Smith's journey to discover the essence of the 'American character' through the diversity of its citizens, she has 'moved across many cultural boundaries' (Smith 2000, 24).

Since the early 1980s, Anna Deavere Smith has been creating a series of one-woman performance pieces, which are part of a much larger personal journey, *On the Road: A Search for American Character.* Her pioneering one-woman documentary theatre includes a collage of individuals who represent diverse backgrounds and levels of authority. Their stories either provide the audience with varied perceptions on urban, racial and class conflicts, which are responsible for shaping and reshaping the nation, or offer multiple angles to explore the human condition. In some of her work, Smith explores the following 'big questions.' Is there a common citizenship? Can we negotiate racial and ethnic difference? Are we paying enough attention to the well-being of our youth? She believes raising these questions will encourage thought about oppression, will lessen the anxiety that comes with talking about difficult issues involving race, community and identity and will generate social action.

In a single performance piece, Anna Deavere Smith may portray several dozen people selected from hundreds of her interviews. Her theatre has a journalistic style and is often referred to as a 'one-person documentary.' Smith begins her creative process by interviewing and recording individuals whose words and physical gestures convey authentic impressions about the psychological and social complexity of their lives. Her unique form of documentary theatre has been referred to as 'verbatim theatre' because she uses the interviewee's *exact* words when she portrays that person's persona on stage. She studies not only what people say, but also how they say it. In her interviews, she looks for the 'extraordinarily communicative moment' when she hears, for example, the breakdown of syntax. Her portrayals not only capture the exact words and speech patterns of her interviewees but also include their intermittent coughs, laughter, pauses and ums. Smith works against Stanislavsky's method of acting, which trains actors to create characters from the inside out; her outside-in approach relies on how the act of speaking reveals something truthful about the individual's identity (See Smith 2000, 7–12).

Performance art emphasises a minimal approach to stage and costume design. In Smith's work, the focus is on the artist who conveys, mostly through language, a plethora of different individuals by simply changing a hat or a piece of clothing and/or moving to a different part of the stage. Smith's performance pieces may include symbolic set pieces, descriptive videos or thought-provoking images that help the audience contextualise the historical event or say something specific about its social ideas. For example, in the 1993 design of the Mark Taper Forum production of *Twilight: Los Angeles*, Smith performed in front of a '25-foot white wall with a bay-window-sized opening.' Daniel B. Wood's review notes that the opening 'serves as graffiti-covered window, living-room backdrop, or video-monitor bank.' He adds that

> perhaps the evening's most poignant moments coincide with her use of the well-known 82-second tape of the King beating [by four L.A. police officers on March 3, 1991], and local news footage of rioting, including the beating of truck driver Reginald Denny [which occurred after the policemen were acquitted at the Simi Valley trial on April 29, 1992].
>
> *(The Christian Science Monitor, July 8, 1993)*

Anna Deavere Smith's performance art creates a theatre experience that encourages artistic democracy—the coming together of people to discuss issues that have community value. Therefore, her performance pieces offer the audience various points of identification to encourage the building of bridges about such topics as race, ethnic, health and gender issues. Smith's theatre presents the audience with questions not answers, multiple opinions on the same topic and dialectical arguments. To facilitate her view of artistic democracy, she often includes post-play conversations so that there is the opportunity for the audience to express their opinions about what they heard and saw on stage.

'Theatre of Real People': today's Living Newspapers

Lastly, students expanded their frame of reference by focusing on the theatre of Rimini Protokoll, an internationally known artists' collective, based in Berlin, Germany, founded by Stefan Kaegi, Helgard Haug and Daniel Wetzel. Trained in Applied Theatre Studies, this team of equals works both individually and collaboratively. They see themselves as authors/directors who produce eclectic contemporary and experimental theatre forms that involve film, audio-installation, mobile technology and new theatrical tools (living interactive installations). Since its inception in 2002, the group has created over 50 productions on social issues that span a wide range of life experiences such as war, global market economy, capitalism, unemployment, old age, dying and death. In *Experts of the Everyday: The Theatre of Rimini Protokoll*, Florian Malzacher explains that while there is joy in their collaborative spirit, Rimini Protokoll is a 'brand name that facilitates communication, an effective working network, an umbrella organisation without an official statement of intent, that maintains separate accounts even today.' With no one artist/genius at the helm, their strengths come from 'their differences not from the similarities that have grown over the years' (Malzacher 2008, 21).

The collective is interested in pursuing experimental dramaturgies that involve documentary and literary methods that encourage deep thinking and interconnectivity. Instead of actors, they enlist non-professional performers, who in the past were called 'experts' and are now called 'protagonists.' They view their performers as 'experts of everyday life.' One of Rimini Protokoll's goals is to develop 'new theatrical realities' that expand traditional definitions of what a stage is and 'allow for unusual perspectives' about how we perceive reality. To that end, many of Rimini Protokoll's theatre experiences do not occur in a traditional theatre, but instead use locations that represent some facet of our life that isn't acknowledged or appreciated (a tour of our home town, someone's living room, a cemetery, the back of a transparent freight truck, etc.). Daniela Hahn suggests that walking tours of a city have been turned into 'sites of artistic investigation' when theatre events moved to streets and buildings and away from 'institutionalized spaces for the arts' (Hahn 2014, 31). Stefan Kaegi and Jörg Karrenbauer, organizers for *Remote Houston* (2016), describe the experience of exploring the city (on foot and by public transportation) for approximately two hours with the help of a guided audio tour:

> With recordings and soundscapes taking over your ears, the cityscape of Houston turns into your personal film. As you move along, the voice in your headphones becomes a more active participant, [an] artificial intelligence exploring human activity, and you are the vehicle for that exploration.
>
> *(Alley Theatre, 'Remote Houston')*

Rimini Protokoll has been credited with inventing a twenty-first-century form of documentary theatre, which Florian Malzacher describes as 'one in which the conventional notion of objective documentary is juxtaposed with very subjective experiences, in which the individual and the social are brought together in a way that expands both objective and subjective perception' (Malzacher 2010, 80–1). More interested in people than in facts, Rimini Protokoll differs from the 1930s Living Newspaper documentary theatre or Anna Deavere Smith's one-person 'verbatim' documentary theatre since this artist collaborative often alters its interview material for presentational purposes. A company member since

2005, Jörg Karrenbauer explains: 'We take over people's stories at a certain moment. We rearrange them, cut them and extend them. It's not always true, what they say onstage. It's not necessarily their experience of life' (as quoted in Trueman 2016). The final narrative becomes a 'literary blending' of actual interviews, research and the behaviour and motivations of people who are involved in a particular situation. Rimini Protokoll has reinvented documentary theatre and the Living Newspaper to respond to the 'digital age.'

Situation Rooms is an extreme example of how Rimini Protokoll 'blends' the narration of the 'participants/inhabitants' and the experiences of the 'audience/players.' Rimini Protokoll calls *Situation Rooms* a 'multi-player video piece' where the 'players,' the audience members, are called 'spectators.' The setting for the piece consists of a labyrinth of rooms and environments that mirror the actual spaces where 20 people ('inhabitants') from a variety of continents tell their stories (on video) of how their lives have been shaped by weapons. The spectators carry iPads which they have been given prior to entering the maze of rooms, on which they listen to the personal narratives of the 'inhabitants' and are guided through their reconstructed environments, as a disembodied voice instructs them on where to go and what to do, and as they view the inhabitants speak from the actual environments where they recorded their stories[1] (Haug, Kaegi, and Wetzel 2013). As the relationships of the participants/inhabitants/audience/players are blurred, as boundaries are crossed, a new and deeper understanding of the weapons industry is developed.

Stage Two

Research and creative exploration

In Stage Two of this course, students built on their knowledge of these three model documentary theatre forms and applied their understanding of the historical and contemporary forms of Living Newspapers as they contemplated and wrote their own version of a Living Newspaper play. Metacognitive experiences, high-impact educational practices and scaffolding experiences (e.g., apprenticeship models) helped students think about, integrate knowledge and participate in discussing, examining and revising their plays. The collaborative learning community provided a creative and supportive environment for students to develop social awareness and interpersonal skills. In this atmosphere, students listened to the insights of others and evaluated their progress. Online reflective writing activities mapped the progress of their learning and personal growth by charting the students' development as undergraduate researchers and collaborative theatre practitioners. Meaningful reflection papers and discussions helped students analyse interdisciplinary problem-based approaches and experiences in order to discover new ways of thinking about the development of their creative signature projects.

Early on in Stage Two, students selected a topic that had social implications and composed a Purpose Statement about why they elected this specific issue to describe, explain, examine, break down and assess through their contemporary Living Newspaper play. At this point, the door to intentional learning opened.

As you might imagine the topics undergraduates were interested in pursuing often had something to do with their majors or their personal lives, such as the environment, social justice, deportation, rape or gender issues. For example, one of my students elected to write a Living Newspaper piece on the objectification of women. In her Purpose Statement, she offered that the intent behind her Living Newspaper project was to bring 'attention to the subtle cases of objectification that women deal with every day, [and] to put audience members in a woman's shoes so they can understand how it *feels* to be objectified.' In her play, she wanted the audience to 'leave knowing how ingrained and ever-present objectification of women is in society and understanding how this makes women feel/the damage this does, which should provoke them to do something about it.'

Social issues are complex; they also can be intimidating and emotionally charged, but they provide powerful opportunities for students to assess their personal thoughts about, for example, race and ethnicity. In one instance, a student discovered her own 'unconscious racial tendencies' during a discussion about an early iteration of her Living Newspaper play. Subsequent conversations with me, as well as her online reflections, revealed this student's discomfort about not only her unexamined thoughts but also her need to learn more about people from different backgrounds by listening more carefully to their experiences. She used her metacognitive experience as a self-regulating opportunity. She reshaped the purpose of her Living Newspaper project from how an unhealthy inner-city environment affects its children to one that helped her audience learn how to be self-reflective about race relations, especially as these issues relate to 'environmental justice, gentrification/displacement, [and] childhood development.'

Once students composed their purpose statements, they turned their attention towards examining the play's world – a process that involved crossing cultural boundaries, researching the play's societal issues, talking with their mentors and conducting field interviews. Writing a Living Newspaper play includes researching and presenting authentic information and compelling images about the project's 'big questions.' In order to enhance the students' research experiences, I connected my students to faculty members across the university whose research interests in the humanities, social sciences and/or sciences matched my students' topics. Students refined their questions, built their bibliographies and developed research networks through engagement with faculty members outside of the class.

Beyond the 'traditional' research methods of reading books and articles, these non-theatre majors were introduced through coaching and modelling to, what I call, 'strategies for creative research.' In several intensive workshops on theatre practice, they were encouraged to sketch rudimentary storyboards that illustrated how each scene in their Living Newspaper project said something specific about the play's social message; to produce portrait galleries, a collection of images that explored their characters' inner psychological qualities; to design soundscapes that uncovered something revelatory about the play's emotional subtext; to create improvisational scenes that revealed how characters' physical actions were sparked by psychological motivations. These explorations involved novice artists actively engaged within 'the context of a community of thinkers...where more than one person is working on the solution of a particular problem or within the particular genre of expression' (Rogoff 1990, 198). Their visual, audible and physical explorations were presented and discussed in class. Strategic imaginative thinking helped students identify and think about complex human relationships. For example, when a music student presented her soundscape with an accompanying selection of abstract images that explored the emotional context of rape, the class immediately was engrossed by the music and the images to the extent that we all invested in a personal journey about the psychological ravages of rape. Because the class experienced such strong empathic responses to an individual caught in the world of sexual assault, students became more interested in the social implications of the subject. Subsequent online discussions referred back to the power of this soundscape and, in some instances, students advocated for this exploration to remain in the final draft of the Living Newspaper Treatment. The soundscape exercise became a significant indication that imaginative exploration generates reflective thought about complex subjects.

At this point of the course, the learning environment moved from the classroom and the library to field study. Inspired by what they were learning through both traditional and non-traditional research activities, students began to pursue another way to think about their play's social issues. This part of their learning journey involved the process of formulating questions for individuals on their interview list. Going into the field and talking to strangers was an intimidating learning activity for some of my students. Jonathan Kaufman, Pulitzer-prize-winning journalist at Northeastern University, provided an additional scaffolding experience. He joined our learning community and addressed their concerns and questions through a learning strategy that resembles apprenticeship. Following Vygotsky's model in which 'social interaction facilitates cognitive development,' this example of a joint problem-solving activity extended the 'existing knowledge and skills' of my students (Rogoff 1990, 141). For example, Kaufman talked about the responsibility of being an 'unbiased journalist' who looks for the 'truth' by asking questions that elicit detail, capture controversial points of view and articulate the intricacies of the issue. Students considered Kaufman's opinions about interviewing, especially in the dramaturgical context of what they already knew about the 1930s Living Newspaper play as well as the theatre of Anna Deavere Smith and Rimini Protokoll. The results of the discussion with a journalism professor made a strategic difference in how my students thought about their signature projects. One student posted that Professor Kaufman made it clear that although interviews include the words of other people, 'it is ultimately the person writing the story (i.e., putting the pieces together) whose voice and perspective is truly being represented.' Another student offered that this encounter with a renowned journalist affected her prior patterned responses to what she thought she knew about journalism. She posted her reflections: 'The best thing about this meeting for me was seeing journalism through fresh eyes.' [I was able] 'to look at journalism as human interaction and storytelling as opposed to what it has become in my eyes: a horrifying burden of dark stories, betraying people, awkward social interactions, and an abusive relationship with the truth.' Ultimately, those students who engaged in original field interviews found the experience to be filled with rich informational material, which provided them with reflective opportunities to develop a deeper understanding of the social and psychological implications inherent in the creation of a Treatment for their authentic Living Newspaper plays.

Traditional inquiry-based methods of research, field interviews and creative explorations provided possibilities for reflective thinking and collaborative learning. Students were coached to select what kind of presentational model best fit the intent of their Living Newspaper project from the three examples offered in Stage One of the class. The possibilities included a contemporary version of a traditional 1930s Living Newspaper play, a form of journalistic theatre inspired by Anna Deavere Smith's 'verbatim theatre' or one of the two Rimini Protokoll examples: a 'Remote X' site-specific walking tour or a living installation as staged in *Situation Rooms*. Several students were inspired by the work of Rimini Protokoll's *Situation Rooms* and devised projects that had audience members move through rooms representing different social situations that captured the essence of the play's topic. One student whose project focused on the objectification of women posted:

> I chose this style because it allows for the push and pull between an audience member being completely immersed in emotion and being able to step back and reflect upon/consider what they just saw. This is crucial in any theatre trying to provoke social change.

Another student chose a combination of Rimini Protokoll's 'Remote X' site-specific theatre and Bertolt Brecht's Lehrstück (learning play) ideas to encourage people to appreciate and think about their current and future relationship with nature. She posted: 'I realized early on in the process of writing this Contemporary Living Newspaper that if I wanted to write a piece about nature then it would be best to get people outside and engaging with nature.' She added that her project 'resembles, a Brechtian Lehrstück' because she would focus on cognitive engagement rather than 'passively sitting and receiving information.'

Throughout Stage Two, students posted relevant material about the substance and organisation of their Living Newspaper play on an ePortfolio website designed especially for the class by my teaching assistant. Instalments included purpose statement, mentor insights, reflections about their conversation with the Journalism professor, questions for their interviewees and audience members, creative explorations (soundscapes), multiple drafts of each project with frequent peer and instructor comments and, eventually, the final Treatment for their Living Newspaper play. The ePortfolio became a metacognitive strategy that was aimed at making cognitive progress through purposeful collaborative monitoring, assessing personal knowledge and producing further metacognitive experiences; as such, this collaborative space facilitated meaningful dialogues and fostered metacognition (See Takayama 2014, 24–6).

Students used the site to post questions to the community of learners, especially about the effectiveness of their current draft. The community responded, offered advice and encouraged risk taking. The collaborative exchange of ideas was successful because, as 'targeted feedback,' it addressed both the content relative to the progress students were making towards achieving their goal of creating a signature work and it directed the students' subsequent practice of incorporating that feedback into further practice (See Ambrose et al. 2010, 141). Feedback gave rise to successive questions, helped students think about changes and additions and mapped the progress of their reflections through each of their iterations. Online discussions became the topics for in-class discussions; both types of exchanges were effective incentives for reflection and multiple drafts.

Students discovered they were not just completing an assignment; instead, they were engaging in the process of learning in resourceful ways. One student commented:

> This is a really unique and wonderful opportunity to learn how to present and understand information in many different ways. It may seem intimidating to come up with an idea for your own piece of documentary theatre, but the class is very collaborative and makes a huge project seem not so intimidating at all. I would encourage everyone to step out of their comfort zone, because this class will change the way you think about theatre and telling stories.

Comments such as this demonstrate the value of HIPs strategies that lead to reflective practice and mindful learning.

Stage Three

The workshop presentation

Students presented a final Treatment Paper of their Living Newspaper projects to the community of learners who critically evaluated the effectiveness of the works in conveying social messages. The final projects consisted of an individual student's 30-minute PowerPoint presentation followed by a 15-minute group discussion.

Their Treatments included information addressing the following sections: Purpose of the Project, Research Questions, Intended Audience, Production Form, Key Characters, Scene or Episode Breakdown, Breakout Scene, Reflection Statement and the Bibliography.

The Breakout Scene (which could be roughly staged) offered the presenter the opportunity to provide the learning community with an in-depth look at one scene, showing how the play's social ideas were created through some of the following examples: photographs that referenced the actual sites of the walking tour, visual diagrams of the site-specific rooms, performed dialogue, videos, slides and sounds. The final Treatment Paper as well as PowerPoint presentations, videos, written dialogue and soundscapes that were mentioned in the students' final oral presentation was posted online. Final presentations were held in a classroom except for one presentation that occurred in the actual location specified in the student's Breakout Scene.

Concluding thoughts

The Living Newspaper project combined HIPs—undergraduate research, inquiry-based, goal-directed exploratory activities, writing-intensive assignments and cross-disciplinary learning environments—with three specific metacognitive strategies—frequent opportunities for reflection, collaborative problem-solving activities within an active environment of learners and guided mentorship. The process of developing a Treatment for a contemporary Living Newspaper involved cognitive and metacognitive strategies as well as the ability to identify diverse (sometimes difficult) social perspectives, to question, to be flexible and to recognise the value of learning how to learn.

Through this process my students gained a new respect for complex social issues by employing some of the professional tools and metacognitive strategies of the journalist and the theatre artist. Furthermore, because my students became intentional, informed and socially responsible learners, they created unmistakably lucid and creatively nuanced Treatments for their Living Newspaper plays. Throughout the course, they were provided with numerous opportunities to take risks, move beyond their comfort zones and reflect on their progress. Therefore, it seems plausible to suggest that by engaging students in HIPs and metacognitive strategies they will continue to incorporate some or many of these methods of learning as a way of thinking and working when approaching life's unscripted problems.

Note

1 For a full description of *Situation Rooms*, see Helgard Haug, Stefan Kaegi, Daniel Wetzel, Situation Rooms: A Multiplayer Video Piece, www.riminiprotokoll.de/website/media/situationrooms/programmhefte/Situation_Rooms_englisch.pdf.

References

Alley Theatre. (2017). 'Remote Houston.' Accessed May 24, 2017. www.alleytheatre.org/plays/remote-houston.

Ambrose, S. A., Michael W. Bridges, Michele DiPietro, Marsha C. Lovett, and Marie K. Norman. (2010). *How Learning Works: Seven Research-Based Principles for Smart Teaching*. San Francisco, CA: Jossey-Bass.

Association of American Colleges and Universities. (2008). *College Learning for the New Global Century: Executive Summary with Employers' Views on Learning Outcomes and Assessment Approaches*. Washington, DC: Association of American Colleges and Universities.

Association of American Colleges and Universities. (2015). *The Leap Challenge: Education for a World of Unscripted Problems*. Washington, DC: Association of American Colleges and Universities.

Brown, J. S., Allan Collins, and Paul Duguid. (1988). 'Situated Cognition and the Culture of Learning.' *Educational Researcher* 18 (1): 32–42.

Dreysse, M., and Florian Malzacher, eds. (2008). *Experts of the Everyday. The Theatre of Rimini Protokoll.* Berlin: Alexander Verlag.

Flanagan, H. (1940). *Arena.* New York: Duell, Sloan, and Pearce.

Flavell, J. (1979). 'Metacognition and Cognitive Monitoring: A New Area of Cognitive-Developmental Inquiry.' *American Psychologist* 34 (10): 906–11.

Hahn, D. (2014). 'Performing Public Spaces, Staging Collective Memory: 50 Kilometres of Files by Rimini Protokoll.' *TDR: The Drama Review* 58 (3): 27–38.

Haug, H., Stefan Kaegi, and Daniel Wetzel. (2013). *Situation Rooms: A Multiplayer Video Piece.* www.riminiprotokoll.de/website/media/situationrooms/programmhefte/Situation_Rooms_englisch.pdf.

Kuh, G. D. (2008). *High-Impact Educational Practices: What They Are, Who Has Access to Them, and Why They Matter.* Washington, DC: Association of American Colleges and Universities.

Malzacher, F. (2010). 'The Scripted Realities of Rimini Protokoll.' In *Dramaturgy of the Real on the World Stage*, edited by Carol Martin, pp. 80–7. Basingstoke: Palgrave Macmillan.

Rogoff, Barbara. (1990). *Apprenticeship in Thinking: Cognitive Development in Social Context.* New York and Oxford: Oxford University Press.

Smith, A. D. (2000). *Talk to Me: Listening Between the Lines.* New York: Random House.

Takayama, K. (2014). 'A Metacognitive Approach to Mapping Collaborative Inquiry through E-Portfolios.' *Peer Review* 16 (1): 24–6. http://aacu.org/peerreview/2014/winter/takayama.

Trueman, M. (2016). 'International: The New Documentary Theatre of Germany's Rimini Protokoll.' *The Stage*, July 7. www.thestage.co.uk/features/2016/international-rimini-protokoll-germany/.

Vygotsky, L. S. (1978). *Mind in Society: The Development of Higher Psychological Processes.* Cambridge, MA: Harvard University.

Wood, D. B. (1993). 'Many Voices in One Mouth.' *The Christian Science Monitor*, July 8. www.csmonitor.com/1993/0708/08121.html.

14

FROM BANANA PHONES TO THE BARD

The developmental psychology of acting

Thalia R. Goldstein

Acting is uniquely human. There is no theatre of chimpanzees, or drama for reptiles. But acting is not a monolithic skill. There are a large number of cognitive, social and emotional skills that are recruited and required in order to act. While some of these skills may be uniquely human and some are shared with other species, all evolved from animal origins. In this chapter, I explore a variety of developing psychological capabilities and their associations with the skills necessary to act and perform theatrical works. I briefly discuss how each skill develops both ontogenetically (i.e., during the child's lifespan) and phylogenically (i.e., how it developed through evolutionary means and what purpose it serves). I also examine whether these skills are increased by affiliation or exposure to engaging in theatre. Finally, I propose a coordinated framework of what it takes, cognitively and emotionally, to be an actor, and how each of these achievements is built upon (or not) in development. My goal is breadth, not depth. Any single skill discussed in this chapter is the topic of extensive research in psychology. Some skills also have large research histories on their connections to play, pretend, imagination, drama and theatre. But as theatre is a complex, intricate, human, cultural endeavour, so is developmental psychology. Matching semantic terms and descriptions is a first step to creating a dialogue between the two fields.

What are the skills required to act? Each acting teacher, theorist and performer may have a different list, and even within those lists, requisite skills may differ depending on role, play, venue and mood of the individual day. Although we do not have a systematic analysis of what happens, psychologically, in acting, and there have been very few experimental studies of the acting classroom (see Chapters 15 and 27), there are some candidate psychological abilities. These are present in every child, and many may have their bases in pretend play, a universal milestone in typical development (Weisberg, 2015). Broadly, psychological skills are typically divided into cognitive skills and social-emotional skills. 'Cognitive' skills include thought processes, such as learning, memory, language and problem solving. 'Social' skills involve engaging with others, understanding of others, behaviour towards others and understanding of self. 'Emotional' skills involve control and understanding of personal behavioural and emotional states. While each type of skill affects all other skills—development does not happen as isolated abilities—each skill is approached here more or less independently for ease of theorising.

The skills presented are not meant to be a complete catalogue of the psychological skills used while acting (and this chapter does not discuss physical or vocal skills), but without these capacities, performance would look very different than the kind of naturalistic, emotional enactment we expect actors to develop and train for performance in Western plays and films. These abilities are presented in the order of preparation for performance: first, the cognitive skills that set the scene, then the social skills to understand character and culture and finally the emotional skills to create and present a work to an audience. All of these capacities change over development: newborns have no concept of separating a fictional world from daily life, for example; toddlers are highly emotionally deregulated; watching young children lie can be quite humorous, as they cannot do so convincingly; and an in-depth understanding of other people's emotions is still incomplete even in adolescence.

Real and pretend: cognitive skills

Cognitive skills allow actors to understand scripts and characters, and to think about the task of acting itself. To begin, theatre cannot exist without an understanding of fiction. Acting requires a cognitive quarantining of the real world from the world of the play, a split of real and fictional, self and character. A lack of this ability to quarantine is dangerous. Even very young children (as well as dogs and kittens, playing) understand quarantine, and engage in play-fighting and soft biting (Smith, 1982). At its most basic level, understanding the reality-fictional boundary begins at the age of 12 months and develops from there. An assumption is often made, but not (yet) supported empirically, that pretend play and acting are psychologically similar. Both involve quarantine (Leslie, 1987), imaginative creation and a separated character. But pretend involves demarcated behaviour (Lillard and Witherington, 2004; Ma and Lillard, 2013, 'What Makes'), while acting is not behaviourally tagged (Goldstein and Bloom, 2011). Pretend play involves over-exaggerated emotions, behaviours and actions. For example, children quickly learn when people are pretending to smile when they do not need to, or when it is incongruous with the situation being pretended (Ma and Lillard, 2006). In much of contemporary acting, the behavioural goal is naturalism or realism.

Pretense does not need to be taught. Even in cultures where pretend is discouraged or banned (Carlson, Taylor, and Levin, 1998), children still begin pretending around 12–18 months old. In fact, a lack of pretend is considered an early symptom of later developmental disorders (Jarrold, 2003). While most developmental textbooks and theories claim pretend play ends around the ages of 6–8 years, other research has found children continue to engage in pretend play through age 11 years or even later (Smith and Lillard, 2012). The nature of play changes and shifts over time. Children begin by using real objects in their world, as well as themselves as literal players, such as pretending to brush their own hair with a plastic hairbrush, or playing 'mommy' with a realistic baby doll. But over the preschool years, children begin to engage in symbolic substitution, using props and objects that may not be as characteristic of the real object, and themselves as characters further from their own lived experiences. At its most advanced meta-representational level, pretend play involves children undergoing transformations into animals or characters, or playing multiple characters in one pretend play session, using 'air' as props (e.g., imagined fish and pole while fishing, possibly a precursor to mime) (Hanline et al., 2008).

Research across psychology, anthropology and sociology has found that pretend play is associated with a variety of other developing skills, including social understanding (e.g., Youngblade and Dunn, 1995), executive functioning (e.g., Carlson and White, 2013; Taylor et al., 2009) and language (e.g., Bergen, 2002). However, whether pretense causes such gains still requires more careful research (Lillard et al., 2013). The young of many other species play as a way to learn important skills, such as hunting and fighting (Bekoff and Byers, 2011). Evolutionarily, play may be the best recourse for learning critical survival skills (Sandseter and Kennair, 2011) as well as cultural norms and rituals (Legare and Harris, 2016; Tomasello, 2016). But pretend play is different from acting. Acting is sustained, and typically involves narrative, plot and arc. There is a goal of final or formal performance, which pretend play lacks (Weisberg, 2015). Pretend play seems more likely to be a way for children to learn or explore new ideas (Henricks, 2015; Gopnik and Walker, 2013; Weisberg et al., 2013). Acting may build on skills and constructs in pretend play, but it is not pretend play.

As part of pretend play and playing a character in a show, actors must engage in a more basic cognitive task: task switching between the needs of the play and audience, and the needs of the self as performer. They must engage in meta-representation of the world of the play and the world outside the play. This ability, whether in pretend play or a theatrical play, relies on a developing executive function. Executive function is the ability to resist acting impulsively, select where to focus attention and adapt to changing circumstances (Diamond, 2013). In rehearsal and performance, actors must be able to keep in mind their lines, the emotional state of the character, their behaviour, their movements around the stage, the placement of the audience, and so forth. Children develop their executive function beginning in preschool, and continue to gain through adolescence and adulthood. This ability underlies most higher-order cognitive skills leading to better school readiness (Blair, 2002), academic success (Fitzpatrick et al., 2014), job success (Bailey, 2007) and even physical health (Stautz et al., 2016). There is some evidence that pretend play, and particularly engaging in high-fantasy pretend play (i.e., less realistic and more fantastical content), can help train and is associated with higher levels of executive function, particularly the ability to task switch (Berk and Meyers, 2013; Carlson, White, and Davis-Unger, 2014; Thibodeau et al., 2016). Without executive function, almost any task in acting would become impossible. Obviously, these types of planning and attention functioning are critical for evolutionary development, as they allow animals to adapt to changes in the environment. The focus of executive functioning for creative pursuits and art, however, is uniquely human.

Another set of cognitive skills necessary to act, which build on executive functions but are more cognitively complex, include memory, language and the ability to lie. Any engagement with scripted work is going to require memory and language abilities, core cognitive capacities that progress over development. However, most actors would not consider memory to be a distinct skill that makes them better at acting, but rather a skill required in order to engage with scripted work. There is extensive research on actors' cognitive techniques to best remember their scripts, as well as how these techniques can be used to help older adults increase their memories. (See Noice and Noice, Chapter 27, 'A Theatrical Intervention to Lower the Risk of Alzheimer's and Other Forms of Dementia' in this volume.) The basic findings from this body of work are that it is the embodiment of concepts – a focus on goals, objectives and tactics, rather than just focusing on memorisation – that causes higher levels of learning. However, this work does not investigate the developmental trajectory of memory for lines within children's pretend play or acting. Children's memory, as one would expect, gets better over time, correlated with their improved executive functioning (Zelazo et al., 2003). At the same time, children have a better memory for rhymes and songs than even adults (Király et al., 2017)!

More than memory, interpretation and understanding of the meaning of language—its subtleties and exact definitions—is central to performing. It is also the area in which educational theatre and curricula-integrated drama is most well researched, with the strongest evidence for transfer (where experience within theatre and drama leads to skills outside of theatre and drama). In a large meta-analysis (a study of studies), children who participated in drama over a range of ages were found to have higher levels of verbal skills across a wide range of verbal outcomes, including reading comprehension, vocabulary level and writing, than children who did not participate in drama (Podlozny, 2000). Verbal development enables understanding of scripts, how best to perform both the words on the page and, critically, the subtext, or meaning underneath the lines. Yet children's understanding of metaphor develops slowly and similarly to their understanding of lying, as both are forms of representational language (Winner, 1997). Children comprehend the non-literal use of language, whether lying, metaphor or sarcasm, at around eight years old (Demorest et al., 1983). As young children cannot understand non-literal text, they might have difficulties interpreting scripts.

There is no question that all forms of language are a singularly human ability. Language requires representation, as words do not usually sound like what they mean; only onomatopoetic sound words such as 'crash' and 'bang' have that distinction. While there have been a few animals who can understand language (such as Koko the signing Gorilla [Patterson, 1978], or dogs who understand commands [Kaminski, Call, and Fischer, 2004; Markman and Abelev, 2004], no other animal produces language or communicates through spoken or written phrases).

Finally, lying as a cognitive skill is directly related to meta-representation and shares several important cognitive qualities with acting. When lying, the liar's goal is to convince the people watching them that the emotional and behavioural experience they are having is real. But a major difference between lying and acting is that lying involves only the liar knowing that what is happening is not true, while in acting, both audience and actor understand that what is happening on stage is part of a play (Goldstein and Bloom, 2011). Lying requires three separate cognitive components: an understanding of the liar's own beliefs and desires (also perhaps necessary in acting), an understanding of the beliefs and desires of the other person (similar to understanding a character) and the ability to engage in the type of behaviour that will make the other person believe what the liar is telling them (the actual performance of acting). Even semantically, we often confuse the terms lying and acting in everyday use.

There are multiple other animals (and even plants) in our evolutionary line that engage in deception. When discussing the evolution of lying, however, the question is whether 'lying' requires conscious and purposeful deception of others, or if automatic deceptive behaviour can be considered lying (Whiten and Byrne, 1988). One might argue that camouflage or mimicry (e.g., a stick insect that looks like a stick) is a form of deception, but is not lying (Wickler, 1968).

Children learn to lie slowly and in steps, depending on the type of lie. Antisocial lies develop early, at around the age of three, although it takes a long time for children to be able to lie without giving away their lies (until at least seven years old) (Talwar and Lee, 2002 'Development of lying'). Prosocial lies (e.g., white lies), which are more culturally imbued, develop later, although children still cannot back up false statements with additional false statements, engaging in 'semantic leakage' in which they just cannot help but tell the truth (Xu et al., 2010; Talwar and Lee, 2008 'Little liars'; Talwar and Lee, 2008 'Social and cognitive correlates').

Lying depends on a theory of mind (the ability to understand another's beliefs, desires or intentions), so much so that training a child on theory of mind will then cause them to try out lying (Ding et al., 2015)! However, theory of mind is a cognitive capacity more central to social abilities. Thus, I now turn to the social and emotional skills that allow actors to produce characters, and how the development of such skills allows or prevents children from being able to act.

Self and other: social skills

Actors must understand and interact with others, understand themselves in social contexts, portray facial, vocal and physical states read by an audience and understand the emotions, goals, intentions, beliefs, personalities and backgrounds of their characters. Psychologically, before an actor can engage socially, the first step to perform a character is the separation of presentational self from authentic self. This can occur either through a separation of real world from pretend world, and then further character creation within the pretend world, or as a decision to perform the self for presentation (such as comics and hosts who are not inhabiting a character, but are inhabiting a performative self). Actors must understand themselves before they can portray someone else: they must understand that they exist as a separate self from other people, and how to control their physical and vocal self in order to portray states and traits to others. This begins developmentally around the age of 18 months: when toddlers can first begin to recognise themselves in the mirror (Amsterdam, 1972). Being able to recognise oneself in a mirror means you see yourself as a unique individual, a 'self' separate from the world around you, that remains consistent in appearance over time (Asendorpf and Baudonnière, 1993). However, this ability is not unique to humans. Chimpanzees (Gallup, 1970), bottlenose dolphins (Reiss and Marino, 2001), magpies (Prior, Schwarz, and Güntürkün, 2008) and elephants (Plotnik, De Waal, and Reiss, 2006) all recognise themselves.

Beyond mere recognition, creating an identity, a sense of self as part of the larger culture, is a developmental progression that continues through adolescence and into adulthood (Erikson, 1968; Marcia, 1966). Beginning in adolescence, individuals go through an exploration process, in which they reevaluate past goals and explore other options for future goals, before committing to an identity going forward. Both elements of exploration (i.e., looking at the various options for identities) and commitment (firmly deciding the facets of identity they want to create) are required for healthy understanding of self (Dunkel and Anthis, 2001). While animals may have a basic sense of self as an individual within a group, we do not know if animals have a sense of self as a distinct personality or identity.

All individuals inhabit different roles in different areas of their lives, and some may feel more or less authentic to an actual self (Sheldon et al., 1997; Goffman, 1978). While acting, it is psychologically unknown whether a separation between self and character is necessary, but at the very least, an actor who cannot separate the character he or she is portraying from his or her own life at all is going to have trouble acting. Even with psychological separation, it is possible that the actors' self-perceptions of their personalities change as a result of engaging with the personality of a character (Hannah et al., 1994). The supposed merging of self and other when acting is regularly part of interviews with actors. Audience members, child and adult, believe this merging is happening (Goldstein and Bloom, 2015 'Characterizing characters'; Goldstein and Bloom, 2015 'Is It Oscar-Worthy?'; Goldstein and Filipe, 2017). Yet we have very little systematic research. The idea of blending or changing of selves may be unique to humans, evolutionarily. While other animals can deceive, the act of putting on a presented personality, consciously, for the enjoyment of an audience, is uniquely human.

As children gain a firmer understanding of the self, they also begin to understand, orient and react to others. Acting requires an understanding of the character being portrayed, deeply enough that a whole portrayal can be constructed. This requires a psychological way of thinking about others: a theory of mind. Theory of mind is the ability to understand another's beliefs, desires and intentions (Wimmer and Perner, 1983). This is related to the concept of cognitive empathy, which is the ability to understand what someone else is feeling (Strayer, 1987), as well as emotional empathy, which is the matching of someone else's feelings (Shamay-Tsoory, Aharon-Peretz, and Perry, 2009). But theory of mind is separate from compassion and sympathy, feeling badly or sorry for someone else (without necessarily matching what they are feeling) (Davis, 1983; Coplan, 2011). Parallel developments of executive function, arousal regulation and language in preschool-aged children underlie broader social orientations and the development of appropriate responding to others (Blair, 2002). Acting requires creating meaning out of behaviours—understanding and expressing the subtext of a script. A better understanding of the real social world and people, therefore, should produce a better understanding of a play's social world and characters (although there is not psychological research to support such a statement).

Fiction generally has been theorised as an abstraction and simulation of the social world (Mar and Oatley, 2008), and actually embodying and interpreting such simulation may allow actors to practice and gain social skills. In fact, research with 8- and 14-year-olds has shown that a year of acting classes (as compared to a year of visual arts or music classes) can increase empathy and theory of mind (Goldstein and Winner, 2012). College-aged actors also show higher levels of theory of mind than psychology majors (Goldstein, Wu, and Winner, 2009) and professional actors recall higher childhood orientation to other people than lawyers (Goldstein and Winner, 2009) and self-report higher levels of empathy (Nettle, 2006). The idea that actors have higher social skills than non-actors has received theoretical support (Levy, 1997; Verducci, 2000; Metcalf, 1931), while the opposite conclusion of actors as selfish and narcissistic also persists (Dufner et al., 2015; Young and Pinsky, 2006).

Yet acting requires more than theory of mind about the characters' current state or an emotional matching with a character. Understanding characters requires a broad view of human desire, belief, intention and behaviour, beyond momentary emotions and intentions. Good actors (as compared to bad) may have a broader understanding of personality, how personalities work within situations and both immediate and superordinate goals. Comparatively, non-human primates, our closest evolutionary relatives, have a relatively impoverished view of conspecifics (i.e., other primates of the same species). While even very young children (perhaps as young as 18 months) can form 'representational' awareness of another's beliefs (i.e., that others may have a false belief about the status of the world; Baillargeon, Scott, and He, 2010; Onishi and Baillargeon, 2005), non-human primates cannot do this. Chimpanzees can only track what another chimpanzee knows about the world through where they have looked and what they have seen. Rhesus macaques show the same specific ability. This means that apes and monkeys can represent true beliefs in others, but not false ones (Martin and Santos, 2016). It is this distinction that may be critical for pretend play, fiction, theatre and even language. And while children's understanding of false beliefs in others may be developed by age 5 (Wellman, Cross, and Watson, 2001), a full understanding of how others' behaviours are representations of their inner states continues to develop over the lifespan (Kuhn, 2000).

Once actors have a clear understanding of themselves, the character they must portray and the other characters in the play, they must then discern how to present themselves to be read by an audience – the understanding of self as a social skill. This requires behavioural control and emotional control and regulation, as well as physical and vocal control.

The truth of acting: self-regulation and emotion regulation

How does an understanding of a character become a fully realised portrayal? Through physical, vocal, behavioural and emotional regulation and change. Actors must be able to regulate (the psychological term for both internal and external control) themselves in two ways: managing their own emotions and behaviours to become a character, and managing the character's emotions and behaviours to engage in portrayal. While theorising about the nature and strength of this regulation from the perspective of acting theory is beyond the scope of this chapter, there are a few theoretical possibilities that lead to different psychological predictions. In the most extreme cases, the actor has to completely suppress his or her own emotional state and up-regulate the emotional state of the character in order to fully experience and therefore portray the character, using whatever methods (e.g., sense memory, imaginative 'as if,' dissociation; Hull, 1985; Hagen, 2009; Thomson and Jaque, 2011) she or he can employ. On the other end of the continuum, the actor needs to be aware of his or her current emotional state and where it intersects with the character, but actual emotional experience of the character is unnecessary. Somewhere in the middle, then, is both some down-regulation of the emotional and physical state of the actor and an up-regulation of the emotional and physical state of the character.

Emotional and self-regulation may be the factor that precludes children from being able to act well. When children are determining their own topics and emotional contents of pretend play, they can simply play how they are already feeling (Fein, 1981). But acting has external requirements of script and scene. Actors must change their emotional state, and within the world of a play, from scene to scene, frequently must do so rapidly. In interviews, actors often speak of the emotional challenges of acting, and discuss feeling exhausted or worn out by the end of a performance (e.g., Gross, 2014). Even for an actor who is operating only with physical and behavioural regulation, with no emotional attachment to their performance at all, the task of making the audience believe that the character is undergoing the emotional events of the play is difficult. Adding in emotional connection to the material makes it more difficult.

Children's emotion regulation develops from being other-oriented (i.e., dependent on a caregiver to help with regulation) to self-oriented (Cole, Martin, and Dennis, 2004). Individual abilities in self-regulation show continuity throughout the lifespan (Rothbart et al., 2011), and as cognitive skills develop, children are able to engage in more complex (Shields and Cicchetti, 1997) and adult-like emotion regulation techniques. The use of different regulation strategies, dependent on situation, also develops through middle childhood (Saarni, 2000), but adolescents are still developing their abilities to control their emotions and emotional behaviours, as well as learning to use more healthy and adaptive strategies (Zeman et al., 2006).

Previous work focused on lifetime strategies of emotion regulation have concluded it is more healthful to engage in cognitive reappraisal (i.e., reframing) of emotional experiences, rather than suppression (i.e., non-expression) of emotional experiences (Gross and John, 2003). When children and adolescents are enrolled in acting classes, they show lower levels of suppression of emotions (although not higher levels of cognitive reappraisal) than children and adolescents in visual arts or music classes (Goldstein, Tamir, and Winner, 2013). It may be that as children and adolescents create symbolic events in acting classes that cause them to have emotions, they can learn to modify those emotions in the symbolic experience of the play, which then transfers to the real world (Bretherton, 1989).

There is some causal evidence that engaging in drama games and play leads to better emotional control and self-understanding. Goldstein and Lerner (2017) found that for four-year-old children enrolled in Head Start preschool, eight weeks of guided dramatic pretend play games (i.e., Spolin-type warm up games) led to better levels of emotional control, when compared to engaging in either block play activities or story time. In another study, DeBettingnies and Goldstein (under review) found that ten-year-old children who engaged in Spolin-like drama games increased their self-concept, when compared to ten-year-old children who spent time in a study hall. However, these results were specific only to those children who began with the lowest level of self-concept. It may be that drama games help emotional control and self-awareness, but only if the participants have lower than average levels of functioning (e.g., as children from lower-income households often are [McClelland, Acock, and Morrison, 2006]).

Tying these areas together, emotion regulation is dependent on social skills such as emotional understanding, and cognitive skills, such as executive function. Each set of skills develops in conjunction with other skills, leading to a coherent whole.

Development and evolution: a coordinated framework of acting

When a child picks up a banana, puts it to her ear as a phone and says 'hello, Gramma!' how similar are the cognitive, social and emotional skills she is using to those needed by the West End actor who steps to centre stage and begins 'To be, or not to be – that is the question'? In the same way that all humans develop a set of cognitive and social skills such as understanding of the self, understanding of others, language and memory to reach adulthood, so do actors refine and develop their required skills over a lifespan. But until psychology has a firm grasp on the cognitive, social and emotional skills necessary to become a talented adult actor, it is hard to pinpoint which skills in childhood are most conducive to being an actor, and whether children can act well and similarly to adults, given enough experience and ability with these skills. In developmental psychology, social, emotional and behavioural abilities are coordinated and a global level of ability in these skills is often associated with positive impact on development (Durlak et al., 2011). A child who is gifted in one area (e.g., cognitive) will often also be gifted in other areas, such as social and emotional development (Reis and Renzulli, 2004).

Beyond ontological development, it is important to remember that it is hard to provide strong evidence for evolutionary theories (Buss, 2005). However, humans are the only species to create theatre. The evolution of meta-representational skills, linguistic skills and pretense seems to be uniquely human. Many species are social. Dogs, apes, monkeys, fish and even some snakes and birds all live socially. Therefore, the skills that make humans socially adept, understanding others' intentions, engaging with behaviours of others and managing their own behaviour to fit the social patterns of others, must either have come from common ancestry with many other species, or, all of these examples of social groups and social skills are evidence of convergent evolution (van Schaik and Burkart, 2011).

Can any child with a standard set of cognitive and social abilities learn how to act? Theoretically, yes. Acting requires sets of skills. First, cognitive, to understand the language and symbolism of a play, have memory for lines and understand the representational nature of fiction. Then, social, to understand the people of the play (after all theatre is about people). And finally, emotional, to create a performance. All of these must be brought together with vocal and physical control, to create a presentation of emotional life to be read as real by an audience. Each of these areas includes multiple interrelated skills, including theory of mind, meta-representation and behavioural regulation. None are sufficient for acting, and with perhaps the exception of the ability to understand language, none may be necessary either. But some combination of some of these skills *is* necessary for acting. As interdisciplinary conversations between psychology, cognitive science and theatre and the humanities move forward, distinguishing which skills, when and how will be a primary task to allow for conversation and understanding of what it means to act.

Acting training and theatre classes, of course, go through their own developmental progressions. Acting teachers, through training and instinct, learn to adjust their classes to the developmental abilities and levels of the children in front of them, without necessarily knowing about the psychological capabilities that go into the classes. Drama games for early childhood classes look very different from drama games for adolescent actors, which are then also different from complex scene study and character study (Spolin, 1986). Having access and understanding to the multiple levels of developmental complexity and knowledge within acting is important.

To conclude, one question completely unexplored by this chapter is the psychological difference between a bad actor and a good one. Is it simply training? Underlying capabilities such as executive function or imitation? A combination of time, place and social skills? Ability to dissociate from the current world and engage in the world of the play? Certainly, someone who cannot remember lines, does not understand the character or cannot express emotion onstage is at the far end of the continuum of a 'bad' actor. Yet characteristics that make up a great actor, other than 'believability' or 'expression,' are under-defined. While giftedness in math or reading may be relatively easy to isolate, psychology does not have a definition of what meta-representational talent (for example) could look like. Without working knowledge of how good acting happens, it is hard to pinpoint which developmental processes may lead to being a good actor. This field is ripe for study, definition and experimentation. Theorists and practitioners interested in children's developing skills, and how such skills are applied to theatre and theatre-adjacent activities throughout the lifespan have a fertile field to work in. In whichever area of theatre people are interested, they can find a psychological concept to explore. Because in the end, theatre and psychology are interested in the same questions: 'Why do we do the things we do?' and 'What makes us human?'

References

Amsterdam, Beulah. 'Mirror Self-Image Reactions before Age Two.' *Developmental Psychobiology* 5 (1972): 297–305.

Asendorpf, Jens, and Pierre-Marie Baudonnière. 'Self-Awareness and Other-Awareness: Mirror Self-Recognition and Synchronic Imitation among Unfamiliar Peers.' *Developmental Psychology* 29.1 (1993): 88.

Bailey, Charles. 'Cognitive Accuracy and Intelligent Executive Function in the Brain and in Business.' *Annals of the New York Academy of Sciences* 1118.1 (2007): 122–41.

Baillargeon, Renée, Rose Scott, and Zijing He. 'False-Belief Understanding in Infants.' *Trends in Cognitive Sciences* 14.3 (2010): 110–18.

Bekoff, Marc, and John Byers. *Animal Play*. Cambridge: Cambridge University Press, 2011.

Bergen, Doris. The Role of Pretend Play in Children's Cognitive Development. *Early Childhood Research and Practice: An Internet Journal on the Development, Care and Education of Young Children*. 4.1 2002.

Berk, Laura, and Adena Meyers. 'The Role of Make-Believe Play in the Development of Executive Function: Status of Research and Future Directions.' *American Journal of Play* 6.1 (2013): 98.

Blair, Clancy. 'School Readiness: Integrating Cognition and Emotion in a Neurobiological Conceptualization of Children's Functioning at School Entry.' *American Psychologist* 57.2 (2002): 111.

Bretherton, Inge. 'Pretense: The Form and Function of Make-Believe Play.' *Developmental Review* 9.4 (1989): 383–401.

Buss, David. *The Handbook of Evolutionary Psychology*. Hoboken, NJ: Wiley, 2005.

Carlson, Stephanie, and Rachel White. 'Executive Function, Pretend Play, and Imagination.' In Marjorie Taylor (ed). *The Oxford Handbook of the Development of Imagination*. New York: Oxford University Press. (2013): 161–74.

Carlson, Stephanie, Marjorie Taylor, and Gerald Levin. 'The Influence of Culture on Pretend Play: The Case of Mennonite Children.' *Merrill-Palmer Quarterly* (1998): 538–65.

Carlson, Stephanie, Rachel White, and Angela Davis-Unger. 'Evidence for a Relation between Executive Function and Pretense Representation in Preschool Children.' *Cognitive Development* 29 (2014): 1–16.

Cole, Pamela M., Sarah E. Martin, and Tracy A. Dennis. 'Emotion Regulation as a Scientific Construct: Methodological Challenges and Directions for Child Development Research.' *Child Development* 75.2 (2004): 317–33.

Coplan, Amy. 'Will the Real Empathy Please Stand Up? A Case for a Narrow Conceptualization.' *The Southern Journal of Philosophy* 49.s1 (2011): 40–65.

Davis, Mark H. 'Measuring Individual Differences in Empathy: Evidence for a Multidimensional Approach.' *Journal of Personality and Social psychology* 44.1 (1983): 113–26.

DeBettingnies, Brooke, and Thalia Goldstein. Improvisational Theatre and Children's Self Concept. *Psychology of Aesthetics, Creativity and the Arts* (under revised review).

Demorest, Amy, Lisa Silberstein, Howard Gardner, and Ellen Winner. 'Telling It as It Isn't: Children's Understanding of Figurative Language.' *British Journal of Developmental Psychology* 1.2 (June, 1983): 121–34. *CoINs.*

Diamond, Adele. 'Executive Functions.' *Annual Review of Psychology* 64 (2013): 135–68.

Ding, Xiao, Henry Wellman, Yu Wang, Genyue Fu, and Kang Lee. 'Theory-of-Mind Training Causes Honest Young Children to Lie.' *Psychological Science* 26.11 (2015) 1812-1821.

Dufner, Michael, Boris Egloff, Christoph Hausmann, Lisa-Maria Wendland, Franz Neyer, and Mitja Back. 'Narcissistic Tendencies among Actors: Craving for Admiration, but Not at the Cost of Others.' *Social Psychological and Personality Science* 6 (2015): 447–54.

Dunkel, Curt, and Kristine Anthis. 'The Role of Possible Selves in Identity Formation: A Short-term Longitudinal Study.' *Journal of Adolescence* 24.6 (2001): 765–76.

Durlak, Joseph, Roger Weissberg, Allison Dymnicki, Rebecca Taylor, and Kriston Schellinger. 'The Impact of Enhancing Students Social and Emotional Learning: A Meta-Analysis of School-Based Universal Interventions.' *Child Development* 82.1 (2011): 405–32.

Erikson, Erik. *Identity: Youth and Crisis*. New York: WW. Norton, 1968.

Fein, F. 'Pretend Play in Childhood: An Integrative Review.' *Child Development* (1981): 1095–118.

Fitzpatrick, Caroline, Rachel McKinnon, Clancy Blair, and Michael Willoughby. 'Do Preschool Executive Function Skills Explain the School Readiness Gap between Advantaged and Disadvantaged Children?' *Learning and Instruction* 30 (2014): 25–31.

Gallup, Gordon. 'Chimpanzees: Self-Recognition.' *Science* 167.3914 (1970): 86–7.

Goffman, Erving. *The Presentation of Self in Everyday Life*. London: Harmondsworth, 1978.

Goldstein, Thalia R., and Paul Bloom. 'The Mind on Stage: Why Cognitive Scientists Should Study Acting.' *Trends in Cognitive Sciences* 15 (2011): 141–142.

Goldstein, Thalia R., and Paul Bloom. 'Characterizing Characters: How Children Make Sense of Realistic Acting.' *Cognitive Development* 34 (2015): 39–50.

Goldstein, Thalia R., and Paul Bloom. 'Is It Oscar-Worthy? Children's Meta-Representational Understanding of Acting.' *PloS One* 10.3 (2015): e0119604.

Goldstein, Thalia R., and Aline Filipe. 'The Interpreted Mind: Understanding Acting'. *Review of General Psychology* 22 (2018): 220–229.

Goldstein, Thalia R, and Matthew Lerner. 'Dramatic Pretend Play Games Uniquely Improve Emotional Control in Young Children.' *Developmental Science* 21 (2018): e12603.

Goldstein, Thalia R, Maya Tamir, and Ellen Winner. 'Expressive Suppression and Acting Classes.' *Psychology of Aesthetics, Creativity, and the Arts* 7.2 (2013): 191.

Goldstein, Thalia R., and Ellen Winner. 'Living in Alternative and Inner Worlds: Early Signs of Acting Talent.' *Creativity Research Journal* 21.1 (2009): 117–24.

Goldstein, Thalia R, and Ellen Winner. 'Enhancing Empathy and Theory of Mind.' *Journal of Cognition and Development* 13.1 (2012): 19–37.

Goldstein, Thalia R., Katherine Wu, and Ellen Winner. 'Actors Are Skilled in Theory of Mind but Not Empathy.' *Imagination, Cognition and Personality* 29.2 (2009): 115–33.

Gopnik, Alison, and Caren Walker. 'Considering Counterfactuals: The Relationship between Causal Learning and Pretend Play.' *American Journal of Play* 6.1 (2013): 15.

Gross, T. Philip Seymour Hoffman on Acting: An 'Exhausting' and 'Satisfying' Art. *Fresh Air* (2014). www.npr.org/2014/02/03/270954011/philip-seymour-hoffman-on-acting-an-exhausting-and-satisfying-art.

Gross, J., and Oliver John. 'Individual Differences in Two Emotion Regulation Processes: Implications for Affect, Relationships, and Well-being.' *Journal of Personality and Social Psychology* 85.2 (2003): 348–62.

Hagen, Uta. *Respect for Acting*. New York: Wiley, 2009.

Hanline, Mary Frances, Sande Milton, and Pamela Phelps. 'A Longitudinal Study Exploring the Relationship of Representational Levels of Three Aspects of Preschool Sociodramatic Play and Early Academic Skills.' *Journal of Research in Childhood Education* 23.1 (2008): 19–28.

Hannah, M., G. Domino, R. Hanson, and W. Hannah. 'Acting and Personality Change: The Measurement of Change in Self-Perceived Personality Characteristics during the Actor's Character Development Process.' *Journal of Research in Personality* 28 (1994): 277–86.

Henricks, Thomas. 'Play as Experience.' *American Journal of Play* 8.2 (2015): 18–49.

Hull, S. *Strasberg's Method as Taught by Lorrie Hull: A Practical Guide for Actors, Teachers, and Directors*. Woodbridge, CT: Ox Bow Pub., 1985.

Jarrold, Christopher. 'A Review of Research into Pretend Play in Autism.' *Autism* 7.4 (2003): 379–90.

Kaminski, Juliane, Josep Call, and Julia Fischer. 'Word Learning in a Domestic Dog: Evidence For 'Fast Mapping'.' *Science* 304.5677 (2004): 1682–3.

Király, Ildikó, Szilvia Takacs, Zsuzsa Kaldy, and Erik Blaser. 'Preschoolers Have Better Long-Term Memory for Rhyming Text Than Adults.' *Developmental Science* 20.3 (2017) e12398.

Kuhn, Deanna. 'Theory of Mind, Metacognition, and Reasoning: A Life-Span Perspective.' In Peter Mitchell and Kevin J Riggs (eds). *Children's Reasoning and the Mind* (2000): 301–326.

Legare, Cristine, and Paul Harris. 'The Ontogeny of Cultural Learning.' *Child Development* 87.3 (2016): 633–42.

Leslie, Alan. 'Pretense and Representation: The Origins of Theory of Mind.' *Psychological Review* 94.4 (1987): 412–26.

Levy, J. 'Theatre and Moral Education.' *Journal of Aesthetic Education* 31.3 (1997): 65–75.

Lillard, Angeline, and David Witherington. 'Mothers' Behaviour Modifications during Pretense and Their Possible Signal Value for Toddlers.' *Developmental Psychology* 40.1 (2004): 95–113.

Lillard, Angeline, Matthew Lerner, Emily Hopkins, Rebecca Dore, Eric Smith, and Carolyn Palmquist. 'The Impact of Pretend Play on Children's Development: A Review of the Evidence.' *Psychological Bulletin* 139.1 (2013): 1–34.

Ma, Lili, and Angeline Lillard. 'Where Is the Real Cheese? Young Children's Ability to Discriminate between Real and Pretend Acts.' *Child Development* 77.6 (2006): 1762–77.

Ma, Lili, and Angeline Lillard. 'What Makes an Act a Pretense One? Young Childrens Pretend-Real Judgments and Explanations.' *Child Development Research* (2013) e467872. doi:10.1155/2013/467872.

Mar, Raymond, and K. Oatley. 'The Function of Fiction Is the Abstraction and Simulation of Social Experience.' *Perspectives on Psychological Science* 3.3 (2008): 173.

Marcia, James. 'Development and Validation of Ego-Identity Status.' *Journal of Personality and Social Psychology* 3.5 (1966): 551.

Markman, Ellen, and Maxim Abelev. 'Word Learning in Dogs?' *Trends in Cognitive Sciences* 8.11 (2004): 479–81.

Martin, Alia, and Laurie Santos. 'What Cognitive Representations Support Primate Theory of Mind?' *Trends in Cognitive Sciences* 20.5 (2016): 375–82.

McClelland, Megan, Alan Acock, and Frederick Morrison. 'The Impact of Kindergarten Learning-Related Skills on Academic Trajectories at the End of Elementary School.' *Early Childhood Research Quarterly* 21.4 (2006): 471–90.

Metcalf, John. 'Empathy and the Actor's Emotion.' *The Journal of Social Psychology* 2.2 (1931): 235–8.

Nettle, D. 'Psychological Profiles of Professional Actors.' *Personality and Individual Differences* 40.2 (2006): 375–83.

Noice, Helga, and Anthony Noice, 'A Theatrical Intervention to Lower the Risk of Alzheimer's and Other Forms of Dementia' in this volume, 280–90.

Onishi, K., and Renee Baillargeon. 'Do 15-Month-Old Infants Understand False Beliefs?' *Science* 308.5719 (2005): 255.

Patterson, Francine 'The Gestures of a Gorilla: Language Acquisition in Another Pongid.' *Brain and Language* 5.1 (1978): 72–97.

Plotnik, Joshua, Frans DeWaal, and Diana Reiss. 'Self-Recognition in an Asian Elephant.' *Proceedings of the National Academy of Sciences* 103.45 (2006): 17053–7.

Podlozny, A. 'Strengthening Verbal Skills through the Use of Classroom Drama: A Clear Link.' *Journal of Aesthetic Education* 34.3/4 (2000): 239–75.

Prior, Helmut, Ariane Schwarz, and Onur Güntürkün. 'Mirror-Induced Behaviour in the Magpie (Pica Pica): Evidence of Self-Recognition.' *PLoS Biol* 6.8 (2008): e202.

Reis, Sally, and Joseph Renzulli. 'Current Research on the Social and Emotional Development of Gifted and Talented Students: Good News and Future Possibilities.' *Psychology in the Schools* 41.1 (2004): 119–30.

Reiss, Diana, and Lori Marino. 'Mirror Self-Recognition in the Bottlenose Dolphin: A Case of Cognitive Convergence.' *Proceedings of the National Academy of Sciences* 98.10 (2001): 5937–42.

Rothbart, Mary, Brad Sheese, Rosario Rueda, and Michael Posner. 'Developing Mechanisms of Self-Regulation in Early Life.' *Emotion Review* 3.2 (2011): 207–13.

Saarni, Carolyn. 'Emotional Competence: A Developmental Perspective.' In R. Bar-On and J. D. A. Parker (eds). *The Handbook of Emotional Intelligence: Theory, Development, Assessment, and Application at Home, School, and in the Workplace.* San Francisco, CA: Jossey-Bass. (2000): 68–91.

Sandseter, Ellen Hansen, and Leif Ottesen Kennair. 'Children's Risky Play from an Evolutionary Perspective: The Anti-Phobic Effects of Thrilling Experiences.' *Evolutionary Psychology* 9.2 (2011): 147470491100900212.

Shamay-Tsoory, S., J. Aharon-Peretz, and D. Perry. 'Two Systems for Empathy: A Double Dissociation between Emotional and Cognitive Empathy in Inferior Frontal Gyrus versus Ventromedial Prefrontal Lesions.' *Brain* 132.3 (2009): 617.

Sheldon, Kennon, Richard Ryan, Laird J. Rawsthorne, and Barbara Ilardi. 'Trait Self and True Self: Cross-Role Variation in the Big-Five Personality Traits and Its Relations with Psychological Authenticity and Subjective Well-being.' *Journal of Personality and Social Psychology* 73.6 (1997): 1380.

Shields, Ann, and Dante Cicchetti. 'Emotion Regulation among School-Age Children: The Development and Validation of a New Criterion Q-sort Scale.' *Developmental Psychology* 33.6 (1997): 906.

Smith, E., and Angeline S. Lillard. 'Play on: Retrospective Reports of the Persistence of Pretend Play into Middle Childhood.' *Journal of Cognition and Development* 13.4 (2012): 524–49.

Smith, Peter 'Does Play Matter? Functional and Evolutionary Aspects of Animal and Human Play.' *Behavioural and Brain Sciences* 5.1 (1982): 139–55.

Spolin, Viola. *Theater Games for the Classroom: A Teacher's Handbook.* Chicago, IL: Northwestern University Press, 1986.

Stautz, Kaidy, Rachel Pechey, Dominique-Laurent Couturier, Ian Deary, and Theresa M. Marteau. 'Do Executive Function and Impulsivity Predict Adolescent Health Behaviour after Accounting for Intelligence? Findings from the ALSPAC Cohort.' *PLoS One* 11.8 (2016): e0160512.

Strayer, J. 'Affective and Cognitive Perspectives on Empathy.' In Nancy Eisenberg and J. Strayer (eds). *Empathy and Its Development.* Cambridge: Cambridge University Press. (1987): 218–44.

Talwar, V., and Kang Lee. 'Development of Lying to Conceal a Transgression: Children's Control of Expressive Behaviour during Verbal Deception.' *International Journal of Behavioural Development* 26.5 (2002): 436.

Talwar, V., and Kang Lee. 'Little Liars: Origins of Verbal Deception in Children.' In Itakura S., and Fujita K. (Eds). *Origins of the Social Mind.* Tokyo: Springer. (2008): 157–78.

Talwar, Victoria, and Kang Lee. 'Social and Cognitive Correlates of Children's Lying Behaviour.' *Child Development* 79.4 (2008): 866–81.

Taylor, M., M. Rhodes, and Susan Gelman. 'Boys Will Be Boys; Cows Will Be Cows: Children's Essentialist Reasoning about Gender Categories and Animal Species.' *Child Development* 80.2 (2009): 461–81.

Thibodeau, Rachel, Ansley Gilpin, Melissa Brown, and Brooke Meyer. 'The Effects of Fantastical Pretend-Play on the Development of Executive Functions: An Intervention Study.' *Journal of Experimental Child Psychology* 145 (2016): 120–38.

Thomson, P., and S. Jaque. 'Testimonial Theatre-Making: Establishing or Dissociating the Self.' *Psychology of Aesthetics, Creativity, and the Arts* 5.3 (2011): 229.

Tomasello, Michael. 'Cultural Learning Redux.' *Child Development* 87.3 (2016): 643–53.

van Schaik, Carel P., and Judith M. Burkart. 'Social Learning and Evolution: The Cultural Intelligence Hypothesis.' *Philosophical Transactions of the Royal Society of London B: Biological Sciences* 366.1567 (2011): 1008–16.

Verducci, S. 'A Moral Method? Thoughts on Cultivating Empathy through Method Acting.' *Journal of Moral Education* 29.1 (2000): 87–99.

Weisberg, Deena Skolnick. 'Pretend Play.' *Wiley Interdisciplinary Reviews: Cognitive Science* 6.3 (January 29, 2015): 249–61.

Weisberg, Deena, Jennifer Zosh, Kathy Hirsh-Pasek, and Roberta Golinkoff. 'Talking It Up: Play, Language Development, and the Role of Adult Support.' *American Journal of Play* 6.1 (2013): 39.

Wellman, Henry, David Cross, and Julanne Watson. 'Meta-Analysis of Theory-of-Mind Development: The Truth about False Belief.' *Child Development* 72.3 (2001): 655–84.

Whiten, A., and R. Byrne. 'Tactical Deception in Primates.' *Behavioural and Brain Sciences* 11.2 (1988): 233–44.

Wickler, Wolfgang. *Mimicry in Plants and Animals*. London: Weidenfeld & Nicolson, 1968.

Wimmer, H., and J. Perner. 'Beliefs about Beliefs: Representation and Constraining Function of Wrong Beliefs in Young Children's Understanding of Deception.' *Cognition* 13.1 (1983): 103–28.

Winner, Ellen. *The Point of Words: Children's Understanding of Metaphor and Irony*. Cambridge, MA: Harvard University Press, 1997.

Xu, Fen, Xuehua Bao, Genyue Fu, Victoria Talwar, and Kang Lee. 'Lying and Truth-Telling in Children: From Concept to Action.' *Child Development* 81.2 (2010): 581–96.

Young, Mark, and Drew Pinsky. 'Narcissism and Celebrity.' *Journal of Research in Personality* 40.5 (2006): 463–71.

Youngblade, L., and Judith Dunn. 'Individual Differences in Young Children's Pretend Play with Mother and Sibling: Links to Relationships and Understanding of Other People's Feelings and Beliefs.' *Child Development* 66.5 (1995): 1472–92.

Zelazo, Philip, Ulrich Müller, Douglas Frye, Stuart Marcovitch, Gina Argitis, Janet Boseovski, Jackie Chiang, Donaya Hongwanishkul, Barbara Schuster, and Alexandra Sutherland. 'The Development of Executive Function in Early Childhood.' *Monographs of the Society for Research in Child Development* 68.3 (2003): i–151.

Zeman, Janice, Michael Cassano, Carisa Perry-Parrish, and Sheri Stegall. 'Emotion Regulation in Children and Adolescents.' *Journal of Developmental & Behavioural Pediatrics* 27.2 (2006): 155.

15

'I'M GIVING EVERYBODY NOTES USING HIS BODY'

Framing actors' observation of performance

Claire Syler

> This is for everyone, okay? This is our first time where we're doing performances and I'm coaching. Now. I'm telling him stuff (*pointing to a student onstage*), but am I only telling him stuff? (*the students in the audience shake their heads no*) Who am I really telling? (*the students murmur 'me' and 'all of us'*) Everybody else! So, I'm giving *everybody* notes using *his* body.

This is how a university-level acting instructor, Professor D,[1] began the first performance coaching session in his basic acting class. And as the college students bobbed their heads in agreement, I leaned forward to scribble in my fieldnotes, *metacomment*. At its most basic, metacommunication concerns 'communication about communication' (Bateson 1951, 209), but, in a more nuanced sense, metacommunication can act as a frame to structure attention, relationships and interactions (Goffman 1974, 1981). Anthropologist Gregory Bateson (1951, 1972) was among the first to recognise that metacommunication is also crucial to *play* (and therefore performance), because it can signal the opening of an additional, counterfactual frame of activity. It was for these reasons that I became interested in Professor D's use of metacomments when observing his acting class as part of a larger ethnographic project. What began as a seemingly casual utterance became a pedagogical tool Professor D routinely used to develop his students' thinking about performance.

This chapter examines how an experienced acting instructor, Professor D, repeatedly deployed metacomments to frame his students' observation of performance. Rather than emphasise the *doing* of performance – the action, impulse and spontaneity that so often represent performance participation – I focus on how Professor D framed the students' *observation* of performance, which, in turn, cultivated their thinking and promoted the transfer of performance knowledge. Applying research from the learning sciences, a field that emerged in part from the cognitive sciences (Sawyer 2014), as well as research from cognitive theatre studies to evidence collected in an ethnographic case study, this chapter characterises metacomments as a cognitively rich instructional practice. I do so with the assumption that if more acting instructors realised the significance of using metacomments to frame their students' observation of classroom performance, perhaps the discursive tool would be more widely and purposefully used.

Professor D's basic acting class exemplified the traditional, Stanislavski-based course offered at most colleges and universities across the United States.[2] Although common, this site of performance learning is significant because it involves a perennial population of novice actors engaged in foundational training activities, such as solo performance work, dyadic scene study and actor coaching sessions. Although the learning environment and participants I describe are specific, the case study provides a familiar scenario and proposes general issues concerning how acting instructors, like Professor D, can frame actors' observation of performance during classroom activities.

A professional fight choreographer and actor, Professor D has taught university-level acting courses for a decade. His teaching practice is shaped by a regional theatre career, a BFA in Acting from a leading U.S. conservatory, an MFA in Performance Pedagogy and, at the time of the study, two years of doctoral study in theatre and performance studies. Although external researchers are rarely allowed access to systematically study performance training *in situ*, Professor D granted me (and my video camera) entrance to his course. This was because, as a professional practitioner and colleague in the same graduate programme, I was not entirely an outsider. But, Professor D's willingness also stemmed from a general indifference to the privacy and protection so often exerted by acting teachers. Our partnership resulted in a rich opportunity for me to study the multifaceted nature of performance training, one aspect of which concerns the importance of framing young actors' observation of performance during classroom activities.

The cognitive complexities of metacomments

In a post-course interview, I asked Professor D why he began each performance coaching session with a statement to include everyone in the class. His answer was straightforward and pragmatic: 'From the moment I start [teaching], every minute needs to get used up. Every moment needs to be towards learning...; that means that when I'm coaching [one actor], I'm coaching everybody.' Listening to Professor D's thoughtful answer, I suspected his pedagogic practice might be richer than he knew. And this was appropriate, given that his role in the classroom was to teach and not theorise. But, since the 1970s, learning scientists and educational anthropologists have extended the science of learning by venturing into classrooms and informal learning settings with the goal of explaining how learning happens and, when appropriate, explicating the instructional practices that give rise to learning opportunities (Sawyer 2014). Knowing this, it seemed that Professor D's metacomments deserved further attention. In particular, I was interested in the least 'active' participant in the coaching framework — the student spectators. What cognitive opportunities did Professor D's metacomments prompt for the audience?

To begin, Professor D used metacomments to establish two overlapping frames of performance learning activity. Using Goffman's (1974) classic understanding of a frame as the 'organizational premises' that are 'sustained both in the mind and in activity' (247), Professor D identified one frame in which he would coach the onstage actor, as well as a second frame in which he would imaginatively coach the student spectator. Moving to stand between the actor(s) onstage and the student spectators seated in a loose proscenium orientation, Professor D faced the audience and bracketed each coaching session with a statement like: '[Audience] I'm coaching them, *(pointing to the students onstage)*, but you are learning through what my comments are to them.' After this framing move, Professor D took a seat in the audience while the actor(s) onstage made their final adjustments before launching into a rehearsed performance and the audience members shifted in their seats to better view their instructor's imminent coaching.

Actor coaching involves a familiar (if not ritualised) event structure in which an instructor (or coach) contingently intervenes into an actor's performance to clarify intentions, relationships and, in general, improve practice as a means to train the actor.[3] Coaching represents a significant instructional event within the broader phenomenon of actor training, but the significance of *observing* coaching is rarely acknowledged as equally important. This is perhaps ironic, given that observation has long been considered a tool of the actor. Performance training, as movement specialist Mark Evans (2015) notes, often guides actors 'to look anew at the world they live in and at the theatre practice that surrounds them' (xxviii). Evans goes on to cite Brecht to emphasise his point: '[o]bservation is a major part of acting. The actor observes his fellow-men (*sic*) with all his nerves and muscles in an act of imitation which is at the same time a process of the mind' (xxviii). Thus, Professor D's choice to frame classroom coaching sessions as significant for the student spectators carved out important learning opportunities dependent upon active observation.

One way Professor D's metacomments did this was to frame the audience members' participatory purpose or, as he called it, 'buy in.' By telling the spectators things like, 'I want you to be in the type of headspace where I'm talking to one person but I'm really talking to *you*,' Professor D assigned the students a role – or 'headspace' – to guide their viewing. More specifically, Professor D's metacomments asked the observing students to project themselves into their peers' performances to receive Professor D's coaching as their own. Cognitive performance theorist Bruce McConachie (2015) reminds us that spectating involves the same kind of psychological projection, or conceptual blending of self and other, which is foundational in performance practice. Yet, in this learning opportunity, the spectator's blending of self and other did not aim to foster empathy or develop a character. Rather, the spectator's self-projection created a kind of performance filter, which one student, Mark, described as a 'mental checklist of dos and don'ts that I pick up from watching other people [perform during coachings].'

Still, Professor D's metacomment could only prove productive if the student spectators agreed to play along and align their frame of learning (while spectating) with Professor D's frame of activity. Acting pedagogue Dick McCaw (2015) notes that a key component of performance training concerns 'accepting the offer to play,' and, in this class, students seemed happy to take up Professor D's framing invitation (176). When I asked students about what it was like to view their peers' coachings in post-course interviews, they replied enthusiastically, as illustrated by these excerpts from three students:

> [Professor D] prefaced [coachings] with an 'I'm not just doing this for you, but this is for everyone to take a part in.' So, I liked that because if I see something I like, I log it in my head. I'm like, 'I like the way that he's emoting during this scene'—or 'I really didn't like the way he moved from upstage to downstage.'
>
> [During coachings if] people were doing something that I liked or I didn't like—like some people would shift their weight a lot, which I found like super distracting—I would consciously think in my head 'ok, don't shift your weight onstage.'
>
> [Observing a classmate being coached is] cool because then you get to see what you did, but through like third view.

Of interest, here, is the students' cognitive characterisation of observing their peers' coaching sessions. Phrases like 'log it in my head,' 'consciously think in my head' and 'third view' demonstrate the complex thinking the students experienced while watching their peers' perform.

The playful space from which the students observed coaching sessions can be described as 'liminal, inter-subjective' or, as Nicola Shaughnessy (2012) has noted, 'the ludic third' (38). Play is a phenomenon that has been rigorously researched by scholars from varying fields and, while resistant to firm definition, it is generally considered cognitive in nature.[4] This is because, to exist within a zone of play, individuals must uphold multiple (sometimes contradictory and counterfactual) viewpoints simultaneously. In this class, Professor D's metacomment encouraged the spectators to intentionally foreground their peer's stage performance as a means to better understand their own practice. Observing from this 'third view' can provide a significant opportunity to learn because, as one student, Ava, noted, 'it's always easier to catch other people's flaws than to catch your own.' When pressed to explain what she meant, Ava stated, 'if you see that a person could have done [something] a little bit better [during a coaching], it makes me question, "Oh, do I do that, too? Yeah, I guess I kinda do." So, it kind of makes me reflect on my own flaws and try to, the next time, improve on that specific thing.'

Ava's comment points towards two cognitive characteristics of performance learning that are significant. First, Ava acknowledges the mostly tacit nature of performance practice when she states, 'it's always easier to catch other people's flaws than to catch your own.' Actor training tends to develop performance knowledge through activity-based experiences, which, scholar of acting Ian Watson (2015) notes, are often 'absorbed and applied tacitly rather than as an explicit conscious process' (11). Indeed, McConachie (2015) notes that performance activity typically transpires in a form of 'primary consciousness,' which is bound to the present moment (33). Yet, it is often the practical activity experienced in primary consciousness, which sets the stage for higher-order thinking. In other words, Ava's ability to recognise her own performance flaw (via observation of a peer's performance) is possible, in part, because of her own foundation in performance practice. Yet, Ava's self-awareness of this fact (or flaw) may not have been possible if Professor D had not framed Ava's viewing experience from the outset.

The second cognitive characteristic embedded in Ava's comments concerns her growing ability to think *across* coaching contexts to plan her future performances ('next time [I'll] improve on that specific thing'). The ability to transfer knowledge across contexts is arguably the end goal of most instruction. Learning scientist, Randi Engle (2006), contends that metacomments can be used to foster knowledge transfer by creating *intercontextuality*, which occurs 'when two or more contexts become linked with one another...[so that] the content established during learning is considered relevant to the transfer context' (456). By using metacomments to frame the audience's observation of coaching sessions, Professor D increased the likelihood that the student spectators might understand the relevance between their peers' coachings and the coachings they would take part in as actors. Thus, one reason the young actors were able to successfully harness particular concepts of performance knowledge (e.g., 'don't shift your weight') and transfer it to their own upcoming coaching sessions was because Professor D framed their viewing as relevant to their own future performance work.

A final cognitive complexity Professor D's metacomments set in motion relates to his decision to repeatedly use them. Before *every* coaching session Professor D stood up and positioned himself, once again, between the young actors onstage and the student audience to bracket the coaching session: 'Okay, remember. I'm coaching them, but you are learning through what my comments are to them.' The accumulation of these statements encouraged the audience members to learn from particular moments of performance (i.e., 'flaws'), but it also invited them to notice their peers' improved practice over time. Angie, a student in the class, stated 'seeing [my peers] make improvements was really fun,' because their performances became 'better than [what] I had seen the day before or the week before.' Angie's comment indicates that observing a peer's developing performance can provide an opportunity to build a shared sense of what improvement looks like.

Cognitive scientists Colleen Seifert and Edwin Hutchins (1992) might refer to Angie's ability to identify a peer's improved practice, alongside an awareness of her own performance work, as a 'horizon of observation' (427). The authors note, '[t]he utility of shared knowledge can be enhanced by providing access to other's performances. This *horizon of observation* is the functional workspace that each participant can monitor in addition to his own task' (427). Although Angie was not directly engaged in her own performance task while observing her peers' performances, her own upcoming coaching session remained close at hand due to Professor D's metacomment. Moreover, Angie's horizon of observation was focused on a *peer's* improvement, which allowed her to view improvement at a level of skill comparable to her own. In this regard, Seifert and Hutchins note 'an unlimited horizon of observation might prove too distracting' (428). Here, it seems significant for young performers, like Angie, to view their peers' practice to establish an observational frame that resides at, or just beyond, their own performance ability.

Although Professor D's metacomments were only one part of his broader classroom pedagogy, they did important cognitive work. By structuring his students' participatory purpose during actor coaching sessions, Professor D opened up a frame of observation that overlapped with his primary activity of coaching actors on their solo and scene work. But, these seemingly simple framing statements pressed the student spectators towards new ways of seeing, which allowed them to identify, trace and re-purpose performance in important ways.

Sustaining an observational frame

After thinking about metacomments for a good while, I wondered about Professor D's mental focus during his coachings. Who was he really addressing? When I asked him, he told me, 'I don't really focus on [just the actors]—I'm thinking about everybody learning this thing that I'm trying to teach.' Professor D's statement suggests his attention was on the classroom *system*, just as an actor's focus must extend beyond a scene partner. Acting teachers, like stage actors, must attend to the broader system of performance learning to monitor the efficacy of its varying frames. In this regard, Professor D's metacomments initiated a frame of observation, which he nurtured throughout the class.

In addition to bracketing coaching sessions, Professor D repeatedly checked in with the observing students to ensure they were paying attention to his coaching practice or to underscore a particular performance concept. For example, in the midst of one coaching session, Professor D stopped a student's performance to emphasise the importance of repetition in dramatic language and performance: 'Alright hold. (*turning to address the entire class*) Now anytime in a monologue that we have the same word said more than once, it's repetition [and]…we can't have them sound the same.' Then, in subsequent coaching sessions, Professor D re-circulated the performance concepts and, at opportune moments, asked students to re-voice their significance ('So, during the last coaching that you and I had, I talked to you about [repetition]. Explain it to us.'). Known-answer questions, of course, often push learners to memorise certain concepts, but, in this case, they also allowed Professor D to draw attention to particular observations in which he wanted the spectators to engage.

To help the spectators accumulate their performance observations, Professor D frequently directed the students to reflect on their viewing experience. After coaching several actors,

Professor D invited the entire class to discuss what they learnt from coaching sessions: 'So what are the takeaways in terms of how I was coaching everyone?' In response to his query, the students usually called out the varying performance concepts highlighted during coachings, for example, 'repetition,' 'beats' and 'subtext.' In a similar fashion, Professor D frequently asked the students to keep track of their 'takeaways' in writing during in-class journaling. Both efforts pressed students to articulate their observations of coaching sessions, which, in turn, helped to sustain the observational frame of activity launched via his metacommentary.

Like many performance-based courses, Professor D's basic acting class comprised a series of layered activities. Within this dynamic flow of action, Professor D dedicated a good amount of time to working with actors *and* spectators during coaching sessions. By asking everyone in the class to experience a peer's coaching as their own, Professor D invited spectators to engage in cognitively rich learning opportunities dependent upon mental self-projection and play. Metacomments offer acting teachers a way to animate an observational frame of activity that can support students in mindfully observing coaching sessions, which leads students to reify performance concepts and promotes the transfer of performance knowledge.

Notes

1 In keeping with the qualitative research paradigm as practiced in the United States, the names in this chapter are pseudonyms. After selecting his pseudonym, Professor D told me his chosen name denotes a character, comprising all his past performance instructors, who he performs while teaching. I have written about Professor D's noteworthy teaching practice elsewhere (Syler 2017).

2 Indeed, Patricia Downey's (2013) dissertation, which reviews a data set of 75 course syllabi utilised in introductory acting courses at 61 different U.S. universities between 2001 and 2007, confirms the dominant performance tradition taught is a version of Stanislavski's System.

3 In this case, Professor D's emergent evaluation of what constituted improved practice aligned with a Stanislavski-based tradition of realistic performance. As with all modes of performance, what determines improvement aligns with the culturally constructed aesthetics and performance techniques codified in sociohistorical contexts.

4 For more on the intersections between play, performance and cognition, see Chapter 1 in McConachie's *Evolution, Cognition, and Performance* (2015) and Chapter 2.1 in Shaughnessy's *Applying Performance* (2012).

References

Bateson, Gregory. 2008 (1951). "Information and Codification: A Philosophical Approach." In *Communication: The Social Matrix of Psychiatry,* by Jurgen Ruesch and Gregory Bateson, 168–211. New York: Routledge.

Bateson, Gregory. 1972. *Steps to an Ecology of Mind: Collected Essays in Anthropology, Psychiatry, Evolution, and Epistemology.* Chicago, IL: University of Chicago Press.

Downey, Patricia K. 2013. "An Exploratory and Descriptive Inquiry into the Relationship between the Goals of General Education and Disciplinary Content in Acting for Non-Majors Courses in Colleges and Universities in the United States." PhD diss., University of Missouri – Columbia.

Engle, Randi A. 2006. "Framing Interactions to Foster Generative Learning: A Situative Explanation of Transfer in a Community of Learners Classroom." *Journal of the Learning Sciences* 15(4): 451–98.

Evans, Mark. 2015. "Introduction." In *The Actor Training Reader,* edited by Mark Evans, xix–xxxii. New York: Routledge.

Goffman, Erving. 1974. *Frame Analysis: An Essay on the Organization of Experience.* Cambridge, MA: Harvard University Press.

Goffman, Erving. 1981. *Forms of Talk.* Philadelphia: University of Pennsylvania Press.

McCaw, Dick. 2015. "Introduction to Part V: Presence, Physicality, Play and Communion." In *The Actor Training Reader,* edited by Mark Evans, 171–81. New York: Routledge.

McConachie, Bruce A. 2015. *Evolution, Cognition, and Performance*. Cambridge: Cambridge University Press.

Sawyer, R. Keith. 2014. "Introduction." In *The Cambridge Handbook of the Learning Sciences*, edited by R. Keith Sawyer, 1–20. New York: Cambridge University Press.

Seifert, Colleen M. and Edwin L. Hutchins. 1992. "Error as Opportunity: Learning in a Cooperative Task." *Human–Computer Interaction* 7(4): 409–35.

Shaughnessy, Nicola. 2012. *Applying Performance: Live Art, Socially Engaged Theatre and Affective Practice*. New York: Palgrave Macmillan.

Syler, Claire. 2017. "Conceptualising Actor Coaching: Talk Moves as Tools." *Theatre, Dance and Performance Training* 8(3): 317 328.

Watson, Ian. 2015. "Introduction to Part II: The Purpose of Actor Training." In *The Actor Training Reader*, edited by Mark Evans, 9–20. New York: Routledge.

16

ACTING TECHNIQUE, JACQUES LECOQ AND EMBODIED MEANING

Rick Kemp

Introduction

The idea that actors employ technique can be surprising to many outside the fields of theatre and performance. While the visibility of technique varies according to style and medium, in the naturalistic style of most Western drama, actors seem to appear to respond spontaneously to events, creating an apparent naturalness of behaviour in fictional circumstances. In fact, this naturalness is generally the result of many years of training, and long and painstaking preparation for individual roles. Actors preparing and playing a role engage in most, if not all, of the cognitive processes that humans conduct in daily life – with a crucial difference. Actors consciously elicit and regulate phenomena that generally arise spontaneously for most people. It is probably easy to recognise that an actor will seek to consciously regulate features such as vocalisation, facial expression, gesture and movement, because these are audible and visible. But actors also address many other phenomena that are not so readily apparent. Imagination is necessary to embody a fictional character in fictional circumstances. Empathy with the fictional character may be stirred by this process, and the actor certainly seeks to stimulate empathy in an audience. Actors also seek to stimulate and regulate emotion – the believability of emotion in fictional circumstances is highly valued by Western audiences.

A heightened faculty of memory is necessary not only for the obvious task of precisely memorising dialogue, but also for less evident tasks, such as reproducing the planned movement and behaviour in a specific production, often at the micro-level of 'actions' that communicate unspoken thoughts and feelings. In preparing a role by analysing a script or through improvisatory creation, an actor needs a knowledge of linguistic and narrative forms, as well as a psychological and sociological understanding of human behaviour. This list is not exclusive, but it gives an idea of the realm of extended cognitive-affective activities within which an actor's technique operates. Some of the activities are conscious, many operate below the level of conscious awareness most of the time and many can never be consciously accessed. So actors are frequently seeking to consciously elicit otherwise unconscious phenomena. The means by which they do this are the assorted exercises and practices that I collectively call technique. It is these activities that are taught in actor training.

In Western theatre, actor training as a dedicated activity arose predominantly in the twentieth century, with varied forms of apprentice systems largely responsible for the development of technique prior to this. The most influential formulator of contemporary acting technique in the West was Russian actor and director Konstantin Stanislavski (1863–1938), who developed a system for preparing a role that is widely taught in European, North American and Western-influenced theatre conservatoires. His work is analysed from a cognitive perspective by Sharon Carnicke in Chapter 1 of this book. Stanislavski's approach is largely applied to text-based performance and associated mainly with the style of psychological realism that has been prevalent on stage and screen in the second half of the twentieth century and early part of this century. By contrast, the performance pedagogy of Jacques Lecoq (1921–1999) deals mainly with improvisation, deliberately avoids language in its early stages, engages with multiple styles of performance and explicitly seeks to stimulate the creation of new forms of performance. It is a training system for actors that is radically different from Stanislavski's, yet the two are widely used alongside one another in Western actor training programmes, suggesting that in practice they complement each other.

In this chapter I propose that Lecoq's conceptualisation of acting technique is implicitly congruent with the principles of embodied cognition, and often explicitly forecasts its precepts. When viewed in the theoretical framework of embodied cognition, many of Lecoq's training practices consciously elicit and define what would otherwise be unconscious activity, articulating expressive techniques that demonstrate the dynamic and constructive nature of meaning as broadly posited in the concept of embodied simulation. I will give a brief overview of some of the relevant principles of embodied cognition and will then relate these to key themes of Lecoq's pedagogy. The chapter concludes with a case study that gives an example of how Lecoq's rigorous training not only benefits actors, but can also provide information for research beyond the field of acting. The case study focuses on a video of Lecoq teaching a class on gesture and language. The extract that I describe uses gestures associated with 'taking' executed by different language groups among Lecoq's students. The phrase 'je prends' (normally translated as 'I take') provokes a variance of gesture among the different language groups. When viewed from the perspective of embodied cognition, this suggests that the concepts associated with this action differ significantly among language groups, indicating the culturally situated nature of meaning. This also suggests that the linguistic translation of 'take' for 'prendre' overlooks the cultural nuances of meaning that are revealed by Lecoq's training in acting technique.

Embodied cognition

Embodied cognition proposes that thinking and behaviour are properties of the whole human organism, not the brain alone, and that body, brain and cognition are 'situated' – engaged with the surrounding environment. This presents a radical challenge to the Cartesian separation of mind from body that has long influenced traditional Western psychology. The field of embodied cognition incorporates research from many related disciplines – psychology, linguistics and neurobiology among many others – and inevitably contains varied opinions. Certain concepts, however, are considered foundational. One of these is the principle that has emerged from neuroscientific findings that sensorial and motor experiences form the neural foundations for mental concepts.

As cognitive linguist George Lakoff and cognitive philosopher Mark Johnson explain:

> Our abilities to move the way we do and to track the motion of other things give motion a major role in our conceptual system. The fact that we have muscles and use them to apply force in certain ways leads to the structure of our system of causal concepts. What

> is important is that the peculiar nature of our bodies shapes our very possibilities for conceptualization and categorization.
>
> *(Lakoff and Johnson 1999, 19)*

This concept challenges assumptions that underlie much of Western thought, since it demonstrates that mental concepts are shaped by physical experience in the material environment and use many of the same neural pathways that are involved in physical action and sensorial experience. Based on this understanding, phenomena such as consciousness, empathy, intersubjectivity, affect and aesthetic responses 'come from having a body with various sensorimotor capacities [that] are themselves embedded in a more encompassing biological, psychological and cultural context' (Varela, Thompson and Rosch 1991, 173). Within the field of embodied cognition there is a growing consensus that meaning results intersubjectively from our situated interactions with the world. Thus, meaning can be quite personal, as it depends on our particular experiences in particular environments. By extension, meaning is also variable across cultures. This theory of meaning depends on the theory of embodied simulation – the experience of perception and action without their physical manifestation. According to the embodied simulation hypothesis, meaning does not arise from the deployment of abstract mental symbols but is constructed from the neural experiences triggered by various stimuli and is thus dynamic and constructive (Gallese 2007a). I will describe other aspects of theory from embodied cognition later. For now, I would like to trace the links between these fundamental precepts of embodied cognition and three key precepts of Lecoq's training: 'Tout bouge' ('everything moves'), 'Le fonds poetique commun' (translated below) and 'Dynamiques' ('Dynamics').

Jacques Lecoq – Tout bouge

The influence of Jacques Lecoq on modern theatre is significant. He founded an international school of performance training in Paris in 1956 where he taught until a few days before his death in 1999. The school continues to thrive under the direction of Pascale Lecoq, Lecoq's daughter, and has trained over 5,000 students from at least 84 countries. Many of these students have formed their own companies, such as Le Théâtre du Soleil in Paris, Complicite in London, Mummenschantz in Switzerland, Footsbarn in France and Pig Iron in Philadelphia as well as many others worldwide. Graduates have also found success as directors of theatre and film (e.g. Julie Taymor, Luc Bondy, Simon McBurney, James McDonald), or as actors in mainstream movies (e.g. Geoffrey Rush, Toby Jones and Sergi Lopez). Many others teach in actor-training programmes or have founded their own schools in countries ranging from Chile to Germany to the USA to Belgium, Italy and Spain (for more information on Lecoq and his school, see Lecoq 2001, 2006; Murray 2003; Kemp 2012; Evans and Kemp 2016).

Lecoq's guiding principle was 'Tout bouge' – everything moves. His rigourous analysis of movement in humans and their environments formed the foundation for a refined and nuanced repertoire of acting exercises rooted in physical action. These exercises develop a heightened somatic awareness in the actor of the relationship between thought, feeling, gesture and language, preparing him or her to communicate with movement in a variety of styles, to employ physical actions that both provoke and define emotion and to invest spoken language with meaningful gesture. Given the primacy that embodied cognition places on sensorimotor experience and its role in shaping meaning, Lecoq's focus on movement immediately resonates with its principles and the involvement of physical activity with communication. It must be remembered, however, that the ideas of embodied cognition are still in the process of gaining widespread recognition, and many people in mainstream theatre consider Lecoq's teaching to be less sophisticated an approach to acting than Stanislavski's script-oriented process – precisely because Lecoq's originates in movement. As recent scholarship shows (e.g. Carnicke 2009; Kemp 2012) the linking of Stanislavski with a purely mental and linguistic approach has been overstated, as his 'Active Analysis' work demonstrates. Nevertheless, the widespread idea of Stanislavski as 'psychological' and Lecoq as 'physical' persists, indicated by the tendency of many actor-training programmes to teach Stanislavski practice in 'Acting' classes and Lecoq work in 'movement' classes. This curricular arrangement underestimates the scope and nuance of Lecoq's pedagogy – something that an analysis from the perspective of embodied cognition will show.

The training programme that Lecoq created at his own school drew on many sources, is rooted in physical and verbal improvisation and engages with several theatrical styles – what he called 'dramatic territories': Greek Tragedy, Commedia dell'arte, red nose clown, melodrama, the grotesque parodies of 'bouffons.' Work on these styles is rooted in movement analysis and initial training with the Neutral Mask, created by Lecoq with sculptor and mask-maker Amleto Sartori (1915–1962). The Neutral Mask is a full-face mask, made of leather, with a neutral facial expression. Through wearing it in a variety of exercises and observing others doing the same, actors develop a heightened awareness of the communicative potential of the body. Since we are habituated to pay attention to facial expressions in daily life, the meaningful content of posture, gesture and gait becomes much more apparent to the observer when the face is covered – information that can then be used when wearing the mask oneself. The term Neutral Mask refers to the mask itself, the type of exercises conducted with it and the persona that is apparent when a performer wears it. Exercises are always conducted in an ensemble mode, with groups of 5–7 students assaying a particular exercise while being observed by the instructor and other students in the larger group of 20 or more. This mode is an important part of the training. Through the process of wearing the mask *and* observing others in it, actors develop many sensitivities, not least an awareness of the communicative potential of the body.

Lecoq's use of the mask to train actors is prescient of the discoveries of cognitive science in many ways. For example, a 2005 study demonstrates that bodily posture is highly significant in determining the meaning of emotional facial expressions for observers (Meeren, van Heijnsbergen and de Gelder 2005). By heightening an actor's awareness and expressive control of postural communication, Lecoq's Neutral Mask exercises assist in clarifying emotional expression when the mask is removed (Lecoq used the Neutral Mask only in training, not for performance). For the wearer of the mask, an interesting phenomenon arises – as one doesn't need to concern oneself with facial expression as communication, prolonged use of the mask encourages a relaxation of the facial muscles that in turn seems to prompt a sense of calm and focus. Psychologist Paul Ekman's work on facial expressions has shown that consciously arranging the facial musculature in the patterns associated with various primary emotions provokes the affective state of the emotion. This indicates that there is a reflexive proprioceptive and interoceptive relationship between facial musculature and emotion (Ekman, Davidson, and Friesen 1990; Ekman 1999, 2003) – a phenomenon that may explain the sense of calm that arises from the relaxation of the facial muscles under the mask.

Other research points to a similar reflexive relationship between larger bodily activity and the experience of emotion (e.g. Tom, Pettersen, Lau, Burton and Cook 1991; Stepper and Strack 1993, Duckworth, Bargh, Garcia and Chaiken 2002). Movements of the whole body are involved in Neutral Mask 'identifications' work, in which actors consciously embody the rhythms of movement found in natural, social and fabricated environments. This activity forms the foundation of the remarkable synchrony between Lecoq's pedagogy and the precepts of embodied cognition, as it articulates Lecoq's conviction that the starting point for theatre is not a scripted play, but the actor's engagement with the sensorimotor experience of her environment:

> In my method of teaching I have always given priority to the external world over inner experience. … It is more important to observe how beings and things move, and how they find a reflection in us. … People discover themselves in relation to their grasp of the external world.
>
> *(Lecoq 2001, 19)*

This principle of Lecoq's correlates with the foundational concept of embodied cognition that I described earlier; that sensorial and motor experiences form the neural foundations for mental concepts. Beginning with sensorimotor investigations and moving through spoken, then written language, Lecoq's training grounds performance in an explicit re-experiencing of human cognitive development: He states

> [T]he laws of movement govern all theatrical situations. A piece of writing is a structure in motion. Though themes may vary (they belong to the realm of ideas), the structures of acting remain linked to movement and its immutable laws
>
> *(Lecoq 2001, 24)*

Lecoq's idea of 'the laws of movement' refers to the affordances and constraints of the typical able-bodied human anatomy and its relation to the physics of movement in the material environment. The concepts of affordances and constraints are ones that arise in significant theories of embodied cognition. I use the term 'affordance' in the sense that was originated by J.J. Gibson (1979): 'The affordances of the environment are what it offers the animal, what it provides or furnishes, either for good or ill' (Gibson 1979, 127). So a flat rock affords sitting, a tree affords climbing, a river affords swimming. These examples demonstrate how the concept of affordances acknowledges a symbiosis between body and environment – action arises from what a body is able to do in engagement with features of its physical environment. Gibson was prescient about the integration of body and environment that is involved in the theories of situated cognition:

> An affordance is neither an objective property nor a subjective property; or it is both if you like. An affordance cuts across the dichotomy of subjective–objective and helps us to understand its inadequacy. It is equally a fact of the environment and a fact of behavior. It is both physical and psychical, yet neither. An affordance points both ways, to the environment and to the observer.
>
> *(Gibson 1979, 129)*

Lecoq's Neutral Mask 'identifications' work embodies this concept through activities stimulated by the observation and conscious imitation of phenomena. In this work, the affordances are both those of the environment that enable and provoke an action of the body, and those of the body in the extent to which it can imitate rhythms of the environment. In this way, Lecoq's 'identifications' work actively demonstrates Gibson's idea that an affordance is 'equally a fact of the environment and a fact of behavior' (ibid.). This 'both/and' quality of affordances is useful in understanding the theory of enactivism, which proposes that cognition is a process of continuous reciprocal interactions between brain, body and world.

While the concept of enactive affordances is implicitly present in Lecoq's exercises, that of constraints is explicitly identified and applied: 'Different ways of using exercises are employed at each stage: [including] the method of constraints (spatial, temporal and numerical)' (Lecoq 2015, 15). Simply put, in Lecoq's pedagogy, a constraint is a limit – the boundaries of a space in which one can move, the duration of time allotted to an exercise or presentation, the number of people available to execute a task. Cognitively, constraints are also enactive in that they help us make sense of the environment and are involved in the shaping of affordances – for example, how high a body can reach or how far it can step. The idea of a constraint also helps cognitive scientists create models of how organic beings behave and interact. It is a significant feature of Dynamical Systems Theory (DST).

Originating in mathematics and developed significantly by Esther Thelen (1941–2004) in the field of Developmental Psychology, DST provides models that describe the flow of relationships among the components of a whole phenomenon. These models acknowledge the real-time, reciprocal cause-and-effect action of elements upon each other within a system. Consequently, DST models are well-suited to describe the ways in which characters, events and circumstances simultaneously affect one another in the context of a dramatic interaction. Given Lecoq's focus on movement, the following quote from a group of physiotherapists aptly describes the way in which we can understand the application of DST to the analysis of movement in conjunction with environment:

> The fundamental premise of *dynamic systems theory* when applied to coordination and control is that movement patterns emerge from the interplay of the *constraints* between and within the elements of the *system*.... The patterns that emerge are a reflection of these changing organismic, task and environmental *constraints*.
>
> *(emphasis in original)*
> *(Holt, Wagenaar and Saltzman 2010, 448)*

In Lecoq's 'identifications' work, the limitations placed on the movement of beings and objects by their physical qualities serve as both constraints and affordances. Mud, for example, oozes, but does not spark. Fire sparks, but does not ooze. The human body can be considered a system that interacts with the imagined affordances and constraints of natural phenomena like these in order to effectively recreate their rhythms of movement. As Lecoq develops the actor's work, different levels of systems arise. Lecoq refers, for example, to dramatic styles as 'constraints.' From the perspective of embodied cognition, this invites a view of style as a system in which the interactive elements are performers. Lecoq's interest in the relationship between the movement of the human body and its environment through the physics of movement is also expressed in the use of architecture as a companion discipline in his school. He and architect Krikor Belekian developed the Laboratoire d'Etude du Mouvement which is taught as a parallel activity to the school's main pedagogy. The course is now led by Pascale Lecoq, who trained as an architect before becoming the school's director.

'Le fonds poetique commun'

As might be expected from Lecoq's statement about the importance of the 'external world,' much of the content of his training programme arises from his perception of the ways in which humans interact with their material environment. This perception is expressed both in specific exercises and in broad conceptualisations that create a framework for these exercises.

One of the most significant of these conceptualizations is Lecoq's idea of 'le fonds poetique commun.' A full appreciation of this concept benefits from some reviewing of previous translation. David Bradby, who overall did an admirable job of translating Lecoq's book '*Le Corps Poetique*' (1997) for its English publication as *The Moving Body* (2001), rendered this phrase as 'the universal poetic sense.' However, he has acknowledged in a later editorial comment that 'the word "fonds" conveys something more real and concrete than a "sense"' (Lecoq 2006, xiii). In an earlier piece (2016), I suggested that there is a more nuanced way of understanding this concept than is provided by Bradby's free translation, and here I build on that proposal in order to gain a fuller sense of Lecoq's meaning. This is significant, as the concept of 'le fonds poetique commun' (properly defined) relates directly to the concept of situated cognition and the sensorimotor links between physical activity and semantic meaning. The association of these features can be drawn out from the way the idea evolved over time in Lecoq's pedagogy and by detailing the semantic associations of the French words that form the phrase.

Lecoq's pedagogy grew organically throughout his life. While 'le fonds poetique commun' is a term that he only began to use in the last third of his teaching career, practical elements of what he came to define as this concept were present from the inception of his school in 1956. From the beginning, his curriculum incorporated exercises that drew on his own training in sport, as a physical therapist, and with French theatre practitioners Jean-Marie Conty and Jean Dasté. These were combined with the results of practical research that he had conducted into the historical styles of Greek tragedy and Commedia dell'arte during an eight-year stint in Italy. As these were both masked styles, they placed emphasis on physical communication in addition to verbal and vocal communication (Commedia used both speech and gibberish). As Lecoq trainee and scholar Ismael Scheffler describes, Lecoq's training incorporated 'exercises of movements of identification and expression of natural elements and phenomena' (Scheffler 2016, 182) within its idea of mime; the school's original name was L'École Internationale de Théâtre et de Mime -The International School of Theatre and Mime. It is clear from Lecoq's practices that he used the word mime to suggest physical re-enactment rather than the formalised movement styles associated with Etienne Decroux and Marcel Marceaux. Lecoq's form of mime engaged students in activities that consciously repeated humans' corporeal engagement with the rhythms and sensations of their environment. He struggled to find an explicit way of distinguishing his idea of mime from the stylised and codified approaches of Decroux and Marceau. As Scheffler points out, '[a]nalyzing Lecoq's publications before 1969 … one can perceive the difficulty Lecoq had until then to define his comprehension of mime' (Scheffler 2016, 182). It was not until Lecoq read the work of anthropologist Marcel Jousse (1886–1961) in the early 1970s that he found both a vocabulary and a conceptualisation that resonated with his own practice. Jousse considered that humans had an unconscious and intuitive ability to reproduce the physical stances, rhythms and movements of beings and objects – a capacity that he called 'mimisme' (Jousse 1932). These ideas enabled Lecoq to refine and articulate what he called his 'mimodynamic' approach in which students would first consciously absorb the dynamics of their environment and then 'replay' them – a word ('rejeu' in French) that he had adopted from Jousse. By the late 1980s, when Lecoq compiled a book of his own and others' writings that expressed his views on theatre (*Theatre of Movement and Gesture*, 1987), he had assimilated the ideas of Jousse to the point where he quotes him as saying:

> Miming differs from mimicry in this respect; it is not imitation, but a way of grasping the real that is played out in our body. A normal human being is 'played' by the reality

> that reverberates in him. We are the receptacles of interactions that play themselves out spontaneously within us. Human beings think with their whole bodies, they are made up of complexes of gestures and reality is in them, without them, despite them.
>
> *(Jousse in Lecoq 2006, 9)*

Lecoq recapitulated this concept in his own words shortly before his death:

> The 'miming body' is the body that has the faculty to take on the dynamics that surround it, the world that surrounds it – like a child. [...] All children are mimes. People [adults] too, but they don't know it.
>
> *(Lecoq, speaking in Roy and Carasso 1999, n.p.)*

The term 'mime' with its associations of silent white-gloved performers trapped in glass cases can easily trivialise Lecoq's approach. However, his anthropologically influenced conception of mime has considerable congruity with influential ideas in the field of embodied cognition. Neuropsychologist Merlin Donald proposes that mimesis was the evolutionary adaptation that generated a distinctly human culture, defining mimesis as a way of representing knowledge through consciously chosen motor activity. The capacity for symbolic thought has arisen from this cognitive evolution of the brain in symbiosis with culture. This has significance for present-day human activities, as evolutionarily new brain activities are 'scaffolded' on existing cognitive capacities. As a consequence of this biological phenomenon, Donald states: 'If mimesis was the adaptation that generated a distinctly human culture, it follows that the deepest communicative framework of human culture must still be mimetic' (Donald 2005, 293).

Lecoq's concept of mime, shaped by his phenomenological experience imbued with Jousse's analysis, anticipates Donald's scientifically oriented analysis:

> Children gain their understanding of the world around them by miming it: they mimic what they see and what they hear. They replay with their whole body those aspects of life in which they will be called on to participate. In this way they learn about life and, little by little, take possession of it.
>
> *(Lecoq 2006, 1)*

It is within the context of these ideas that we need to understand Lecoq's idea of 'le fonds poetique commun.' As described earlier, Bradby used the term 'universal' for the word 'commun.' Literally translated, this would be 'common' – in this context meaning 'in common' or 'shared.' The term 'fonds commun' is normally translated as 'mutual fund' or 'common fund.' Removed from a financial context, we can understand the use of 'fonds' as 'fund' in the sense of 'fund of knowledge.' The word 'fonds' also has strong associations of 'foundation' or 'base' because of its homonym 'fond,' as in the widely used phrase 'au fond'; literally 'at the bottom of.' So Lecoq believes that in addition to having principles (or 'laws') of movement in common, humans also have a 'fund' (or 'source') of foundational shared knowledge gained from a mimetic absorption of rhythms and sensorial experience of the physical world. Within this concept, Lecoq's use of the adjective 'poetic' is not romantic or vague. His practical investigation of poetry in his pedagogy demonstrates his awareness of the ways in which poetry has effect – the importance of rhythm and sensorial and perceptive stimuli and the centrality of metaphor in human experiencing. (See Gilrain 2016 for a description of how Lecoq's teaching encourages the conscious awareness of metaphoric thinking to the point of induced synesthesia.) This is significant when we think of Lakoff and Johnson's analysis of metaphor as a fundamental feature of human cognitive activity, given that physical experience provides the source domain for abstract thought: 'Conceptual metaphor is pervasive in both thought and language. It is hard to think of a common subjective experience that is not conventionally conceptualized in terms of metaphor' (Lakoff and Johnson 1999, 45). So 'fonds poetique commun' can be understood as a fund of sensorial knowledge that humans have in common, developed through physical engagement with the material world. This foundational principle of Lecoq's actor training accords strongly with the foundational precept of embodied cognition that sensorial and motor experiences form the neural foundations for mental concepts.

Dynamiques

In many of the quotes that I have used, Lecoq uses the term 'dynamics' ('dynamiques') to describe the situated environment that humans engage with through physical activity and mimesis. This term is a significant one in his pedagogy, as he defines 'dynamiques' as combinations of rhythm, force and space. Again, this is an astute conceptualisation when considered through the lens of embodied cognition. The foundational proposition of embodied cognition that I've just reiterated shows us that the mind is inherently embodied, not simply because the brain operates in a body, but because physical experience shapes conceptual thought. Furthermore, thought employs many of the same neuronal pathways as physical action (a process often called 'neural exploitation' – see Gallese 2007b). Kinaesthetic and perceptual experiences of the material world generate cognitive systems that reflect our physical environments and interpersonal experiences and form patterns for higher cognitive activity. As a result, cognitive processes like language and conceptual thought use partial re-activations of sensory, motor and affective systems. As Vittorio Gallese proposes:

> [K]ey aspects of human social cognition are underpinned by brain mechanisms originally evolved for sensorimotor integration. It is proposed that these mechanisms were later on adapted as new neurofunctional architecture for thought and language, while retaining their original functions as well. By neural exploitation, social cognition and language can be linked to the experiential domain of action.
>
> *(Gallese 2007b, 317)*

So Lecoq's description of lived experiences in the physical world as 'dynamics' -combinations of rhythm, force and space – identifies them at their sensorimotor level, the actional level at which we engage with our environments before we start to consciously reflect on them or describe them in language. It is at this level that his Neutral Mask training re-sensitises actors to the sensorimotor sources of mental concepts.

For example, they work with the dynamics of water in different states (bubbling spring, meandering river, stormy sea), of fire, of air, also those of materials like paper and cellophane and of humans experiencing varied physical environments. These activities demand a sustained mental and physical discipline to accurately observe and physically embody different dynamics in ways that make them specific and expressive. For instance, the dynamics of waves in a storm are distinct from those of a bubbling brook. The performer must distinguish how the forces within the varied physical constraints affect the water, and change its rhythm, tempo and directional tendencies. These activities provide the actor with a reinvigorated awareness of the relationship between these rhythmic patterns and the concepts that they generate and shape through the relationship between motor cortex activity, thought and language. These activities also stimulate emotion, a phenomenon that I've addressed in an earlier publication (Kemp 2016).

Lecoq extends this process beyond simple physical mimicry by subsequently combining the experiential physical activities with varied modes of linguistically expressed meaning – single words, poems, improvised dramatic narrative. Through these sequences of exercises, the 'dynamics' of nature and materials are invested in and correlated with the 'dynamics' of communicative expression. Lecoq considers that this training develops lasting patterns of behaviour in the performer: 'The main result of this identification work are the traces that remain inscribed in each actor, circuits laid down in the body, through which dramatic emotions also circulate, finding their pathway to expression' (Lecoq 2001, 45). Using consciously articulated patterns of muscular activity, actors develop what is known in theatre parlance as 'muscle memory' of movement schema that are linked to concepts and emotions. Using concepts from cognitive neuroscience, we could describe this as a refined awareness of the proprioceptive and interoceptive dimensions of emotion and thought. Lecoq's phrase 'circuits laid down in the body' also evokes another primary assumption of embodied cognition – 'that any type of recall includes a sensorimotor simulation of the processes involved in the original encoding of the experience' (Koch, Fuchs, Summa and Mulller 2016, 2, referencing Barsalou, Niedenthal, Barbey and Ruppert 2003, Niedenthal 2007).

Actions and language

Another result of Lecoq's training is the specificity of expressive physical action that arises from this level of observational and physical rigour. When engaged with language, this physical precision offers us a view of meaning in interpersonal communication that accords with findings that language and gestural actions are intertwined in communication. As cognitive linguist David McNeill asserts, 'gestures are an integral part of language as much as are words, phrases and sentences – gesture and language are one system' (McNeill 1992, 2). Lecoq's work on language and gesture can also be usefully appreciated through Glenberg and Gallese's action-based theory of language. They point to 'findings that strongly support the existence of mirror neurons in the human motor system and [that] have led to the notion of a mirror neuron system involving areas in the frontal lobes (notably, Broca's area) and parietal lobes' (Glenberg and Gallese 2011, 8). The significance here is that Broca's area has traditionally been associated in brain research with speech and language. Glenberg and Gallese are careful to point out that they consider that both emotion and perception systems are active in language in addition to action systems, but have assimilated data from multiple findings to form a theoretical framework that roots language in action. They state that:

> parietal mirror neurons not only code the goal of an executed/observed motor act, like grasping an object, but they also code the overall action intention […] The 'motor vocabulary' of grasping-related neurons, by sequential chaining, reorganizes itself [so] as to map the fulfillment of an action intention. The overall action intention (to eat, to place the food or object) is the goal-state of the ultimate goal-related motor act of the chain.
> *(Glenberg and Gallese 2011, 3)*

This concept of an 'action intention' meaning that is definable through motor actions is another concept of embodied cognition that Lecoq's practice both anticipates and illuminates. Earlier work (Murray 2003, Kemp 2012) has described Lecoq's analysis of, and work with, the action verbs of 'push' and 'pull.' This work arises from Lecoq's recognition that these actions are essential in the mechanics of movement and have extensive metaphorical applications in both verbal and nonverbal communication. To conclude this chapter, I will point out another aspect of Lecoq's work with language that anticipates the view in embodied cognition that culture is one of the factors that is involved in the situated nature of meaning.

Case study: 'Je prends'

In the video *Les deux voyages de Jacques Lecoq* (Roy and Carasso 1999), Lecoq can be seen teaching part of a lesson on language and gesture. He's working on the action verb 'je prends' – normally translated as 'I take.' In the extract of the lesson that is shown in the video, he has divided a large group of about 20 students into nationality groups and asked each group to define the gesture that expresses this action for them. Each group is given a few minutes to confer and practice (the groups work simultaneously) and is then asked to show the gesture while saying the phrase 'I take' in their own language. The first group to show their work is composed of five or six Americans (of mixed gender) who stand in a circle. The gesture that accompanies the phrase is a motion with the hand and arm that reaches directly forward in front of the torso, grasps an imagined item and sharply draws it towards the torso. The movement is sudden and forceful, and the phrase 'I take' is repeated multiple times. Lecoq is surprised; from his French perspective the American version of 'take' is described as 'arrache' – 'grab.' The Americans are followed by a group of four British English speakers (again of mixed gender) who stand in a square formation. They say the phrase 'I take' once and make a similar gesture to the Americans – reaching out in front, grasping an imagined item, then pulling the item towards themselves. However, the British version, while also including the action of pulling towards the torso, seems less acquisitive than the American because there is less force, a slower tempo and the vocal expression is considerably softer – a contrast that provokes laughter among the observers and the comment from Lecoq that the gesture is 'more diplomatic.' The British are followed by a group of three Scandinavian women whose symmetrical gesture involves using both hands to reach out rapidly, grasp and then pull an imagined item smoothly towards the torso. The sustained tempo of the pulling action is in marked contrast to the sudden and rapid action of the American gesture, but is nevertheless decisive in quality. The next example is given by an individual – a woman who is the only speaker of Serbo-Croat in the class – whose gesture involves placing both hands on an imagined item in front of her and then pulling this item with a sustained tempo to the right side of her torso. This lateral movement is significantly different from the pulling action of the three previous groups that drew the 'taken' item directly towards the torso. The gesture associated with the Serbo-Croat verb does not bring what is taken to the individual but places it to the side. This prompts Lecoq to comment that the gesture evokes the idea of 'putting aside for winter.' The final group shown in the video is French – again of mixed gender. Their gesture involves both arms and hands reaching out in front of the torso and then resting on an imagined object. This completes the gesture. Intriguingly for English speakers, for whom the word 'take' has connotations of 'acquire,' the French gesture has no activity of pulling the imagined item towards the self. Lecoq seems satisfied that the gesture communicates the meaning of the French phrase 'je prends,' saying 'It's mine [...] I have my hand on it, I stay there' as an elaboration of the meaning of the gesture. These five gestures are all prompted by the concept 'I take,' but have significant differences in what they communicate.

Lecoq's training in the dynamics of movement enables both the clear execution and description of gesture in terms of direction, force, tempo, tension and so on. This level of specificity highlights the visible nuances of meaning among the different nationality groups. When considered in the context of Glenberg and Gallese's action-based language theory, one can see that Lecoq's approach identifies how the word 'take' – apparently equivalent in meaning when translated linguistically – has different action intentions in different cultures. Differences in force, tempo and degree of suddenness between the American, English and Scandinavian groups make gestures that have the same basic components (grasping something in front of oneself, drawing it towards oneself) evoke different qualitative associations. The similarity of the basic action components may relate to the fact that the English 'take' has its etymological roots in 'taka' – a word from Old Norse, which is the ancestor of Scandinavian languages. As I hope my description has made clear, both the Serbo-Croat and the French gestures associated with the phrase 'I take' suggest not just qualitative differences, but markedly different *action intentions*. The two-handed drawing motion to the side of the body enacted with the Serbo-Croat 'ja uzimam' evokes connotations of 'putting aside' – and we see that the gesture enacted with the French 'je prends' has no component of pulling the imagined item towards the self. The video of Lecoq's lesson shows the difference in action intentions manifested by the varied gestural patterns of the nationality groups. Further differences might well be observed were the nationality groups engaged in such an exercise further sub-divided, so that the cultural nuances attendant upon gender, socio-economic status, ethnicity and race could be expressed.

These concrete examples of meaningful behaviour that arise from Lecoq's teaching have significant implications. They demonstrate that movement tends to carry more specific meanings about physical activity than language, supporting Merlin Donald's proposal (quoted earlier) that '[I]f mimesis was the adaptation that generated a distinctly human culture, it follows that the deepest communicative framework of human culture must still be mimetic.' When we connect this idea with Lakoff and Johnson's axiom that mental concepts are shaped by physical experience, the visibly different gestures can be understood to express variations of the concept of 'take' among the nationality groups. This suggests that the concept involved in 'je prends' is more accurately translated into English as 'I take hold of' or even 'I grasp.' Conversely, the English concept of 'take' has connotations of 'acquire' – 'acquérir' in French. Lecoq's exercise gives us a vivid example of the culturally situated and embodied nature of meaning – and demonstrates ways of clarifying meaning in the symbiosis of movement and language.

Conclusion

While Lecoq's pedagogy is certainly designed as actor training, graduates of his school have achieved success in many ways. They work as actors in various styles, as clowns and as practitioners of devised theatre and also as playwrights, painters, sculptors and directors of film and theatre. In addition to their presence in acting programmes, Lecoq's techniques are being used in the training of doctors and architects, to teach English as a foreign language and as part of reconciliation programmes in South Africa (Evans and Kemp 2016). This wide diversity of application arises because Lecoq's concepts of 'Tout bouge,' 'le fonds poetique commun' and 'dynamiques' have shaped his training in ways that are congruent with a core proposition of embodied cognition: that meaning results intersubjectively from our situated interactions with the world. The exercises that Lecoq has developed clarify and articulate this proposition, enabling conscious eliciting of what is often unconscious behaviour in daily life. Empirically supported theories of embodied cognition show that mental, physical and emotional activities are all intertwined. Lecoq's training benefits actors and other artists by working in a way that both acknowledges and consciously utilizes this phenomenon of human cognition. Consequently, his training can provide a valuable forum of research for those from various disciplines outside of theatre who wish to study human behaviour from the perspectives of embodied cognition.

Acknowledgement

This chapter was developed from a paper presented at the Third Symposium on Embodied Cognition, Acting and Performance at the 2016 Convention of the Society for the Study of Artificial Intelligence and Simulation of Behaviour (AISB). The paper was then published in *Connection Science* (Vol. 29, Issue 1, 2017) the Journal of AISB.

References

Barsalou, L. W., P. M. Niedenthal, A. Barbey and J. Ruppert (2003), 'Social embodiment.' In *The Psychology of Learning and Motivation,* B. Ross (Ed.). San Diego, CA: Academic Press, Vol. 43, pp. 43–92.

Carnicke, S. (2009), *Stanislavsky in Focus* (2nd Ed.). Abingdon and New York: Routledge.

Donald, M. (2005), 'Imitation and mimesis' in *Perspectives on Imitation, Vol. 2. From Neuroscience to Social Science: Imitation, Human Development and Culture,* Susan L. Hurley and Nick Chater (Eds.). MIT Press, pp. 283–300.

Duckworth, K., J. Bargh, A. Garcia and S. Chaiken (2002), 'The automatic evaluation of novel stimuli.' *Psychological Science,* Vol. 13, No. 6, pp. 513–9.

Ekman, P. (1999), "Basic emotions." in *Handbook of Cognition and Emotion,* T. Dalgleish and M. Power (Eds.). New York: Wiley, pp. 46–60.

Ekman, P. (2003), *Emotions Revealed: Recognizing Faces and Feelings to Improve Communication and Emotional Life.* New York: Henry Holt and Company.

Ekman, P., R. J. Davidson and W. V. Friesen (1990), "Emotional expression and brain physiology II: The Duchenne smile." *The Journal of Personality and Social Psychology,* Vol. 58, pp. 342–53.

Evans, M. and R. Kemp (Eds.) (2016), *The Routledge Companion to Jacques Lecoq.* New York: Routledge.

Gallese, V. (2007a), 'Before and below "theory of mind": Embodied simulation and the neural correlates of social cognition.' *Philosophical Transactions of the Royal Society B: Biological Sciences,* Vol. 362, No. 1480, pp. 659–69.

——— (2007b), 'Mirror neurons and the social nature of language: The neural exploitation hypothesis.' *Social Neuroscience,* Vol. 3, No. 3, pp. 317–33.

Gibson, J. J. (1979), *The Ecological Approach to Visual Perception.* Boston, MA: Houghton Mifflin.

Gilrain, J. (2016), 'The Mimo-dynamics of poetry, short story and music: Lecoq on Bartok' in *The Routledge Companion to Jacques Lecoq,* M. Evans and R. Kemp (Eds.). New York: Routledge, pp. 127–34.

Glenberg, A. M. and V. Gallese (2011), 'Action-based language: A theory of language acquisition, comprehension, and production.' *Cortex,* Vol. 48, No. 7, pp. 1–18. Available online: doi:10.1016/j.cortex.2011.04.010 (accessed on 5/13/16).

Holt, K. G., R. O. Wagenaar and E. Saltzman (2010), 'A dynamic systems/constraints approach to rehabilitation.' *Brazilian Journal of Physical Therapy,* Vol. 14, No. 6, pp. 446–63. São Carlos.

Jousse, M. (1932), Mimétisme et mimisme. 4ème conférence. École d'Anthropologie. 28 nov. 1932. Le cours de Marcel Jousse. CD-rom 1/2. Association Marcel Jousse, 2003.
Kemp, R. (2012), *Embodied Acting: What Neuroscience Tells Us about Performance*. Abingdon and New York: Routledge.
——— (2016), 'Lecoq, embodied cognition and emotion' in *The Routledge Companion to Jacques Lecoq*, M. Evans and R. Kemp (Eds.). New York: Routledge, pp. 199–207.
Koch, S., T. Fuchs, M. Summa and C. Mulller (Eds.) (2012), *Body Memory, Metaphor and Movement*. Philadelphia, PA and Amsterdam: John Benjamins Publishing Company.
Lakoff, G. and M. Johnson (1999), *Philosophy in the Flesh: The Embodied Mind and Its Challenge to Western Thought*. New York: Basic Books.
Lecoq, J. (1997), *Le Corps Poetique*. Paris: Editions Actes Sud.
Lecoq, J. (2001), transl. D. Bradby. *The Moving Body*. New York: Routledge.
Lecoq, J. (2006), ed. D. Bradby. *Theatre of Movement and Gesture*. London and New York: Routledge.
McNeill, D. (1992), *Hand and Mind: What Gestures Reveal about Thought*. Chicago, IL: University of Chicago Press.
Meeren H. K. M., C. C. R. J. van Heijnsbergen, and B. de Gelder. (2005). 'Rapid perceptual integration of facial expression and emotional body language.' *Proceedings of the National Academy of Sciences of the United States of America*. November 8, 2005. Vol. 102, No. 45, pp. 16518–23.
Murray, S. (2003), *Lecoq*. London: Routledge.
Niedenthal, P. (2007), 'Embodying emotion.' *Science*, Vol. 316, No. 5827, pp. 1002–5.
Roy, J.-N. and J.-G. Carasso (1999), *Les deux voyages de Jacques Lecoq*. (Video). Paris: La Septe ARTE-On Line production ANRAT.
Scheffler, I. (2016), 'Laboratory of movement study' in *The Routledge Companion to Jacques Lecoq*, M. Evans and R. Kemp (Eds.). Abingdon and New York: Routledge, pp.179–86.
Stepper, S. and F. Strack (1993), 'Proprioceptive determinants of emotional and nonemotional feelings.' *Journal of Personality and Social Psychology*, Vol. 64, No. 2, pp. 211–20.
Tom, G., P. Pettersen, T. Lau, T. Burton and J. Cook. (1991). 'The Role of Overt Head Movement in the Formation of Affect.' *Basic and Applied Social Psychology*. Vol. 12, No. 3, pp. 281–89.
Varela, F. J., E. Thompson and E. Rosch (1991), *The Embodied Mind: Cognitive Science and Human Experience*. Boston, MA: MIT Press.

PART III: SCHOLARSHIP

Introduction

Bruce McConachie

This part of nine essays examines the many ways that cognitive science has assisted and informed theoretical, critical and historical scholarship in theatre and performance studies. While most of the chapters in our book thus far have relied on insights from **embodied** areas of cognitive science, several of these essays venture into the fields of **enactive** and **distributed** cognition. The first essay, in fact, underlines the importance of systems theory for the epistemology of enactivism and the second primarily explores insights from the philosophy of phenomenology into understanding theatrical spectating as enactive. The science on how spectators actually attend to a performance – the focus of the third essay – nicely complements the second chapter. Spectator attention necessarily leads to meaning-making in the next essay. This is followed by two chapters oriented towards emotional experience – the first about the enjoyment of *rasa* by audiences at performances of traditional Indian theatre and the next about the emotion of shame that shapes the experience of many traditional tragedies. Performances about ageing bodies and ethical responsibility centre the next two essays, in which a son struggles to help his old father with his bowel problems as an impassive Christ looks on and a dancer has a lively conversation with her complaining legs, arms and feet. A final chapter uses distributed cognition to answer historical questions about the likely staging of four different kinds of theatre, ranging from ancient Roman comedy to contemporary puppet performance.

In 'Systems theory, enaction, and performing arts,' Gabriele Sofia contextualises the paradigm of **enaction** in cognitive science within similar paradigm shifts that have occurred in other disciplines (including sociology, biology and theatre) when those scholars recognised that **dynamic systems theory** altered key relationships between its parts and the whole. Just as a whole 'society' in sociology is more than the sum of its parts ('social groups,' 'language,' 'norms,' etc.) under systems theory, so cognition as a whole system is more than neurons, memory, perception and its other constitutive parts. Sofia points out that systems theory in these and other areas has epistemological ramifications that break down the usual distinctions between objective and subjective knowledge. An audience member, for example, can never gain completely objective knowledge of any performance she is watching because her presence in the theatre necessarily alters – if only in small ways – the relationship between performers and spectators. From an enactive and systems perspective, one phenomenological result of this insight is that the interactions of spectators and actors are the primary co-constituents of all performances, within an ecology that is bounded by specific constraints of space, air and time.

Stanton Garner attends to one aspect of the phenomenological consequence of enaction in his essay, 'Watching movement: Phenomenology, cognition, performance.' He investigates what **mirror neurons** and motor cognition generally can tell us about the cognitive and emotional transactions that couple spectators and actors and the implications that this science has for the experience of spectating. Garner enters the ongoing dialogue between science and phenomenology warily, acknowledging the long history of mutual suspicion that has made this conversation difficult. Part of the problem has to do with the scientific understanding that many cognitive processes are pre- or unconscious, while phenomenology privileges the possibilities of **consciousness.** Nonetheless, Garner cites several phenomenologists and scientists who have productively borrowed from each other to forge a rapprochement, including Merleau-Ponty, Varela, Sheets-Johnstone and Thompson. Garner ends his essay with a cognitive-phenomenological analysis of the climactic scene between Willy Loman and his son, Biff – emphasising their changing physical and emotional relationships and his responses to them – in the 1999 production of *Death of A Salesman* with Brian Dennehy and Kevin Anderson in New York.

Given the importance of cognitive attention in watching *Salesman*, we follow Garner's essay with a chapter by James Hamilton, 'Attention to theatrical performances.' Looking first at the debates about attention in cognitive science and philosophy, Hamilton then comments on the demands that the theatre typically places on spectators for their attention. He recognises that the past theater-going experience of spectators shapes their attention, that regularities in the theatrical environment matter for attentiveness and that directors and performers often create moments of focus to capture spectatorial attention. Hamilton uses the **predictive processing** ideas of Andy Clark and related theories to understand how spectators seek to minimise prediction errors while watching plays. In accord with **enaction** theory, he notes that spectator attention is often drawn to performer action as well as their preparation for action.

Amy Cook begins with the example of the New York Public Theatre's controversial production of *Julius Caesar* in 2017 to ask what it means to make meaning in the theatre in her 'Emergence, meaning, and presence: An interdisciplinary approach to a disciplinary problem.' Meanings **emerge** in the interactions between performers and spectators within the constraints of theatrical space and time; meanings are never given in the symbols on stage or in the minds of spectators. Turning to the theory of **conceptual integration** (aka, blending), Cook explains that it challenges conventional semiotic approaches to meaning-making. She also remarks that 'blending' has become an unfortunate synonym for conceptual integration because it connotes the operation of a kitchen blender, which merely mixes ingredients together instead of creating new integrations and meanings. Cook adds that she, like many spectators, wants more than spoon-fed meanings when she goes to the theatre; she wants to experience the mystery that derives from 'presence' as well.

Part of that presence may derive from what ancient Indian aestheticians called rasa, which Erin Mee explores from the angle of **enactivist emotion** theory in 'Relishing performance: Rasa as participatory sense-making.' Simply put, rasa is what audiences 'taste' while experiencing a performance; it occurs in the interactions between performers and spectators, whom Mee terms 'partakers.' While rasa as 'taste' is a useful metaphor, Mee insists that this experience can be more fully realised if it is understood as the emotional dimension of intersubjectivity joining performers and partakers. She turns to the ideas of Giovanna Colombetti, whose enactivist approach to empathy and the emotions has been widely cited, to better understand the kinds of actor-audience interactions that occur in performances today as well as those that happened in the traditional theatre of India.

In 'The self, ethics, agency, and tragedy,' David Palmer begins by asking how Darwin's materialist understanding of evolution and cognition could possibly have anything to do with our understanding of tragedy, which is necessarily premised on the possibility of agency for its tragic hero. Without some ability to know and choose his fate, the hero's fall is reduced to melodrama; he becomes the victim of circumstances beyond his control. Palmer gets around this problem by dividing the hero in two, such that he is both a material self and an image of and for himself – a **conscious** image in a life narrative about himself that secures his self-respect. If the hero violates the ethical expectations of this narrative, he often falls into shame and possible death. When Macbeth violates his better judgement and self-respect by killing Duncan, for example, he suffers from guilt and remorse, which only deepen his tragic entanglements for the rest of the play. Palmer's essay unites a cognitive understanding of the emotions with a Platonic understanding of tragedy.

John Lutterbie examines the ethical and political possibilities of theatrical *ostranenie* (Shklovsky's term for making familiar matters strange) in his 'Aesthetics and the sensible.' He borrows 'the sensible' from Rancière, who uses it to mean the institutions and beliefs that organise our social world as normative and acceptable; like Shklovsky, Rancière believes that disrupting 'the sensible' and rendering it strange through performance can open up possibilities for progressive political action. According to Rancière, two types of images have this power to disrupt normative spectator expectations: the intolerable and the pensive. Lutterbie finds potent examples of both images in *The Concept of the Face: Regarding the Son of God*, Romeo Castellucci's three-part production intended to challenge the power of the Catholic Church. A pensive head of Christ devoid of compassion dominated the background of the production as three different dramatisations featuring intolerable images occurred in the foreground. Lutterbie focuses his attention on the first piece, which featured a middle-aged son trying to help his incontinent old father, who fouls his diaper three times. Lutterbie reports that the intolerable images (and smells) of these three defecations forced him to look away, but the impassive image of Christ intentionally offered no consolation.

'Talk this dance: On the conceptualization of dance as fictive conversation,' by Ana Margarida Abrantes and Esther Pascual, foregrounds the importance of conversation as a flexible and powerful frame for performative communication. The authors demonstrate their thesis through an exploration of *Tchaikovsky's Left Leg*, a dance with dialogue written by Tiago Rodrigues for prima ballerina Barbara Hruskova, who danced and spoke it in Lisbon, Portugal, in 2015. In the performance, Hruskova talked to different parts of her body – her legs, feet, arms and so forth – and then related what her body parts said and did back to her; her feet, for example, tell her they are tired and want to rest and then run off (seemingly by themselves) when she tries to get them to dance again. Staging the conflict between the ageing dancer and her body through a dialogue of fictive conversation and action evoked more **empathy** for her than would a conventional monologue, the authors assert. Rodrigues' script also invited Hruskova to mark several of her moves rather than fully dancing them, a strategy Abrantes and Pascual relate to Edwin Hutchins's understanding of **distributed cognition**.

Distributing cognition in the environment is the focus of our final essay in this part by Evelyn Tribble and Robin Dixon, '**Distributed cognition**: Studying theatre in the wild.' Their title echoes Hutchins's path-breaking 1995 book, *Cognition in the Wild*, which makes the case that cognition is best studied where it is normally practiced – in work places, everyday life and nature, not in psychology labs. In their chapter, Tribble and Dixon primarily study how various groups of actors offloaded their cognitive challenges to opportunities offered in three different historical theatres and their technologies as a means of stabilising its primary conventions for themselves and communicating them to their spectators. How did Shakespeare's performers memorise all those lines in such a short time? How did the actors in ancient Rome communicate the fixed locales of Plautus's comedies to their audience? How did the improvising performers of commedia dell' arte keep track of where they were in the plots of their fast-moving narratives? These are just some of the cognitive problems that actors at the time had to solve and that historians today must try to understand in order to figure out how such theatre succeeded with their spectators. For their fourth example, the authors look specifically at how the puppeteers behind (and underneath) the Handspring Puppet Company's production of *War Horse* endowed their constructions of wood, plaster and cloth with the breath of life.

17

SYSTEMS THEORY, ENACTION AND PERFORMING ARTS

Gabriele Sofia

The revolution of systems theory

Systems theory has undoubtedly been one of the most revolutionary turning points of epistemology in the twentieth century. We could hardly find a field of studies that has not been influenced by this theory. Its paradigms have affected several research areas, from biology to philosophy, from sociology to theatre cultures.

One of the insights that has promoted such a *paradigm shift* is the awareness that, in order to analyse living systems, we cannot merely describe their single components, but have to consider how they interact with each other. Adopting a well-known expression, the 'whole is greater than the sum of its parts.' The added value is due to their organisation. One of the first scientific fields to embrace this paradigm was the biological sciences. On the basis of the understanding that anatomy itself was not sufficient for a description of organisms, the biologist Ludwig von Bertalanffy, founder of the *General System Theory*, stated:

> Since the fundamental character of the living thing is its organization, the customary investigation of the single parts and processes cannot provide a complete explanation of the vital phenomena. This investigation gives us no information about the coordination of parts and processes.[1]
>
> *(Von Bertalanffy 1972, 410)*

The epistemological consequences of such considerations caused a deep questioning of the processes of knowledge organisation in Western cultures. Edgar Morin tried to synthesise it, by comparing the paradigm of 'simplification' with that of 'complexity':

> The paradigm of simplification is based on the separation of the different areas of knowledge; in this sense, the objects of knowledge are separated from their context. Therefore, we think that we 'know,' when we isolate the object. The first aspect of simplification is separation, the second is reduction: the knowledge of a set of elements of a whole is reduced to the knowledge of each of its single parts, without considering

> that there are qualities of the whole which cannot be detected when each of its parts is analysed singly. On the contrary, the paradigm of complexity consists in maintaining the interdependency between objects: we observe all of them, but maintain them as united.
>
> *(Morin 2002, 25)*

One of the most important innovations in this sense has been the central role played by the notion of *environment*. For hundreds of years the cultural hegemony of scientific objectivity has imposed the experimental practice of isolating the analysed phenomenon from the physical, environmental, cultural context within which it naturally takes place. In contrast, systems sciences have highlighted the connection between the different elements and therefore reintroduced the study of the relations to which a phenomenon gives rise within its environment. 'Complex' stems from the Latin root *cum-plexus*, 'to weave with.'

One of the most important consequences of this approach, which have sometimes been defined as *ecological*, is the focus on the inevitable influence of the observer on the phenomenon, with which he shares the same environment. The research on quantum physics provided impressive examples.[2] In this account, there can be no presumption of purely objective measurements, which are considered as resulting from the relationships among the observed phenomenon, the observer, the adopted tools and the environment in which the observation takes place. In other words, the scientific world has been forced to overcome the distinction between subject and object that had been fundamental in the past.

Theatre cultures have been widely affected by this revolution. In the 70s, Grotowski had already proposed the following definition of theatre: 'We can thus define the theatre as what takes place between spectator and actor. All the other things are supplementary' (Grotowski 1968, 32). With these words, Grotowski introduced a systemic logic into the study of the theatrical event, which is no longer considered as an *object*, but as a constantly developing *relationship*, that *takes place* between (at a minimum) two human beings.[3] By his mere presence, the spectator is therefore a co-creator of the theatre event. The actor-audience relationship is no longer characterised by a unidirectional logic, but rather by a circular system of stimulations.

Undoubtedly a great contribution to the spread of systems theories within the performing arts has been those new theories that propose patterns of cognition no longer isolated in an individual's brain, but embodied and located within an environment. The model that has probably facilitated the most complete synthesis of these principles is the *enaction* model, proposed by Francisco Varela in the late 80s.

Enaction: a new model of human cognition

The notion of enaction is the result of a long process of research and reflection that Francisco Varela had carried out since his youth, when he started to collaborate with his mentor Humberto Maturana. In fact, this notion is based on what the two scientists defined as *autopoiesis*, which is a life system's capacity for self-production through constant interaction with its environment, despite the ongoing changes of its components (Varela, Maturana, and Uribe 1974; Maturana and Varela 1973, 1984). The notion of *autopoiesis* permitted the definition of a living system not through its elements but through its dynamical organisation. In this sense, the ontology of a living system is not determined by what it *has got*, but by what it *does*, by its way of interaction and self-organisation. Therefore, *action* plays a central ontological role that, once applied to the cognitive sciences, led Varela and his colleagues to propose a radically new model:

> Since this analytical perspective is specifically interested in highlighting the notion of action with respect to that of representation, it is suitable to name this new approach of cognitive sciences *enaction*.
>
> *(Varela 1989, 93)*

In fact, enaction opposes representational models, on which almost all the studies of the cognitive sciences had so far been based. The representational models described the relationship of the human being to its environment, according to a logic based on a strict distinction of inner and outer. Initially, the so-called 'cognitivist' models stated that the brain produced 'inner' symbolical representations of the 'outer' world and that they had to be deciphered through cognitive operations. These models were then replaced by the so-called 'emergentist' ones, which proposed that the world cannot be experienced directly but only through emerging representations that the brain *projects* onto the environment.

As an alternative to these perspectives, the enaction model was not based on representations but on *embodied action*:

> These two extremes both take representation as their central notion: in the first case representation is used to recover what is outer; in the second case it is used to project what is inner. Our intention is to bypass entirely this logical geography of inner versus outer by studying cognition not as recovery or projection but as embodied action.
>
> *(Varela, Thompson, and Rosch 1991, 172)*

The latter is then defined according to two characteristics:

> By using the term embodied we mean to highlight two points: first, that cognition depends upon the kinds of experience that come from having a body with various sensorimotor capacities, and second, that these individual sensorimotor capacities are themselves embedded in a more encompassing biological, psychological, and cultural context. By using the term action we mean to emphasize once again that sensory and motor processes, perception and action, are fundamentally inseparable in lived cognition.
>
> *(Varela, Thompson, and Rosch 1991, 172–3)*

Underlying the impossibility of separating the perceptive from the motor processes (the same will be confirmed at a physiological level by the studies on the mirror neuron mechanism, cf. Rizzolatti and Sinigaglia 2006), the enaction model based cognition on the circular relationship between human beings and their environment. This understanding can be synthesised in the following few lines. Each time we 'perceive' the world, we are actually *acting* upon it, and it upon us. In other word, we are modifying the object that we intend to perceive. If every perceptive act modifies the very object of perception, then there is no pre-determined world that can be perceived 'as it is.' Instead, the world itself is established and modified by the individual who perceives it. What we refer to as 'perception' is not a passive process, it is not data collection, but it is the result of a circular interaction between the individual and the world.

Situating performance

As I have already said, the performing arts have been influenced by enaction theories, both directly and indirectly.[4] However, this influence has often been limited to the adoption of some successful key words (*embodied mind*, *autopoiesis*, etc.), without embracing the full paradigm shift required by enaction theories. For example, many analyses continue to employ a representational logic, or a strict dichotomy between inner and outer is often maintained.

Moreover, most scholarship in the performing arts continues to operate according to the paradigm of simplification, proposing a separation of diverse genres (dance, theatre, circus, etc.), competencies (lighting design, dramaturgy, scenography, etc.), steps of production (training, rehearsals, performance, etc.), components (text, acting, direction, etc.) and even the phenomenon itself (performance vs. reception). In other words, performance has often been considered as an *object* rather than as a constantly changing and dynamic *relationship*. If at a certain level of the work such a method of analysis can appear functional, it is because one does not analyse the interactions and the organisation of the examined elements.

Precisely as occurs in systems sciences or enaction theories, the first step to subvert this trend is the act of situating the performance event within its environment. If studies on reception have often focused on the way the spectator perceives the performance, the opposite influence has often been neglected. A strict systemic approach forces us to imagine how the mere presence of the spectator affects – or better, interferes with – the performer's cognitive organisation. Doing this, we become aware of two remarkable *interferences*:

1 The former concerns the organisation of the performer's action. If an actor has to drink from a glass on the stage, every motor act he is going to activate ('grabbing the glass,' 'bringing the glass to his mouth,' 'swallowing the water,' 'putting the glass again on the table') will be aimed at performing the act of 'drinking a glass of water.' but, though the action is factually the same and follows *the same series of motor acts* and is performed *in the same time*, it will be different, since the actor is here supposed to stimulate the spectator's attention. The actor's action is not only directed to an on-stage aim (drinking the glass of water) but also to the audience (stimulating the spectators' attention). The same action has two different aims. This gives rise to a 'double intention' of the actor or rather to a *dilated intention* that broadens from the performed action out to the audience. We can reasonably suppose that such a 'broadening' of the intention concerns a peculiar neuromotor dimension.

2 The latter concerns the planning of actions. The spectator is not merely someone who is present in that particular moment; he is rather someone who 'expects' something from the performer. The Latin etymology of the word 'spectator' is *spectare*, which is the same stem of the Italian word *aspettare* (expecting/waiting for). In this sense, the spectator's aim is always to anticipate, with more or less consciousness, the actions that will take place on the stage.[5] On the other side, the effect of spontaneity of the actor is based on the ability of performing every action, albeit usually planned and rehearsed, as if it took place there for the very first time, without making any unconscious anticipations. Let's take an example. Let's think about something on the stage that is supposed to appear as 'unpredictable' for the spectator, like the sound of someone knocking at the door. If this moment is incoherent because the actor knows exactly when that would happen and moves towards the door before the sound, the spectator can anticipate the knock and the scene is unbelievable. This means that the actor needs to modify his daily and

> natural tendency to anticipate the sequence of his actions and rebuild it with a different awareness of his stage behaviour.

This example shows the difference between the everyday and the on-stage action and allows us to remark that, considered in neurophysiological terms, this difference is probably quite significant, since it may concern the processes of action planning, postural control, decision-making, body schema and so on. On stage, through specific training, the human being has then to develop neuromotor dynamics different from those adopted in his everyday life. The great exponents of the theatre of the twentieth century defined this alternative organisation of the performer's thought-action in different ways: as second nature (Stanislavski), as achieved spontaneity (Copeau), as biomechanical (Meyerhold), as Über-marionnette (Craig), as an actor's extra-daily techniques (Barba) and so forth. If we analyse how the pedagogical practices of these directors operate on the actor's organism, we find several correspondences. Despite the aesthetic differences between one director and the other, most of the exercises that they proposed operate on mechanisms related to the *body schema*. The studies concerning the notion of body schema teach us that our body-mind system involves widely dispersed operations that cannot be traced back to a 'mechanism of central control' that combines information concerning the outer environment, the position of the actor's body in space, his posture and his possibility of interacting. Some research suggests that actors may develop some deep modification in their body schema dynamics, acquiring what I have elsewhere defined as a *performative body schema* (Sofia 2013a, 2013c; Lippi et al. 2016). That could give to the actor a different awareness of his stage behaviour. Moreover, a *performative body schema* would permit the actor to develop new motor routines that allow for the *dilated intention* I have mentioned before.

Co-constitution and intersubjectivity

In the last years of his life, Francisco Varela underlined how the enactive paradigm could provide a new methodology to approach the study of human consciousness. Varela looked for evidence as to how each subjective experience is connected to the experience of the others:

> Experience is clearly a personal event, but that does not mean it is *private*, in the sense of some kind of isolated subject that is parachuted down onto a pre-given objective world.... An investigation of the structure of human experience inevitably induces a shift to considering the several levels on which my consciousness is inextricably linked to those of others and the phenomenal world in an empathic mesh.
>
> *(Varela 1996, 340)*

Questioning the actor-spectator 'empathic mesh' could be one of the key problems of a systemic approach to the performing arts. If an investigation of human experience is possible, the study of human experience when involved in an actor-spectator relationship should be possible as well.[6] Recent studies on intersubjectivity are essential in order to approach this problem. Alain Berthoz and Jean Luc Petit (2008) proposed the notion of *co-constitution*. This notion updates what Edmond Husserl (1973) defined as *Mitkonstitution* and recalls enaction theory, according to which there is no 'pre-determined' world, because the world itself is constituted, modified and perceived by the subject through a circular process of perception-action. Thus, the world is not separated from the subject, but rather composes the latter; it is part of him. In this perspective, two interacting subjects do not 'share' a same pre-determined world but actively co-constitute it. The world perceived as 'already constituted' is an effect due to the fact that the subject has always been experiencing a context shared with other subjects, who interact with the world through the same motor routines of the subject. This is how we can realise that my way of perceiving the world depends on the way in which the other subjects interact with me, co-constituting the world. More recently, some research (Costantini et al. 2011; Fini et al. 2015; Committeri and Fini 2016) has reproposed this paradigm, highlighting some important aspects. The perception of an object does not depend only on my potential action on the object, but also on the potential action of the other individuals who, being present within my field of vision, might interact with the same object. We do not simulate only our interaction with space, but also those with whom we *co-constitute* the space. Let us suppose that I am a spectator and that the individual with whom I co-constitute space is an actor. The actor would inhabit the co-constituted space according to neuromotor routines *different* from the usual ones: he would actually adopt a *performative body schema*. His extra-daily motor techniques would also make my simulation of his potential actions extra-daily. Moreover, in this way, the perception of the space that I, as spectator, have co-constituted with the actor, changes radically and provides me with an extra-daily experience of the environment. My experience would then acquire different and unique *structure* not only with respect to all the other daily experiences, but also with respect to all the other spectator's experiences that do not entail a co-constitution of space with another human being (cinema, TV, internet, etc.). This could be, therefore, one of the fundamental phenomenological characteristics of the *performative* experience of the spectator in a theatre.

From the phenomenological level to the sense-making process

What we have discussed up to now could give us a new way of thinking about the motor-cognitive dimension of the actor-spectator relationship. At a first glance, this seems really far from the spectator's activity of making sense of the performance, and it probably is. But one of the most important insights of systems theory has been to make complex the notion of biological organisation, which reveals how direct and quick the perturbation between one level and another could be, despite their distance. Human motor behaviour is actually strongly connected to the process of making sense throughout the assemblage of motor schemas.

> For example, in a particular situation, a person has to recognize many things – the people sitting around the room, the furniture in the room, the location of a particular object the person is looking for – and this means that one has different schemas for recognizing the object, the furniture, and the people. Furthermore, such schemas may have to be combined in order to represent a totally novel situation. One thus calls upon the appropriate knowledge for making sense of that situation. I call this schema assemblage. At any particular time there is a network of interacting schemas pulled together to represent the situation.
>
> *(Arbib and Gallagher 2004, 54–5)*

Making a precise description of all the layers engaged in the spectator's experience is clearly impossible, but it is important to understand that the motor-cognitive layer is always in resonance with all the layers of the experience (emotional, cultural, biographical, etc.) above it. We could actually envision that the performative event engages the spectator in an ongoing reassembly of his schemas, giving birth to new meanings and new performative experiences.

The more we refine our theoretical tools, the more an exhaustive description of the spectator's experience becomes difficult. This puzzle doesn't make the systemic or enactive approach useless. The aim is not actually the quest for an answer to this 'hard problem' but the radical change of the epistemological platform we use to deal with it.

Notes

1 I discovered this quote, thanks to the PhD thesis of Victor Jacono (2012), who remains one of the researchers who have consistently questioned the relationship between complexity sciences and theatre acting pedagogy (cf. also Jacono 2009, 2014 and 2016).

2 On the epistemological consequences of quantum physics on humanities, there is an excellent essay by one of the founders of subatomic physics, Niels Bohr (1937).

3 We have to remember that Grotowski was knowledgeable in quantum physics. He had often defined the institute of Niels Bohr as a great inspiration for his laboratory theatre. A quite useful essay on the scientific influences on Grotowski's work is that by Jean Marie Pradier (2013).

4 For example, in the 90s, Humberto Maturana wrote a dialogue-book along with Susana Bloch, neurophysiologist and theatre theoretician, founder of the *Alba Emoting* method (Maturana and Bloch, 1996). In contrast, Francisco Varela, despite his close contact with the milieu of theatre (his wife was a dancer, while his daughter Leonor is a famous actress), gave only one interview on cognitive sciences and performative arts to Maria Leao for her PhD thesis (2002).

5 For more considerations of the spectator's tendency to anticipate the performer's actions in a pre-reflective way, cf Sofia (2013a, 2013b).

6 With this aim some reflection on 'embodied theatrology' (Cf. De Marinis 2016, Sofia 2016).

Bibliography

Arbib, Michael A. and Shaun Gallagher. 2004. "The minds, machines, and brains of a passionate scientist: An interview with Michael Arbib." *Journal of Consciousness Studies* 11(12): 50–67.

Berthoz, Alain and Jean-Luc Petit. 2008. *The physiology and phenomenology of action*. Oxford University Press. [Original version: 2006. *Phénoménologie et physiologie de l'action*. Paris: Odile Jacob].

Bohr, Niels. (1937) 1972. "Biologie et physique atomique." In *Physique atomique et connaissance humaine*. Paris: Gauthier-Villars.

Committeri, Giorgia and Chiara Fini. 2016. "Body presence and extrapersonal space perception." In *Theatre and Cognitive Neuroscience*, edited by in Clelia Falletti, Gabriele Sofia and Victor Jacono, pp. 23–34. London: Bloomsbury Methuen Drama.

Costantini, Marcello, Giorgia Committeri and Corrado Sinigaglia. 2011. "Ready both to your and to my hands: Mapping the action space of others." *PLoS One* 6: e17923.

De Marinis, Marco. 2016. "Body and corporeity in the theatre: From semiotic to neuroscience. A small multidisciplinary glossary." In *Theatre and Cognitive Neuroscience*, edited by in Clelia Falletti, Gabriele Sofia and Victor Jacono, pp. 61–74. London: Bloomsbury Methuen Drama.

Fini, Chiara, Marcel Brass, and Giorgia Committeri. 2015. "Social scaling of extrapersonal space: Target objects are judged as closer when the reference frame is a human agent with available movement potentialities." *Cognition* 134: 50–56.

Grotowski, Jerzy. 1968. *Towards a Poor Theatre*. Holstebro: Odin Teatrets Forlag.

Husserl, Edmund. (1929–1935) 1973. *Zur Phänomenologie der Intersubjektivität*. The Hague: Martinus Nijhoff.

Jacono, Victor. 2009. "La scienza dell'attore nel teatro della complessità." In *Dialoghi tra teatro e neuroscienze*, edited by Gabriele Sofia, pp. 142–153. Roma: Edizioni Alegre.

Jacono, Victor. 2012. "Questioning how knowledge acts the relationship between the performer's pedagogy and cognitive neuroscience." PhD diss. Sapienza University of Rome.

Jacono, Victor. 2014. "Letter for an exchange between scientists of cognition." *Brazilian Journal on Presence Studies* 4(2): 205–222.

Jacono, Victor. 2016. "Complexity, cognition and the actor's pedagogy." In *Theatre and Cognitive Neuroscience*, edited by in Clelia Falletti, Gabriele Sofia and Victor Jacono, pp. 103–116. London: Bloomsbury Methuen Drama.

Leão, Maria. 2002. "Le prémouvement anticipatoire, la présence scénique et l'action organique du performeur: méthodes d'entraînement à travers la méthode Danis Bois." PhD diss. Université Paris VIII.

Lippi, Daria, Corinne Jola, Victor Jacono, and Gabriele Sofia. 2016. "Step towards the art of placing science in the acting practice. A performance-neuroscience perspective." In *Aesthetics and Neuroscience. Scientific and Artistic Perspective*, edited by Zoi Kapoula and Marine Vernet, pp. 141–164. Springer International Publishing.

Maturana, Humberto and Francisco Varela. 1973. De Maquinas y Seres Vivos. Una caracterizacion de la organizacion biologica. Editorial Universitaria. Santiago [Eng. tr. 1980. *Autopoiesis and Cognition: The Realization of the Living*. Boston, MA: Riedel Publishing Co.]

Maturana, Humberto and Francisco Varela. 1984. *El Arbol del Conocimiento: Las Bases Biologicas del Conocer Humano*. Editorial Universitaria, Santiago: La Edicion [Eng. tr. 1998. *The Tree of Knowledge: The Biological Roots of Human Understanding*. Boston, MA and London: Shambala].

Maturana, Humberto and Susana Bloch. 1996. *Biologia del emocionar y Alba Emoting. Bailando juntos*. Santiago del Chile: Dolmen.

Morin, Edgar. 2002. "Le complexus, ce qui est tissé ensemble." In *La Complexité, vertiges et promesses. 18 histoires de sciences*, edited by Réda Benkirane. Paris: Le Pommier.

Pradier, Jean-Marie. 2013. "Grotowski et les Sciences." In *L'Anthropologie Théâtrale selon Jerzy Grotowski*, edited by Jarosław Fret and Michel Masłowski, pp. 163–180. Paris: Édition de l'Amandier.

Rizzolatti, Giacomo and Corrado Sinigaglia. 2006. *So quel che fai. Il cervello che agisce e i neuroni specchio*. Milano: Raffaello Cortina Editore [En. Tr. 2008. *Mirrors in the brain. How our minds share action and emotions*. Oxford University Press].

Sofia, Gabriele. 2013a. "Achieved spontaneity and spectator's performative experience. The motor dimension of the actor-spectator relationship." In *Moving Imagination: Explorations of Gesture and Inner Movement*, edited by Helena De Preester, pp. 69–86. Amsterdam and Philadelphia, PA: John Benjamins Publishing Company.

Sofia, Gabriele. 2013b. *Le acrobazie dello spettatore. Dal teatro alle neuroscienze e ritorno*. Roma: Bulzoni Editore.

Sofia, Gabriele. 2013c. "The effect of theatre training on cognitive functions." In *Affective Performance and Cognitive Science*, edited by Nicola Shaughnessy, pp. 171–182. London: Methuen Drama.

Sofia, Gabriele. 2016. "Towards en embodied theatrology?" In *Theatre and Cognitive Neuroscience*, edited by in Clelia Falletti, Gabriele Sofia and Victor Jacono, pp. 40–60. London: Bloomsbury Methuen Drama.

Varela, Francisco. 1989. *Connaître. Les sciences cognitives, tendences et perspectives*. Paris: Édition du Seuil.

Varela, Francisco. 1996. "Neurophenomenology: A methodological remedy to the hard problem." *Journal of Consciousness Studies* 3: 330–50.

Varela, Francisco, Humberto Maturana, and Ricardo Uribe. 1974. "Autopoiesis: The organization of living system, its characterization and a model." *Biosystems* 5: 187–96.

Varela, Francisco, Evan Thompson, and Eleanor Rosch. 1991. *The Embodied Mind. Cognitive Science and Human Experience*. Cambridge and London: The MIT press.

Von Bertalanffy, Ludvig. 1972. "The history and status of general systems theory" *The Academy of Management Journal* 15(4) : 407–26.

Translation Gennaro Lauro

18

WATCHING MOVEMENT

Phenomenology, cognition, performance

Stanton B. Garner, Jr.

Peter Brook famously wrote, 'A man walks across [an] empty space whilst someone else is watching him, and this is all that is needed for an act of theatre to be engaged' (1968, 9). In recent decades, science has taught us much about the cognitive mechanisms engaged by this minimalist scene and more elaborate ones that include performers moving and acting on stage. We know that our perceptual system operates on and through movement – in other words, that we understand our environment through our potential to interact with it in embodied, situated ways. As part of our evolutionary legacy, we are cognitively primed to detect movement in this environment, especially animate movement. Though the precise mechanisms underlying this association are subject to disagreement, we know that we engage in similar neural operations when we perceive intentional movements – or actions – as we do when we execute these movements ourselves. It is also clear that our ability to do so in specific situations is influenced by cultural conventions, individual movement repertoires and factors such as training. Finally, we know that the acts of imagining, remembering and verbally recounting actions engage cognitive mechanisms that we employ in action observation and execution. This is true when we read about an action or hear an action recounted by someone else.

As I consider the implications of movement and movement perception for theatrical and other forms of spectatorship, I will return to some of these insights and the cognitive research that supports them. Drawing together cognitive science, neuroscience, psychology, cybernetics and linguistics, the expanding field of motor cognition has much to tell us about the cognitive transaction between audience and performer. While neuroscientists and other cognitive scientists illuminate the cognitive operations governing movement perception, they have little to say about the experiential dimensions of this and other cognitive processes: 'what it is like' (in Thomas Nagel's phrase) to engage in cognitive acts such as moving or observing movement. To raise this issue is to introduce concepts such as consciousness, subjectivity, givenness and awareness that science has historically excluded from its epistemological and methodological domains. The relationship between what can be described scientifically and what we experience has been termed the 'hard problem' of consciousness. In philosopher and cognitive scientist David J. Chalmers's words, 'How can we explain why there is something it is like to entertain a mental image, or to experience an emotion? It is widely agreed that experience arises from a physical basis, but we have no good explanation of why and how it so arises. Why should physical processing give rise to a rich inner life at all? It seems objectively unreasonable that it should, and yet it does' (Chalmers 1995, 201). This relationship is complicated by the fact that the cognitive operations underlying this inner life are largely sub-personal, operating below the threshold of awareness. Shaun Gallagher uses the term 'prenoetic' to designate those aspects of the mind that influence 'how the body anticipates and sets the stage for consciousness' (2005, 2). But cognitive mechanisms often have experiential correlates and outcomes, whether we attend to them or not. Neural mechanisms that respond to the movements of others by vicariously enacting their execution operate automatically; in this sense, we are cognitively wired for movement responsiveness. It is often also the case, though, that we experience this resonance in our bodies, swaying uncomfortably in our seats, for instance, while watching a tightrope walker balance herself 30 feet above us. There may be an epistemological gap between third-person and first-person perspectives, but cognition retains its subjective registers.

Consciousness, subjectivity and experience are the domain of phenomenology, a philosophical discipline that attempts to understand the givenness of experience and the elements that constitute it (*phenomenon* comes from a Greek word meaning 'to appear'). For most of its 125-year history, this tradition has developed apart from and in varying stances of opposition to the empirical tradition of the human and physical sciences. Practitioners of the sciences of mind, in turn, have historically discounted phenomenology for what they perceive as its subjectivism and lack of scientific rigour. In recent decades, however, a number of cognitive scientists, philosophers of mind and phenomenologists have embraced a dialogue between these two approaches. In the pages that follow, I will discuss this dialogue and some of the methodological terrain it has attempted to negotiate. I will then consider the convergences and divergences of these two approaches on the subjects of movement and movement perception. Finally, I will analyse a scene from a recent production of Arthur Miller's *Death of a Salesman* in order to demonstrate the insights that a phenomenologically inflected cognitive reading – or a cognitively inflected phenomenological reading – can provide when applied to movement perception in performance.

Phenomenology and cognitive science

When philosophers, scholars and the occasional layperson use the term 'phenomenology,' they often refer to different things. In its loosest sense, the term denotes introspection, the act of paying attention to one's experience. Health-care professionals often use the term in this way when they advocate listening to patients' accounts of their symptoms in addition to analysing these symptoms diagnostically. A more specialised sense of the term is employed by practitioners in a variety of disciplines, who pursue first-person insight while simultaneously combatting the subjectivism that 'untreated' experiential accounts fall prey to. In fields such as psychology, sociology, anthropology and geography, specialised methods have been developed for obtaining phenomenological data, including carefully prescribed interview protocols, and for analysing the experiential structures that characterise psychological states such as jealousy and regret, initiation rituals and cultural spaces.

Behind these practical uses of 'phenomenology' is the philosophical movement inaugurated by Edmund Husserl at the turn of the twentieth century and the methods associated with this philosophy. Husserl's phenomenology was an attempt to ground philosophy in consciousness's engagement with the world. It did so by challenging what Husserl called the 'natural attitude': the everyday assumption that the world exists as we think it does when we take it for granted, that it exists outside and independent of us. Subject to measurement, experiment and calculation, this is the realist world that science takes as its subject. If we bracket this assumption, however, and suspend all certainties as to the world's objective existence or non-existence (Husserl's terms for this process are *epoché* and *reduction*), we enter the phenomenological attitude, which discloses the world as it manifests within our experience. Husserl writes, 'It is not until one consistently and completely carries out the *phenomenological reduction* [...] that one obtains *pure* lived experience, as the object of the phenomenological perception, and, for the first time, achieves genuine phenomenological perception in its radical distinctiveness from empirical perception' (Husserl 2006, 40–41). While this bracketing is sometimes mistaken as a retreat from or denial of the world, in phenomenological terms it discloses the world as a correlate of our experience rather than accepting it unreflectively as those in thrall to the naturalist attitude do. Evan Thompson describes this shift, '[O]nce we adopt the phenomenological attitude, we are interested not in *what* things are in some naïve, mind-independent or theory-independent sense, but rather in exactly *how* they are experienced, and thus as strict relational correlates of our subjectivity' (Thompson 2007, 18–19). Gravity appears very differently when we consider it as an experiential phenomenon than it does when we contemplate it according to Newton's or Einstein's laws.

Husserl's successors in the continental phenomenological tradition – Edith Stein, Martin Heidegger, Jean-Paul Sartre, Maurice Merleau-Ponty, Emmanuel Levinas and others — challenged and refined Husserl's phenomenological project and the methods he developed for implementing it. More recent philosophers and scholars have deepened this appropriation and critique by setting the phenomenological tradition in dialogue with contemporary critical theory and identity studies.[1] This interest in applying phenomenology to rapidly evolving theoretical fields — and thereby testing its explanatory reach and limitations — extends to theatre, dance and performance studies, which have seen an upsurge of phenomenological interest in recent years. The formal application of phenomenology to dance studies was inaugurated with Maxine Sheets-Johnstone's *Phenomenology of Dance* (1966), and this approach was introduced to theatre studies with Bruce Wilshire's *Role-Playing and Identity: The Limits of Theatre as Metaphor* (1982) and Bert O. States' *Great Reckonings in Little Room: On the Phenomenology of Theatre* (1985).[2] In response to new forms of technologically mediated, immersive and participatory performance, and taking advantage of new ways of thinking about experience, subjectivity and the body, phenomenological scholarship since 2000 has kept pace with the expanding field of performance studies.[3]

Surprisingly little of this work has acknowledged or engaged with cognitive studies, which has expanded over the same period in ways that often complement the work of performance phenomenology. For their part, cognitive theatre scholars have only recently begun acknowledging the important role that phenomenology has played in the theoretical development of cognitive science over the last 25 years. A key moment in the broader conversation between phenomenology and cognitive science that followed a long period of mutual neglect and suspicion was the publication of Francisco Varela, Thompson and Eleanor Rosch's *The Embodied Mind: Cognitive Science and Human Experience* in 1991. In this pivotal study, the authors proposed an enactive model of cognition, in which embodied, autonomous agents bring forth their cognitive domains through skilful interaction with their environments. By understanding cognition as 'lived history, whether seen at the level of the individual (ontogeny), the species (evolution), or social patterns (culture)' (Varela et al. 1992, 213), this approach returned experience and its organism-centred perspective to cognitive science. Enactivism as articulated by Varela et al. owes an acknowledged debt to phenomenology, particularly Merleau-Ponty's phenomenology of embodiment, perception and situatedness in the world. 'We hold with Merleau-Ponty,' they write, 'that Western scientific culture requires that we see our bodies both as physical structures and as lived, experiential structures—in short, as both "outer" and "inner," biological and phenomenological' (xv). By incorporating Merleau-Pontean phenomenology with the emerging 'interdisciplinary matrix' of cognitive science (xvi), Varela, Thompson and Rosch consider their project a radically new implementation of the French philosopher's conception of embodiment.[4] (For additional insight into the links between enactivism and phenomenology, see Chapter 17.)

Enactivism became one of a number of areas in which cognitive science, philosophy of mind and phenomenology interact in order to explore their compatibilities and differences. The journal *Phenomenology and the Cognitive Sciences* was founded in 2002 as a venue for investigating this intersection, and its contributors have been at the forefront of those seeking to integrate empirical science with the study of consciousness (and vice versa). Some of the most influential books for cognitive theatre studies have advanced the research opened up by this convergence: Alvin Noë's *Action in Perception* (2004), Gallagher's *How the Body Shapes the Mind* (2005) and Thompson's *Mind in Life: Biology, Phenomenology, and the Sciences of Mind* (2007). In each of these studies, empirical science is challenged to accept the necessary perspective of experience, while phenomenology is asked to address the insights generated by neuroscience and other experimental sciences. Gallagher and Dan Zahavi's book *The Phenomenological Mind* (2008; 2nd ed. 2012) introduces cognitive scientists to the phenomenological dimensions of their research, while the former's introductory book *Phenomenology* (2012) situates the philosophical tradition within this new interdisciplinary matrix.

Theoretical and methodological collaboration between these approaches is not without its challenges. One of these challenges has to do with the nature and scientific status of first-person evidence. Scientists engaged in the experimental study of cognition typically collect data from selected subjects under controlled laboratory conditions, and they take care to eliminate individuating variables that skew representative conclusions. In other words, they employ outside-in research protocols. Phenomenologists, on the other hand, investigate cognitive and perceptual processes from the inside-out, taking the individual's experience as their starting point. It would be a mistake to equate this process, as some have in the past, with subjectivism, the idea that experience and knowledge are merely subjective, hence biased and unable to achieve general explanatory power. As Varela points out, 'Experience is clearly a personal event, but that does not mean it is *private*' (1996, 240). Though phenomenologists take subjectivity as their object of inquiry, they employ rigorous methods to identify the essences (*Wesen*) underlying subjective phenomena, elements that persist when a phenomenon is varied. Identified through this process, phenomenological insights and models become transpersonal, available to others to test, confirm and revise. In this sense, phenomenological understanding is progressive, as previous models and descriptions are re-examined in light of new experiential data and acts of attention. Attention to issues of gender in the late twentieth century, for instance, led feminist philosophers to challenge the models of experience produced by earlier male phenomenologists.

Since phenomenology employs its own procedures and yields its own kind of results, however, the question of how one incorporates its first-person descriptions with the results of experimental science remains an open one. Among the solutions proposed for addressing this issue are strategies for 'naturalising' phenomenology by making its data and methods usable to scientific inquiry (see Petitot et al. 1999). Neurophenomenology, a programme introduced by Varela in the mid-1990s and taken up by researchers in a number of areas, seeks to integrate phenomenological analysis of experience, dynamic systems theory (which is highly compatible with phenomenological models) and experimental brain science. Subjects in neurophenomenological studies are trained to provide reliable and consistent descriptions of their experience, and these descriptions are used to generate variables for further experimentation and to provide models that can be used in interpreting neurobiological data. Central to Varela's attempt to bridge the 'experimental mind-experiential mind gap' is his principle that '[p]henomenological accounts of the structure of experience and their counterparts in cognitive science relate to each other through reciprocal constraints' (Varela 1996, 343). This principle, which provides a useful guideline for anyone working at the intersection of cognitive science and phenomenology, dictates that each approach provides a reference point and check for the other. In practice, the notion of 'constraint' suggests that neurological findings can be validated or questioned in part according to their 'fit' with experiential data and that phenomenological accounts of experience be re-evaluated if they depart markedly from agreed-upon findings in experimental science. Orthodox practitioners in both disciplines will reject such constraints, but the theoretical and methodological reciprocity between these different approaches to the study of mind has demonstrated its power to advance cognitive science, phenomenology and the conversation between them.

One of the areas where this conversation is starting to have an impact is theatre and performance studies. Phillip Zarrilli's work, for instance, illustrates the inroads that phenomenology and enactive theory offer to the actor's creative process and the field of actor training,[5] and it is increasingly common to see phenomenology referred to by scholars and practitioners adopting cognitive approaches to other performance areas. My focus in the remainder of this essay will be on one of the most fertile areas where phenomenology and cognitive science converge: theatrical and other forms of spectatorship. If the idea of enaction and its cognitive/phenomenological corollaries lead naturally to the performer making things happen on stage through embodied, meaning-filled interactions with her environment, they also point to the spectator, who co-constitutes the performance through equally embodied perceptual acts. Some of the most important work on performance cognition — Bruce McConachie's *Engaging Audiences: A Cognitive Approach to Spectating in the Theatre* (2008), for instance — focuses on spectatorship, while theatre phenomenology has a long tradition of approaching performance from the audience's point of view (hence the importance to this tradition of Merleau-Ponty's phenomenology of perception). I use 'spectator' and 'spectatorship' advisedly, since both approaches have challenged the ocularcentrism and sense of disembodiment with which these terms are associated. In phenomenology as in enactive cognitive science, vision forms part of the subject's broader, sensorimotor engagement with its world. Far from being the property of a passive disembodied consciousness, it moves, acts and explores in conjunction with the body's other sensory openings on its environment. What happens to perception when we return it to movement — the movements it enacts and the movements it perceives in the world around it? What happens to spectatorship if we consider vision and the other senses within a dynamic of sensorimotor engagement?

Movement, perception, kinaesthesia

As early as the nineteenth and early twentieth centuries, psychologists and philosophers recognised a connection between movement, perception and action recognition. In 1852, Rudolf Lotze theorised that visual sensations are integrated with an individual's muscular sense in making spatial determinations, while Hermann von Helmholtz argued that the information that triggers a motor command is also recruited to recognise external movement (Berthoz 2000, 9; Viviani [1990], 17). This 'motor theory of perception' was amplified years later in James Gibson's ecological psychology, which posits that '[t]o see things is to see how to get about among them and what to do or not do with them' (Gibson 1979, 223). It was also developed in Noë's *Action in Perception* and in the work of Marc Jeannerod, Alain Berthoz and others in the field of motor cognition. As Jeannerod states in the foreword to *Motor Cognition: What Actions Tell the Self*, this field investigates 'the way actions are thought, planned, intended, organised, perceived, understood, learned, imitated, attributed or, in a word, the way they are represented' (2006, v). It should be noted that 'representation,' in Jeannerod's sense of the term, does not refer to a Cartesian cognitive model in which pictures of the world and one's experiences are manipulated by disembodied mental processes; rather, it encompasses the multiple ways that action manifests itself in an embodied, environmentally oriented cognitive field.

Contemporary interest in this relationship was intensified by the discovery in the early 1990s of mirroring systems in the brains of macaque monkeys and the later discovery of equivalent systems in humans. Neurons in these systems fire when a subject moves, when that subject observes another subject move and when actions are remembered, imagined or read. Action-oriented mirror neurons are associated with other neurons that respond similarly to the observation of emotion. Mirror neurons were heralded in the popular press and by some scientists as providing the neurological key to mind-reading, action understanding, imitation and empathy, and these claims were subsequently challenged by other scientists who disputed the neurological evidence and the far-reaching conclusions drawn from it. Stepping back from the polemics on both sides of this debate (particularly those of mirror-neuron sceptics), we should recognise that mirror-system research is compatible with the work of other scientists studying motor perception, much of which preceded and does not depend on the neuronal claims of mirror-neuron theory. One of the distinguishing features of mirror neurons, for instance – that they respond to goal-directed, or intentional, action but not to random movements – was well established by earlier experimenters such as Gunnar Johansson, who studied movement perception using point-light displays of moving figures in the early 1970s.[6] One need not subscribe to the precise neural mechanisms that mirror-system researchers propose for linking mirror-neuron activity with other cognitive activity – the interaction between mirror neurons and higher-order cognitive functions like theory of mind, for instance, which attributes mental states to others – in order to incorporate their expanding understanding of movement perception with parallel research in this area.

Empirical research on movement perception and motor resonance has yielded an increasingly detailed understanding of theatrical and other forms of spectatorship. Not surprisingly, given the medium's focus on kinetic action, much of the analysis of spectatorship and movement has focused on dance. In the early 2000s, a research team headed by Beatriz Calvo-Merino looked at the brain activities of experts in classical ballet, experts in Brazilian capoeira and inexperienced control subjects when they were shown videos of ballet or capoeira movements (Calvo-Merino et al. 2005). Their findings showed increased activities in areas of the brain associated with mirroring activity in those who viewed movements from the tradition they were familiar with, thereby demonstrating a link between motor expertise and action observation. In a similar study based on interviews and other qualitative research methodologies, Matthew Reason and Dee Reynolds found that the kinaesthetic and emotional responses of spectators asked to watch classical ballet and the South Indian *bharatanatyam* differed in kind and intensity based on their cultural familiarity with these forms (Reason and Reynolds 2010). Such research contributes to a kinaesthetically oriented tradition of dance theory that stretches from John Martin's modern dance criticism of the 1930s to Susan Leigh Foster's *Choreographing Empathy: Kinesthesia in Performance* (2011). But empirical research on movement and movement perception has been applied to theatrical and other forms of performance, as well. In their study of the roles of distance and biological movement in extra-personal space perception, for example, Giorgi Committeri and Chiara Fini speculate that motor resonance (vicariously enacting the movements of other) can be modulated by the physical distance between the observer and the observed body. Our desire to sit closer to the stage, they suggest, could be driven not only by our desire to see better, but also by the attempt 'to have a better "resonance" with the movements and gestures of the actors, in order to be part of this "near," shared common space' (2016, 34).

Phenomenology has played and continues to play an important role in the study of movement and movement perception. Vittorio Gallese, a member of the research team who discovered mirror neurons at the University of Parma, used the philosophy of Merleau-Ponty to frame the team's neuroscientific findings (Iacoboni 2008, 16–17), and Alain Berthoz has collaborated with philosopher Jean-Luc Petit on a physiological and phenomenological study of action (2006). Phenomenology, of course, has a long history of considering movement 'from the inside,' and much of it is centred on kinaesthesia and kinaesthetic experience. The term 'kinesthesia,' which combines the Greek words *kinein* (to move) and *aesthesis* (sensation), was coined in the late nineteenth century by British neurologist Henry Charlton Bastian to describe sensations resulting from or associated with movement (1880, 543). Closely akin to 'proprioception,' the sense I have of my body's posture and the relationship of my limbs to each other, kinaesthesia is my awareness when I move that I am doing so. Motility and the kinaesthetic sense are fundamental to Husserl's phenomenology of consciousness and world-constitution: I come to know my world by moving through and engaging with it, and this interaction deepens my reflexive sense of my own kinaesthetic capabilities. Kinaesthesia is also important to Merleau-Ponty's analysis of motility, motor intentionality and what he calls 'body schema.' In a memorable passage from *The Structure of Behavior* in which he describes a soccer player moving a ball downfield, Merleau-Ponty describes the merging of movement and environment in the player's felt dynamism: 'The field itself is not given to him, but present as the immanent term of his practical intentions; the player becomes one with it and feels the direction of the "goal," for example, just as immediately as the vertical and horizontal planes of his own body' (1963, 168). At this moment, he writes, 'consciousness is nothing other than the dialectic of milieu and action' (169).

No philosopher has contributed more to the understanding of movement experience than Maxine Sheets-Johnstone, whose book *The Primacy of Movement* (1999; expanded ed. 2011) and numerous other publications over the past 50 years offer a kinaesthetically grounded dynamic phenomenology. According to Sheets-Johnstone, we are born into movement, born *as* movement: 'This primal animateness, this original kinetic spontaneity that infuses our being and defines our aliveness, is our point of departure for living in the world and making sense of it. [...] *We literally discover ourselves in movement*' (1999, 117). Most explanations of human movement, she argues – including the work of much traditional phenomenology, cognitive science and other disciplines exploring this area — minimise or neglect the role of kinaesthesia in sensorimotor and higher-order development. Against mirror-neuron theorists who privilege autonomous neural mechanisms in action understanding, for example, she argues that mirror neurons 'are contingent on morphology and corporeal-kinetic tactile-kinaesthetic experience' (2012, 385).[7] Our brains, in other words, form neural connections based on our own kinetic/kinaesthetic dynamics and possibilities of movement. In identifying and detailing the kinaesthetic foundations of human movement, Sheets-Johnstone draws widely and deeply on infant research, paleoanthropology, microbiology and neurology, and her arguments engage important debates within these (and other) fields. But as befits a philosopher who is also a trained dancer, her phenomenology is rooted in the self-exploration of movement. Any time we want to pay closer attention to the kinaesthetic body, she points out, 'there it is' (2003, 75). By insisting on this step, she reminds us that phenomenology is more than a field of insights that can be brought into dialogue with scientific findings; it is a methodological invitation to inhabit these insights oneself and formulate new ones. A 'corporeal turn' in philosophy, she writes, requires that we become aware of movement: 'It thus asks us [...] to be silent, and, in our silence, to witness the phenomenon of movement—our own self-movement and the movement of all that is animate or animated in our surrounding world' (2011, xix).

Attending to movement on the stage

Sheets-Johnstone's words present scientific accounts of motor cognition with a challenge and an invitation. When neuroscientists study movement perception, they do so in laboratory settings where subjects are immobilised to varying degrees by the requirements of the experiment. The functional magnetic resonance imaging (fMRI) technology that many experimenters use to measure neural activity, for example, may require that the subject be strapped down with his or her head secured in a brace. In addition, the actions these subjects are asked to observe often involve isolated body movements (an arm and hand lifting a glass of water) rather than movements that are integrated with the entire body and responsive to other people's movements. Scientists, in other words, tend to study motor cognition outside of the real-world conditions in which embodied subjects move in relation to their environment and other moving subjects. A phenomenologically sensitive approach to movement perception offers the opportunity to return cognition to its natural experiential field. In this animated field, we are always moving, even when we seem to be still. This is as true for theatrical and other forms of spectatorship as it is in non-performative situations. While the myth of the audience's stillness is well-entrenched — 'the spectators hung motionless on every word' — theatrical watching is an intensely kinetic and kinaesthetic activity. Seated in the auditorium while watching a play, I adjust my position, feel my chest rise and fall as I breathe, glance around me from time to time, lean forward at a particularly intense moment and glance at my watch during a particularly tedious one. When I fix my gaze on something, my eyes undergo saccadic, or involuntary, movements every second. In differently socialised theatrical cultures and in immersive and other forms of participatory theatre, I may encounter expanded possibilities for movement, but these possibilities represent an expansion of spectatorship's inherent motility.

In conjunction with these and other movements, my attention ranges over and through my perceptual field, actively selecting perceptual objects within a background awareness of my body as kinaesthetic entity. As George Home-Cook notes in his phenomenological study of theatrical listening, attention is 'a dynamic, intersensorial, bodily engagement with the "affordances" of a given environment' (2015, 2). By foregrounding elements of this environment in my awareness, attention inside and outside the theatre selects the objects of consciousness and cognition. This cognitive faculty plays a determining role in action observation and the kinaesthetic resonances that accompany it. When I walk down a city street or sit on a bench and observe the flow of pedestrians around me, I am faced with myriad subjects walking, pushing strollers, carrying things, talking with others or on cell phones, doing some or all of these at the same time. Individuals move along 'intentional arcs' (Merleau-Ponty 2012, 137), but they do so with environmentally attentive, often improvised coordination—hence the fact that they're not colliding with each other every few steps. As my eyes survey this scene, attention determines what I focus on: the man trying to eat a sandwich, the couple hurrying to cross the street before the light changes, the elderly woman walking her dog while other pedestrians try to pass her. Every time my focus shifts and I observe someone new, I apprehend that subject's action as part of the collective flow and sense it with my kinaesthetic imagination.

Theatre, of course, does not disclose itself to me in the same way that a city street does — even, provocatively, when it takes place on one of those streets. Unlike the movements and gestures of everyday pedestrians, its actions present themselves *for* the eyes and other senses as well as *to* them. These actions are selected, rehearsed and staged in ways that call attention to individual actors, gestures and lines of dialogue over others. Perceptually speaking, though, theatre presents kinetic/kinaesthetic attention with complexities similar to those it encounters in extra-theatrical environments: multiple action points, movement interactions and spectators who direct their attention consciously as well as preconsciously through the movement 'affordances' they are presented with. Motor cognition and mirror system research describe the preconscious mechanisms at work when I observe a complex movement field like this, but without a first-person perspective on environmentally situated movement interactions, this research is limited in its ability to account for the experiential dynamics of this process.

In order to suggest what a phenomenologically informed cognitive analysis of theatrical movement perception might look like, I will briefly discuss the climactic confrontation sequence between Willy Loman and his son Biff Loman in Robert Falls' 1999 production of *Death of a Salesman*, which featured Brian Dennehy as Willy and Kevin Anderson as his oldest son.[8] This sequence, which takes place late at night in the Loman kitchen after their failed restaurant meeting, is initiated by Biff telling his father that he's leaving in the morning and won't return.[9] Refusing to shake his hand, Willy attacks Biff for making a failure of his life, but Biff counters with accusations of his own and an angry, pleading insistence on who they both are ('Pop! I'm a dime a dozen, and so are you!'). Kinetically and emotionally, the scene escalates in intensity until Biff, shattered by his outburst, breaks down, crying with his head in his father's lap.

When one attends to the experience of watching this sequence in performance, its kinaesthetic dynamics reveal themselves in complex, performance-specific ways. Different actors make individual choices when interpreting their characters' movements, and these choices are realised through specific bodies and individual movement styles. Taking advantage of his heavy stature, for example, Dennehy moved his upper torso as a block, leading with his chest and shoulders as if armouring himself while pushing his way through the world. For much of his performance he accentuated what he was saying using his right arm and hand; at other times, he kept one or both hands in his pocket.[10] To watch Dennehy's Willy move on stage was to recognise and feel the embodied, self-containing dynamic of his actions and gestures. As he and Anderson faced off in the kitchen, their physical action tracked the emotional dynamic that fueled their interaction. Cognitive scientists and phenomenologists have both noted the connection between emotion (or affect) and movement. As Alain Berthoz puts it, '[T]here is no perception of space or movement, no vertigo or loss of balance, no caress given or received, no sound heard or uttered, no gesture of capture or grasping that is not accompanied by emotion or induced by it' (2000, 7). When Anderson's Biff recounted to Willy that he ran down the stairs at Bill Oliver's office with a stolen pen in his hand, he gesticulated agitatedly, underscoring deeply felt words with arm and hand movements, then clutched his father when he pleaded that 'all I want is out there, waiting for me the minute I say I know who I am! Why can't I say that, Willy?' When Willy responded to his 'dime a dozen' claim by lashing back ('I am not a dime a dozen! I am Willy Loman, and you are Biff Loman!'),

Anderson rushed at Dennehy, grabbed him and shoved/dragged him to the chair as Dennehy cowered and moaned. It was a kinetic and emotional slug-fest, with each taking the initiative and then registering the impact when his attack is countered.

Neurological and other research on motor cognition clarifies the mechanisms at work when we watch stage actions such as these: our cognitive involvement in goal-directed movements such as Anderson grabbing Dennehy by the shoulders or the latter's fierce arm movements when he declares 'I am Willy Loman.' As the sequence unfolded on stage, though, these mechanisms operated with an integration and variability that phenomenological observation is better suited to illuminate. Actors move with their whole bodies, for one thing; when Anderson grabbed Dennehy's shoulders, he braced his legs, torso and neck to counter the latter's resistance. Actors also move in relation to and at the same time as other actors, which means that spectator processes a field of multiple movements, each with a rival claim to kinaesthetic attention. As I watched this scene multiple times on videotape a number of years after seeing it live, I felt my attention drawn to different actors, sometimes in rapid succession. Though the sequence centres on Willy and Biff, Linda and Happy were also onstage, and while they were relatively motionless during most of the confrontation, they retained their kinetic/kinaesthetic agency. I occasionally glanced over to reaffirm their presence as witnesses to this encounter.

The intricacy and mobility of the spectator's kinaesthetic allegiances in this sequence were foregrounded at its conclusion. Having pushed Dennehy's Willy down onto his chair and towered over him (Dennehy shielding his head with his right arm), Anderson's Biff sank to his knees at the end of the line 'There's no spite in it any more. I'm just what I am, that's all,' his anger broken, and started to sob while he held the side of his father's face with his hand. Raising himself on his toes, he fiercely kissed Willy on the cheek. Dennehy held his hands awkwardly above Biff's head as if not sure what to do with them then put his arm around his son's neck, ran his fingers through his hair with his other arm around his shoulder, pulled him upward and kissed him on the neck. As he held Anderson in an embrace, the latter reached around and held his father firmly and tenderly. Dennehy kissed him again, and Anderson quietly rose and disengaged himself. The intimacy of these movements was inseparable from the intense tactility of their contact with each other, and both contributed to the exchange's powerful emotional/kinaesthetic effect. Watching this sequence, I felt each hugging the other, alternatingly and at the same time. In this sense, I switched between the two men, inhabiting the kinaesthetic perspective of one, then the other. But I also inhabited their mutual embrace in a wider attentional field that transcended the individuality of kinaesthetic points of view. Theirs were not separate actions; they were interactions, each man's gesture intimately caught up in the other's. From this perspective, the object of kinaesthetic attention was mutual movement—sometimes synchronous, sometimes asynchronous, but always coordinated in perceptual terms. The tightness of their embrace made this interactivity clear, but the kinaesthetic inseparability of individual and collective action evident here — and the spectator's attentional agency in shifting between these kinaesthetic levels — could be seen in their heated interactions leading up to this point as the actors moved toward, away from, around and in physical contact with each other. The interactivity of their movements resembled, at moments, the experience of watching dance, where performers engage in individual and ensemble movements and spectators vicariously inhabit these on a continuum between the kinaesthetic I and the kinaesthetic we.

The spectatorial responses I have detailed in my account of the confrontation sequence in *Death of a Salesman* are clearly influenced by 'mirroring' and other cognitive mechanisms that scientists have made great strides in understanding. That these responses have empirical corroboration enhances the explanatory authority of phenomenological description. What phenomenology offers, in turn, is the ability to investigate the operations of movement and movement observation in dynamic, intercorporeal situations. With its rigorous experiential investigations, phenomenology provides a method for recognising the kinaesthetic nuances generated in these situations, for understanding the holistic, dynamic body as it moves and is observed in multiple active kinetic environments such as theatre. In doing so, it ventures outside and often beyond empirical studies in order to confirm their insights, complicate their models and offer directions for future research. It will be fascinating to know, when the experimental procedures are developed for measuring this, what the neuroscience of kinaesthetic perception in theatrical and other performance environments looks like. But experience has its own modes of disclosure, and these illuminate the felt dimension of cognitive life. One perspective supports the other. When phenomenology and cognitive science engage in dialogue, the study of movement, movement perception and other cognitive actions reaches above and beneath the threshold of consciousness.

Notes

1 See, for example, Young (1980), Ihde (1993), and Ahmed (2006).
2 For a history of phenomenological approaches to theatre in the late twentieth century, see Garner (2001).
3 See, for example, Kozel (2008), Home-Cook (2015), and Bleeker et al. (2015).
4 It is worth noting in this context that Merleau-Ponty engaged with neurology and developmental science in his early work. *The Phenomenology of Perception*, for instance, refers to cases of neurological pathology like that of Johann Schneider, who sustained brain injuries during World War I and was the subject of a 1918 study by German neurologists Kurt Goldstein and Adhémar Gelb (see Merleau-Ponty 2012, 105–140; Garner 1994, 33–36).
5 See, for example, Zarrilli (2007, 2009). Another scholar whose work on acting has been influenced by this dialogue is John Lutterbie, whose use of dynamic systems theory in *Towards a General Theory of Acting: Cognitive Science and Performance* (2011) bears evidence of his earlier work on phenomenology and performance.
6 Point-light displays are produced by filming a moving human figure with lights attached to his or her joints. Abstracted from the human body, these lights are set against a dark background. In a still image, the glowing dots are typically not recognisable, but when viewed in motion—in other words, dynamically—they are immediately discernible as a moving figure. Subjects in these experiments show a preference for biological over non-biological movement.
7 As Sheets-Johnstone points out in *The Primacy of Movement*, the sensorimotor system is the first developing perceptual system evident in the foetus (2011, 228).
8 This production of Miller's play, which originated at Chicago's Goodman Theatre in 1988, opened at the Eugene O'Neill Theatre in February 1999.
9 The sequence can be found in Miller (1967, 127–33).
10 Watching Dennehy move as Willy is different from watching Dustin Hoffman in the 1985 television production of *Salesman*. In this production, Hoffman walks stiffly but moves his arms and upper body with an almost histrionic gestural expressivity.

Works Cited

Ahmed, Sara. *Queer Phenomenology: Orientations, Objects, Others*. Durham, NC: Duke UP, 2006.
Bastian, Henry Charlton. *The Brain as an Organ of Mind*. New York: Appleton, 1880.
Berthoz, Alain. *The Brain's Sense of Movement*. Trans. Giselle Weiss. Cambridge, MA: Harvard UP, 2000.
Berthoz, Alain and Jean-Luc Petit. *The Physiology and Phenomenology of Action*. Trans. Christopher Macann. Oxford: Oxford UP, 2008.
Bleeker, Maaike, Jon Foley Sherman, and Eirini Nedelkopoulou, eds. *Performance and Phenomenology: Traditions and Transformations*. New York and London: Routledge, 2015.

Brook, Peter. *The Empty Space.* New York: Atheneum, 1968.

Calvo-Merino, Beatriz, Daniel E. Glaser, Julie Grèzes, Richard E. Passingham, and Patrick Haggard. 'Action Observation and Acquired Motor Skills: An fMRI Study with Expert Dancers.' *Cerebral Cortex* 15, no. 8 (2005): 1243–49.

Chalmers, David J. 'Facing Up to the Problem of Consciousness.' *Journal of Consciousness Studies* 2, no. 3 (1995): 200–19.

Committeri, Giorgia and Chiara Fini. 'Body Presence and Extra-Personal Space Perception.' *Theatre and Cognitive Neuroscience.* Ed. Clelia Falletti, Gabriele Sofia, and Victor Jacono, 23–34. London: Bloomsbury Methuen, 2016.

Foster, Susan Leigh. *Choreographing Empathy: Kinesthesia in Performance.* London and New York: Routledge, 2011.

Gallagher, Shaun. *How the Body Shapes the Mind.* Oxford: Oxford UP, 2005.

Gallagher, Shaun. *Phenomenology.* Basingstoke: Palgrave Macmillan, 2012.

Gallagher, Shaun and Dan Zahavi. *The Phenomenological Mind.* 2nd ed. London and New York: Routledge, 2012.

Garner, Stanton B., Jr. *Bodied Spaces: Phenomenology and Performance in Contemporary Drama.* Ithaca, NY: Cornell UP, 1994.

Garner, Stanton B., Jr. 'Theater and Phenomenology.' *Degrés: Revue de synthèse à orientation sémiologiques* 107–108 (Autumn–Winter 2001): B1–17.

Gibson, James J. *The Ecological Approach to Visual Perception.* Boston, MA: Houghton Mifflin, 1979.

Home-Cook, George. *Theatre and Aural Attention: Stretching Ourselves.* Basingstoke: Palgrave Macmillan, 2015.

Husserl, Edmund. *The Basic Problems of Phenomenology: From the Lectures, Winter Semester, 1910–11.* Trans. by Ingo Farin and James G. Hart. Dordrecht: Springer, 2006.

Iacoboni, Marco. *Mirroring People: The Science of Empathy and How We Connect with Others.* New York: Picador, 2008.

Ihde, Don. *Postphenomenology: Essays in the Postmodern Context.* Evanston, IL: Northwestern UP, 1993.

Jeannerod, Marc. *Motor Cognition: What Actions Tell the Self.* Oxford: Oxford UP, 2006.

Kozel, Susan. *Closer: Performance, Technologies, Phenomenology.* Cambridge, MA: MIT P, 2007.

Lutterbie, John. *Toward a General Theory of Acting: Cognitive Science and Performance.* Basingstoke: Palgrave Macmillan, 2011.

McConachie, Bruce. *Engaging Audiences: A Cognitive Approach to Spectating in the Theatre.* Basingstoke: Palgrave Macmillan, 2008.

Merleau-Ponty, Maurice. *Phenomenology of Perception.* 1945. Trans. Donald A. Landes. London and New York: Routledge, 2012.

Merleau-Ponty, Maurice. *The Structure of Behavior.* 1942. Trans. Alden L. Fisher. Pittsburgh, PA: Duquesne UP, 1963.

Miller, Arthur. *Death of a Salesman.* 1949. New York: Penguin, 1976.

Nagel, Thomas. 'What Is It Like to Be a Bat?' *Philosophical Review* 83, no. 4 (1974): 435–50.

Noë, Alvin. *Action in Perception.* Cambridge, MA: MIT P, 2004.

Petitot, Jean, Francisco J. Varela, Bernard Pachoud, and Jean-Michel Roy, eds. *Naturalizing Phenomenology: Issues in Contemporary Phenomenology and Cognitive Science.* Stanford, CA: Stanford UP, 1999.

Reason, Matthew and Dee Reynolds. 'Kinesthesia, Empathy, and Related Pleasures: An Inquiry into Audience Experiences of Watching Dance.' *Dance Research Journal* 42, no. 2 (2010): 49–75.

Sheets-Johnstone, Maxine. 'Kinaesthetic Memory.' *Theoria et Historia Scientiarum* 7, no. 1 (2003): 69–92.

Sheets-Johnstone, Maxine. 'Movement and Mirror Neurons: A Challenging and Choice Conversation.' *Phenomenology and the Cognitive Sciences* 11, no. 3 (2012): 385–401.

Sheets-Johnstone, Maxine. *The Phenomenology of Dance.* Madison: U of Wisconsin P, 1966.

Sheets-Johnstone, Maxine. *The Primacy of Movement.* 1999. Expanded second ed. Amsterdam: John Benjamins. 2011.

States, Bert O. *Great Reckonings in Little Rooms: On the Phenomenology of Theater.* Berkeley: U of California P, 1985.

Thompson, Evan. *Mind in Life: Biology, Phenomenology, and the Sciences of Mind.* Cambridge, MA: Harvard UP, 2007.

Varela, Francisco J. 'Neurophenomenology: A Methodological Remedy for the Hard Problem.' *Journal of Consciousness Studies* 3, no. 4 (1996): 330–49.

Varela, Francisco J., Evan Thompson, and Eleanor Rosch. *The Embodied Mind: Cognitive Sciences and Human Experience*. Cambridge, MA: MIT P, 1991.

Viviani, Paolo. 'Motor-Perceptual Interactions: The Evolution of an Idea.' *Cognitive Science in Europe: Issues and Trends*. Ed. Massimo Piattelli Palmarini, 11–39. Ivrea: Golem, 1990.

Wilshire, Bruce. *Role Playing and Identity: The Limits of Theatre as Metaphor*. Bloomington: Indiana UP, 1982.

Young, Iris Marion. 'Throwing Like a Girl: A Phenomenology of Feminine Body Comportment, Motility, and Spatiality.' 1980. *Throwing Like a Girl and Other Essays in Feminist Philosophy and Social Theory*, 141–59. Bloomington: Indiana UP, 1990.

Zarrilli, Phillip B. 'An Enactive Approach to Understanding Acting.' *Theatre Journal* 59, no. 4 (2007): 635–47.

Zarrilli, Phillip B. *Psychophysical Acting: An Intercultural Approach after Stanislavsky*. London and New York: Routledge, 2009.

19

ATTENTION TO THEATRICAL PERFORMANCES

James Hamilton

How do audiences attend to theatrical performances? An account of attention is useful to theatre and performance studies for the obvious reason that attention is the necessary first step for coming to an understanding of how audiences grasp and respond to a performance.

In the first section of this entry, I situate attention to performances within current debates about attention in psychology, neuroscience and philosophy. From that vantage point, we can survey the demands that must be placed on any model of attention by the specifics of performance and the relevant behavioural goals such a model must be responsive to. I address these tasks in the second and third sections of the entry. Note that I use the words 'audience,' 'spectator' and 'attendant' interchangeably. I favour the latter term although I do not argue for its use (di Benedetto 2007).

I. 'Attention' in empirical and philosophical literature

Attention is usually thought to occur at the initial stages of perception and to cause the delivery of perceptual content to the senses (Zhang and Lin 2013). And it is thought to be important for managing the limited resources available in sensory areas of the neocortex. Some influential theorists regard the phenomena of perception and attention as reducing uncertainty in predictions that we make about what we are going to encounter (Feldman and Friston 2010; Friston et al. 2012). These phenomena are understood within the theory that the brain has evolved as a mechanism for reducing free energy expenditures within the lives of animals (Yu and Dayan 2005a; Friston 2009).

The ideas of 'reducing uncertainty in predictions' and 'free energy' are directly connected:

> Free energy is a quantity from information theory that quantifies the amount of prediction error or, more formally, it is a variational approximation to the surprise or negative log likelihood of some data given an internal model of those data.
>
> *(Friston 2012, 249)*

Some recent debates have concerned whether attention causes the delivery of perceptual content to the senses or is, rather, an effect of an animal's decision-making process (Krauzlis et al. 2014). Those debates have arisen out of the investigation of relationships between attention and action. Other debates concern which uncertainties are reduced – whether it is uncertainties about the locations of objects or about their properties, or somehow about both (Chikkerur et al. 2010).

The state of the relatively young science of perception and attention is illuminated by considering problems with the key distinctions that have guided research in the empirical literature about attention for the past 70 years.

One distinction concerns whether the kind of attention being investigated is guided by what is internal to the animal ('endogenous') or whether it is guided by what is external to an animal ('exogenous'). Endogenous attention is task-driven and exogenous attention is comparatively task-independent. Another distinction concerns whether the form of attention being investigated is tied to eye-movements ('overt') or can shift around independently of where the subject is 'fixated' at any given moment ('covert'). Yet another distinction concerns whether animals are attending to environments only in a 'distributed' fashion, gaining thereby a fuller but less detailed sense of what is around them, or if they are more narrowly 'focused' in attending to a limited number of specific objects or a limited number of the specific features of an object (Mack 2002; Nanay 2016). Each of these frequently overlapping distinctions has been made within a view that the function of attention – no matter whether that function is as cause or as effect – is to 'facilitate target processing while inhibiting distraction and noise' (Chun, Golumb, and Turk-Browne 2011, 77).

All of these distinctions have problems, as is revealed, for example, by an examination of endogenous and exogenous attention when those modes of attention are understood or referred to as 'top-down' versus 'bottom-up' attention control. One problem with that distinction concerns just what the relation between them is. Evidence from studies showing that attention can be captured by low-level salient, task-irrelevant features has convinced some that top-down attention is actually governed by bottom-up processes; however, evidence from studies showing that low-level salient stimuli can effectively be ignored has convinced others that immediately upon bottom-up stimulation, top-down attention is invariably employed according to the goals of the agent (Hickey, van Zoest, and Theeuwes 2010). Theatrical examples of both low-level and top-down attention can readily be found. The first appearance of Hedda, according to the script for the play *Hedda Gabler*, is heralded by sounds she makes in the hallway prior to entering the room. Changes in sounds and movements are salient features that capture attention. But later in the same play, and based on a spectator's understanding of the character of Judge Brack, a perceptive attendant will expect Brack to use what he knows to attempt to exert power over Hedda. And such a spectator will attend closely and with anticipation to Hedda picking up a pistol at that point. Prior understanding is crucial to the top-down management of both expectations and attention (Saltz 2006; Blair 2010).

Another problem is that the top-down versus bottom-up picture seems to have foundered over the fact that some strong selection biases cannot be explained using either side of the distinction. In particular, neither side of the distinction can explain selection biases stemming from the history of selection that seem to govern some current selections nor some biases in current selections that seem to be governed by past rewards (Awh, Belopolsky, and Theeuwes 2012). Finally, the so-called 'top-down control of attention' is deeply involved in the very area that bottom-up attention was supposed to be controlling – namely, in the deployment of attention for spatial selection (Hunt and Kingstone 2003; Zelinsky et al. 2005). If a theatre-goer is already familiar with the play, *Hamlet*, he will expect to look and see something on the parapet and as a result be conducting a visual search at that location, even if the movement or sounds of whatever or whomever represents Hamlet's father's ghost is what triggers the attention.

Reliance on these distinctions is further complicated by the advent of *cross-modal* studies of attention – attention that relies upon visual, auditory, tactile and other sensory inputs (Li et al. 2008; Ide, Zhang, and Li 2014).

Most models of eye-gaze guidance and visual search have been built on the assumption that low-level conspicuousness is to be explained by appeal to features that can be replicated in static pictures, are pre-attentively salient, become salient to an agent in a manner that is task-free and stimulus driven and are aimed at providing an explanation of spatial rather than temporal aspects of viewing behaviour (Tatler et al. 2011, 4–8). This way of understanding eye-gaze guidance and visual search has made it easier to develop computational representations and modeling of attention – a worthy goal of attention studies (Itti and Koch 2001; Borji, Sihite, and Itti 2012; Itti and Borji 2015; yet also see Zhang and Lin 2013, 167–206). But the foregoing disagreements and debates are part of what supports the claim that there may be no single model of attention (Chun, Golumb, and Turk-Browne 2011, 76; Lavie et al. 2004). And, in any case, this is not what is needed in theatre and performance theory, where most attendants already know they are at the theatre, watching and listening to moving and speaking performers, and are actively trying to figure out what it is they are spectating during the time of the performance.

Accompanying the lack of clarity and precision in the uses and applications of the leading distinctions in the empirical literature is also the adoption, by philosophical theorists, of larger but competing metaphysical models of what is going on in our brains when we attend, perceive and cognitively and emotionally grasp the world. Some of those models hold that we directly perceive and are physically involved in a way that shapes our perceptions (Clark 2013, 2015). Others hold we infer the causes of our perceptions, and so we must be distinct from the rest of the world (Hohwy 2012).

Yet, despite the disagreements within the empirical and philosophical communities, there are five important factual claims for which there is converging evidence. First, specification of tasks is important to models of attention-guidance and attention-selection across modalities (Yu and Dayan 2005b; Land and McLeod 2000). For example, the common tendency to look into empty spaces in anticipation of an event is problematic for bottom-up conspicuity models precisely because they rely on task-free skills and low-level feature salience. Whereas, if attention-guidance and attention-selection across modalities are determined by the activities the agent is engaged in – by her tasks – anticipatory searching is readily explained. This is an important key to understanding how a spectator's attention is focused within a theatrical performance as well as how performers create focal points (Chaikin 1980, 59). In *King Lear*, Act III, Scene vi, Lear, deep into madness, is brought by Kent to a farmhouse adjoining Gloucester's castle, accompanied by the King's Fool and by Edgar, Gloucester's estranged son. He begins to arraign his absent daughters, Regan and Goneril. Looking into empty air, he first arraigns Goneril. A typical spectator will search the space in front of Lear for any trace of her and will not settle on a specific place until the Fool says, 'Cry you mercy, I took you for a joint stool'. Moreover, this feature of theatrical performances – that they induce spectators to attend to places before the action in them commences – is also relied upon by animators and cartoonists (Johnstone and Thomas 1995, 51–3; Cohn and Wittenberg 2016).

Second, if the subjects being studied or modeled are aware of the actual task at hand and actively participate in that task, they tend to shape 'their distribution of attention to match the expected events' (Hayhoe et al. 2003, 49–50). If they are not, or are prompted to attempt a different task, previous episodes of attention to a scene or stages of a scene neither are remembered nor guide attention. One phenomenon that brought this to light is often called 'change blindness' (Levin and Simons 1997; Simons and Levin 1997, 1998). The classic examples of change blindness occur when a subject fails to detect changes in a scene when focused upon a non-relevant task. Simons and Levin report cases in which, when being asked directions, subjects failed to notice that the *persons* asking for the directions had actually changed – were in fact different people – while a large object was briefly carried between them, obscuring the changes in personnel, but where the changes were perceptible. The problem is that the tasks the subjects believed they were supposed to undertake, and to which they were attentive, prevented them from recognising the changes that occurred in the scenes they actually experienced. Theatre provides a contrasting case to these in one respect because for the most part theatre and performance theorists *can* simply assume that attendants with a certain level of experience know they are witnessing a theatrical performance and are trying to figure out the same performance that is observed by others in attendance. That is their relevant task.

Third, regularities in the attended environment – which are common in theatrical performances, especially in the behaviour of performers – are important for attention-guidance (Zhao, Al-Aidroos, and Turk-Browne 2013). Every performer works at selecting what to do and how to say things (if there are to be spoken lines) in a manner that is consistent across scenes. And in traditional acting, only one person, with his or her distinctive persona, will play a given character (Hamilton 2013, 42–44). For something of an outlier example, members of Actors from the London Stage, who perform entire plays by Shakespeare using only four-five performers, provide regularities to their audiences in the form of multiple but simple identifying props (http://shakespeare.nd.edu/actors-from-the-london-stage/about/ – accessed 17 November, 2017).

Fourth, Bayesian inference – that shows precisely how individuals challenge their 'prior beliefs' by confronting them with their current experiences and thereby infer new, 'posterior beliefs' – is involved in all current models of attention, perception and cognition, including the recognition and response to affect in other animals (Pollick et al. 2001; Friston 2012). Indeed, it is

> a guiding assumption of prediction error minimization theories of attention and perception … that any system that minimizes long-term prediction error will approximate Bayesian inference.
>
> *(de Bruin and Michael 2017, 59)*

Other terms for 'beliefs' might be 'expectations,' 'credences' or 'predictions;' and I will use those words interchangeably.

Several aspects of Bayesian inference are important to models of attention, perception and cognition. Bayesian inferences are often undertaken without awareness (Perfors et al. 2011). The prior beliefs we have are already informed by prior experience, so the notions of a 'data-stream' or the 'collection of data' do not presuppose anything about the metaphysical nature of that data. Nor should we make such presuppositions, for example, by claiming that the 'data-stream' consists of theory-independent materials for theory construction (Glymour 2000; Noë 2007). Moreover, Bayesian inference does not deny but cuts right across the 'top-down/bottom-up' distinction that has often been invoked in studies of visual perception. All of this is important to understanding the reception of theatre because its events, each of which must be understood (at some level) when they happen, take time to unfold and also are connected over time. So, a spectator's attention must not only adjust to what is happening at any given moment within that moment, but must also adjust both to what happens in the next moments and what has happened in the previous moments as well.

Fifth, and finally, attention is connected to preparation for action and to action. The idea of 'passive' data-collectors and hypothesising observers – if ever it did hold sway among cognitive scientists and philosophers of perception – has certainly collapsed in light of the important connections now drawn between attention, preparedness for action and action (Yu and Dayan 2005b; Creem-Regehr and Kunz 2010; Wu 2011; Friston et al. 2010; Friston 2012; Friston et al. 2013). This is parallel to the fact that theatre practitioners and theorists have rejected the idea that attendants to theatrical performances are 'passive,' a trope that (sometimes deliberately) mis-describes spectator behaviour, even in the least interactive styles of theatre that have been preeminent in the Western tradition.

II. 'Attention' in a theory of theatre and performance

These five agreed-on fact claims – that tasks guide attention, that awareness of the relevant task is crucial to attention guidance, that regularities in the attended environment also assist in guiding attention, that Bayesian inference is involved in all models of attention, perception, and cognition, and that attention is connected to preparation for action – suggest that we can understand theatrical performances in a new way, as signalling games. In a simple signalling game, we have two players – a sender and a receiver. The senders send signals to the receivers who cannot observe the situation directly, but can observe the signal. Receivers then choose and perform actions that affect both the senders and the receivers themselves in the attempt to get a positive payoff for both (Skyrms 2010; Kuhn 2014).

Crucial to the application of this model to theatrical performance is the fact that receivers, like attendants in theatre, are active (Ranciere 2009, 13). They do not, despite appearances to the contrary, passively and directly perceive the world of the play; instead, they actively draw inferences about what is to happen and gradually – guided by the performers' signals – build up a conception of that world and react to it. The discrepancy – between the feeling of direct, passive perception of the theatrical world and the reality of inference-drawing on the basis of sensory cues – is one aspect of the important fact that many of the inferences attendants make are often well below the level of conscious awareness.

A full philosophical theory of theatre modeled as a signalling game must have at least three parts. One part concerns how signals are generated and sent. A second part concerns how attendants pay attention to and have their attention captured by performances.

A third part concerns how attendants draw further inferences, that is, develop and change their expectations or predictions about what is going to happen and what kind of thing is causing what they attend to, in response to changes in what they are attending. For that third part, perhaps a 'meta-Bayesian' account of updating is required (Daunizeau et al. 2010). And this looks promising for two reasons.

First, an interesting example of change from a prior to a posterior belief that depends crucially on the data as it is presented is provided by what happens when an attendant observes two different performances of, say, *Richard II*. Suppose one of those performances has its modern intermission set right after Richard has shown himself to be a brash but clear-eyed and greedy cynic (i.e., after Acts I and II). An attendant to such a performance is unlikely to ever take Richard's 'epiphany' in Act V very seriously. If, in contrast, the intermission is set after Act I and all of II except that last scene – where the Queen declares him 'much put upon' and a 'sweet guest' – an attendant is much more likely to take the later epiphany seriously and perhaps even to come to view Richard as alternating between anger and self-pity (Toole 1978, 166–67). A similar case is to be found when thinking of a spectator to the 1947 performance of *A Streetcar Named Desire* who finds

Blanche 'neurotic and unlikable,' but when seeing the *movie* version in 1951, other critics found a 'nuanced and sympathetic Blanche.' One important difference appears to be that in the play there was an intermission between scenes four and five that solidified the negative conclusions about Blanche, whereas no such intermission occurred in the movie (McConachie 2014).

Second, the fact that attendants develop and change their credences about what *kind* of object is being performed is also a reason for adopting a hierarchical Bayesian theory about changes in credences when modelling the reasoning of audiences (Goodman et al. 2011; Hamilton 2015). For this model shows how it is possible for a spectator to correctly infer that she is witnessing a production in the Naturalistic style (as opposed to a Brechtian style, for example) based only on the acting and *mise en scene* she experiences. She is not limited only to asking and making guesses about what happens next but also can assist herself in that very task because she is also able to infer what *kind* of production she is experiencing from *the same data* she uses to form a prediction about what will happen next.

However, this entry has been primarily a contribution only to the second part of a full theory, namely, an account of how attendants pay attention to and have their attention captured by performances, based on the agreed-upon facts about attention and on the signalling game model. The following two questions remain. To what do spectators attend? and what is their manner of attention? So, in conclusion, I will now venture suggestions for how we might respond to them.

III. The attention of audiences

First, attention to features of performers and to the *mise en scene* generates the collected 'data streams' from which attendants draw their inferences. Features become salient to an attendant just when the learner-attendant, under a suitable common knowledge requirement, can notice those features as regularities in the behaviour of the performer or the presented environment (Hamilton 2007, 91–113). What is salient for the attention of a spectator will, naturally, depend on her level of familiarity and expertise with theatrical performances. She may be able to foresee certain features because she has been appropriately backgrounded for it (Saltz 2006). But she may be entirely a novice, unable to rely on effective prior beliefs and thus unable to foresee those features. Nevertheless, in both cases, and since most of the regularities that attendants are interested in primarily are regularities of actions, I believe it is *primarily* features of actions that audiences attend to (Shepard 1971; Schachner and Carey 2013; Novack and Goldin-Meadow 2016).

Secondly, consider the behaviour of someone attending to the sick or attending to a lover. Those behaviours provide useful analogies for understanding the manner in which a receiver-spectator attends. Even though some of this behaviour is not at all relevant to this issue, some is highly relevant.

1 Attending in these cases involves listening and watching for – *searching for* – certain kinds of features that are appropriately related to the tasks of grasping and responding affectively to the behaviour of others. The results of laboratory studies suggest that the specifics of a subject's task are crucial to the subject's feature searches (Joseph, Chun, and Nakayama 1997; Kiss et al. 2014). And theatre and performance practitioners have long known that attendants must actively search for what to attend to in the behaviour of performers in order to make sense of it. In this, they are guided both by performers and their own past experiences.

2 Attending in these cases requires being prepared to respond to that for which one has been listening and watching. Moreover, the forms of that preparedness to respond and the responses themselves are usually physical or active in some way. Audience members squirm, grimace, guffaw or wince when, for example, they believe a character is about to do something stupid, they often lean forward in anticipation of bad people getting their just desserts and they occasionally cry for the losses characters suffer. Spectators react physically to images, events, characters, their traits and conduct, often without self-awareness. A character may be felt to be in a stronger position in the story simply because the performer is positioned on stage in a particular way. And, unless the attendant is sufficiently cognisant of the relevant bits of stagecraft, she will be completely unaware of how that response to the character was occasioned (Hamilton 2007).

3 Then there is usually a physical effect on the person who is the object of attention. For two simple but powerful examples, being attended to can trigger approval-seeking behaviour or it can determine where people look (van Rompay, Vonk and Frasen 2009; Risko and Kingstone 2011). That is, the effects can also signal back something to the original signaler (Skyrms 2010). Some acting styles – such as Brecht's perhaps – aim to exploit this very natural reaction (Rouse 1984). Others – such as Naturalism – aim to suppress it (Stanislavski 1980, 84). This helps explain what *performers* must do, given their stylistic choices, when they are attended to.

References

Awh, Edward, Artem V. Belopolsky, and Jan Theeuwes. 2012. 'Top-Down versus Bottom-Up Attentional Control: A Failed Theoretical Dichotomy'. *Trends in Cognitive Sciences* 16(8): 437–43.

Blair, Rhonda. 2010. 'Acting, Embodiment, and Text: Hedda Gabler and Possible Uses of Cognitive Science'. *Theatre Topics* 20(1): 11–21.

Borji, Ali, Dicky N. Sihite, and Laurent Itti. 2012. 'An Object-Based Bayesian Framework for Top-Down Visual Attention'. *Association for the Advancement of Artificial Intelligence* 1529–35.

Chaikin, Joseph. 1980. *The Presence of the Actor.* New York: Atheneum.

Chikkerur, Sharat, Thomas Serre, Cheston Tan, and Tomaso Poggio. 2010. 'What and Where: A Bayesian Inference Theory of Attention'. *Vision Research* 50: 2233–47.

Chun, Marvin M., Julie D. Golomb, and Nicholas B. Turk-Browne. 2011. 'A Taxonomy of External and Internal Attention'. *The Annual Review of Psychology* 62: 73–101.

Clark, Andy. 2013. 'Whatever Next? Predictive Brains, Situated Agents, and the Future of Cognitive Science'. *Behavioral and Brain Sciences* 36: 181–253.

Clark, Andy. 2015. 'Radical Predictive Processing'. *The Southern Journal of Philosophy* 53 (Special Supplement): 3–27.

Cohn, Neil, and Eva Wittenberg. 2015. 'Action Starring Narratives and Events: Structure and Inference in Visual Narrative Comprehension'. *Journal of Cognitive Psychology* 27(7): 812–28.

Creem-Regehr, Sarah H., and Benjamin R. Kunz. 2010. 'Perception and Action'. *Reviews in Cognitive Science* 1: 800–810.

Daunizeau, Jean, Hanneke E. M. den Ouden, Matthias Pessiglione, Stefan J. Kiebel, Karl J. Friston, and Klaas E. Stephan. 2010. 'Observing the Observer (I): Meta-Bayesian Models of Learning and Decision-Making'. *PLoS One* 5(12): 1–19.

di Benedetto, Stephen. 2007. 'Guiding Somatic Responses within Performative Structures'. In *The Senses in Performance*, edited by Sally Banes and Andrew Lepecki, 124–134. New York: Routledge.

de Bruin, Leon, and John Michael. 2017. 'Prediction Error Minimization: Implications for Embodied Cognition and the Extended Mind Hypothesis'. *Brain and Cognition* 112: 58–63.

Feldman, Harriet, and Karl J. Friston. 2010. 'Attention, Uncertainty, and Free-Energy'. *Frontiers in Human Neuroscience* 4: 1–23.

Friston, Karl. 2009. 'The Free-Energy Principle: A Rough Guide to the Brain?' *Trends in Cognitive Science* 13(7): 293–301.

Friston, Karl. 2012. 'Prediction, Perception, and Agency'. *International Journal of Psychophysiology* 83: 248–52.

Friston, Karl, Jean Daunizeau, James Kilner, and Stefan J. Kiebel. 2010. 'Action and Behavior: A Free-Energy Formulation'. *Biological Cybernetics* 102: 227–260.

Friston, Karl, Rick A. Adams, Laurent Perrinet, and Michael Breakspear. 2012. 'Perceptions as Hypotheses; Saccades as Experiments'. *Frontiers in Psychology* 3: 1–20.

Friston, Karl, Philipp Schwartenbeck, Thomas FitzGerald, Michael Moutoussis, Timothy Behrens, and Raymond J. Dolan. 2013. 'The Anatomy of Choice: Active Inference and Agency'. *Frontiers in Human Neuroscience* 7(article 598): 1–18.

Glymour, Bruce. 2000. 'Data and Phenomena: A Distinction Reconsidered'. *Erkenntnis* 52: 29–37.

Goodman, Noah D., Tomer D. Ullman, and Joshua B. Tenenbaum. 2011. 'Learning a Theory of Causality'. *Psychological Review* 118(1): 110–9.

Hamilton, James R. 2007. *The Art of Theater.* Hoboken, NJ: Wiley-Blackwell.

Hamilton, James R. 2013. 'Acting'. *The Journal of Dramatic Theory and Criticism* 28(1): 35–59.

Hamilton, James R. 2015. 'Notes toward a Theory of Spectating'. *The Journal of Dramatic Theory and Criticism* 29(2): 105–25.

Hayhoe, Mary M., Anurag Shrivastava, Ryan Mruzek, and Jeff B. Pelz. 2003. 'Visual Memory and Motor Planning in a Natural Task'. *Journal of Vision* 3: 49–63.

Hickey, Clayton, Wieske van Zoest, and Jan Theeuwes. 2010. 'The Time Course of Exogenous and Endogenous Control of Overt Attention'. *Experiments in Brain Research* 201: 789–96.

Hohwy, Jakob. 2012. 'Attention and Conscious Perception in the Hypothesis Testing Brain'. *Frontiers in Psychology* 3: 1–14.

Hunt, Amelia R., and Alan Kingstone. 2003. 'Covert and Overt Voluntary Attention: Linked or Independent?' *Cognitive Brain Research* 18: 102–5.

Ide, Jaime S., Sheng Zhang, and Chiang-shan R. Li. 2014. 'Bayesian Network Models in Brain Functional Connectivity Analysis'. *International Journal of Approximating Reason*, 56(Part 1): 1–26.

Itti, Laurent, and Ali Borji. 2015. 'Computational Models: Bottom-Up and Top-Down Aspects'. *arXiv* 3(7): 1–30.

Itti, Laurent, and Christof Koch. 2001. 'Computational Modeling of Visual Attention'. *Nature Reviews Neuroscience* 2: 1–11.

Johnstone, Ollie, and Frank Thomas. 1995. 'The Principles of Animation'. In Disney Animation: The Illusion of Life, 47–69. New York: Hyperion.

Joseph, Julien S., Marvin M. Chun, and Ken Nakayama. 1997. 'Attentional Requirements in a "Preattentive" Feature Search Task'. *Nature* 387: 805–7.

Kiss, Monicka, Anna Grubert, Anders Petersen, and Martin Eimer. 2012. 'Attentional Capture by Salient Distractors during Visual Search Is Determined by Temporal Task Demands'. *Journal of Cognitive Neuroscience* 24(3): 749–59.

Krauzlis, Richard J., Anil Bollimunta, Fabrice Arcizet, and Lupeng Wang. 2014. 'Attention Is an Effect Not a Cause'. *Trends in Cognitive Science* 18(9): 457–64.

Kuhn, Steven L. 2014. 'Signaling Theory and Technologies of Communication in the Paleolithic'. *Biological Theory* 9: 42–50.

Land, Michael F., and Peter McLeod. 2000. 'From Eye Movements to Actions: How Batsmen Hit the Ball'. *Nature Neuroscience* 3: 1340–5.

Lavie, Nilli, Aleksandra Hirst, Essi Viding, and Jan W. de Fockert. 2004. 'Load Theory of Selective Attention and Cognitive Control'. *Journal of Experimental Psychology, General* 133(3): 339–54.

Levin, Daniel T., and Daniel J. Simons. 1997. 'Failure to Detect Changes to Attended Objects in Motion Pictures'. *Psychonomic Bulletin & Review* 4(4): 501–6.

Li, Junning, Z. Jane Wang, Samantha J. Palmer, and Martin J. McKeown. 2008. 'Dynamic Bayesian Network Modeling of fMRI: A Comparison of Group-Analysis Methods'. *NeuroImage* 41: 398–407.

Mack, Arien. 2002. 'Is the Visual World a Grand Illusion? A Response'. *Journal of Consciousness Studies* 9(2): 102–10.

McConachie, Bruce. 2014. 'All in the Timing: The Meanings of *Streetcar* in 1947 and 1951'. In *The Theatre of Tennessee Williams*, edited by Brenda Murphy and Kevin J. Wetmore Jr., 181–204. London: Bloomsbury Methuen Drama.

Nanay, Bence. 2010. 'Action-Oriented Perception'. *European Journal of Philosophy* 20(3): 430–46.

Nanay, Bence. 2016. *Aesthetics as Philosophy of Perception.* Oxford: Oxford University Press.

Noë, Alva. 2007. 'The Critique of Pure Phenomenology'. *Phenomenological Cognitive Science* 6: 231–45.
Novack, Miriam A., and Susan Goldin-Meadow. 2017. 'Gesture as Representational Action: A Paper about Function'. *Psychonomic Bulletin & Review* 24: 652–65.
Perfors, Amy, Joshua B. Tenenbaum, Thomas L. Griffiths, and Fei Xu. 2011. 'A Tutorial Introduction to Bayesian Models of Cognitive Development'. *Cognition* 120(3): 302–21.
Pollick, Frank E., Helena M. Paterson, Armin Bruderlin, and Anthony J. Sanford. 2001. 'Perceiving Affect from Arm Movement'. *Cognition* 82: B51–B61.
Ranciere, Jaques. 2009. *The Emancipated Spectator.* Translated and edited by Gregory Elliott. London: Verso.
Risko, Evan F., and Alan Kingstone. 2011. 'Eyes Wide Shut: Implied Social Presence, Eye Tracking and Attention'. *Attention, Perception, & Psychophysics* 73: 291–6.
Rouse, John. 1984. 'Brecht and the Contradictory Actor'. *Theatre Journal* 36(1): 25–42.
Saltz, David. 2006. 'Infiction and Outfiction'. In *Staging Philosophy: Intersections of Theater, Performance, and Philosophy,* edited by David Krasner and David Saltz, 203–20. Ann Arbor: University of Michigan Press.
Schachner, Adena, and Susan Carey. 2013. 'Reasoning about "Irrational" Actions: When Intentional Movements Cannot Be Explained, the Movements Themselves are Seen as the Goal'. *Cognition* 129: 309–27.
Shepard, Richmond. 1971. 'What Is Mime About?' In *Mime: The Technique of Silence.* New York: Drama Book Specialists.
Simons, Daniel J., and Daniel T. Levin. 1997. 'Change Blindness'. *Trends in Cognitive Science* 1(7): 261–7.
Simons, Daniel J., and Daniel T. Levin. 1998. 'Failure to Detect Changes to People during a Real-World Interaction'. *Psychonomic Bulletin & Review* 5(4): 644–9.
Skyrms, Brian. 2010. *Signals: Evolution, Learning, and Information.* Oxford: Oxford University Press.
Stanislavski, Konstantin. 1980. *An Actor Prepares,* 2nd ed. Translated and edited by Elizabeth Hapgood. London: Methuen.
Tatler, Benjamin W., Mary M. Hayhoe, Michael F. Land, and Dana H. Ballard. 2011. 'Eye Guidance in Natural Vision: Reinterpreting Salience'. *Journal of Vision* 11(5): 1–23.
Toole, William B. 1978. 'Psychological Action and Structure in *Richard II*'. *The Journal of General Education* 30(3): 165–84.
Yu, Angela J., and Peter Dayan. 2005a. 'Uncertainty, Neuromodulation, and Attention'. *Neuron,* 46: 681–92.
Yu, Angela J., and Peter Dayan. 2005b. 'Inference, Attention and Decision in a Bayesian Neural Network'. In *Advances in Neural Information Processing Systems 17,* edited by L. K. Saul, Y. Weiss, and L. Bottou, 1577–84. Cambridge, MA: MIT Press.
Van Rompay, Thomas J. L., Dorette J. Vonk, and Marieke L. Fransen. 2009. 'The Eye of the Camera: Effects of Security Cameras on Prosocial Behavior'. *Environment and Behavior* 41: 60–74.
Wu, Wayne. 2011. 'Attention as Selection for Action'. In *Attention: Philosophical and Psychological Essays,* edited by Christopher Mole, Declan Smithies and Wayne Wu. Oxford: Oxford University Press, 97–116.
Zelinsky, Gregory, Wei Zhang, Bing Yu, Xin Chen, and Dimitris Samaras. 2005. 'The Role of Top-Down and Bottom-Up Processes in Guiding Eye Movements during Visual Search'. In (NiPS 18) *Advances in Neural Information Processing Systems,* edited by Y. Weiss, B. Scholkopf, and J. Platt, 1569–76. Cambridge, MA: MIT Press.
Zhao, Jiaying, Naseem Al-Aidroos, and Nicholas B. Turk-Browne. 2013. 'Attention Is Spontaneously Biased Toward Regularities'. *Psychological Science* 24(5): 667–677.
Zhang, Liming, and Weisi Lin. 2013. *Selective Visual Attention: Computational Models and Applications.* Singapore: Wiley-IEEE Books.

20

EMERGENCE, MEANING AND PRESENCE

An interdisciplinary approach to a disciplinary question

Amy Cook

During the summer of 2017, The Public Theater's production of *Julius Caesar* created quite a stir: it seemed to depict the assassination of President Donald Trump and thus was perceived by some to be a dangerous piece of propaganda calling for the death of the president. Scholars, critics—anyone who had read or seen the play, really—quickly pointed out how ridiculous this interpretation was: the play does not go on to suggest that this action, taken early in the play, was such a good idea and things don't end well for the conspirators. Nonetheless, Shakespeare's play became ground zero in a political/cultural battle over meaning and representation.

Those who argued that The Public's *Caesar* did not advocate assassination of Trump would have agreed that the actor playing Caesar was costumed and directed to evoke the U.S. President and that the play involves that character being stabbed multiple times by the senators around him. Yet two different groups took two different meanings from the same performance. This might be an extreme example—Fox News and rabid Trump supporters can only be straw men in an argument on Shakespearean performance analysis—but the question remains: What does it mean to make meaning? How do we know when we have achieved it? These are legitimate questions being addressed by scholars across the academy: linguists want to know, for example, how each 'there' in 'there's no there there' can mean different things; computer scientists working on artificial intelligence want to know how to teach a computer to perceive 'meaning;' medical researchers struggle with patients with various serious problems all related to a felt loss of meaning; and neuroscientists, cognitive scientists and psychologists want to know how humans ascribe meaning. A theatrical interplay with the cognitive sciences helps to reimagine what it can mean to mean.

My work started with a desire to understand how complicated poetry affected an audience. I wanted to know, for instance, why spectators sit forward at a particular piece of poetry. Why everyone remembers some lines ('to hold the mirror up to nature') but not others ('twas caviar to the general')? Cognitive linguistics, with its insistence on embodiment and scale of evidence similar to literary theory (both look at comprehended text to understand something behind or beyond the words spoken or written), has been a powerful tool for these kinds of inquiries. Using conceptual integration and metaphor theory, I pulled apart sections of poetry (Hamlet's 'mirror held up to nature,' Richard's 'winter of our discontent' and *Henry V*'s 'crooked figure,' for example) and argued that the networks of meanings evoked and primed in the process of understanding the text are then part of the scaffolding of larger thematic threads throughout the play (Cook 2007, 2010, 2012; Fauconnier and Turner 2002). The mirror Hamlet holds up to nature at the centre of the play, the one that stands in for the 'purpose of playing,' is not an actual mirror. This is not the mirror we use to check our blind spot or correct our lipstick: this mirror is capable of reflecting impartially and also anatomising or expanding. The purpose of playing is not just strict duplication or precise mimetic representation (indeed, the play he has performed for the King is 'something *like*' the murder of his father), it is a slightly edited, amended or distorted reflection that allows the audience to see virtue's feature, for example, in 'Virtue.' Through examining *how* audiences have understood this creative image, rather than what they have understood, I found a web of evoked sources (convex mirror, political tracts, the use of glass in scientific instruments and the small flat glass mirrors newly available from Italy) tied together. Understanding 'the purpose of playing' evokes a number of tools for vision that do different things.

This is only helpful to see, of course, if seeing it allows us to ask new questions about the poetry or to perceive new angles in the performance. It does not prove that Shakespeare 'anticipated' current theories in cognitive science and it does not prove that these theories are accurate because we see evidence in Shakespeare. Cognitive linguists have received many calls to find a way to empirically verify their theories, and interdisciplinary scholars must be attuned to the disciplinary status of studies or theories that we are importing into our field. Across the sciences, findings evolve, theories are challenged and research proves difficult to replicate. This caveat must be reiterated, and care must be taken at the start of any project like this: new research is happening every day, challenging and stabilising interpretations and assumptions of the past. While I cannot use Shakespeare's poetry to prove a question in cognitive linguistics and cognitive linguistics cannot prove the value of Shakespeare, integrating the two enriches both and ignoring the knowledge and research across the disciplines imperils the work in our own. My means are interdisciplinary, but my goals are disciplinary.

This essay will explore the various ways cognitive science challenges how we make sense of theatre and performance. A cognitive approach must foreground the embodied and embedded nature of communication, attending to the importance of time, presence, emotion and learning, areas traditionally overlooked by more semiotic 'readings' of meaning onstage. Research in cognitive linguistics offers a dynamic and embodied perspective on meaning. The theory of conceptual integration networks, also called blending, has been influential outside of its home discipline, and yet the true value in applying it to theatre and literature comes from how it operates specifically and rigorously, not as applied metaphorically, as often happens. Language onstage, of course, comes out of the mouth of an actor playing a character. I will apply integration theory to complicate our understanding of the relationship between the actor's body and the character's story. Finally, I will discuss Alva Noë's ideas about 'presence' and art; my aim is to destabilise, from the beginning, the idea of meaning.

Conceptual integration networks

Cognitive linguists are now almost unanimous in understanding that thinking and speaking are creative and metaphoric. We do not use language as a code with which we translate what is out there; we organise what is out there around metaphors, image schemas and mental spaces that come from embodied and embedded experience in the world. The father points to the illustration in the book and asks his child, 'do you see the blue ball?' and the child connects the visual activity with the words spoken. The father later asks the child, 'do you see how the pan is hot? Can you see how dangerous that is?' and the process of visual perception is abstracted to more general intellection. TO SEE IS TO KNOW has been referred to as a 'dead metaphor,' but it is powerful despite and because it is rarely visible as metaphoric. I do not respond to my interlocutor by insisting that her 'point' is not visible and thus of course I do not 'see' it; we experience intellection and visual perception similarly.

There is growing evidence, however, that the connection between the abstract and the concrete (between understanding and seeing, for example) remains active in our bodies. In other words, the metaphors are not dead but operate zombie-like, shaping our perception of the discourse. Researchers found that subjects will map directions onto verbs, such that 'lifted' is coded as upward, 'pulled' and 'fled' as leftward, 'walked' as rightward and 'owned' as downward (Richardson et al. 2001). Other studies suggest that readers of many languages with left-to-right word order mentally represent consecutive events as laid out from left to right, where speakers of languages with no egocentric directions have been observed to prefer east-to-west (Boroditsky 2010; Fuhrman 2010; Santiago et al. 2007). The state of the body is both an input into language interpretation and an output. When we read 'open the drawer' we are much quicker to perform a movement moving our hands towards our bodies than away, for example, suggesting that the comprehension of the sentence accessed the motor cortex sufficiently to prime one physical action (movement towards) rather than another (movement away) (Bergen 2012). Others have extended this kind of result to show that the hand muscles are primed even by sentences that describe metaphorical exchanges ('You delegate the responsibilities to Anna') (Glenberg et al. 2008). Comprehending language is a full-bodied affair. We cannot rely on readings that disembody language and talk about meaning as a kind of semiotic code. We require a new look at the language that moves us.

Blending theory builds on Gilles Fauconnier's theory of mental spaces (*Mental Spaces* xvii). Fauconnier defines mental spaces as 'constructs distinct from linguistic structures but built up in any discourse according to guidelines provided by the linguistic expressions' (Fauconnier and Turner 2002, 16); these are packets of information constructed and framed on the fly in which information is organised. Mental space theory provides a model for meaning construction that is fluid and expandable, capable of explaining many examples in language that the more complicated logical theories cannot, such as (as Lakoff and Sweetser point out) 'If I were you, I'd hate me,' co-reference and propositional problems (Lakoff and Sweetser 1996). Some words prompt for meaning, these are 'space builders,' such as 'Max believes' or 'In that movie' (Fauconnier and Turner 2002, 17 and 18). These words set up a space that will inform and/or structure the words/information to come. 'Max believes Sarah went to the store,' for example, creates an event as understood in relation to what Max believes. To connect Max's belief system with whether or not someone went to the store creates a complex social scene in one short sentence.

Certain language explodes in my mouth like pop-rocks: I swirl the words around and the effect multiplies. I repeat, 'let slip the dogs of war' or 'Cowards die many times before their deaths; / The valiant never taste of death but once' (2.2.34), and the enjoyment gives way to mystery. How can someone die before death? How can 'death' seem to mean more than one thing in the same sentence? Traditional theories of language comprehension suggest that we first access the definition of the word and only when it does not fit (one cannot die many times) do we explore more metaphoric or figurative meanings. Language comprehension is always dynamic and embedded; there is no vacuum of ideal meaning from which we find variation. One critical insight of George Lakoff and Mark Turner is that abstract concepts like time, life, love and death are understood metaphorically by mapping – or connecting through perceived similarity – elements of a concrete experience (such as heat) with an abstract concept (such as love) (Lakoff and Turner 1989). We can understand a relationship heating up, consuming us in flames and cooling down. This relies on compression, a process by which the relationship between love and heat is compressed and love *becomes* heat. In this way, all that we know about how heat operates can be recycled to think and speak about love. It also means that we can imagine impossible things (basking in the warmth of her love) because the concept of love is dynamic and networked.

Death may seem like a literal thing, but without any experience of it, we rely on other experiences to make sense of it. We often borrow the path or journey metaphor we use to speak of life to speak of death: 'she passed on' or 'he is gone.' Personification allows us to understand abstract states or nouns by turning them into things capable of causing the thing. The state of no longer being alive (death) can become the agent that caused the state ('Death') such that one can then say 'Death, be not proud' and be understood. Something caused the death of our grandmother and so we narrativise the mystery by creating a Death that can 'come for her.' An agent is created where none originally existed. Death in general need not have intention or agency; death occurs because humans are mortal. The only evidence we have of Death is its effect. Fauconnier and Turner call this entrenched compression a causal tautology. As Turner explains elsewhere,

> from the Event, we read off a Cause that is tautologically and exclusively defined in terms of the event category and is referred to by the very terms for that category ... 'Death' here is an 'empty cause:' Lust causes all events of lust, Hunger causes all events of hunger, Death causes all events of death. In the blend, the specific event of dying is caused fundamentally by Death-in-general; the specific manner of death is the means.
> *(Turner 2004, 14)*

The reason these insights are crucial for those of us working in the arts and humanities is that if we examine the network that allows Death to be proud, we can better understand the language that moves us. Because a blend like 'to pass away' evokes many mental spaces necessary to understand it (travel, here vs. there, etc.), something simple can become complex. As Turner says: 'A blend is not a small abstraction of the mental spaces it blends and it is not a partial cut-and-paste assembly, either, because it contains new stuff, new ideas. It is a tight, packed little compression. It contains much less information than the full mental web it serves. From it, we can reach up to manage and work on the rest of the vast mental web' (Turner 2014, 8).

Cognitive linguistics, with its insistence on an embodied experience of and articulation of life, has produced many influential works on Shakespeare and classic texts. Donald Freeman views *Macbeth* as being tightly constructed around the image-schemata of PATH and CONTAINER and notes that the metaphoric structure of the play then becomes the metaphoric structure for the critics who write about the play; the scaffolding then becomes contagious (Freeman 1995). Mary Thomas Crane's *Shakespeare's Brain* examines how the language of the early modern period reflects and illuminates the brain that created the work (Crane 2001). More recently, Raphael Lyne explains and connects theories of rhetoric and contemporary cognitive science in order to explore how the two may be harnessed together to depict Shakespeare's characters as thinking with and through their language. For Lyne, moments of rhetorical failure are often the most exciting; his discussion of Macbeth's 'pity, like a naked new-born babe' simile shows how Macbeth is unable to find the words to compel restraint as evidenced by the conflicted and arresting image of this striding, vulnerable infant (Lyne 2011). In addition to Shakespeare, cognitive linguistics has proven influential in thinking on, for example, the Cold War theatre of America, suspense films and viewpoint (McConachie 2003; Oakley and Tobin 2005; Dancygier and Sweetser 2012).

According to Fauconnier and Turner, as we process language, mental spaces are networked, connected analogically by a shared, generic space, and emergent meaning comes from the integration of the networked spaces. Blends create emergent meaning by combining structural information from many input spaces. This integration is not complex, weird, advanced or literary; it is everyday and omnipresent. This is in contradiction to Noam Chomsky's theory of language wherein we inherit a system of rules and, through exposure to language, we learn how to generate sentences. Many cognitive linguists view language as being compositional and creative from the start. If language was a system of rules and definitions in the brain, then children should have tremendous difficulty understanding stories involving talking animals, and yet these seem to be the bulk of the characters in children's books. We compose and stage the world around us, generating new ideas (like a grumpy donkey or dying more than once) by linking up disparate spaces.

It is unfortunate that 'blend' is such a simple and evocative word. This can make it seem like 'blending' is some special thing we do when we are being poetic or quirky. It can seem like a blurring or combining of ideas. This is not accurate. Conceptual integration is a theory about how we make meaning below the level of consciousness. Its greatest flaw is also its greatest strength: it explains too much. It hasn't been empirically validated and its proponents have not come up with a way to falsify it. This does not mean that it is not accurate or a powerful explanatory paradigm, but it does mean that claiming that something is a blend does not say enough. What is illuminated once the spaces evoked and connected necessary to create a seemingly objective or literal thing (such as a mirror) are displayed? How do those connected spaces allow us to ask and answer new questions about our disciplinary object? There have been extraordinary studies that have been inspired by cognitive linguistics and would not have been possible without it. For example, Rafael Núñez and Eve Sweetser find an alternate conception of the mapping of the past and the future in the gestures of Aymara Amerindians (Núñez and Sweetser 2006). Teenie Matlock did an experiment that showed that people activate the motor cortex when they read 'the trees ran down the driveway' but not when they read 'the trees lined the driveway' (Matlock 2010). While the impact of this research is on demonstrating the profoundly embodied nature of our cognition (we need our non-neural bodies to make sense of trees running down the driveway), it would not have been possible without recognising the conceptual blend that is 'ran' in this sentence. It is not up to those of us in the arts and humanities to prove or disprove the work in the cognitive sciences, but we should understand that work created in that field is meant to communicate with that world. They, too, are answering disciplinary questions.

Actors and characters

In *The Way We Think*, Fauconnier and Turner refer to the blend that is created when an actor takes the stage:

> The character portrayed may of course be entirely fictional, but there is still a space, a fictional one, in which that person is alive. We do not go to a performance of *Hamlet* in order to measure the similarity between the actor and a historical prince of Denmark. The power comes from the integration in the blend.
>
> *(Fauconnier and Turner, 266)*

Embodied by the actor, 'Hamlet' can become Hamlet. The actor remains visible while also evoking this other entity: 'While we perceive a single scene, we are simultaneously aware of the actor moving and talking on a stage in front of an audience, and of the corresponding character moving and talking within the represented story world' (Fauconnier and Turner, 266). Fauconnier and Turner are correct to perceive that the audience is neither in a trance state nor 'suspending disbelief' but as I have said elsewhere, I worry that my earlier work implied a simplification of what this 'blend' is. By referring to the character blend with a slash ('character/actor' or 'Hamlet/Hawke'), as both Bruce McConachie and I have, we are attempting to keep the actor's body present, not obscure a more complicated network of integration. Casting, the process by which actors are selected for characters in a particular production, is everything, as many directors will tell you. The protesters at *Julius Caesar*, for example, were responding to the whole cast, not just the costume and hair of the murdered senator.

Shakespeare was well aware of which actor would play which part, and the conventions of the time meant that spectators used the information they had about Richard Burbage, Will Kempe, the boy player and so forth to anticipate the story through reference to the actor. The plays often display the actors bodies or histories underneath them. As I argued elsewhere, Hamlet playfully disrobes the boy player playing Ophelia when he comments on the 'nothing' between his legs (Cook 2006). As Arden editor Harold Jenkins points out, the original actor of Polonius also played Caesar the year before, sharing a stage with Burbage, who is thought to have played Brutus (Jenkins 1982). This makes the discussion that Hamlet and Polonius have about 'acting' a rich, intertextual event for the knowing audience member. Hamlet asks Polonius about his acting past and Polonius reports, 'I did enact Julius Caesar. I was killed i' th' Capitol. Brutus killed me.' And Hamlet responds, 'It was a brute part of him to kill so capital a calf there.' For spectators at the Globe, Polonius is simultaneously speaking to both the character of Hamlet and the actor, Burbage. Polonius is pointing to his past and future with this particular actor. For spectators in the Globe who had seen *Julius Caesar* the year before, this might be a bit of foreshadowing. It might invite a layer of complexity to the troubled relationship between Polonius and Hamlet if we are also watching Ceasar and Brutus. Caesar and Brutus ghost Polonius and Hamlet.

'Ghosted' is what theatre historian Marvin Carlson calls it when the previous roles of an actor bleed through the current performance – either enriching it or undermining it, as the case may be (Carlson 2001). Oskar Eustis, director of The Public's 2017 *Julius Caesar*, was aware of this when casting his production. Gregg Henry, the actor cast as Caesar, may not have reminded everyone in the audience of Hollis Doyle, the dirty-trick-playing presidential candidate in the television show *Scandal* from the previous year, but his tall physique and blond hair have gotten him many roles before as lawyers, politicians and villains. With a bit of mousse to his blond hair and an extra-long red tie, he was easy to see as a particular politician. Further, the rest of the cast is filled with actors seen most recently in contemporary political dramas like *House of Cards* (Eisa Davis as Decius Brutus, Elizabeth Marvel as Antony, Corey Stoll as Marcus Brutus), *Homeland* (Elizabeth Marvel) and *Madam Secretary* (John Douglas Thompson as Caius Cassius). Eustis used the casting to locate the time and place for *Julius Caesar* as right now in Washington D. C. In the curtain speech Eustis made about the show (and subsequently published online), he notes, 'we didn't write any new lines. It's all Shakespeare.' Apparently, his casting worked so well that he wanted to remind the audience of the authorship.

Familiarity is a positive element of a production in most cases. We build character through a dynamic interplay between a number of conceptual spaces – visible before the actor crosses the stage or delivers a line: the body (age, race, gender, physical attributes); textual information (actions taken or lines said about or by the character); what we know already or anticipate based on historical information, personal information about the character and the actor portraying it; the reputation of the character (and the actor) and what we know about other roles the actor's body has taken on. The body of the actor and his/her history is always onstage with the character. By casting actors who many spectators would recall from other fictional depictions of political intrigue, Eustis knew the audience would perceive this 'Rome' as the contemporary 'D. C.' that these actors are usually seen in. With the cast set, the play was going to be about Washington politics no matter what other choices were subsequently made by the costume designer, set designer, director or actors. With the cast set, a large part of the meaning was already made.

Presence

Theatre scholars and practitioners spend a lot of time thinking about what happens onstage and (relatively) little time talking about what is going on in the audience.[1] There are experiences as a spectator that make us feel like a part of something bigger and then there are those experiences where what is onstage cannot upstage what's happening in the house. Wagner built his Beyreuth Theatre to give each spectator a good view of the stage, without being distracted by other audience members or the musicians, who were now hid in a pit. The audience was supposed to be consumed by the *gesamtkunstwerk*. The work of the theatre was on stage and the spectators were to be transported. This set of conventions continues, largely, into the present day. We enter the theatre, sit in our seat and are told to turn off our phones and be quiet. We will feel for the characters, get caught up in the story, gasp at the spectacle, clap when it ends and leave talking about what it meant. 'I thought the tree represented hope.' 'I thought she was going to die at the end.' Our job, as audience, is to read the meaning and then probe the illness through the symptomatic behaviour of the characters.

There are several problems for me with this scenario. First, this describes relatively few of my experiences in the theatre. I get distracted by the man who is falling asleep one row down, or I'm wondering how long until intermission or I'm trying to recall where I saw the actor playing the waiter. This may be because of bad luck in audiences, my own lack of discipline or poor casting, but being pulled out of my current state is not common. The second problem is that I know that I am supposed to feel real feelings for these fictional characters, to worry about what will happen to them, but I rarely do. The secret will come out, the gun will go off, the door will slam. I can be delighted, I can laugh, I can register sadness and I can appreciate virtuosic acting, but I rarely feel swept up in the drama as this scenario of expectations depicts. Finally, if I leave propelled to talk about what it meant, if it seems to me that the play was supposed to mean something, I wonder why the producer didn't just save us all a lot of time by telling us what it meant in the programme and letting us go home. I don't want meaning and I don't want escape: I want something I can use.

I am building a straw-man theatre experience from which I can launch my next point about presence and art. Clearly, I am describing a particular kind of experience in the theatre and my own very particular biases. I am not being fair and I'm not trying to be. I believe that *A Doll's House* was art in 1879 and *A Streetcar Named* Desire was art in 1947. When I saw *Long Day's Journey Into Night* by Eugene O'Neill in 11th grade, I was moved – though the *meaning* was slow to reveal itself. I struggled for a week afterwards with what happened: why was the story chilling to me? What did I recognise out of the corner of my eye, resisting words or narrative? Eventually I did have that moment when the pieces came together: when my feelings for the Tyrone family linked up with an analysis of the family unit and my own developing self. My teenage self felt that the play provided me with a way of understanding the self as having an internal psychology messed up by one's parents, containing memories that haunt until purged. This genre of theatre stages a Freudian self and a Cartesian mind, which was tremendously valuable when this was the dominant scientific paradigm.

R. Darren Gobert argues that theatre after Rene Descartes reflected a concept of interiority and the importance of passions: actors were prized 'not only for their outward abilities to represent emotion but for their interior perceptual, emotional, and volitional apparatuses, which determine these outward abilities and shape their performative expression' (Gobert 2013, 86). Gobert refutes the reading of Descartes as separating the body and the mind that was popularised by Antonio Damasio's book, *Descartes' Error*. In fact, argues Gobert, Descartes saw the 'animal spirits' as uniting the body, mind and passions. While contemporary scientists might point to a lack of evidence for 'animal spirits,' Gobert reads Descartes as committed to a mind-body union: 'Descartes defines the emotions as bodily perceptions and thus precisely as a source of knowledge.... Failing to understand the workings of mind-body union, critics... misread Descartes's concept of mind but also miss the crucial role that the passions play in the process of reason' (Gobert 2013, 12). The cultural shift Descartes engendered in this area is most vibrantly evident at the theatre. Gobert finds in Corneille's 'deviation from classical form' a turn to 'wonder,' 'the precise emotion that Descartes located at the center of his emotional physics and moral philosophy.' He explores Racine's *Phedre* as a demonstration of the power of the passions to overcome the body, and argues 'perspectival staging seemed to promise the ontological security of the spectator' (Gobert 2013, 89, 131). Gobert argues that scientific conceptions of the self drive and are reflected in what is on stage.

Theatre has to work on an audience and a spectator has to work to be part of it. It doesn't require labour, but it does require something to be shaken, changed, opened. The pleasure, for me, is in the way the work resists the categories I bring to it and challenges me to create new categories. I want theatre that does something to me that I don't even recognise that I need. I want it to show me something I didn't know. An experience like this might fit the definition of art, as defined by Alva Noë in his book *Strange Tools*: 'Art aims at the disclosure of ourselves to ourselves and so it aims at giving us opportunities to catch ourselves in the act of achieving perceptual consciousness—including aesthetic consciousness—of the world around us. Art investigates the aesthetic' (Noë 2015, 71). He describes art that works as philosophy, as a thinking tool: 'it is the domain in which we grapple with what we already know (or think we know). It is the domain in which we try to get clear about the ways we think and respond and assign value' (Noë 2015, 203). For Noe, theatre, all art, should help us reorganise ourselves to better fit the world around us. Through engaging with this strange experience, this moment in the theatre, we are charged to find new ways of picking up, of exploring the ideas involved.

In his earlier book on *Varieties of Presence*, Noë reconceives perception as a 'skillful engagement' (Noë 2012, 2) and argues for the value of honing our 'sensorimotor understanding' (Noë 2012, 20) of the world around us through a continual adjustment of our concepts:

> Don't think of a concept as a label you can slap on a thing; think of it as a pair of calipers with which you can pick the thing up. Seeing something is picking it up using one sort of caliper. Thinking about its absence requires that we pick it up, or at least try to, in a different way. If there is a difference between seeing something and thinking about it,

> it is because of differences in our calipers. Insofar as there are overlapping similarities and kinship between thought and experience, this is because we use some of the same tools in both cases.
>
> *(Noë 2012, 36)*

The theatre I want now will expand my ability to see, it will give me new ways to imagine history, the self, the universe, physics, cause and effect, and how I think and feel as a spectator sitting there. I no longer care about secrets, about Oedipal desires, I don't even care about family dysfunction: I care about humans as mutually dependent organisms with the miraculous ability to communicate. I don't want to be swept away: I want to be called up short. I want to be given new tools to deal with this rapidly changing world. I want to experience myself in the audience differently. The theatre that excites me now does not stage a Freudian or Cartesian self, it stages the expanding conceptions of language, the body and cognition, that is also exciting my colleagues in the cognitive sciences.

Conclusion

The growing consensus within the cognitive sciences is that thinking is not computing in the brain but action with the body in the world. This is not a small change to the general received wisdom about literature, theatre and art. Such an approach offers a method of understanding performance and connecting it with work being done in other disciplines. It is no longer adequate to say about good theatre that, 'we know it when we see it.' Making meaning is not magic; it can be studied, applied and demonstrated. If thinking is 'world-making,' rather than processing stimuli into meaning, then the hermeneutic tradition of literary and art scholarship must adapt; we do not read to attain meaning, we read to enact worlds within which to experience anew. If cognition is embedded in a given environment, always affective, and extended outside of skin and skull, then a spectator's experience, emotion and attention depend less on what is happening 'inside' the actors onstage – as the acting style of Constantin Stanislavsky and others, and the playwriting of Tennessee Williams, and others, seem to suggest – and more on an enacted experience of her own. If thinking means using objects in our environment in order to make changes to our own and extended ecosystem, then an interaction with a work of art can be aesthetic, poetic and autopoetic.

Note

1 There are notable exceptions, of course, such as McConachie (2008); Bennett (1997); and States (1987).

References

Bennett, Susan. *Theatre Audiences: A Theory of Production and Reception* (New York: Routledge, 1997).

Bergen, Benjamin. *Louder Than Words: The New Science of How the Mind Makes Meaning* (New York: Basic Books, 2012).

Boroditsky, Lera and A. Gaby. "Remembrances of Times East: Absolute Spatial Representations of Time in an Australian Aboriginal Community." *Psychological Science* 21, 1635–9 (2010).

Carlson, Marvin. *The Haunted Stage: The Theatre as Memory Machine* (Ann Arbor: University of Michigan Press, 2001).

Cook, Amy. "The Narrative of Nothing: The Mathematical Blends of Narrator and Hero in Shakespeare's Henry V," *Blending and the Study of Narrative*. Eds. Ralf Schneider and Marcus Hartner. (Berlin: de Gruyter, 2012).

Cook, Amy. *Shakespearean Neuroplay: Reinvigorating the Study of Dramatic Texts and Performance through Cognitive Science*. (New York: Palgrave Macmillan, 2010).

Cook, Amy. "Interplay: The Method and Potential of a Cognitive Scientific Approach to Theatre," *Theatre Journal* 59.4 (2007).

Cook, Amy. "Staging Nothing: *Hamlet* and Cognitive Science," *SubStance* #10 35.2 (2006).

Crane, Mary Thomas. *Shakespeare's Brain: Reading with Cognitive Theory* (Princeton, NJ and Oxford: Princeton University Press, 2001).

Dancygier, Barbara and Eve Sweetser (eds), *Viewpoint in Language: A Multimodal Perspective* (Cambridge: Cambridge University Press, 2012).

Gilles Fauconnier and Mark Turner, *The Way We Think: Conceptual Blending and the Mind's Hidden Complexities* (New York: Basic Books, 2002).

Freeman, Donald. "'Catch[ing] the Nearest Way': *Macbeth* and Cognitive Metaphor," *Journal of Pragmatics* 24, 689–708 (1995).

Fuhrman, Orly and Lera Boroditsky. "Cross-Cultural Differences in Mental Representations of Time: Evidence from an Implicit Nonlinguistic Task," *Cognitive Science* 34: 1430–51 (2010).

Glenberg, Arthur M., M. Sato, L. Cattaneo, L. Riggio, D. Palumbo, et al. "Processing Abstract Language Modulates Motor System Activity," *Quarterly Journal of Experimental Psychology* 61.6: 905–919 (2008).

Gobert, R. Darren. *The Mind-Body Stage: Passion and Interaction in the Cartesian Theater* (Stanford, CA: Stanford University Press, 2013).

Jenkins, Harold. *Hamlet* (London: Arden Shakespeare, 1982).

Lakoff, George and Eve Sweetser. "Sorry, I'm Not Myself Today: The Metaphor System for Conceptualizing the Self." In Gilles Fauconnier and Eve Sweetser (eds) Spaces, Worlds, and Grammars (Chicago: University of Chicago Press, 1996).

Lakoff, George and Mark Turner, *More Than Cool Reason: A Field Guide to Poetic Metaphor* (Chicago: University of Chicago Press, 1989).

Lyne, Raphael. *Shakespeare, Rhetoric and Cognition* (Cambridge: Cambridge UP, 2011).

Matlock, Teenie. "Abstract Motion Is No Longer Abstract," *Language and Cognition* 2 (2010).

McConachie, Bruce. *American Theater in the Culture of the Cold War* (Iowa City: University of Iowa Press, 2003).

McConachie, Bruce. *Engaging Audiences* (New York, Basingstoke and London: Palgrave Macmillan, 2008).

Noë, Alva. *Strange Tools: Art and Human Nature* (New York: Hill & Wang, 2015).

Noe, Alva. *Varieties of Presence* (Cambridge, MA: Harvard University Press, 2012).

Núñez, Rafael and Eve Sweetser. "'With the Future Behind Them': Convergent Evidence from Aymara Language and Gesture in the Crosslinguistic Comparison of Spatial Construals of Time," *Cognitive Science* 30.3 (2006).

Oakley, Todd and Vera Tobin, "Attention, Blending, and Suspense in Classic and Experimental Film," *Blending and the Study of Narrative* (Berlin: de Gruyter, 2012).

Richardson, Daniel C., et al., "'Language Is Spatial': Experimental Evidence for Image Schemas of Concrete and Abstract Verbs," *Proceedings of the 23rd Annual Meeting of the Cognitive Science Society*, 873–878 (2001).

Santiago, Julio, et al., "Time (also) Flies from Left to Right," *Psychonomic Bulletin and Review*, 14.3: 512–16 (2007).

States, Bert. *Great Reckonings in Little Rooms: On the Phenomenology of Theatre* (Berkeley: University of California Press, 1987).

Turner, Mark. *The Origin of Ideas: Blending, Creativity, & the Human Spark* (Oxford: Oxford University Press, 2014).

Turner, Mark. "The Ghost of Anyone's Father," in Shakespearean International Yearbook, ed. Graham Bradshaw, Thomas Bishop, and Mark Turner (Rants, UK: Ashgate, 2004).

21
RELISHING PERFORMANCE
Rasa as participatory sense-making

Erin B. Mee

'There is no drama without *rasa*' according to *The Nātyashāstra* (The Science of Drama), the Sanskrit aesthetic treatise attributed to Bharata (1996, 54). Rasa has been variously translated as juice, flavour, taste, extract and essence; it is the 'aesthetic flavour or sentiment' savoured in and through performance. Bharata tells us that when foods and spices are mixed together in different ways, they create different flavours; similarly, the mixing of different emotions and feelings arising from different situations, when expressed through the performer, gives rise to an experience or 'taste' in the partaker, which is rasa (55). Rasa is what is 'tasted' when a performance is 'digested' or 'taken in' by a partaker. The goal of Sanskrit drama was to create rasa, and rasa remains central to genres such as *kutiyattam* (a particular way of performing Sanskrit drama in Kerala, South India) and *kathakali* (a genre of classical dance-drama in Kerala). Rasa exists only as and when it is *experienced*: 'the existence of rasa and the experience of rasa are identical' (qtd Deutsch 1981, 215). Similarly, rasa exists always and only as the result of an *interaction between* performer and partaker. For Abhinavagupta (CE 950–1025), who commented extensively on *The Nātyashāstra*, rasa is not a gift bestowed upon a passive spectator or a commodity bought by a consumer, but an attainment, an accomplishment; someone who wants to experience rasa has to be an active participant – or, to use the dining metaphor, partaker – in the work.

Because Bharata has taken his metaphor from food, I refer to the spectator or audience member as a partaker: someone who has to choose to take in a performance, who has to actively put it in the 'mouth,' chew on it, break it down and roll it around on the 'tongue' to relish it; who has to 'ingest' and 'digest' the performance, incorporating it into the self. Rasa is active, participatory, interactive, social, experiential, sensual, tactile, multi-sensory, internal, emotional, intellectual, embodied and an attainment. Abhinavagupta refers to rasa as an '*act* of *relishing*' (Deshpande 1989, 85 emphasis mine), and as such, rasa is both a noun and a verb: the relishing of the flavour and the flavour that is itself relished.

While the dining metaphor is useful for a first pass at understanding rasa, it eventually breaks down because experiencing and processing a performance is not, at the cognitive level, the same as eating a meal. Ultimately, rasa is a theory of embodied response to a performance – a theory of partakership – that can be understood in light of cognitive science. I will put rasa in conversation with Giovanna Colombetti's discussion of embodied, experiential, participatory sense-making in *The Feeling Body* to position rasa as an 'affective dimension of intersubjectivity, construed as an embodied or jointly enacted practice' (Colombetti 2017, 172) of participatory sense-making.

According to Bharata, 'rasa is the cumulative result of *vibhāva* [a stimulus], *anubhāva* [an involuntary reaction to the stimulus], and *vyabhicāri bhāva* [a voluntary reaction to the stimulus]' (1996, 55). Bharata lists eight rasas, each of which has varying degrees of intensity: *shringāra* (love, affection), *hasya* (joy, laughter, happiness), *raudra* (anger, rage), *karuna* (sadness, grief, depression), *vira* (strength, heroism), *adbhuta* (wonder, awe), *bibhatsa* (disgust) and *bhayanaka* (fear).

To examine how rasa works according to Bharata, let us focus on shringāra. Shringāra has been translated into English as 'love,' but in Bharata's description, shringāra encompasses much more:

> In our daily life whatever is pure, holy, resplendent is referred to as shringāra [...] Moreover, a person enjoying happiness, achieving his desires and helped by proper season, flowers, etc. when he is in a woman's company – that is called [...] This shringāra results in the case of men and women, of healthy youth. It is of two kinds: *sambhoga* (fulfillment), *vipralambha* (non-fulfillment; lit. separation).
>
> *(Bharata 1996, 56, 57)*

Shringāra that is fulfilled can be stimulated by 'season (i.e., spring), garlands, scent (anointment), ornament or experience or by listening to, or seeing desired company, beautiful surroundings, delightful music, [and/or] beautiful parks' (57). Clearly, shringāra can exist in the everyday world as well as on stage, and can be stimulated by many things, so rasa encompasses emotional experience in everyday life and on the stage.

Each rasa is built on a 'sthāyi bhāva,' or underlying emotional state. Shringāra is based on love; hasya is based on humour; karuna is based on compassion; raudra is based on horror; vira is based on the heroic; bhayanaka is based on fear; bibhatsa is based on repulsion; adbhuta is based on wonder. As Vinay Dharwadker points out:

> *The Nātyashāstra*'s boldest implication is that the sthāyi bhāvas comprise an individual subject's fundamental mode of existence in the world – that a self exists only in one or another of these long-lasting states at any given time, persists over time in a succession of such states, and has no other mode of existence.
>
> *(2015, 1384)*

Shringāra is built on the sthāyi bhāva *rati* (love). For shringāra an underlying emotional state is subjected to a stimulus which, in the case of shringāra, might be a smell, a memory, a trip to the park, an interesting conversation, or an interaction with a loved one. This stimulus (vibhāva) would then mix with the sthāyi bhāva to create anubhāva – an involuntary or non-conscious psychophysical reaction to the stimulus, such as 'sweating, thrill, break in voice, trembling, pallor, tears and breakdown' (Bharata 1996, 57). The stimulus and the sthāyi bhāva also create vyabhicari bhāva, a more conscious and somewhat controllable reaction to the stimulus, which might include impatience, bashfulness, excitement, dissimulation or jealousy.

The Nātyashāstra's main goal is to help performers evoke rasa; most of its chapters focus on how to create and present facial expressions, hand gestures and bodily postures that will embody the outward manifestation of an emotion so it can be shared with, and evoke a response in, the partaker. Bharata says that shringāra 'must be expressed [...] by loving looks, lifting (raising) eyebrows, side-glances, graceful steps and gestures, which are all anubhāvas or involuntary (natural)' (Bharata 1996, 57). Rasa occurs when the sthāyi bhāva of the partaker is subjected *ble 21.1* How the rasa shringāra is generated

hayi ava (in e partaker)	+ *Vibhava (stimulus)*	+ *Anubhava (involuntary reaction to the stimulus)*	+ *Vyabhicari Bhava (voluntary reaction to the stimulus)*	*When expressed by the performer through abhinaya and shared with the partaker through:*	= *Rasa*
ati (love)	Season (i.e., spring), garlands, scent (anointment), ornament or experience or by listening to, or seeing desired company, beautiful surroundings, delightful music, beautiful parks.	Feeling stunned, sweating, thrill, break in voice, trembling, pallor, tears and breakdown.	Dejection, lassitude, suspicion, jealousy, infatuation, fatigue, helplessness, anxiety, confusion, remembrance, boldness, bashfulness, fickleness, pleasure, excitement, heaviness, pride, sorrow, impatience, sleep, forgetfulness, dream, awakening, intolerance, dissimulation, ferocity, desire, disease, insanity, death, fear and guessing.	Loving looks, lifting (raising) eyebrows, side-glances, graceful steps and gestures.	Sringara

to the vibhāva of the situation, and the anubhāva and vyabhicari bhāva of the character as performed by the performer, and shared with the partaker through *abhinaya*, the Sanskrit term for acting. Abhinaya[1] literally means 'to carry forward,' and the actor is known as a '*katha patram*,' or vessel for the story and its thematic and emotional content. The actor's primary responsibility, which is built into the terminology, is to be a vessel to carry forward the character-story-situation in order to evoke a rasic exchange with the partaker (Table 21.1). To put it mathematically: sthāyi bhāva + vibhāva + anubhāva + vyabhicari bhāva + abhinaya (4 aspects) = rasa

Colombetti articulates a theory of embodied sense-making that parallels Bharata's articulation of rasa. She argues that skills such as imitation, along with a responsiveness to others' facial expressions and physical gestures, 'embody [...] a *pragmatic* form of understanding others' (172) that is not based on internal simulation or mentalising, but constitutes an embodied *practice*. She refers to this as '*participatory sense making*, which is enacted in the concrete interaction between two or more autonomous agents coupled via reciprocity and coordination' (172). These skills, present in daily life and in partakers of performance, create rasa – an emotional taste – through pragmatic, participatory and embodied sense-making in the coordinated and reciprocal interaction between performer and partaker.

An understanding between self and other involves empathy, which is, in Colombetti's view, 'an *experiential* access to the other's subjectivity,' a 'feeling in' (174). She stresses the '*sensual* nature of our experience of others,' and refers to empathy as a process of sensing-in (174). For example,

> I do not experience the other's bodily sensation [as my own]. Hence when I see a hand tensely contracted in a fist, I do not experience this tenseness in my own hand, as if my hand were itself tensely contracted in a fist. At the same time, however, I do not just see the other's hand and judge that it is tense [or mentalize about its tenseness]; rather, I experience the tenseness in the other's hand.
>
> *(174)*

Crucially, for a discussion of rasa as partakership, this direct body-to-body empathy occurs in the *relationship between* self and other: 'I neither "lose myself" in the others nor incorporate the others' experience into mine in a sort of extended awareness of myself' (181). Empathy, then, is not self-referential: I do not convert the other person's experience into my own in order to understand it (e.g., 'I feel your pain'). 'Rather, I retain an awareness of myself and the others as distinct subjects. At the same time, however, I am also aware, via basic empathy, that the others' feeling is the same as mine' (181). Colombetti's analysis of empathy functions as a description of rasa as *sensual* and *experiential*, as a *relationship* between performer and partaker and as a *process* of sensing-in. This opens up an understanding of rasa as a *practice* of empathy in performance.

Colombetti points out that 'one need not be able to name the emotion that one empathizes in the expression – even though, arguably, one's emotional vocabulary can affect how one perceives expression' (177). In other words, this connection is often, even usually, 'prereflective' (181). Colombetti also points out that there can be a 'mismatch between the feeling that is empathized and the one that the observed person actually experiences' (177). For example, an actor portraying anger may evoke fear; an actor portraying fear may evoke pity. Nonetheless, the partaker experiences the performer 'as a source of feeling' (181).

If rasa is a way of experiencing another's emotions through the embodied relationship between self and other, if rasa is an affective dimension of intersubjectivity, it is an act of empathy, a practice of empathy and an empathetic response. Colombetti argues that the *awareness* of sharing a feeling (which is different from empathy) leads to a 'higher unity' between self and other (181). Rasa, which is an awareness of a shared feeling in that the partaker attends a performance to have a shared feeling and to be aware of that shared feeling – is then a mode of social bonding. Although rasa is most often discussed as an exchange between performer and partaker, performers partake of each other's performances, and partakers experience each other's responses, meaning that rasa becomes a flow of intersubjectivity between performer and partaker, partaker and partaker, and performer and performer.

As a fundamental mode of existence, as embodied sense-making, as a way of responding to others (whether fictional or real), as a response to performance and as an act or performance in and of itself, rasa is incorporated in(to) the self. Which is to say it participates in constituting the constantly becoming self. Crucially, rasa constitutes the constantly becoming social self.

Note

1 Bharata discusses four kinds of abhinaya: bodily expression (facial expressions, gestures, foot work), known as āngika abhinaya; vocal expression (vāchika abhinaya); expression through costumes, scenery and props (āharyā abhinaya); and the outward expression of the performer's emotions and feelings (sāttvika abhinaya).

References

Bharata (1996). *The Natyasastra*. Translated by Adya Rangacharya. Delhi: Munshiram Manoharlal Publishers Pvt. Ltd.

Colombetti, Giovanna (2017). *The Feeling Body: Affective Science Meets the Enactive Mind*. Cambridge: MIT Press.

Deshpande, G.T. (1989). *Abhinavagupta*. Delhi: The Sahitya Akademi.

Deutch, Elliott (1981). "Reflections on Some Aspects of the Theory of Rasa." In *Sanskrit Drama in Performance*. Honolulu: University of Hawaii Press. Rachel van M. Baumer and James R. Brandon, eds.

Dharwadker, Vinay (2015). "Emotion in Motion: The *Nātyashāstra*, Darwin, and Affect Theory." *PMLA* 130.5: 1381–99.

22

THE SELF, ETHICS, AGENCY AND TRAGEDY

David Palmer

A consequence of the Darwinian revolution is the enhancement of a materialist theory of the universe. Darwin showed that order can arise simply from the interaction of physical elements within a system. A conscious designer with a teleological plan is not needed to explain the evolution of species any more than a conscious designer is needed to create the recurring patterns in a stack of marbles that have been run down an inclined peg board: the pattern is a function simply of the interaction of the physical elements of the system, such as the diameters of the marbles and the arrangement of the pegs. Physical systems can order themselves driven completely by laws of cause and effect (Dennett 1995, 125–33; Richards 2000, 11–18).

Extending this view, the brain is merely one more bodily element, like the stomach or the foot, that takes its current form from the evolutionary history of how individuals with particular physical traits engaged with their environments. People's minds, their personal experience of themselves and the world, are simply the mental phenomena that emerge from the physical brain/body system's interaction with the environment. Human experience and action result strictly from physical interactions and the laws of cause and effect. This has led many people to argue that free will – commonly understood as action initiated by the person – is an illusion along with the notion of a self as the experiencer of the environment and the guiding agent of our actions (See Hood 2012; Wegner 2002; Wegner & Gray 2016).

These general ideas are not new. Similar ones were expressed by William James at the end of the nineteenth century in his chapters on the stream of consciousness, the self and the will in *The Principles of Psychology* (1890) and *Psychology: Briefer Course* (1892). For James, the self is not a mental entity having our experiences and directing our actions but a character in a narrative our brains create as a way of giving our sensations meaning so that they are not, as Kant said, merely a chaos of disjointed experiences but rather a coherent context to which we can respond. As James says in *Psychology: Briefer Course*, 'the thoughts themselves are the thinkers' (1892, 209. See also James 1890, 401).

Contemporary psychological research supports these views. Many animals have what Daniel Kahneman has called System 1 thinking: a kind of automatic response to the environment that is either innate or developed through learning. These creatures function primarily by instinct rather than deliberation. Compared to other animals, humans are especially good at deliberative System 2 thinking: conscious analysis of a situation and the weighing of options. This allows an ability to plan rather than merely to react automatically on instinct, which gives humans an evolutionary advantage. System 2 thinking requires a narrative brain where the human imagines various scenarios for the future and selects the one that appeals most strongly. (2011, 19–30. See also Seligman, Railton, Baumeister & Sripada 2016.)

This selection process, however, does not involve free will. Consider David Eagleman's example of choosing one flavour over another when ordering an ice cream cone (2015, 104–18). The brain creates competing anticipatory narratives of possible future experiences: how will having chocolate feel, strawberry, macadamia nut …? These narratives are built on memories of past experiences along with imaginings of what future situations might be. However, what we experience as a free choice is in fact the result merely of chemical reactions in the brain/body physical system to these competing narratives constructed in the brain as temporary connections of neurons: the neural connection that gives rise to the greatest release of dopamine, a neurotransmitter associated with pleasure, wins and drives our action. This is a completely physical process in the brain; there is no space here for free choice as we commonly understand it (Eagleman 2015, 114–18; Frith 2007, 92–100; Sripada 2016, 91–206).

Given this materialist account of brain functions and the mind, how are we to understand the human experience of agency, of being in control and responsible for what we do? Further, what does this account of agency mean for our understanding of tragedy? In 'Tragedy and the Common Man' (1949), Arthur Miller drew the distinction between tragedy and pathos by invoking the difference between being an agent and being a mere sufferer of external circumstances. Tragedy for Miller requires at least the possibility of warring against and defeating the forces seeking to drive us (10). But the accounts of deliberation in recent neuroscience make it seem that no such freedom and rebellion are possible.

Have psychological theory and recent findings in neuroscience shown our concepts of character, freedom and tragedy to be so confused that they must be abandoned? I think not, but these concepts do need to be re-examined and reconfigured in light of clearer ideas that have emerged following William James's insight into the nature of the self and its role in organising our experience.

One of James's central insights is the difference between experiencing the self as an 'I' or a 'Me.' We experience ourselves as an I when we conceive of the self as a subjective entity having the experience of thought or directing the action. This is the self James has in mind when he says the thinker itself is a thought: in this conception, the self is a thought we have, not an actual mental entity the brain creates that then has thoughts and experiences. The Me, on the other hand, is presented to this subjective I as an object for examination, like other objects in the world outside the I that the I experiences. The Me is a narrative we tell ourselves and other people about who we are and how we fit into the world. It is our way of understanding ourselves, of making our experience of ourselves coherent (James 1892, 174–84). Samuel Beckett depicts the interplay of these two aspects of the self vividly in his play *Ohio Impromptu* (1981), where he has the subjective I (the Listener) confront itself as an objective story (the Me) being presented back to the I as a book of memories read aloud by the Reader as the I listens.

The Me is the foundation of our sense of dignity. As the story of who we are, the Me must stay intact – seem true and real – or it ceases to function and the person goes into crisis, losing his or her sense of direction in life. Miller, again in 'Tragedy and the Common Man,' makes this notion of dignity central to his conception. According to Miller, the distinctive element in tragedy, what captures our interest and 'shakes us', is our 'underlying fear of being displaced, the disaster inherent in being torn away from our chosen image of what and who we are in the world' (9). Tragedy for Miller depicts people's responses to assaults on their dignity, assaults on the narrative Me that is essential to how we direct our actions in life. To understand both the experience of agency (freely chosen action) and the nature of tragedy we need to understand how the Me is constructed and the way in which it contains a moral vision of what is valuable. This takes us to Plato and his invention of virtue ethics in *Republic*.

Republic as a continuous dialogue properly begins at the start of Book 2, where Glaucon and Adeimantus challenge Socrates to show that a rational person would follow the precepts of morality if these precepts could be avoided without harm to oneself. The key point to note is that Glaucon and Adeimantus assume morality is a system of rules that prohibits people from pursuing their self-interest and directs them towards the interests of others (357a–367e). The central move in Plato's argument here is to reject that conception of morality and to consider instead the view that ethics is the study of what genuinely is in a person's self-interest and the types of beliefs, attitudes and traits of character a person needs in order to pursue this genuine self-interest effectively. In making this move, Plato rejects common conceptions of morality, conceptions that dominated Western moral theory from the Hebrew's commandments from Yahweh, through the Middle Ages, into the Enlightenment, and beyond. Plato invents a different conception that has come to be called 'Virtue Ethics' because of its focus on the virtues: the traits of character that lead a person to a flourishing human life, which the Greeks called a state of *eudaimonia*, a word we translate as 'happiness.' Although people choose different personal paths for their lives, the goal of every human life is the same: people seek lives that will enable them to have a sense of *eudaimonia*, a sense that they are doing a good job of living a human life. *Eudaimonia*, then, can perhaps best be understood as the particular kind of happiness or satisfaction that accompanies a person's feeling of self-respect. The opposite of *eudaimonia* is not unhappiness but shame: a pervasive sense of having betrayed something that was essential to the person's understanding of her Me, her proper place in the world and the values that give her life meaning.

In this approach, the construction of a self requires a moral vision, a set of personal values. Glaucon and Adeimantus had argued that people pursue only the appearance of being moral, a reputation for moral uprightness. People, they claim, have no interest in morality itself and no motivation to pursue it if morality inhibits pursuit of their own self-interest. Socrates's reply is that this is a wrong-headed approach to ethics. In order to act in the world, people require a vision of what is valuable and worthwhile, which Plato calls their 'vision of the Good.' People want this vision to be grounded in reality; they want to have a correct vision of what actually is in their self-interest. They do not want to pursue a false vision only to discover later that their vision has been misguided, that their actions have in fact been destructive and that they have wasted their lives (*Republic* 504e–506c).

What James calls the Me must involve some vision of the Good, of what is valuable, if this self-narrative is to function as a guide for action. Initially, my personal vision may be misguided; the goals I value may change as I try following them and learn the ways they must be re-envisioned so that they will lead me to a genuinely flourishing life. This is how Plato gets from this section of the *Republic* on the Good to the Cave Parable (514a–518b), his discussion of the importance of being willing to 'turn around' and examine the concepts we use to guide our lives in an attempt to be certain they are correct.

Plato's ideas here are supported by recent research into the role of emotion in decision-making. Emotions provide our psychological connection to what we value. People who have damaged the neural connections between emotion and reasoning have difficulty making even simple decisions: they have no bodily feeling of what is good to guide them. (For a vivid and poignant account of this, see Eagleman 2015, 110–14.)

These ideas about the role of values in constructing a Me also underlie Harry Frankfurt's conception of personhood, and Frankfurt's ideas enable us to understand how our experience of agency, of acting with a free will, is possible despite the materialist consequences of Darwin's theory and the recent findings of neuroscience about decision-making as a strictly physical process.

Frankfurt distinguishes between first-order and second-order desires (1971, 11–25). First-order desires are desires for particular outcomes; for example, in Eagleman's case of the person ordering ice cream, each of the possible flavours has some first-order desire attached to it. Each of these first-order desires is represented in the brain by particular temporary connections of neurons, and the connections that lead to the greatest release of dopamine drive us to select that flavour. This is a strictly physical process; there is no room for an experience of agency or free will at this first-order level.

However, Frankfurt argues, humans, unlike other animals, have not only first-order but also second-order desires. A second-order desire is a desire to have a certain first-order desire win the battle for dopamine release with other first-order desires and actually drive us to action. To understand this, consider Frankfurt's account of a drug addict who wants to end his addiction. When the addict is tempted to take the drug, he has conflicting first-order desires: take the drug, do not take the drug. His action will be driven by whichever of these first-order desires turns out to be the strongest. There is no place for a concept of agency or free will here. He simply will do whatever the physical systems of dopamine release in his brain drive him to do.

But this addict also has a second-order desire not to take the drug, to have that particular first-order desire be the one that wins the battle. Therefore, if this addict takes the drug, he has the experience of being overwhelmed by his addiction. If he does not take the drug, he has the experience of having acted freely, of being an agent who is guiding his own actions. Our second-order desires embody our conceptions of what Plato calls 'the Good' and James calls 'the Me': these are the narratives we live by, the stories of what we think is valuable and whom we want to be. The addict who has the second-order desire to end his habit and nonetheless takes the drug experiences his action as something shameful, a betrayal of the values that make him who he really is, a betrayal of his Me. That is why this addict experiences his act not as an act of agency but as a shameful experience of being overwhelmed by temptation, a weakness of will. The interrelation between first- and second-order desires creates the psychological space in which agency and free will can be experienced.

The interrelation between these two orders of desire also creates the psychological space in which we can understand the concept of tragedy and see the way it differs from melodrama. Plato shows this in *Symposium*, which I suggest is more than a dialogue about love. Plato uses love (*eros*) here as a way of exploring how values (what we love) are endemic to a person's sense of self, to what James calls 'the Me': quite literally, we are what we love; we are constructed as particular people by the values that drive us. Tragedy, as opposed to melodrama or the absurd, occurs when people are forced to confront the fact that they irreparably have betrayed their Me, their second-order desires.

Symposium was written in the late 380s BCE and is set roughly 35 years earlier in 416. It portrays a small dinner party in celebration of Agathon's winning a prize for tragedy that year. The dialogue has three major sections: the introduction and opening speeches about love by Agathon and his dinner guests (172a–198a), Socrates' speech about love based on his lessons from Diotima (198a–212c) and Alcibiades' speech about Socrates, a topic he chooses to speak on instead of love (212c–223d).

Alcibiades was a brilliant Athenian general who had spent time in his youth as a follower of Socrates. He was driven from Athens by political rivals in 415, one year after the date in which the dialogue is set, and then betrayed his city by colluding with the Spartans against

Athens in the Peloponnesian War. These events would have been familiar to Greek readers of the dialogue, and Plato is using this third section of *Symposium* to explore Alcibiades' character.

Alcibiades enters the party late and drunken. Seeing Socrates, he immediately feels ashamed, and rather than continuing the earlier speeches on love, he speaks about this shame: shame being the opposite of *eudaimonia*, a feeling that can be attained only by fulfilling what we love, the vision we have of our better Me. Plato presents Alcibiades as a man whose actions are driven by uncontrolled first-order desires that take him away from his second-order desires of the Me he wants to be. Confronting Socrates, Alcibiades recognises this failure and expresses remorse (216b–c), which for Plato is the emotion central to tragedy.

Remorse is not disappointment, which arises when things go wrong for reasons we cannot control, such as when a friend cancels a visit to which we had been looking forward for reasons that have nothing to do with us. Nor is remorse anger, which is directed outwardly at forces that we feel thwart us unjustly. Anger, directed at these external forces, often is depicted in melodrama. Remorse, the tragic emotion, turns inward. It arises when we must admit that we in some way colluded with the forces that undermine us. We may have been overwhelmed by these forces, but we are not merely innocent victims. We had a role in our own destruction.

This Platonic approach to tragedy, and its similarities to ideas about the self found in James and Frankfurt, can be seen explicitly in Shakespeare's *Macbeth* (1606). Sometimes commentators see the tragic moment in this play in act 5, scene 5, when Macbeth is told that Birnam Wood is moving towards Dunsinane and he must confront the fact that he has been deluded by the witches' prophecy. These commentators claim this is Macbeth's moment of recognition and reversal. But on the view of tragedy presented here, the tragic moment occurs much earlier. In Act 1, scene 7, Macbeth has told Lady Macbeth of the prophecy that he will be king, has discussed with her the possibility of murdering Duncan (act 1, scene 5) and now is weighing how to proceed. In the language of Frankfurt and recent neuroscience, he has conflicting first-order desires – seek the kingship or maintain his dignity and sense of self as a loyal vassal – each represented by a different neural network leading to a release of dopamine:

> If it were done when 'tis done, then 'twere well
> It were done quickly. If the assassination
> Could trammel up the consequences and catch
> With his surcease, success, that but this blow
> Might be the be-all and end-all …
> We'd jump the life to come. But in these cases,
> We still have judgement here that we but teach
> Bloody instructions, which being taught, return
> To plague the inventor …
> … I have no spur
> To prick the sides of my intent, but only
> Vaulting ambition which o'erleaps itself
> And falls on th'other.

(act 1, scene 7, lines 1–28)

Macbeth's action will be driven by strictly physical events in his brain as the neural networks incorporate various elements building their narratives of the future based on each possible path, leading to varying releases of dopamine. This physical process in his brain at this point leads him to decide against the murder: 'We will proceed no further in this business' (line 31), he tells Lady Macbeth when she enters. Note that this is not an act of free will as that term normally is understood. There is no choice here by a self; this is merely a physical process in a bodily organ, the brain, analogous to any other bodily action, such as the stomach's action in digesting food.

Lady Macbeth, however, chides Macbeth for the cowardliness of this decision (lines 35–44), and he replies: 'I dare do all that may become a man; Who dares do more is none' (lines 46–7). This may seem to end the question, but the possible cowardliness of this action had not been factored into the competing neural networks in Macbeth's earlier decision, and its addition now tips the dopamine releases from the network representing remaining loyal to the one depicting pursuing the kingship. Macbeth is a warrior; he does not want to consider himself a coward, so he now experiences this new most-powerful first-order desire to pursue the kingship as fitting with his second-order vision of the Me he wants to be. He experiences himself as acting freely.

By act 2, scene 2, however, all this has changed. Macbeth now has murdered Duncan and emerged from the royal bedchamber holding the bloody daggers. Lady Macbeth tells him he must return the daggers so that Duncan's attendants can be framed for the crime, but Macbeth refuses to go. Lady Macbeth wrongly thinks this again is an instance of cowardice, but Macbeth explains, 'I'll go no more. I am afraid to think what I have done; Look on't again, I dare not ... To know the deed, 'twere best not know my self' (lines 53–5, 76). Macbeth had thought that in killing Duncan he would fulfill his second-order desires: he did not want to be seen as a coward. But that is not the experience he now has. He feels his Me, the narrative he wanted to tell about who he is, has been destroyed. His Me had been that of a loyal vassal and valiant warrior, and now he is neither. He has lost his self and will have no way to go forward in the world until he finds another as a compass to guide him. The action of the last part of the play depicts Macbeth's struggle to find that self, a struggle that leads him only into more venal pursuits.

The story of that loss of self and the struggle against shame is the essence of tragedy. It is what Arthur Miller meant in saying that tragedy is 'the disaster of being torn away from our chosen image of what and who we are' (1949, p. 9). Tragedy depicts the irreparable collapse of the characters' sense of dignity: the realisation that the Me expressed in their second-order desires never will be actualised. Tragedy is experienced in the discontinuity between our second-order desires expressing who we think we are and want to be and the first-order desires that we are forced to admit actually have driven our actions. The distinction between those two orders of desire also explains the human experience of free will or agency. Like tragedy, our experience of agency arises from humans' need to incorporate a vision of the Good in their construction of a self, all of which is grounded in the evolution of the narrative brain in humans as a way of organising our experience and interaction with the world.

Works Cited

Beckett, S 1981, *Ohio Impromptu*, reprinted in *Samuel Beckett: The Grove Centenary Edition*, vol. 3, edited by Paul Auster, pp. 471–6, New York: Grove Press, 2006.

Dennett, D 1995, *Darwin's Dangerous Idea: Evolution and the Meanings of Life*, New York: Touchstone.

Eagleman, D 2015, *The Brain*. New York: Pantheon Books.

Frankfurt, HG 1971, 'Freedom of the Will and the Concept of a Person', reprinted in his *The Importance of What We Care About*, pp. 11–25, New York: Cambridge UP, 1988.

Frith, C. 2007, *Making Up the Mind: How the Brain Creates Our Mental World*, Oxford: Blackwell.

Hood, B 2012, *The Self Illusion: How the Social Brain Creates Identity*, New York: Oxford UP.

James, W 1890, *The Principles of Psychology*, vol. 1, New York: Dover, 1950.

James, W 1892, *Psychology: Briefer Course*, reprinted in *William James: Writings 1878–1899*, New York: Library of America, 1992.

Kahneman, D 2011, *Thinking, Fast and Slow*, New York: Farrar, Straus, and Giroux.

Miller, A 1949, 'Tragedy and the Common Man', reprinted in *Arthur Miller: Collected Essays*, introduction by Abbotson, SCW, pp. 7–10, New York: Penguin Books, 2016.

Plato 1997a, *Republic*, translated by Grube, GMA, and Reeves, CDC, in *Plato: Complete Works*, edited by Cooper, JM, and Hutchinson, DS, Indianapolis, IN: Hackett Publishing Company.

Plato 1997b, *Symposium*, translated by Nehamas, A, and Woodruff, P, in *Plato: Complete Works*, edited by Cooper, JM, and Hutchinson, DS, Indianapolis, IN: Hackett Publishing Company.

Richards, JR 2000, *Human Nature after Darwin: A Philosophical Introduction*, New York: Routledge.

Seligman, MEP, Railton, P, Baumeister, RF & Sripada, C 2016, *Homo Prospectus*, New York: Oxford UP.

Shakespeare, W 1606, *Macbeth*, edited by Gill, R, Oxford School Shakespeare, Oxford: Oxford UP, 1977.

Sripada, C 2016, 'Free Will and the Construction of Options', in Seligman, MEP, Railton, P, Baumeister, RF & Sripada, C, *Homo Prospectus*, pp. 191–206, New York: Oxford UP.

Wegner, DM 2002, *The Illusion of Conscious Will*, Cambridge: Massachusetts Institute of Technology Press.

Wegner, DM & Gray, K 2016, *The Mind Club: Who Thinks, What Feels, and Why It Matters*, New York: Viking.

23
AESTHETICS AND THE SENSIBLE

John Lutterbie

Time, in Gertrude Stein's essay 'On Composition,' has two registers (Stein 1993, 502). The first is the time of the composition, or the context in which the work was created from the perspective of the artist or experienced from that of the beholder. The second, the time in the composition, has an equally double valence: the time required for the artist to create the work of art, and the time taken by the spectator to behold it. Stein's two temporalities do not exist in isolation but are inextricably intertwined. For the purposes of this essay, however, I focus on the spectator who views the work, one who comes with a set of expectations and who engages a composition (whether literature, visual art or performance) for a period of time, the time *in* the composition. This is a first-person, phenomenological encounter that can last a second or continue to engage us for years after the event. It is in this temporality, however long it takes, that aesthetics comes into play. By aesthetic I don't mean the pleasure of encountering something beautiful, although that can be part of the experience. Rather, I mean an encounter with the unexpected, Shklovsky's *ostranenie* or a rendering of the familiar unfamiliar:

> The technique of art is to make objects 'unfamiliar,' to make forms difficult, to increase the difficulty and length of perception because the process of perception is an aesthetic end in itself and must be prolonged. *Art is a way of experiencing the artfulness of an object: the object is not important.*
>
> *(Lodge 1988, 20)*

Shklovsky's use of the term object is unfortunate when talking about actors; rather, its use is better understood as referring to the work of art as a whole. A book is an object, as is a painting. Similarly, a performance can be viewed as an object, such as a production of *King Lear.* Nonetheless, Shklovsky's theory continues to be critically vibrant, surfacing in Brecht as the *Verfremdungseffekt*, the work of the surrealists, and more recently in the aesthetics of Jacques Rancière.

For Rancière, we encounter a work of art with a set of expectations derived from our life experience, which he calls the distribution of the sensible. When he uses the sensible literally, Rancière refers to the world that our senses have access to. Like Merleau-Ponty's flesh, it extends and delimits us as individuals. But he also uses it metaphorically to define the local and global environments we engage as subjects, defining for Rancière both a politics and an aesthetics. In this sense, the sensible is a fabric organised and supervised by institutions through the exercise of power, and breaching the sensible is configured by Rancière as a bottom-up democratic action designed to empower and strengthen communities through collective awareness, if not immediate action. Cognitively, puncturing the sensible has the potential for reorganising our understanding of it. For Catherine Malabou, this results in a transformation of synaptic connections, a characteristic of neural plasticity, with far-reaching implications.

> The concept of plasticity has an aesthetic dimension (sculpture, malleability), just as much as an ethical one (solicitude, treatment, help, repair, rescue) and a political one (responsibility in the double movement of the receiving and the giving of form).
>
> *(Malabou 2008, 30)*

Politically, it is the domain in which the marginalised can actively affirm a community, and intervene to challenge and destabilise policies and institutions. Aesthetically, it provides the work of art with the potential for creating an emerging understanding that 'emancipates' the spectator. The use of the singular 'spectator' indicates that the individual must be emancipated as a step towards the realisation of collective action. Or, following Malabou, when the sensible is disrupted, the transformation in neural networks has the potential for creating forms that enact the political.

Rancière sees in the arts the potential for disrupting the sensible through the use of images that create a tension between perception and expectation. His approach resonates with the Situationist International, and their goal to disrupt the everyday as a means of making people aware of the pervasiveness of the spectacle, thereby revealing forms of oppression and surveillance. For Rancière, there are images/forms that reinscribe existing relations between peoples and those that resist the reproduction of oppressive values. The politically productive ones are those that resist pre-existing expectations. 'Images change our gaze and the landscape of the possible if they are not anticipated by their meaning and do not anticipate their effects' (Rancière 2009, 105). In *The Emancipated Spectator*, he identifies two types of image that he believes can have political efficacy: the intolerable image and the pensive image. The intolerable, as the name suggests, causes the viewer to 'close one's eyes or avert one's gaze' because, in his example of an anorexic model, 'the beautiful appearance' is offset by the 'abject reality' (83, 85). The pensive image operates through ambiguity, by confronting the spectator with 'social banality in the impersonality of art' (119) and thereby creating a 'zone of indeterminacy between thought and non-thought, activity and passivity, but also between art and non-art' (107). The pensive, in this sense, does not suggest a state of cognitive inactivity, but an unreflective acceptance of the status quo. In the need to avert the eyes and/or through the undecidability of the image there is the potential to open a temporal gap in which the viewers' expectations are confronted by the force of the image, creating the possibility for new meanings, understandings to emerge.

I begin this exploration of the image and cognitive processes that underlie it through a case study, a production I saw in 2013. The analysis following the description uses the concepts of fluency and disfluency developed in psychology, a theory of cognitive prediction, and the recent discovery of the default mode network. Aesthetics in this construct is a temporal process in which the ambiguities and juxtaposition of images defer the possibility of spectators arriving at an interpretation. The viewer's expectations are challenged, forcing a reformulation of the engagement with intolerable and/or pensive images before arriving at an emergent understanding of the experience.

The concept of the face: regarding the Son of God

Romeo Castellucci's 2010 production of *Sul concetto di volto nel figlio di Dio* (*The Concept of the Face: Regarding the Son of God*) is a triptych.[1] The first section portrays the tribulations of a middle-aged son dealing with his father's incontinence. The second involves a gang of teenagers throwing hand grenades at a Renaissance portrait of Christ and the third the dissolution of that painting revealing words that clarify the artist's intent. For this discussion, the focus is on the first segment, with reference to the second and third. First, a description of the set (Figure 23.1).

The scene is the home of an old man and his middle-aged son. It is divided into three parts. Stage right there is a living room, in the middle the semblance of a kitchen (represented by a table and two chairs) and stage left the bedroom consisting of a bed, table and lamp. A white carpet covers the floor. The furniture is made of a light-colored wood (possibly birch) with white upholstery, and in the bedroom, white bedding. The legs of the table and chairs are chrome. Ikea-simplicity might be an appropriate name for the style. Everything is placed parallel to the front of the stage, reducing the dynamic tension often associated with modernist scene design. At the back of the stage, from the floor to the flies is a reproduction of Antonello da Messina's *Ecce Homo*, a painting of the passion of the Christ during his scourging at the hands of Pontius Pilate. It is a beautiful, seductive and in Rancière's terms a pensive image; one that is passive, completely devoid of compassion or suffering.

The narrative of the first part of the triptych is of a man taking care of his father. There is very little dialogue, with most of the story communicated through physical action. The acting, for all intents and purposes, is psychological realism. The son, dressed in white shirt and black slacks, appears to be getting ready to go out, when his father enters in a white, terry cloth bathrobe and an adult diaper, although the latter is not at first apparent. As the son scurries around getting ready, the father, in the living room, turns his back to

Figure 23.1 The set for the first part of *The Concept of the Face*.

the audience and one notices immediately a brown stain on the robe. Clearly the father has fouled himself. The son then cleans the father, changing the diaper. This scenario is repeated two more times, once in the kitchen and then in the bedroom. Each incident is more intense and graphic, with the final incontinence being best described as projectile diarrhea. Each time the son gets more frustrated and the father more humiliated. The scene concludes with the father pouring what appears to be disinfectant over himself and sitting on the bed. Evident throughout the performance is the deep and compassionate love the two men have for each other. This is apparent in the care with which the son tends to the father's needs, and the parent's repeated moans: 'Scusi! Scusi! Scusi!' This loving story is in stark contrast to the intolerable image of the incontinence, including the subtle aroma of feces pumped into the theatre.

Significantly, it was impossible for me to keep my eyes on the last two defecations. Each time I had to avert my eyes, to look away, my gaze was drawn to the oversized image of the Christ. The movement from father and son to the Ecce Homo created another triad, a triangulation between the two men and the Son of God. There was, in the movement of the eye, a growing awareness of the impassiveness of the portrait set against the impassioned relationship between the two characters. I found myself, quite literally, regarding the Son of God, with a growing understanding of the way the face was looking down on the action under its gaze. There was no understanding, no compassion, no concern for the human suffering that was being depicted. With each repetition, Castellucci's critique of the Catholic Church became more evident. An interpretation that was confirmed in the final moments of the performance when (unnecessary in my opinion) the words appeared as the portrait dissolved and was shredded: 'the lord is not my shepherd.'

The touching relationship between the human father and son pulled me into the stream of the narrative, only to be cruelly pulled out of it by the graphic depiction of the defecation. Once the immediate shock of the incontinence dissipated, my gaze left Messina's painting and returned to the two men and the changing of the father's diaper. The intolerable image of the man's incontinence was juxtaposed with the pensive image of the Ecce Homo, commenting without commenting. The shifting of the eyes did not lead immediately to understanding, but to a disturbing ambiguity that remained at the margins of awareness as the narrative recaptured my attention. Understanding the cognitive dynamics of this process requires an understanding of fluency and disfluency.

Fluency, disfluency and the aesthetic experience

Scientists are applying their growing knowledge of the brain and how it works to the aesthetic experience. They focus primarily on understanding the pleasure experienced in the encounter with works of art and the beautiful. V.S. Ramachandran and William Hirstein, for example, identify aspects of various works of art that activate pleasure centers in the brain, arguing that organisation, distortions and abstractions in works of art give rise to aesthetic experiences, at least as defined by them (Ramachandran and Hirstein 1999). Psychologist Rolf Reber takes a different tack by looking at the processing of perceptual experiences and theorising that the flow of experience or fluency is the source of aesthetic pleasure. His approach bears at least a resemblance to Mihaly Csikszentmihalyi's *Flow* (1990). 'Processing fluency, or simply fluency, is defined as the ease with which information flows through the cognitive system (which includes both perceptual and conceptual components)' (Reber 2014, 225). The distinction between perceptual and conceptual is significant, and will be addressed later. At the moment the idea of the ease of movement of information through the cognitive system needs to be developed.

The ease with which data are processed cognitively depends on prior exposure to the same or similar stimulations. In this sense, Reber writes: flow depends 'on both the stimulus and the person's prior experience. The better the fit between stimulus and processing capabilities, the more fluently the stimulus is processed' (Rebera 2014, 226). The habituation or at least familiarity with perceptual information facilitates processing through patterns of recognition because we tend to pay attention to familiar circumstances. The same is true of conceptual information: prior experience with related ideas allows for greater fluency in processing. Ease of recognition is understood to be pleasurable because there is minimal disruption of experience, and patterns of thought arise that enhance the emergence of familiar forms of meaning. Such events reinforce expectations of the same and tend to conserve habitual patterns of behavior. Reber's study led him to conclude that flow in aesthetic experience is what gives rise to pleasure in art.

Not everyone agrees with Reber's conclusions. To his credit, Reber acknowledges disfluency, or moments when the flow of perceptual and conceptual experience is disrupted. Claudia Muth and her colleagues decided to test the effect of disfluency in responses to modern art. They discovered that the interest and attention of beholders of art increases when there is ambiguity in the engagement. 'Ambiguity refers to multiple meanings attributed to an object and varies with information, context and interaction between an observer and an object...It is thus more a subjective than an objective variable' (Muth, Hesslinger & Carbon 2015, 2). Their investigation looked at aspects of the art objects to which subjects responded:

> In contrast to previous reports, we found no evidence for a preference for flow or moderate levels of ambiguity but a clear positive relation of high levels of ambiguity with liking, interest and powerfulness of (perceptual and cognitive) affect. We revealed the largest effect for interest—which indicates that this dimension is especially crucial concerning the aesthetic appreciation of ambiguity in modern art.
>
> *(6–7)*

If Muth and her colleagues are correct, greater interest and attention are paid when there is a tension between the manifest qualities of the work of art and the flow of the familiar. My interest in Jackson Pollock shifted radically when I moved from viewing the work from a distance to a close examination of the way the different paints interacted with the canvas. Some 'drips' were absorbed by the canvas, while others formed impasto, adding a third dimension to what appeared to be a two-dimensional painting. Suddenly there was depth, and my understanding of action painting underwent a transformation that I remember and continue to think about. The fluency based on previous perceptual and conceptual experiences was broken, and my attention and interest were directed to information that defied expectation and required a restructuring of my understanding.

Andy Clark, in *Surfing Uncertainty: Prediction, Action, and the Embodied Mind,* replaces fluency and disfluency with prediction and prediction error. Viewing the body as an open system subject to the influence of perceptual experience, he puts less emphasis on what is perceived and more on 'the ceaseless anticipatory buzz of downwards-flowing neural prediction that drives perception and action in a circular causal flow. Incoming sensory information is just one further factor perturbing those restless pro-active seas' (Clark 2016, 52). The fluency of Reber, in Clark's universe, is replaced with the top-down movement of anticipation derived from attention and previous experiences in similar circumstances. The circularity of which he speaks is the integration of perceptions into the appropriate and predicted behaviour, involving the facilitation of percepts that support the current action and ignoring those incoming signals that are not relevant to the flow of experience. Prediction error, or disfluency in the theories of Reber and Muth, occurs when expectations established by the 'top-down movement of anticipation' are disrupted, requiring the formation of a new understanding. 'But should things fail to fall into place (should the results of the perceptual "experiment" appear to falsify the hypothesis) those error signals can be used to recruit a different hypothesis' (70). The result in extreme perturbations is the necessity to formulate 'revisions in our model of the world' (80). For Rancière and Shklovsky, such disruptions define the aesthetic event, for it involves both the viewer's reformulation of the sensible after the encounter with the work of art (Rancière) and the essential function of art – 'The technique of art is to make objects "unfamiliar"' (Shklovsky). We can see in the confluence of the two theorists the intersection of politics and aesthetics, of the beholding of art and the creativity of the artist.

In the flow of fluency, as defined by Reber, perceptual and conceptual understanding can be seen to move in tandem because previous experiences have created associations between what is perceived and the meanings derived. My early engagement with Pollock established a certain way of seeing and thinking that established a set of expectations that predetermined my future encounters with his art. Moving towards the canvas broke that circularity because my perceptual expectations were disrupted, requiring new conceptual formations, a redistribution of the sensible, to reconcile what his work had meant to me and what it could mean to me. The provocation of the new information initiated a 'complex way of seeing' (Brecht 1956, 44) that required a period of reflection because the familiar was suddenly unfamiliar. The complacency of expectation had been ruptured, requiring a recalibration of what art means and can do.

There is experimental evidence for this recalibration. Neuroscientists working with imagining technologies (fMRI, PET Scans, etc.) can identify the default mode network (DMN), that is, a baseline of mental activity, once perceptual and somatic neural activities are extrapolated from the image. The neural activity that persists once interference has been removed suggests the DMN has a function. The 'prevailing view is that the default mode or "conscious resting state" involves the retrieval and manipulation of past events, both personal and general, in an effort to solve problems and develop future plans' (Greicius & Menon 2003, 257). To Michael Greicius and Vinod Menon this suggests, 'that the retrieval of episodic memories and semantic knowledge are likely candidates.' Not everyone believes the separation of the DMN and sensory information is complete. Michael D. Fox and Abraham Snyder believe the 'underlying intrinsic organization encourages shifting one's perspective of brain function from the view of a system simply responding to changing contingencies to one operating on its own, intrinsically, with sensory information modulating rather than determining the operation of the system' (Fox et al. 2005, 9677). The implication is that the DMN marks a period of reflection following perceptual activity:

> One possibility is that the default network directly supports internal mentation that is largely detached from the external world... Another possibility is that the default network functions to support exploratory monitoring of the external environment when focused attention is relaxed...[and] the default network operates in opposition to other brain

> systems that are used for focused external attention and sensory processing. When the default network is most active, the external attention system is attenuated and vice versa.
> *(Buckner, Andrews-Hanna & Schacter 2008, 18–19)*

In other words, when attention in the external world is the strongest, the DMN is less relevant, and when internal thought is strongest, attention to external stimuli is the weakest.

If this is true, when there is fluency in aesthetic interactions, the DMN is less active, and more active when something in the work of art breaks the flow of attending to the world, giving rise to a disfluency. When the external sensory and motor perceptions are strong, less attention is paid to evaluating the present, problem solving and planning for the future because top-down expectations and the focus of attention prescribe the sensory inputs to which I am attending. The opposite seems equally true. There is convincing evidence that cross-fertilisation occurs between the perceptual apparatus and the DMN, with one influencing the other, suggesting a 'dialectical' relationship between the two. Moments of ambiguity in the aesthetic experience create a disfluency that requires reflection and the construction of representations outside of normative expectations, allowing for the emergence of structures of meaning that attest to the relevance of the disruption. Or, in Clark's terms, new hypotheses are recruited. Brecht, reformulating Shklovsky's defamiliarisation for political ends, writes that 'Some exercise in complex seeing is necessary—though it is perhaps more important to be able to think above the stream than to think in the stream.' His *Verfremdungseffekt* creates an ambiguity that breaks with the narrative flow of experience, giving rise to a period of reflection and the attempt to construct a coherent understanding based on perceptual experience. This state of disfluency lasts an indeterminate amount of time before the new hypotheses emerge or some aspect of the event pulls the beholder back into 'the stream' – the rhetoric of the story, in the case of narrative theatre. Whether or not the new hypotheses are satisfactory is something only time will tell, because it may be that meaning will only emerge in a coherent understanding in some future time and the redistribution of the sensible will be complete.

The concept of the face: regarding the Son of God *Redux*

The empathy created with the suffering of the two men, one in frustration, the other humiliation, is broken by the visual/olfactory experience of the three defecations. When I describe this production to colleagues, there is an involuntary withdrawal by the listener; their repulsion recreates a disfluency similar to what I experienced in encountering those intolerable images. My need to avert my eyes and their coming to rest on the face of the Christ created a pensive dialectic between what was happening on stage and the emotionless passivity of the painting. The disfluency was intensified by the beauty of the painting in contrast with the graphic representations of feces. The son's activity of cleaning the father drew my attention back to the narrative. The repetition of the violence through the three instances of incontinence, the triadic structure of the set,and the triangulation between the two men and the image of Christ led to an emerging understanding. This understanding of the characters' dilemma, the first of the three part structure of the performance, was pitted against the throwing of hand grenades at the image of Christ, without impacting the image, by teenagers (who were devoid of any apparent anger or other intense emotion), created another triad between the painting, the action and the father and son narrative. In the final segment, the father is onstage in the bedroom, and one of the young men is on the other side of the stage as the painting dissolves, creating a final triad with the inscription 'the lord is not my shepherd.'

None of the three sections of the work provided an answer to *The Concept of the Face*; rather, the audience is left with an unsettling critique of the Church. The three divisions are marked by disfluency, breaking our engagement through an abrupt change in the flow and direction of the performance. The incontinences of the father, the cacophony of the amplified sound of the projectiles thrown at the painting and the total dissolution of the painting are intolerable images that created a pensive state in me as an audience member, which led to understanding sometime later. Partial interpretations arose but did not sequence themselves into a coherent thematic. What eventually emerged, reinforced by the words projected on the back wall, was a politics that denounced the impersonal hypocrisy of the Catholic Church and its claims to a compassionate spirituality. That understanding arose over time, and was the result of the intolerable and pensive images as defined by Rancière and created by Castellucci.

Jacques Rancière conceptualises the distribution of the sensible as a 'fabric' that defines the parameters within which we normally relate to each other. The potential of art is its ability to puncture the surface of the sensible by employing intolerable and/or pensive images giving rise to a disfluency or prediction error. The repetitions of disfluency that pulled me out of the flow of the production put me in a temporary state of reflection on a set of disturbing perceptual and affective experiences. I was compelled to seek a deeper understanding, the creation of a new hypothesis. Before a final conclusion can be reached, the production seduces me once again into the stream of events. In the moment of reflection evoked by the intolerable images, I deal with my repulsion and inability to look by shifting my gaze from the distress of incontinence to the beauty of the painting and the passivity of the image of Christ. In the disruption of perceptual experience, as partial understandings emerge with each perturbation, I remain trapped in ambiguities while the production returns to performing its structure, bringing my attention once again to the matters at hand. But my expectations are no longer stable despite the return to the norms of narrative. Instead, they are in a continual state of flux, the experience with the pensive image lingering in working memory. Each disfluency is dependent on what has preceded, reminding me of earlier attempts at meaning-making, which, as I experience the newest intolerable image, complicate my attempts at understanding.

Through grappling with the perturbations to the field of expectations, there arise larger social images, imbued with a politics that call for resistance to the institutions that demand conformity, seemingly without empathy for its subjects. The fissures created by the intolerable and the pensive permit the spectator to grapple with ambiguities that lead to an oscillation between my attention to the performance and the reflections induced by images that don't promise a resolution. There is in this aesthetics the potential for a redistribution of the sensible, rendering the familiar unfamiliar, constructing a new set of expectations, and a reformulation of the hypotheses on which we act. In the intersection of Shklovsky and Rancière, Castellucci and Pollock, lies an aesthetics with the potential for the redistribution of the sensible, and in that perhaps a step towards a new way of being together.

Note

1 The performance I attended was at Montclair State College in 2013.

References

Brecht, Bertolt. 1964. *Brecht on Theatre*. Trans. John Willet. New York: Hill and Wang.

Buckner, Randy L., Jessica R. Andrews-Hanna, and Daniel L. Schacter. 2008. 'The Brain's Default Network: Anatomy, Function, and Relevance to Disease.' *Annals of the New York Academy of Sciences*. 1124: 1–38.

Clark, Andy. 2016. *Surfing Uncertainty: Prediction, Action and the Embodied Mind.* Oxford: Oxford University Press.

Csikszentmihalyi, Mihaly. 1990. *Flow: The Psychology of Optimal Experience.* New York: Harper Collins.

Fox, Michael D., Abraham Z. Snyder, Justin L. Vincent, Maurizio Corbetta, David C. Van Essen, and Marcus E. Raichle. 2005. 'The Human Brain Is Intrinsically Organized into Dynamic, Anticorrelated Functional Networks.' *PNAS.* 102, no. 27: 9673–8.

Greicius, Michael D., and Vinod Menon. 2004. 'Default-Mode Activity during a Passive Sensory Task: Uncoupled from Deactivation but Impacting Activation.' *Journal of Cognitive Neuroscience.* 16, no. 9: 1484–92.

Malabou, Catherine. 2008. *What Should We Do with Our Brain?*, New York: Fordham University Press.

Muth, Claudia, Vera M. Hesslinger, and Claus-Christian Carbon. 2015. 'The Appeal of Challenge in the Perception of Art: How Ambiguity, Solvability of Ambiguity, and the Opportunity for Insight Affect Appreciation.' *Psychology of Aesthetics, Creativity, and the Arts.* doi:10.1037/a0038814.

Ramachandran, Vilayanur S., and William Hirstein. 1991. 'The Science of Art: A Neurological Theory of Aesthetic Experience.' *Journal of Consciousness Studies.* 6, no. 6–7: 15–51.

Rancière, Jacques. 2009. 'Aesthetic Separation, Aesthetic Community.' *The Emancipated Spectator.* Translated Gregory Elliot. London and New York: Verso.

Reber, Rolf. 2004. 'Processing Fluency, Aesthetic Pleasure, and Culturally Shared Taste.' In *Aesthetic Science: Connecting Minds, Brains, and Experience,* edited by Arthur P. Shimamura, and Stephen E. Palmer, 223–49. Oxford: Oxford University Press.

Shklovsky, Victor. 1998. 'Art as Technique.' In *Modern Criticism and Theory: A Reader,* edited by David Lodge, 16–30. London and New York: Longman.

Stein, Gertrude 1993. 'Composition as Explanation.' In *A Stein Reader,* edited by Ulla E. Dydo, Evanston: Northwestern University Press.

24

TALK THIS DANCE

On the conceptualisation of dance as fictive conversation

Ana Margarida Abrantes and Esther Pascual[1]

As a means of expression, the unequivocally physical experience of dance is typically construed and spoken about in communicative terms, as a 'conversation between body and soul,' a 'dialogue between dancers' or a means to 'tell a story' to an audience (Brandt 2015; Pascual and Brandt 2015). This chapter analyses this metaphor through a case study of Tiago Rodrigues' choreography *A Perna Esquerda de Tchaikovski* ('Tchaikovsky's Left Leg'). In this performance a ballerina engages in different interactions: (i) with the audience, to whom she tells her life story, (ii) with the pianist on stage, who silently complies with her requests, and, most remarkably, (iii) with her own body, which she presents as the locus of autobiographical memory and of 'thinking with the body' (Kirsh 2011).

The conversation structure of the performance

Tchaikovsky's Left Leg, written and staged by playwright Tiago Rodrigues, premiered on 5 February 2015 at Teatro Camões, the headquarters of the Portuguese National Ballet, in Lisbon. The performance is about a prima ballerina, Barbora Hruskova, who plays herself, dancing the remembrance of episodes in her life and career. By revealing the hard work, the pains and pleasures of dance, the piece also tells the story of any dancer. Hruskova is accompanied by pianist Mário Laginha, the composer of the piece's original music.

As the viewers arrive, the dancer warms up on stage and the pianist tunes the piano. The lights go down and the dancer walks to the front and greets the audience[2]:

> [0:7:10] Good evening. The show has not yet started. <u>This is the moment before the show</u>. I like to come early to the stage. I like to be on stage before the show. I like to be on stage when only the piano tuner is here [*points to pianist*]. [...] I enjoy listening to the piano being tuned. It is not yet music. It is the promise of music. I warm up my body to the sound of the promise of music. Warming up is not yet dance. It is the promise of dance.

The choreography is self-referential, a form of meta-theatre: it is a performance about performing that grants an unusual view of pre-show work. The first overall pattern of enunciation is thus set up: the ballerina performs before an audience and addresses them in speech.

Why she does so gradually becomes clear: this is a confessional piece about her story as a prima ballerina at the end of her career. The dancer has an autobiographical narrative intent, which she delivers both in the performed gestural modality of dance and in the verbal enunciation modality of theatre. Moreover, the story she is about to tell and perform, her own, metonymically represents stories of other dancers, the story of a career in dance.

This opening also presents the identity of the pianist, no longer the established well-known composer, but a mere piano tuner. She addresses him as 'tuner Siegfried,' intentionally blending his dual identity on stage:

> [0:45:05] [*Dancer turns to pianist*] Music! [*Pianist runs to piano and begins to play*] [...]
> [0:50:35] Thank you, tuner Siegfried. Now your solo.

The ballerina recalls that she once danced *Swan Lake*, and asks the pianist to play Prince Siegfried, the male protagonist in this ballet, and accompany her performance. Her purpose is to demonstrate the inherent difficulty of classical choreography, the challenges it imposed on her body.

Apart from the direct actual communication with audience and pianist, the dancer engages right from the beginning in an imagined interaction with her body, presented as an independent organism:

> [0:10:54] My body and I, we are two different <u>people</u>. We live together, we work together but we are different. When I warm up I <u>talk</u> to my body and my body has several parts. I <u>talk</u> to each of these parts.

Throughout the piece, the ballerina's body is staged as her interlocutor:

> [0:20:25] Before my last performance, I <u>spoke</u> at length with my feet. They <u>said</u>: "Enough! We want to rest. We want retirement!" I <u>said</u>: "Just one more time! Please hang in there, my dears, just one more time. Hang in there! And then we will run!", and now here I am. And my feet <u>say</u>: "Again? You had promised it was the last time!" And I <u>say</u>: "Just this last one, my dears. And then we can run forever." And my feet <u>say</u>: [*Music resumes, dancer begins to walk intently and run; music intensifies as she gains speed, running in circles. Running changes to dance*]

This is a clear example of *fictive interaction* (Pascual 2014; Pascual and Sandler 2016), 'the use of the conversation as a cognitive frame to structure mental, discursive, and linguistic processes' (Pascual 2014, 9). In this performed conversation, the feet not only reply to the dancer in a turn-taking sequence, but are also endowed with emotional states, which they express verbally. They vent their discontent and reinforce this with action: as the dancer begins to assertively walk and gradually turns to running, the effect on stage is as though her feet have acquired a life of their own, dictating what is to happen next. Since bodily motion is central to dance, the feet are metaphorically given their own voice, protesting the regime of their owner through enunciation.[3] The speech ascribed to the ballerina's feet is naturally not actual, and it is not fictitious either— the dancer does not set up a fantasy world in which body parts can speak. Rather, the feet's words are *fictive* in the sense of Talmy ([1996] 2000); they need to be construed as non-genuine, inhabiting a realm between reality and fiction, but serving the purpose of the actual narrative in the here-and-now. By animating her body with its own will and voice, the dancer demonstrates the dynamics of physical forces that have defined her life and now dominate the end of her career. If by dancing she expresses emotion to the audience, by fictively conversing with her body she gives these emotional states an explicit narrative context. The quoted fictive conversation between dancer and body is staged for discursive purposes, for the benefit of the audience, the ultimate receivers of the performance's message.[4]

Critically, this piece is *not* one more instance of dance *as* dialogue (Brandt 2015; Pascual and Brandt 2015); it is literally about dialogue *in* dance. It shows the centrality of the Conversation Frame for making sense of autobiographical experience; it becomes communicable and suitable for mnemonic reconstruction. But why should the staged dialogue between an *I* and a *you* be more engaging, perhaps more effective, than a descriptive narrative in the first person? Note that the very process of memory retrieval may be considered a negotiation between different perspectives on (selected pieces of) the past events reconstructed – the situated perspective of the subject that experiences them as they unfold and the same subject that recalls them from the viewpoint of a later time (Abrantes 2010a, 137ff.). The coexistence of perspectives in this piece is consistent with work in cognitive science and cognitive linguistics on the non-pathological consideration through speech of two irreconcilable realities. For instance, we often talk to ourselves when there is nobody in the room – but ourselves – to listen (cf. Mead [1934] 1955; Dennett 1996, 147–52; Rosenthal 2012), which involves a split-self engaged in conversation. The conceptualisation and linguistic presentation of the self and the self's consciousness, or different parts of the self, as different parts of one's personality (or opinions) as separate – and interacting with each other – is in fact extremely common in the constant perspective shift of discourse (e.g., Dancygier, Lu and Verhagen 2016), in both literary and everyday language (Pascual 2002, 2014, chapter 1; Pascual and Sandler 2016, chapter 1). Linguistic examples include expressions such as 'If I were you, I would...' or 'I'm not myself today' (Lakoff 1996; Fauconnier and Turner 2002; Abrantes 2010b); 'I said to myself,' 'This is the linguist in me speaking,' 'A part of me says ..., part of me...' (Pang 2005).

The life story of a dancer can best be told in this manner. It is both the story as lived and the story as seen by others. If in an autobiographic narrative tension arises from the coexistence of the Olympic perspective of the older telling *I* and the recalled, younger, experiencing self, in the case of a performer, the latter self is already split between an 'inside out' perspective confronted with the 'outside in' look of others. These are intertwined, as described by the performer herself:

> [30:18] The mirror. The mirror helps you. The mirror is with you every day. [...] The mirror becomes part of your thinking. You think mirror. <u>You think about yourself as another looking at you</u>. <u>You are yourself and also the one in the mirror</u>.

This passage shows the performing self and the conceptualisation of the self from an outside and an inside perspective simultaneously (cf. Pascual 2014, 3–5; Pascual and Sandler 2016, 5–6). Here, the dancer's self is conceptually separated from her own body, with which she can then interact, and it is her body that imposes its own rhythms on the self's will. The Conversation Frame models the experience of multiple perspectives and demonstrates the ballerina's life as this permanent conceptual intersubjective tension between inner experience and external viewpoint. It helps make sense of it and conveys it as a story that is both told and performed. Thus, while the basic principles of embodied cognition cannot be denied, it seems that cognition is not only shaped by our experiences with the physical world, but also by our social lives.

The dancer's body in *Tchaikovsky's Left Leg* is the ideal interlocutor for triggering the autobiographical narrative. The ballerina's legs, feet, arms and back take turns in 'arguing' with her, and from it accessing significant moments in her past, reaching as far back as her childhood. Choreographer Tiago Rodrigues says of the piece that it is about 'the memory of the body.' As an 'interlocutor' in this fictive interaction, the body allows mediation, because it both *carries* the memory and performs it by *marking* relevant stages in the recall.

The way this ballerina condenses significant moments and events of her career in the performance and in the narrative is similar to the technique of marking movement sequences in dance. This involves schematically 'performing' movement at much smaller scale (e.g., using gestures and only some parts of the body to represent full-fledged choreographies). Apart from the obvious advantage of sparing the body strenuous effort, this technique may be used to convey movement sequences to others and negotiate them; it helps dancers situate themselves in the choreography and memorise sequences. Marking is thus the production of a compressed model of the movement, situated between regular practice of the real-movement sequence and the mental simulation of that movement in the dancer's mind. As a rehearsal technique, marking relies on embodiment, that is to say, the bodily basis of thought and the way we make sense of the world by relying on the experience of navigating the world in a material body. Marking also relies on distributed or extended cognition, that is, the ability to transfer or offload mental content to external forms of symbolisation (cf. Hutchins 1995; Kirsh 2011). The body thus becomes a material anchor (Hutchins 2005), a metonymic marker of the performer's memory, akin to the use of fingers for counting (Hutchins 2005). Marking seems even most similar to so-called 'constructed action' in signed languages, in which different body parts may symbolically stand for different referents, showing mixed viewpoints simultaneously (cf. Liddell 1995, 2003; Aarons and Morgan 2003; Dudis 2004; Jarque and Pascual 2015).

Throughout the performance, the ballerina stages segments of choreographies she has danced, granting the lay viewer privileged access to the strenuous technique, the body's proprioception and the dancer's emotions, which normally remain hidden. Occasionally she uses marking to convey sequences of movements she does not dance, repeating them with her hands instead of her feet:

> [33:15] [*Dancer names and performs miniature classical ballet steps with her hands; pianist reproduces, echoes on the piano the melody of her speech*]: Réverence. Posé, rond de jambes. Piqué, soutenu, developée devant, bourré, bourré, bourré...

The whole performance could be regarded in the same terms: a sequenced alternation of movements and segments from the choreographies she has performed and the body as the vehicle of this autobiographic marking. This understanding relies on the strategy of conceptual metonymy or referential pinpointing, already present in the title of the piece: the left leg is a reference to the particular strength and effort demanded of this limb in the classical choreographies of Tchaikovsky's ballet, and as such it is representative of the effort demanded from the body of a classical ballerina. Sensations of pain, scars, and permanent injuries are the traces of prior experience, and as such the path to recall a different performance and with it a different time, space, and audience:

> [1:18:00] Each pain in my body corresponds to a show. I've been dancing for 30 years now, so I have a great collection of pains. This is *Prokovief*'s foot. This is the hip of *A*

> *Midsummer Night's Dream*. This is the finger of *Carmina Burana*. This is Händel's coccyx. This is the knee of *Sylphide*.

The collection of pains is the collection of memories that the dancer reconstructs in dialogue as the episodes of her own story, in essence, they represent cause-effect compressions of her career. In the performance, both marking and metonymy are supported by indexical hints, marks of spatial and modal deixis:

> [1:03:35] Makarova does <u>this</u> with her leg. [*Dancer demonstrates the movement*] <u>This</u> is *Swan Lake*. It's perfect.
>
> [39:28] And the arms must close without effort, like wings. But mine only close until <u>here.</u> [*Dancer demonstrates the movement*] So I practiced for hours and hours to be able to lift my arm and keep the neck free. Because <u>this</u> is the position of the swan. [*Dancer demonstrates*]

She demonstrates in dance what she tells in speech, thereby integrating dance and dialogue. As an interlocutor in the fictive interaction, the body is often animated into a discordant conversation partner. As the carrier of memory, it bears the traces of past events, recounted verbally and demonstrated with movement, marking with and through the body the core moments of the ballerina's remembered life.

Engaging with a performance: the staged conversation as a trigger for empathy

The Conversation Frame as an underlying conceptual tool may be employed for discourse purposes. In this case, the fictive interaction involves dancer (the locator) and her body (a fictive counterpart), and the unfolding turn-taking occurs before the audience, a third participant that is not directly involved but is invited to take a position. This triadic channel of fictive communication, which is reminiscent of the typical interaction in a court of law (Pascual 2014, chapters 6, 7), makes the case for the demonstrative and argumentative nature of this conceptual structure.

Narrative is foundational in the way we construe reality, experience and memory (Bruner 1991), and it is moreover a condition for empathic involvement (Breithaupt 2009, chapter 4). Stories are basic structures for making sense of experience, as they allow for the retrospective causal linking of past events and multiple perspective-taking. A narrative, whether autobiographical or fictional, a private one or one for an external audience, often unfolds as the dynamic interplay of opposing forces, inviting the taking of sides (Breithaupt 2009, 152 ff.). Following Breithaupt (2009), empathy involves perspective-taking for one of two antagonistic elements or participants in a story. Consequently, fictive interaction is the conceptual strategy that renders it concrete in discourse and performance. In *Tchaikovsky's Left Leg*, the triadic construction calling for empathic involvement becomes especially clear towards the end, when the dancer engages once more in an imagined conversation with her body:

> [1:15:29] I don't want to stop controlling my body. I used to <u>tell my body</u>: "Do this, do that." And it always <u>complied</u>. But now my body has started to go its own way. Now <u>my body tells me</u>: "Don't do this, don't do that." <u>My body tells me</u>: "Tides rise and fall, the sun rises and sets, nature is made of beginnings and endings. And you may dance, but you are just another animal in nature." I <u>reply to my body</u>: "Alright, as you wish. Now you are in charge. But I

> warn you: We will die together. And when I die it will be like one of those old ladies, mad, wearing my ballet shoes, my tutu, and my tiara. And we shall pirouette, pirouette."
>
> *[She pirouettes]*

The contrast between then and now is underscored by the alternate dominance of the ballerina over her body in the past, her body over her in the present, and again her over her body in the future. Also, the precise distinction between first and third person (*I* vs. *my body*) is neutralised in a shared *we*, eventually submitted to the will of the dominant *I*: "We will die together [...] And we shall pirouette" (cf. Lakoff 1996; Lakoff and Johnson 1980, 1999).

The narrative experience is thus metaphorically transported to a reality alternative to one's own. Performing that reality (Gerrig 1993) adds to the narrative of one's own experience in a process of 'narrative world-making' (Hermann 2009). Then, in the case of a dance performance with integrated story-telling, there is one further dimension, as the story is not only mediated by language, but is actually demonstrated by expressive gesture. Viewers are engaged in a double and complementary way. Relying on the fundamental ability of theory of mind (Zunshine 2006), they anticipate through their own experience the performer's thoughts, behaviours and feelings from what the performer *tells* them; also, they are engaged in a 'feeling of body' (Wojciehowski and Gallese 2010), the embodied simulation of what it might be like to be that performer, triggered by what she *dances* before them. The involvement in this narrative is thus twofold, leading to an empathy (in Breithaupt's sense) that is both cognitive and affective, invited by language and by action.

The particular format and style of this dance-drama renders unusual visibility to the conceptualisation of performance as fictive interaction between (at least) performer(s) and silent audience. This allows for a multimodal engagement of the viewer with the narrative and a process of perspective-taking that results in the empathic involvement with the performer and her story.

Conclusion: more than a metaphor, dance as dialogue, dialogue in dance

In our case study, *Tchaikovsky's Left Leg*, the Conversation Frame structures the piece on multiple levels. The dancer engages in fictive interaction with: (i) the silent audience viewing the piece, (ii) the silent pianist complying with her requests, and (iii) her own body 'responding' and occasionally 'rebelling' against her, thereby revealing its pain. The performance stages dialogue in dance as a mechanism for triggering and sustaining autobiographical thinking throughout the performance.

As a choreography about the memory of the body, the piece makes use of marking as a strategy for simulated movement, with the body as a carrier of memories (scars, pains, etc.). Moreover, in *Tchaikovsky's Left Leg*, a solo dancer's body encloses a double perspective; she both looks at the audience and becomes the object of the audience's gaze ('You think about yourself as another looking at you. You are yourself and also the one in the mirror.'). This dual perspective is further extended to the double perspective in autobiographical recall, of a younger experiencing *I* as remembered by its later self.

The performance's multimodality, combining movement, performed gesture, music and dialogue, allows a rare view into what remains mostly hidden in a classical ballet performance: strenuous effort, rigorous discipline, the dancer's emotions. In this piece, construing this experience as a fictive dialogue of antagonistic forces before an audience allows for the autobiographical narrative to be told and demonstrated. Further, it prompts a triadic construction of participants and viewers, and perspective-taking by the latter. This leads to engagement with the piece and a response of cognitive and affective empathy towards the dancer and her story, which could well be that of any dancer.

Construing the physical experience of dance in conversational terms is extremely common (Pascual and Brandt 2015; Brandt 2015) and illustrates the frequent understanding of our inner and outer world as a communicative exchange. Indeed, the Conversation Frame is a powerful cognitive model for organising our thoughts and conceptualising reality (see overview in Pascual 2014, Pascual and Sandler 2016, Pascual and Oakley 2017). More specifically, our case study adds to the growing body of evidence, suggesting that everyday face-to-face interaction serves as a model for cognition (Brandt 2015), language use (Brandt 2004; Oakley 2009; Kövecses 2015) and even grammar, as in the obligatory use of fictive direct speech for the expression of thoughts and emotions (de Vries 1990, 2010; van der Voort 2016), evidential markers (Jarque and Pascual 2015; Spronck 2016) or the future tense (van der Voort 2016) in some languages. Thus, our intense experience with intersubjectivity models our cognition and language in a manner similar to our lifelong experience with our bodies and the physical world (cf. Lakoff and Johnson 1980, 1999; Sweetser 1990). The socio-communicative experience of conversation can be construed and spoken about as physical movement, just as much as the purely physical experience of dance can be construed and spoken about as a socio-communicative experience like conversation.

Notes

1 We thank then artistic director of the Portuguese National Ballet, Luísa Taveira, for providing us with an internal video recording of the official performance. Esther Pascual also acknowledges the 'Hundred Talents Program' at Zhejiang University for generous funding. Address for correspondence: Dept. of Spanish Language and Culture, and Dept. of Linguistics, Zhejiang University, 866 Yuhangtang road, 310058 Hangzhou, P. R. China. E-mail: pascual@zju.edu.cn, esther@estherpascual.com.

2 All quotes from the original Portuguese script were translated into English. Stage directions – in italics and between square brackets – were added, after visualisation of the video. Underlining is used to highlight parts in examples we wish to direct readers' attention to.

3 Giving fictive voice to a body or body part is not restricted to art or creative discourse, but also occurs in ordinary language use (Clark and Gerrig 1990, 794). The conceptualisation of the human body as a conversational partner as a means of expressing its physical state may also become conventionalised, as in these common expressions in Catalan (Pascual and Oakley 2017, 349):

i *El cos diu prou.*
Lit. 'The/My/Your body says 'stop'.'
'X [the owner of the 'speaking' body] is exhausted.'

ii *córrer/marxar cames ajudeu-me*
Lit. 'to run/leave 'legs, help me''
'to run away very quickly and desperately'

4 This interaction is reminiscent of the debate with Kant, a didactic fictive exchange between a modern philosophy professor and the long-deceased Kant, for the benefit of students (Fauconnier and Turner 2002; Brandt 2008).

References

Abrantes, Ana Margarida. 2010a. *Meaning and Mind*. Frankfurt: Peter Lang.

———. 2010b. "Consciousness and Self in Language." *TECCOGS* 4: 7–24. http://www4.pucsp.br/pos/tidd/teccogs/artigos/2010/edicao_4/1-consciousness_and_self_in_language-a_view_from_cognitive_semiotics-ana_margarida_abrantes.pdf

Aarons, Debra and Ruth Zilla Morgan. 2003. "Classifier predicates and the creation of multiple perspectives in South African Sign Language." *Sign Language Studies* 3(2): 125–56.
Brandt, Line. 2008. "A semiotic approach to fictive interaction as a representational strategy in communicative meaning construction." In *Mental Spaces in Discourse and Interaction*, edited by Todd Oakley and Anders Hougaard, 109–48. Amsterdam: John Benjamins.
———. 2015. "Dance as Dialogue." *Signata* 6: 231–49.
Brandt, Per Aage. 2004. *Spaces, Domains and Meanings*. Bern: Peter Lang.
Breithaupt, Fritz. 2009. *Kulturen der Empathie*. Frankfurt: Suhrkamp.
Bruner, Jerome. 1991. "The narrative construction of reality." *Critical Inquiry* 18: 1–21.
Clark, Herbert H. and Richard J. Gerrig. 1990. "Quotation as demonstration." *Language* 66(4): 784–805. doi:10.2307/414729.
Dancygier, Barbara, Wei-lun Lu and Arie Verhagen (eds.). 2016. *Viewpoint and the Fabric of Meaning*. Berlin: Mouton de Gruyter.
Dennet, Daniel C. 1996. "Talking to ourselves." In *Kinds of Minds: Towards an Understanding of Consciousness*, 147–152. Basic Books: New York.
Dudis, Paul G. 2004. "Body partitioning and real-space blends." *Cognitive Linguistics* 15(2): 223–38.
Fauconnier, Gilles and Mark Turner. 2002. *The Way We Think*. New York: Basic Books.
Gerrig, Richard J. 1993. *Experiencing Narrative Worlds*. New Haven, CT: Yale UP.
Herman, David. 2009. "Narrative ways of worldmaking." In *Narratology in the Age of Cross-Disciplinary Narrative Research*, edited by Sandra Heinen and Roy Sommer, 71–87. Berlin: Walter de Gruyter.
Hutchins, Edwin. 1995. *Cognition in the Wild*. Cambridge, MA: MIT Press.
———. 2005. "Material anchors for conceptual blends." *Journal of Pragmatics* 37(10): 1555–77.
Jarque, Maria Josep and Esther Pascual. 2015. "Direct discourse expressing evidential values in Catalan Sign Language." *eHumanista/IVITRA* 8: 421–45.
Kirsh, David. 2011. "How marking in dance constitutes thinking with the body." *Versus* (113–115): 179–210.
Kövecses, Zoltan. 2015. *Where Metaphors Come from*. Oxford: Oxford University Press.
Lakoff, George. 1996. "Sorry, I'm not myself today." In *Spaces, Worlds and Grammar*, edited by Gilles Fauconnier and Eve Sweetser, 91–123. Chicago, IL: University of Chicago Press.
Lakoff, George and Mark Johnson. 1980. *Metaphors We Live by*. Chicago, IL: University of Chicago Press.
———. 1999. *Philosophy in the Flesh*. New York: Basic books.
Liddell, Scott K. 1995. "Real, surrogate and token space: Grammatical consequences in ASL." In *Language, Gesture and Space*, edited by Karen Emmorey and Judy Reilly, 19–41. Hillsdale, NJ: Lawrence.
———. 2003. *Grammar, Gesture, and Meaning in American Sign Language*. Cambridge: Cambridge University Press.
Mead, George H. [1934] 1955. "The self." In *Mind, Self, and Society from the Standpoint of a Social Behaviorist*, 135–226. Chicago, IL: The University of Chicago Press.
Oakley, Todd. 2009. *From Attention to Meaning*. Bern: Peter Lang.
Pang, Kam-yui S. 2005. "'This is the linguist in me speaking': constructions for talking about the self." *Functions of Language* 12(1): 1–38.
Pascual, Esther. 2002. *Imaginary Trialogues*. Utrecht: LOT.
———. 2014. *Fictive Interaction*. Amsterdam: John Benjamins.
Pascual, Esther and Line Brandt. 2015. "Embodied fictive interaction metaphors." In *Addenda*, edited by Sander Lestrade, Peter de Swart, and Lotte Hogeweg, 321–34. Nijmegen: Radboud University.
Pascual, Esther and Sergeiy Sandler (eds.). 2016. *The Conversation Frame*. Amsterdam: John Benjamins.
Pascual, Esther and Todd Oakley. 2017. "Fictive interaction." In *Cambridge Handbook of Cognitive Linguistics*, edited by Barbara Dancygier, 347–60. Cambridge: Cambridge University Press.
Rosenthal, Victor. 2012. "La voix de l'intérieur." *Intellectica* 58(2): 53–89.
Spronck, Stef. 2016. "Evidential fictive interaction (in Ungarinyin and Russian)." In *The Conversation Frame*, edited by Esther Pascual and Sergeiy Sandler, 255–75. Amsterdam: John Benjamins.
Sweetser, Eve. 1990. *From Etymology to Pragmatics: Metaphorical and Cultural Aspects of Semantic Structure*. Cambridge: Cambridge University Press.
Talmy, Leonard. [1996] 2000. "Fictive motion in language and 'ception'." In *Toward a Cognitive Semantics*, 99–175. Cambridge, MA: MIT Press.

Voort, Hein van der. 2016. "Recursive inflection and grammaticalized fictive interaction in the southwestern Amazon." In *The Conversation Frame*, edited by Esther Pascual and Sergeiy Sandler, 277–99. Amsterdam: John Benjamins.
Vries, L. J. de. 1990. "Some remarks on direct quotation in Kombai." In *Unity in diversity*, edited by H. Pinkster and I. Genee, 291–309. Dordrecht: Foris.
———. 2010. "Direct speech, fictive interaction, and bible translation." *The Bible Translator* 61(1), 31–40.
Wojciehowski, Hannah C. and Vittorio Gallese. 2010. "The mirror neuron mechanism and literary studies." *California Italian Studies* 2(1). http://escholarship.org/uc/item/56f8v9bv.
Zunshine, Lisa. 2006. *Why We Read Fiction*. Columbus: Ohio State University Press.

25
DISTRIBUTED COGNITION
Studying theatre in the wild

Evelyn Tribble and Robin Dixon

The model of distributed cognition or cognitive ecology that we advance in this chapter takes its inspiration from Edwin Hutchins's ground-breaking book *Cognition in the Wild* (1995). 'In the Wild' refers to the proposition that thought is not best studied in artificial laboratory settings, but rather in real-world environments that have been designed to think with. Experimental settings by definition seek to reduce complexity and eliminate variables; in Hutchins's view, identifying cognitive mechanisms through these means alone provides only a partial picture of human intelligence. Such methods are generally predicated upon individualistic and atomistic models of thought, which cannot account well for the emergent properties of complex group activities. Hutchins argues for the study of 'human cognition in its natural habitat – the material and social surrounds that enable and constrain thought:

> Humans create their cognitive powers by creating the environments in which they exercise those powers. At present, so few of us have taken the time to study these environments seriously as organizers of cognitive activity that we have little sense of their role in the construction of thought.
>
> *(169)*

Hutchins did his fieldwork for the book on board a naval vessel, where he studied the highly complex task of coordinating social, technological and material systems. His central claim was that the thinking needed to manage the taskworld lies not in the individual nor the structure nor the material, but that all these elements are in dynamic interaction with one another. Activity is stretched across and shuttles among cognitive artefacts such as navigation charts and instrumentation, the spatial organisation of the vessel, including the ways that objects are disposed in space, the social structures and hierarchies and the embodied skills and dispositions of the crew. A key result of such a structure is what Hutchins calls the 'social formation of competence,' as the novice learns 'to organize his behaviour so that it produces a competent performance' (280).

Thus, the model of 'distributed cognition' posits that a complex activity such as performance is spread or smeared across resources such as attention, perception and memory; the experience of training as it is sedimented in the body; social structures; and the material environment. As Kourken Michaelien and John Sutton write, the emphasis upon what might seem like a motley collection of resources is central to the model. The elements 'need not be alike: rather, their distinctive features and capacities complement each other in combination so as to realize the relevant processes collectively' (Michaelian and Sutton 2013: 6). The closely related idea of a 'cognitive ecology' (Hutchins 2010: 705) emphasises the *interplay* of internal cognitive mechanisms and social and physical environments.

It is of course a truism that theatre is a collaborative art coordinating multiple agents in real time to produce its 'magic.' Distributed cognition provides a systems-level account of such coordination, and, importantly, a way of integrating social, material, environmental and technological factors into a robust account of extended and embodied mind that encompasses skilled practice. A key advantage of the model is that it privileges neither agent nor structure *in advance* and thus avoids falling into a binary account of artistic practice of the sort that can plague cultural theory. As Hutchins writes in a later articulation of his theory: 'distributed cognition does not assume a center for any cognitive system' (Hutchins 2014). Indeed, distributed cognition predicts historical change and variation as new configurations of material and social practices emerge. Such models hold promise for analysing collective enterprises that are otherwise difficult to account for. Moreover, the models we sketch here are not just applications; they also offer a way for theatre historians to speculate fruitfully about areas of historical practice that are only partially understood. Historians depend upon documentation, but rehearsal, training and composition practices are very poorly documented in the pre-Modern era. For this reason, we adopt an argument by analogy, using our good understanding of some practices to illuminate our poor understanding of others. By understanding what performers accomplished—for example, the punishing schedule of early modern English playing companies—and by careful attention to the practices that can be documented, distributed cognition allows us to build a model of group intelligence in action, and to explain obscure aspects of theatre history using applications of this model.

In what follows, we present four brief case studies of historical theatrical practice drawn from different eras and cultural contexts. In each case, applying aspects of distributed cognition theory to the historical practice offers fresh insights into that practice, while simultaneously demonstrating the utility of this developing body of ideas.

Early modern English playing companies

Our first case study is the early modern English theatre. As this example has been discussed at length elsewhere (Tribble 2005, 2011), we only trace the outlines here. By the 1590s, several purpose-built theatres dotted the landscape to the south and north of the London city walls, and playing companies competed for customers seeking novelty. Commercial pressures resulted in what seem today like crushing workloads: up to six different plays were performed a week, with relatively little repetition. The cognitive demands of such a schedule seem overwhelming, which prompted some past theatre historians to imagine a rote-like system in which actors simply went through the motions (structure) or were redeemed by a single brilliant playwright (agent).[1] However, the framework of distributed cognition can illuminate how the task-world was managed, without resorting to a deficit model. For example, the practice of learning lines from parts that contained only the actor's lines and his cues worked not just as a means of saving labour and paper, but was an effective way of reducing the 'noise' of extraneous information and allowing the actor to throw focus only upon the most vital task at hand. In turn, playwrights used devices such as iambic pentameter and rhyme, forms that afforded memorisation and also provided a structure for occasional improvisation within the verse form, or 'fluent forgetting.' The plays also used a basic spatial dynamic that allowed actors to map the fictional world of the play onto the physical world of the stage on the fly, with little need for elaborate group rehearsal. The apprentice system, in which boys were apprenticed to senior players, embedded young players into the company by giving them increasingly complex roles. Like the young midshipmen integrated into the smart environment of the naval vessel, novice actors benefitted from intelligent structures. Distributed cognition posits neither central control nor chaos; it locates thinking not only in the individual head of the actor but also across the entire system, depending upon the particular demands at hand.

Place and space relationships in the comedies of Plautus

The analogy of the insights that textual analysis based on distributed cognition has yielded regarding Elizabethan performance suggests that applying these ideas to the plays of Titus Maccius Plautus (c.254–c.184 BCE) and Roman comic performance generally may prove illuminating. Performed during the regular religious festivals of the Roman Republic, these comedies provide a useful case study into theatrical production under circumstances of particular cognitive pressures on actors. Despite centuries of scholarship on these plays, several aspects of their composition and production remain mysterious. Even the most detailed and persuasive recent treatments of Roman stagecraft – for example, Marshall (2006) and Moore (1998) – do not consider training, rehearsal and specific dramaturgical conventions in depth. There are still many unanswered questions about the plays and their performance, writing and rehearsal practices and the theatrical economy and ecology. In this context, a new methodological approach is justified.

Many of the production conditions of Plautine drama are not atypical of pre-nineteenth-century Western theatre. Trained male slave actors performed, and over four weeks of rehearsal time was probably available to the performers for each festival.[2] While the analogy of contemporary theatre practice has led scholars to assume that the production of Roman comedies would have required a central 'directorial' figure, there is no persuasive evidence for this assumption.[3] Instead, the complex staging demands of Roman comedy may have been met by similar cognitive structures to those described earlier for Early Modern theatre: a 'scaffolded' system of apprentice actor training, actors using only their own part to prepare lines,[4] implicit stage directions encoded in those parts and, in particular, a set of fundamental spatial conventions that reduced the need for rehearsal and distributed the responsibility for performance decisions among all performers, rather than locating it in one agent.

This latter structure is of particular interest for an investigation of Roman comedy using concepts drawn from cognitive ecology. One constraint on theatre production in the Roman Republic is highly anomalous in the broader context of Western theatre history. Roman drama before 55 BCE was performed exclusively in purpose-built temporary structures, erected and dismantled for specific theatrical festivals;[5] there were apparently legal and moral prohibitions on the construction of permanent spectator buildings in force until the first century BCE.[6]

These temporary structures were built in a limited number of public locations, primarily the Forum and near various temples.[7] Since all of these locations had other functions incompatible with the long-term presence of a theatre building, the period of time that such buildings were in place must have coincided closely with the duration of the festival in question. Crucially, this suggests that actors would not have had any prolonged access to the performance venue before the performance event, which would have placed sustained pressure on their cognitive resources. Each performer had to make decisions about entrances, exits, stage positioning and movement that had not been exhaustively rehearsed under the instruction of a director nor governed by explicit 'stage directions.'

Close examination of the extant plays of the Plautine corpus suggests an effective and pragmatic solution to this contextual constraint: the employment of a series of conventions for stage movement, encoded implicitly in dialogue through verbal formulae and applied in performance as a consequence of the memorisation of lines. These conventions may be grouped into three categories, the first of which will be considered below: those governing the relationship between physical space and fictional place;[8] those governing entrance and reentrance into, or exit from, the stage space; and those governing performer movement across the stage space.

Roman comedy followed a straightforward system for the 'mapping' of fictional dramatic place onto physical stage space. None of the 'scene changes' familiar to contemporary audiences, or the 'wipe and reset' spatial dramaturgy identified by Fitzpatrick in Early Modern plays (2011: 38) would have been observable in performances of Plautine comedies; instead, after the fictional location of each play was established, it was 'fixed.' This method of depicting fictional places has several dramaturgical advantages, but also generates additional cognitive pressures on actors: inconsistencies in use of space by performers are rendered more obvious to the audience, and can lead to serious problems for comprehension of plots.

In the Plautine corpus, the presented fictional location is remarkably consistent: in all extant plays, the scene is a street in a Greek city, in front of one, three or (most commonly, in around two-thirds of extant comedies) two buildings.[9] This street leads to the unrepresented forum or town centre in one direction, and to the unrepresented harbour or countryside in another. A rectilinear stage featuring three doors, two wing-entrances and a long open space would fulfil the representational requirements of every Plautine drama, and it is no coincidence that the later permanent theatres built across the Roman Empire had these fundamental characteristics, nor that nearly every scholarly reconstruction of the temporary stages follows this basic model.[10]

An efficient spatial dramaturgy influenced the allocation of unrepresented 'offstage' places to the elements of this stage space. The fictional buildings would each be represented by one stage door of the three doors on the typical temporary stage. The relative position of these buildings is often uncertain: only six plays in the corpus contain explicit references to the locations of buildings in the onstage/presentational space, and a simple convention for the arrangement of houses may account for this general absence of explicit allocations. For example, the door on stage right may typically have been used to represent the house of the relatively lower-status or less wealthy character in a given play, as Wiles proposes for Hellenistic New Comedy (1991: 44–5). Following a similarly binary logic, the wing-entrance on stage left was always understood to lead to the forum/city centre and associated places, and that on stage right to the harbour or countryside and other distant places.[11]

These binary arrangements of offstage/unrepresented places opposed the local and foreign, closer and further locations, public and private spaces and the private places of some characters with those of others. Most importantly for practical dramaturgy, this system allowed for clear association of character and place, which in turn would have created conditions of 'cognitive thrift' for the performers. Actors who had learnt only their individual 'parts' and had minimal opportunities for rehearsal in the performance venue could nonetheless adopt a relationship to a particular entrance/exit point that would underpin the majority of their stage movement. Across the corpus of Plautine drama, there is a clear default pattern identifiable: characters tend to enter from and exit to the unpresented fictional place with which they are most closely associated, or that has the most obvious dramaturgical relevance even if not explicitly nominated. Thus, characters foreign to the presented location, or returning to the presented location from foreign places, typically enter from the gate/harbour; actors therefore used the stage-right wing.[12] Local characters, including bankers, moneylenders and parasites, will tend to come from and go to the forum; actors therefore used the stage-left wing.[13] Characters associated with a house onstage enter and exit from that house, and so actors will use the relevant stage door.

Several other implicit conventions are identifiable throughout the Plautine corpus through analysis of repeated verbal formulae and patterns of movement. However, this association of character and place/space is a useful foundation for reconstructions of Roman stagecraft. For the actor unfamiliar with the specific temporary space of performance, adopting a general practice of connecting unrepresented fictional place, access points to the physical stage and the character currently being portrayed would have offered a basic and highly practical default position for exits and entrances even when none of the other conventional verbal formulae were present as cues, and so significantly reduced the cognitive burden of making decisions about movement in performance. The likelihood of problematic inconsistencies in place/space allocations would have been significantly reduced, and the thematic qualities of place, particularly the binary oppositions crucial to many of the plays, emphasised.

Cognitive artefacts in the commedia dell'arte

Distributed cognition does not account only for script-driven models of theatre. In the mode of theatrical performance that has dominated since Aeschylus, recall of previously memorised scripted material and delivery of that material in the presence of an audience has been the fundamental cognitive demand placed on actors; other cognitive pressures (choreography, blocking, gesture, entrances and exits) are closely linked to this dramatic text and the largely pre-rehearsed decisions it motivates or controls in performance.

However, a few theatrical performance genres across Western history have not centred around a dramatic text, instead offering greater or lesser scope for flexible, improvisatory composition-in-performance by the actors. The Atellan farce of second-century B.C.E. central Italy, the *mimus* of first-century C.E. Rome, pre-nineteenth-century *ortaoyunu* in Turkey, contemporary Theatresports™ and some forms of stand-up comedy all exhibit flexible features and encourage composition-in-performance.[14] In several cases, performance works in these forms lack a memorisable dramatic text altogether. The cognitive demands inherent in these performance genres therefore seem to involve less of the recall of specific fixed textual elements and movements fundamental to dramatic theatre, and more the delivery of memorised or partially memorised verbal material, rehearsed physical action and apparently spontaneously generated verbal or physical elements, all arranged into new configurations in every performance. Flexible performance forms therefore provide a fertile field for applications of distributed cognition theory that move beyond memory and consider other aspects of cognition in performance. The following case study draws heavily from Fitzpatrick's landmark 1995 monograph on flexibility, literacy and orality in the *commedia dell'arte*.

The *commedia dell'arte* of Renaissance Italy is perhaps the most distinctive, successful and well-attested historical improvisatory theatre form. *Commedia dell'arte* was essentially a flexible performance form produced by professional companies of actors in Italy between the early sixteenth century and the early eighteenth century. *Commedia* troupes also toured throughout Europe, and the characters and scenarios of *commedia* exercised a defining influence on subsequent generations of actors and playwrights in the literary theatre, including Moliere, Goldoni, Gozzi and Anton Chekhov.

Even during the earliest period of the form's development, *commedia dell'arte* occupied an unusual space between primarily literary and primarily oral cultural practices. The traces of scripted drama, in particular the classical comedies of Plautus and Terence and their revival by the playwrights and actors of the *commedia erudita* in the early Renaissance, were typically present in *commedia dell'arte* performance, particularly in the later period of the form's popularity;[15] the lineage of improvisatory, highly audience-responsive entertainments by *buffoni* and mountebanks in public squares was also crucial (Henke 2002: 50–68). This fertile combination of oral and literate practices and strategies for generating performance material was key to the development of the form, and also led to the use of complementary cognitive artefacts by actors in order to reduce the pressures of moment-to-moment performance decisions. The most important of these artefacts were the *scenario*, a written document that has typically been viewed as equivalent to the dramatic script (Fitzpatrick 1995: 74–8; Henke 2002: 13–15) and the masks, the defining feature of *commedia* and an essential aspect of characterisation and improvisation in the form.

While *commedia dell'arte* is even more ephemeral than many other historical theatrical genres, due to the absence of fully scripted dramatic texts, the extant scenarios constitute the largest and most potentially useful corpus of documents associated with the form (Fitzpatrick 1995: 81). These documents are clearly analogous to the plots utilised in Elizabethan theatrical performance, and like those manuscripts are usefully viewed as essential cognitive artefacts which summarised the fundamental structural aspects of the performance (Bradley 1992; Tribble 2011: 44–54). Much like an Elizabethan plot, a *scenario* was a single document containing an outline of the performance: a summary of the narrative, characters and setting (the *argomento),* various scenes in order of presentation, the characters present in each scene, the defining event/s of each scene, entrances and exits and, occasionally, other information (Henke 2002: 13) The document was displayed backstage, acting as a prompt and guideline for the actors during the performance.

In *commedia*, this outline would be 'fleshed out' by the abovementioned arrangement and rearrangement of rehearsed and unrehearsed material. The actor was therefore spared the cognitive burden of remembering either a script or a complex plot (Henke 2002: 14), allowing them to concentrate on the more important business of flexible composition-in-performance. The physical presence of a document backstage for reference before an entrance could help to maintain a baseline level of coherence in the performance as a whole, balancing what Fitzpatrick describes as the 'elastic' tendency of improvised routines to expand and contract (Fitzpatrick 1995: 225, 263–93). Actors appearing in one scene and then exiting before their next appearance only had to keep the event or events of that single unit of action in mind at a time, and an actor with more continuous stage action could distribute at least some of the cognitive burden involved to their fellow actors.

Crucially, the scenario provided the necessary parameters for improvisation. While a superficial understanding of improvisatory performance processes might suggest that fewer restrictions on decisions made in the moment of performance lead to more successful improvisations, research at both the theoretical and practical ends of the spectrum has clearly indicated a counter-position: improvisation, particularly group improvisation, works best when at least one aspect of the performance genre is relatively fixed, and not open to alteration even if other aspects are constantly being adjusted in performance (Fitzpatrick 1995: 56–60 and 73–8; Henke 2002: 49). *Commedia dell'arte* was no exception; the *scenario* offered actors the necessary outline of essential action and was presumably relatively inviolable, while simultaneously providing a context in which the performer could make a range of impromptu decisions.

However, the relatively fixed structural outline inherent in the scenario was not the only such comparatively inflexible aspect of *commedia dell'arte* performance. The masks synonymous with the form were another crucial cognitive artefact that provided the trained actor with a clear baseline of physical and verbal features from which to improvise. With the exception of the *innamorati* characters that were ostensibly the focus of the plots, and some minor servant roles, each stereotyped character of the *commedia* was associated with a full- or half-faced leather mask of conventional design. These are immediately recognisable to even the casual practitioner of *commedia*: Pantalone's vulture-like beak of a nose and bushy eyebrows, the canine or crocodilian snout of Il Capitano, the drooping eyes and pug nose of Tartaglia, Arlecchino's impish round face and forehead wart and Pulcinella's crooked nose and heavy cheeks.

Mask work, whether in the context of *commedia* or of the modern methods begun by Copeau and continued by Lecoq and others, is still a fundamental element of much actor training, and a key aspect of this training is the idea that the mask shapes character. Instead of the interior, psychologically framed models of characterisation following Stanislavsky, the typical understanding of a theatre mask and the associated stock role is as an exaggerated, external and comparatively simple *persona* that an actor can employ without the need for realism.[16] The situation was similar for the historical *commedia dell'arte*: the masks were a valuable tool that again 'fixed' certain elements of the performance, placing limits on the range and type of physical, verbal and emotional expression available to the actor in a particular role.

These roles were not psychologically complex, instead being based on a defining characteristic (Pantalone's avarice, Arlecchino's hunger, Il Dottore's intellectual pretension) or at most a binary opposition of emotional states (Il Capitano's bluster and cowardice). The visual design of the mask conjured associations with physiognomy, the pseudoscience of connecting moral and emotional traits with facial features,[17] and contemporary experimentation with the stock roles of *commmedia* suggests that each mask encouraged a particular external physicalisation by the actor. The prominent beak of the Pantalone mask promotes a hunched, forward-leaning posture perfect for the portrayal of avarice; the heavy eyelids of Tartaglia motivate a sagging stance ideal for that sleepy, stammering role; the small eyes and proud proboscis of Il Capitano encourage either an aggressive, snarling or a cringing, servile position of head and shoulders.[18]

These are not codified rules of physical presentation for these characters, instead constituting possible starting points for the individual actor to establish a working relationship with a particular mask during training. After training, an actor wearing a mask and assuming the associated role would have performed within conventional parameters for the gestural and movement-related aspects of the performance. Similarly, each mask/persona was correlated with particular verbal patterns, from accent (Pantalone's Venetian or Arlecchino's Bergamask dialect) through to more complex speeches comprising pre-learnt material arranged flexibly (Il Dottore's long-winded lectures, Il Capitano's bravura boasts).[19] A memorised and embodied repertoire of movements, from simple gestures to repeatable decontextualised gag routines (*lazzi*), as well as verbal formulae, was thus associated with each mask type. In performance, this repertoire of actions would provide a baseline of options for the actor regardless of the situations mandated by the scenario, further reducing the cognitive burden of composition-in-performance by reducing the number of variables in play. Understanding the scenario and masks as cognitive artefacts thus offers a clear rationale for their utility in *commedia dell'arte* performance.

Puppet theatre: handspring

Distributed cognition predicts different ways of stretching across environments, social structures, skilled agents and training, rather than treating such non-traditional modes of theatre as aberrations from the norm. Because distributed cognition models do not specify a locus of central control in advance – in other words, because they do not rely upon a pre-determined model of agency, structure or organisation – they may be particularly valuable when thinking through experimental or non-traditional modes of theatre. As shown earlier, *commedia dell'arte* employs the invariable elements of the mask and the scenario as key constraints that in turn guide and enable the more fluid and variable elements of performance. This example also demonstrates the importance of assemblages – the ways that performers integrate material, social and somatic elements into their systems.

We conclude with a contemporary example of such an assemblage: puppet theatre. Puppet theatre has a long history, a vast literature and takes many different forms, and we focus here on just one: the contact puppetry characteristic of the Handspring Puppet Company, which is known for its integration of puppet and human characters within fictions that combine traditional narrative theatre, dance and music.

The integration between humans and puppets includes the way the puppets are designed, their expressive capabilities in relation to the skilled human body, the coordination both among puppeteers and among and between the rest of the cast and the way puppets both mimic and diverge from human movement. Like the masks in *commedia dell'arte*, puppets afford certain kinds of physicalisation. As Basil Jones notes,

> The designer/maker of the puppet is partially responsible for this life the puppet possesses in performance. The jointing (or lack of it) and the structure of the puppet allow for certain forms of expressiveness and not others. The expert designer is acutely sensitive to the movement required by the puppet. So a large part of the liveliness of the puppet is the responsibility not only of the puppeteer but of the puppet's designer/maker as well.
>
> *(2014: 61)*

In the TED talk 'The Genius Puppetry Behind *War Horse*,' (2011), Handspring's Adrian Kohler and Basil Jones describe the 'evolution of a puppet horse' (2011: 0:21).[20] Puppets, they argue, always 'struggle for life' on stage, and the art of constructing a puppet is a 'piece of emotional engineering' (2011: 2:50). Audiences must become convinced of the puppet's life to engage in the fiction. This project is inherently multi-modal and distributed across a network of agents. The puppeteer designs the puppet in such a way that makes certain movements relatively comfortable and others more difficult. He or she works with materials such as wood, plaster, cord and cloth, all of which have their own 'grains' and affordances. As Kohler notes, the 'single biggest practical consideration with wood is weight. The scale would be determined by the amount of weight a puppeteer could hold above his head for any length of time' (Taylor 2009: 71). So the puppet-maker builds into the puppet certain ways of moving designed to work with the ergonomics of the human body. In turn, the puppeteer(s) or manipulators must flesh these out, literally. Nothing animate is ever completely still; animals, including human animals, constantly engage in micro-movements; even when seemingly frozen, they subtly shift balance, quiver and, above all, breathe. And it is the assent of the audience in that act of breath that above all animates the puppets. As Toby Malone and Christopher Jackman write: 'belief in the horse must be performed and enlivened as much by the spectator as by the actors and puppeteers' (2016: 3).

'The authority of breath,' as Kohler terms it (2009: 66), yokes designer, manipulator and audience. The 'breathed stillness' of the puppet creates a join between puppet and audience: 'The audience, in noticing the tiny in breath and out breath of the puppet, enters into an empathetic relationship with the object that is being brought to life' (Kohler 2009: 66). Because the breath is the sign of animate life, without it the stillness becomes uncanny, and the puppet dies. If a puppet does not breathe, 'the tension created becomes uncomfortable. Eventually the audience breathes out and the bond of trust between audience and puppet breaks down' (Kohler 2009: 99). Kohler and Jones write that they first articulated the primacy of breath when attempting a particularly complicated assemblage: adapting Monteverdi's opera *Il Ritorno d'Ulisse de patria* to the puppet theatre. This production, performed in 1998 and again in 2008, used a triad of performers: the puppet itself, the puppeteer, who controlled the head and one hand, and the singer, who manipulated the puppet's other hand while singing. Only by coordinating the movement of the puppet with the breath and movement of the singer could the audience engage in the lifeworld of the opera. As William Kentridge suggests, such a process involved multiple levels of engagement and processing:

> the principle [was] that the manipulator focuses on the puppet and the puppet looks at the audience and the audience has to look at the manipulator, but then follows the manipulator's gaze (as you do when somebody is focusing on something) to the puppet and then back as they become aware of themselves watching the puppet.
>
> *(Jones 2014: 192)*

Kentridge's discussion underlines the importance of tying together puppet, puppeteer and audience through the manipulation of eye direction and gaze. Humans are highly attuned to gaze direction. As Jones notes, audiences watching *War Horse* become fixated on the horse-puppets almost to the exclusion of everything else, displaying 'an affinity and fascination with the horses' (2014: 65). The power of the puppet is such that it can create complications in integrating language, including traditional scriptwriting, and human actors into the larger fiction:

> The scriptwriter was almost powerless to author scenes where the horse was central. Without an intimate knowledge of the capabilities of the puppet and without weeks of watching the puppet in action, it was impossible to write these scenes in any but the sketchiest ways.
>
> *(Jones 2014: 63)*

So the introduction of puppetry into the cognitive ecology of the theatre profoundly alters the other elements in the production. The affordances of the puppets become the key constraint around which the work must be organised.

In the case of the large puppets such as those used in *War Horse*, coordination is made even more complex by the need to integrate the movements of multiple puppeteers. As Jones remarks,

> Although Adrian's horses are capable of a wide range of expression, *realising* that expression through movement requires of the puppeteers the development of a complex set of coordinative skills both personally and as a group. The two main horses each require groups of three operators. A convincing individual horse with a character of its own can be created only by a formidable act of 'group mind'—a level of coordination far beyond what a scriptwriter could produce.
>
> *(Jones 2014: 64)*

War Horse found a particularly interesting solution to the challenges of such coordination. Horse handlers and partially visible puppeteers alike were costumed as English soldiers, which had the effect of integrating the multi-modal elements of the production. In the TED video, Jones and Kohler brought one of the puppet-horses on stage, managed by a uniformed handler. Coordinating the horse-puppet demanded an extremely high level of concentration for the puppeteers, including both those embedded within the horse and the handler, whose work bridged the roles of actor and puppeteer. The handler maintained a focused gaze on the 'horse' at all times, moving in sync with it, seemingly reining it in when it became agitated, manipulating its head. The concentration needed to manipulate the puppet – the attentive gaze – performed double duty in keeping the puppet 'alive' and in orchestrating its effects, including coordination with the other puppeteers and its connection with the audience's avid gaze.

The examples we have reviewed are deliberately diverse, encompassing historical performance work as well as contemporary and emerging practices. 'Thinking' is sometimes a dirty word in theatre, associated with excessive rationalism and 'being in the head.' But such a critique misconstrues the act of thinking itself – thought is embodied and socially and materially extended. Human cognition is itself naturally eclectic; 'in the wild' it seeks out constraints, hints, shortcuts, scaffolds. Analysis of the workings of such systems in no way diminishes the intelligence and training of those who perform in milieus which test the boundaries of human skill in time-pressured settings. Rather, these models illuminate the workings and mechanics of such accomplishments.

Notes

1 See, for example, Bradbrook (1932) and Stern (2000).

2 In the absence of direct evidence for rehearsal practice, it may be assumed that the popularly elected magistrates responsible for funding and organising festivals were not able to make arrangements for festival events prior to their election, and that the accounts of one festival needed to be settled before contracts for the next one could be offered. This would provide between five and eight weeks of rehearsal time for the regular theatrical festivals in April, July, September and November.

3 For example, Richlin (2005: 64).

4 One such part survives: P. Oxy. 4546. See Marshall (2004).

5 See, among others: Bieber (1939: 326–7); Saunders (1913: 96); Duckworth (1952: 79); Campbell (2003: 67); Beacham (1991: 55ff); Goldberg (1998: 1–4); and Welch (2003: 54–5). These scholars tend to draw on the same few references in the primary sources, which are usually comments in the annals of historians like Livy about the construction of temporary theatres.

6 Valerius Maximus 2.4.2; Tacitus *Annales* 14.20; Velleius Paterculus 1.15.3; Appian *Bellum Civile* 1.28. See also Gruen (1992: 205–10).

7 Cicero, *de Harusicum Responso* 24; Livy 40.51.3; Plautus, *Curculio* 466–85; Valerius Maximus 2.4.6; Campbell (2003: 72–3); Goldberg (1998); Hanson (1959; 13–22); Marshall (2006: 36–43); Moore (1991: 126–39); Saunders (1913: 91–2); Sear (2006: 54–5); Wiles (2003: 32–5, 100–3).

8 The useful terminological distinction drawn in McAuley (1999).

9 Plautus' *Asinaria, Bacchides, Casina, Menaechmi, Mercator, Miles Gloriosus, Mostellaria, Persa, Poenulus, Truculentus,* and *Rudens*, and probably *Cistellaria, Epidicus, Pseudolus* and *Trinummus*, all feature two buildings in the represented fictional place. His *Amphitruo* and *Captivi* feature one building, and *Aulularia, Curculio* and *Stichus* three. For the sake of comparison, Terence's *Andria, Adelphi* and *Eunuchus* feature two buildings and his *Heautontimorumenos, Hecyra* and *Phormio* three.

10 For example, Beacham (1991: 60–1 and 216–7); Beare (1955: 166–70); Duckworth (1952: 82–3); Johnston (1933, 15); Marshall (2006: 49–56); Richlin (2005: 7–8); and many others.

11 There is considerable consensus in the scholarship regarding this convention, although the precise articulation of the binary differs between critics. See Beare (1955: 238–45); Johnston (1933: 68–106); Rambo (1915: 415–30); Marshall, (2006: 5); and others.

12 This pattern is most evident in *Menaechmi*, where the actor playing the foreign twin always uses the stage-right wing and never the left, while his local counterpart does the opposite, thus reminding the audience at every entry which twin is which. See Rambo (1915: 421), and most subsequent editions of the play. Other examples of foreign characters' entries otherwise unmarked by explicit verbal formulae occur at *Asin.* 380; *Bacch.* 170, 384 and possibly 573, 842; *Epi.* 526; *Poen.* 930; *Truc.* 482.

13 The following are otherwise unmarked entries which follow this pattern: *Capt.* 69, 191, 401, and see in particular 478–89, which make this association explicit; *Curc.* 375, 526; *Epi.* 620, 647; *Most.* 532, 653; *Pers.* 53, 164, 329; *Stich.* 155.

14 The useful term 'composition-in-performance' is borrowed from Parry (1971) and Lord (1987). It avoids the associations of 'improvisation' while retaining the sense of invention simultaneous with delivery.

15 Henke (2002: 12–14). As Fitzpatrick notes (1995: 14–15) the influence of the literary *commedia erudita* on the *commedia dell'arte*, particularly early in the evolution of the form, has been overstated by many scholars.

16 See, for example, Eldredge (1996); Johnstone (1981: 143–205) offers an extreme version of this idea.

17 A definitive work of Renaissance physiognomy, Giambattista della Porta's *de humana physiognomonia*, was published in 1586; della Porta was himself a playwright contemporary with the first great flowering of *commedia dell'arte*.

18 These observations are drawn primarily from my own experience as a student, teacher and practitioner of *commedia dell'arte*. There are many textbooks and handbooks for contemporary performers in the form, of which Rudlin (1994) is possibly the best; these can be overly prescriptive about the types of gesture and movement associated with each mask, but the basic principle remains sound.

19 Fitzpatrick (1995: 18–19); Henke (2002: 14–15, 45–9). The verbal material memorised by the actors was the element of *commedia dell'arte* most likely to be drawn from literary sources.

20 Malone and Jackman (2016: 19–20) discuss the orchestration and presentation of the TED talk in the context of the '*War Horse* franchise.'

References

Beacham, R. (1991). *The Roman Theatre and Its Audience*. Cambridge, MA: Harvard University Press.

Beare, W. (1965). *The Roman Stage: A Short History of Latin Drama in the Time of the Republic*. Methuen: London.

Bieber, M. (1939). *The History of the Greek and Roman Theater*. Princeton, NJ: Princeton University Press.

Bradbrook, M.C. (1932). *Elizabethan Stage Conditions: A Study of Their Place in the Interpretation of Shakespeare's Plays*. Cambridge: Cambridge University Press.

Bradley, D. (1992). *From Text to Performance in the Elizabethan Theatre: Preparing the Play for the Stage*. Cambridge: Cambridge University Press.

Campbell, C. (2003). "The Uncompleted Theatres of Rome". *Theatre Journal* 55:1, 67–79.

Duckworth, G.E. (1952). *The Nature of Roman Comedy: A Study in Popular Entertainment*. Princeton, NJ: Princeton University Press.

Eldredge, S. (1996). *Mask Improvisation for Actor Training and Performance: The Compelling Image*. Evanston, IL: Northwestern University Press.

Fitzpatrick, T. (1995). *The Relationship of Oral and Literate Performance Processes in the Commedia Dell'Arte: Beyond the Improvisation/ Memorisation Divide*. Lewiston, ME: Edwin Mellon Press.

Goldberg, S.M. (1998). "Plautus on the Palatine". *Journal of Roman Studies* 88, 120.

Gruen, E. (1990). *Studies in Greek Culture and Roman Policy*. Leiden: Brill.

Handspring Puppet Company. "The Genius Puppetry behind *War Horse*." Ted talk 30 March 2011. www.youtube.com/watch?v=h7u6N-cSWtY.

Hanson, J. (1959). *Roman Theater-Temples*. Princeton, NJ: Princeton University Press.

Henke, R. (2002). *Performance and Literature in the Commedia Dell'arte*. Cambridge: Cambridge University Press.

Hutchins, E. (1995). *Cognition in the Wild*. Cambridge, MA: MIT Press.

Hutchins, E. (2010). "Cognitive Ecology." *Topics in Cognitive Science* 2 (2010): 705–710.

Hutchins, E. (2014). "The Cultural Ecosystem of Human Cognition." *Philosophical Psychology*. 27:1, 34–49.

Johnston, M. (1933). *Exits and Entrances in Roman Comedy*. Geneva, NY: W.F. Humphrey Press.
Johnstone, K. (1981). *Impro: Improvisation and the Theatre*. London: Eyre Methuen.
Jones, B. (2014). "Puppetry, Authorship, and the Ur-Narrative," in D. N. Posner, C. Orenstein, and J. Bell., eds., *The Routledge Companion to Puppetry and Material Performance*. Abingdon and New York: Routledge.
Kohler, A. (2009). "Thinking through Puppets," in J. Taylor, eds., *The Handspring Theatre Company*. New York: David Krut Company.
Lord, A. (1987). "The Nature of Oral Poetry," in J.M. Foley, ed., *Comparative Research on Oral Traditions: A Memorial for Milman Parry*. Columbus, OH: Slavica Publishers.
Marshall, C.W. (2004). "Alcestis and the Ancient Rehearsal Process (P. Oxy. 4546)". *Arion* 11:3, 24–45.
Marshall, C.W. (2006). *The Stagecraft and Performance of Roman Comedy*. Cambridge: Cambridge University Press.
McAuley, G. (1999). *Space in Performance: Making Meaning in the Theatre*. Ann Arbor: University of Michigan Press.
Malone, T. and Jackman, C. (2016). *Adapting War Horse: Cognition, the Spectator, and a Sense of Play*. London: Palgrave.
Michaelian, K. and Sutton, J. (2013). "Distributed Cognition and Memory Research: History and Current Directions." *Review of Philosophy and Psychology* 4:1, 1–24.
Moore, T.J. (1991). "*Palliata Togata*: Plautus, *Curculio* 462–86". *American Journal of Philology* 112:3, 343–62.
Moore, T.J. (1998). *The Theater of Plautus: Playing to the Audience*. Austin: University of Texas Press.
Parry, M. (1971). *The Making of Homeric Verse: The Collected Papers of Milman Parry*. Oxford: Clarendon Press.
Rambo, E. (1915). "The Significance of the Wing-Entrances in Roman Comedy". *Classical Philology* 10:4, 411–31.
Richlin, A. (2005). *Rome and the Mysterious Orient*. London: University of California Press.
Rudlin, J. (1994). *Commedia Dell'arte: An Actor's Handbook*. London: Routledge.
Saunders, C. (1913). "The Site of Dramatic Performance at Rome in the Times of Plautus and Terence". *Transactions and Proceedings of the American Philological Association* 44, 87–97.
Sear, F. (2006). *Roman Theatres: An Architectural Study*. Oxford: Oxford University Press.
Stern, T. (2000). *Rehearsal from Shakespeare to Sheridan*. Cambridge: Cambridge University Press.
Taylor, J. (2009). In Dialogue: William Kentridge with Jane Taylor [Interview]. *The Handspring Theatre* Company. New York: David Krut Company, pp. 176–224.
Tribble, E. B. (2005). "Distributing cognition in the Globe". *Shakespeare Quarterly* 56:2, 135–155
Tribble, E. B. (2011). *Cognition in the Globe: Attention and Memory in Shakespeare's Theatre*. New York: Palgrave Macmillan
Welch, K. (2003). *The Roman Amphitheatre*. Cambridge: Cambridge University Press.
Wiles, D. (1991). *The Masks of Menander: Sign and Meaning in Greek and Roman Performance*. Cambridge: Cambridge University Press.
Wiles, D. (2003). *A Short History of Western Performance Space*. New York: Cambridge University Press.

PART IV: TRANSLATIONAL APPLICATIONS

Introduction

Rick Kemp

Recent years have seen instances of trans-disciplinary activity combining concepts and techniques from theatre and performance with scientific methodologies to advance knowledge and understanding in a variety of arenas. Our intent in this part is to provide detailed examples of the development of 'two-way traffic' between the cognitive sciences and theatre and performance studies. The part's title deliberately echoes the field of Translational medicine, which is a rapidly developing inter-discipline in biomedical research that involves collaboration between multiple scientific disciplines. The aim of the field is to facilitate the discovery of new diagnostic tools and treatments by using what is known as a 'bench-to-bedside' approach. The bench in this description is a laboratory bench; our goal here is to demonstrate that theatre and performance can be thought of as laboratories for life – places where humans can safely experiment with provisional versions of social interaction. In contrast to the mediated reality of screen drama, rehearsal studios and performance venues create sets of conditions in which planned or improvised behaviour create embodied meaning. As I've described elsewhere, performers employ most of, if not all, the cognitive activities that are involved in daily life. Theatre and performance-training methods have developed techniques that enable performers to consciously generate behaviour that is normally involuntary in daily life. They do so within multiple intersecting constraints, such as the narrative of a written play, the personality of a fictional character, the theme of a devised or dance piece or a particular style of performance. These constraints enable audience members to focus their attention on 'what if' scenarios of life, perceiving real bodies in ways that stimulate their cognition in ways very similar to social encounters in daily life. One of the barriers to the empirical examination of the living skills embodied in theatre and performance has been the difficulty of communicating exactly what those skills are to investigators in other fields. Both terminology and concepts in training and analysis are highly specialised and vary between different methods and schools of thought. Consequently, they obscure fundamental principles of the human behaviour that is the subject of theatre and performance. The rapidly growing research field that this book represents has begun to address this issue through applying cognitive science to theatre and performance. In this part, we hope to demonstrate how the knowledge gained from doing so can be transposed to other fields.

Our first chapter in this part provides a model of how a cognitive understanding of theatrical expertise can support valuable interventions in social welfare. Cognitive psychologist

Helga Noice and actor Tony Noice report on a theatrical intervention that offers professional acting instruction to ageing populations. They describe a series of related studies that they have conducted in Switzerland and the USA to discover whether acting techniques can assist in maintaining cognitive functioning in older people. The Noices focused on an acting process known as 'living truthfully under imaginary circumstances' that is at the core of the style of psychological realism. This process has been taught to residents in multiple retirement homes and community senior centres in the USA and its effects measured by reliable scientific assessment instruments. For over 20 years, this project has generated evidence that participants have experienced significant cognitive benefits, including increases in episodic memory, list recall, creativity and problem-solving ability. The practice thus lowers risk factors for Alzheimer's disease and other forms of dementia in mentally healthy older adults.

In contrast to the project described by the Noices, the work that my colleague, Rachel DeSoto-Jackson, and I describe is still in its early stages. We both teach performance in the Department of Theater and Dance at Indiana University of Pennsylvania and have been developing training in empathetic communication. Multiple behavioural studies over the last 30 years have indicated that empathy is desirable in the interactions between patients and healthcare professionals. However, there has been little agreement over how empathy training should be designed. Several years ago, our department initiated a collaboration with the nursing training programme at our university to offer a course that would assist trainee nurses in successful interpersonal communication with their patients. DeSoto-Jackson has been teaching this course for the last two years and has introduced techniques from Augusto Boal's Forum Theatre and Image Theatre into the content as well as using her own model for structuring communication between patient and caregiver. Self-reporting from students in the course has shown that these techniques have proved successful in developing what the students themselves describe as empathy. In our chapter, we contextualise a case study of this course with information about empathy derived from the cognitive sciences, which suggest potential explanations for the self-reported benefits of the training by its participants.

Experience Bryon also teaches performance practice, and in her chapter applies her framework of Integrative Performance Practice (IPP) to a field known as Situated Awareness (SA), also known as Situational Awareness. This field seeks to systematise the cognitive activities which one uses to evaluate and predict the actions of multiple elements in one's environment through processes such as pattern recognition, interpretation and evaluation. SA is used in a variety of contexts by professionals like police officers, fire fighters, air traffic controllers, nurses and military commanders. Currently, many approaches to SA depend on a dualistic framing, defining features of SA as either a product or a process. Some theorists propose that situational awareness is a product in the mind of the human operator, while others consider it to be the process of acquiring awareness. Those familiar with concepts of **enactivism** and **situated cognition** will rapidly recognize that these concepts can fruitfully be applied to SA to elide the dualistic separation of an 'agent' from his or her environmental perception. Bryon proposes that her performance practice of 'middle field' awareness can assist theorists and practitioners of SA to improve both its theorisation and its practice.

Our next chapter describes a valuable intervention in a vulnerable population. This is the 'Imagining Autism' project developed by Nicola Shaughnessy and Melissa Trimingham at the University of Kent. Their chapter focuses on an interdisciplinary collaboration between the fields of theatre and psychology to benefit children on the autism spectrum. This project originated in the field of applied and socially engaged theatre, terms that refer to theatre practices that are applied in social, educational and community contexts with a therapeutic or educational purpose. The original aim of the project was to involve autistic children in activities that engaged them socially, physically and creatively. Theatre practitioners designed and delivered theatrical exercises, with psychologists evaluating the effect of the exercises on the children. As the project progressed, however, the success of the exercises in raising assessment scores in the autistic participants prompted adaptations of the research design, creating a 'transdisciplinary' approach. The success of this project has rippled out beyond the immediate participants and has prompted learning in all those involved, children, parents and researchers alike. It is an inspiring example of the transformative power of theatre.

The final chapter in this part describes the concept of 'consilience' – the 'jumping together' of phenomena. The author of this chapter, my co-editor Bruce McConachie, is well known in the field of theatre and performance studies for pioneering the application of cognitive neuroscience to the understanding of theatre and performance, and has written extensively on topics ranging from audience reception to the evolution of performance in play and ritual. Here, he participates in taking up the challenge posed by biologist E.O. Wilson to researchers in the sciences and humanities to create a common groundwork of empirically based knowledge that can link C.P. Snow's 'Two Cultures' of the sciences and the humanities. McConachie takes on the topic of climate change by investigating how potential areas of consilience between history and evolution can inform our understanding of how humans perform politics. He identifies the cognitive phenomena that underlie performance in political contexts that directly embody and practice versions of authority and solidarity, and also in stage performances that comment on the exercise of governmental power. McConachie's ultimate goal in doing this is to contribute to an improvement of democratic citizenship and governance in response to the challenges of global climate change and resource depletion.

26

A THEATRICAL INTERVENTION TO LOWER THE RISK OF ALZHEIMER'S AND OTHER FORMS OF DEMENTIA

Tony Noice and Helga Noice

This chapter describes a theatre intervention that uses professional acting instruction to lower risk factors for Alzheimer's disease and other forms of dementia. Please note that this is not a form of drama therapy, which treats those already afflicted, but is a preventive programme designed to help older adults remain cognitively healthy. The intervention has been supported by four multi-year grants from the National Institutes of Health as well as a number of private funders. The rationale behind this programme is that the practice of acting offers intense mental stimulation because it requires participants to retain theatrical dialogue verbatim without deliberate memorisation, and to perform dramatic scenes *truthfully* in front of an audience of their peers. This latter aspect of acting has been variously referred to as the 'reality of doing,' 'being in the moment' and 'living truthfully under imaginary circumstances,' and is a process that lies at the heart of professional acting. This chapter presents evidence that this approach to performance confers unique cognitive benefits to older non-actors, including increases in episodic memory, list recall, creativity and problem-solving ability, as measured by reliable and valid scientific assessment instruments. The programme has been carried out successfully for over 20 years in dozens of independent-living retirement homes and community senior centres in the USA.

Recently, the authors of this chapter were commissioned to review all existing studies on participatory arts programmes designed to enhance healthy ageing (Noice, Noice, & Kramer, 2014). We examined programmes on acting, dancing, writing, music (vocal and instrumental) and various visual arts (sculpting, painting, collage, etc.). Although we set no chronological limits for the search, we found only 31 extant studies that met our criterion of publication in peer-reviewed professional journals. Of those, fewer than a dozen were rigorous, pre-post studies with random assignment to condition. The latter group included reports on our own 20-year acting programme designed to lower risk factors for Alzheimer's and other forms of dementia. Prior to that, we were *expertise* researchers examining the cognitive processes of professional actors (e.g., Noice & Noice, 1993, 1994, 1996, 1997a). It was in the late 90s that we started to explore the value of acting instruction to enhance healthy cognitive/affective ageing. We still return occasionally to theoretical inquiry (e.g., Noice & Noice, 2007).

Our programme for older adults started with a chance meeting at a memory conference with two prominent Swiss cognitive and developmental researchers, Walter and Pasqualina Perrig (like us, a husband and wife team). We mentioned our finding that college-age student-actors exhibited superior memory for text compared to controls (Noice & Noice, 1997b) and speculated that perhaps older adults might also benefit from studying actors' techniques. The next day over breakfast, the Perrigs asked if we would be interested in coming to Switzerland to pilot test our approach. We acquiesced within seconds, and the Perrigs asked us to generate a brief grant proposal, which was approved in short order by the Schweizer Nationalfond (the Swiss equivalent of the US National Science Foundation). The Perrigs were in the middle of a longitudinal memory inquiry assessing the effectiveness of various interventions for older adults. They asked us to devise and administer a short course in acting as a component of their multi-year study, using the recall and recognition measures they employed for the entire series.

Before going into detail on this and our subsequent healthy ageing experiments, we'd like to sketch in the background of this type of research. The early phase involved specific training programmes such as teaching older adults mnemonic techniques. For example, Robertson-Tchabo, Hausman and Arenberg (1976) found gains of 79% in list-learning ability after five days of instruction using the method of loci. This is a memorisation technique in which words are converted into mental images and then associated with specific areas of a familiar location such as one's home. However, in a later meta-analysis, Verhaeghen, Marcoen and Goossens (1992) concluded that the benefits of such mnemonic instruction do not transfer to other types of cognitive tasks.

The most extensive study investigating techniques to enhance healthy cognitive performance, Advanced Cognitive Training for Independent and Vital Elderly (ACTIVE), employed randomised controlled trials, with 2,802 participants in a total of six US cities (Ball et al., 2002). The study produced evidence for the efficacy of three types of cognitive training: verbal memory, speed of processing and inductive reasoning. However, the results revealed two limitations: the training did not transfer (i.e., those trained on reasoning improved in that aspect of cognition but not on speed of processing or memory) and, despite the targeted gains, no increases were found on performance of activities of daily living. A follow-up study more than ten years later (Rebok et al., 2014) showed that some of the gains in speed of processing had been maintained (especially after booster training) but that the other original gains (memory and reasoning) had not.

Another widely used approach to investigating cognitive enhancement techniques consisted of correlating such activities with markers of healthy cognitive ageing (for a review, see Hertzog et al., 2009; see also, Verghese et al., 2006). However, as all beginning psychology students know, correlation is not causation, and experimental interventions under controlled conditions are necessary to persuasively document gains. One encouraging one (Kattenstroth et al., 2013) employed 35 participants, randomly assigned to either a dance condition or a usual activities (control) condition. Half of them learnt a complex choreographed dance over a six-month period; the other half continued with their usual activities. (None of the participants had engaged in dancing or physically effortful sports for at least five years before the experiment.) Both groups were pre-post-tested on cognitive and physiological measures such as Raven's Standard Progressive Matrices. At the end of the six months, the experimental (dance) group demonstrated significant cognitive and physiological improvements compared to controls. Unfortunately, many older adults cannot or will not engage in such rigorous physical activity for six straight months. Therefore, there appears to be a strong need for a non-strenuous, cognitively stimulating activity, preferably one whose benefits have been validated over a number of studies. A search of the literature revealed that our theatre programme appears to be the only series of studies systematically tested with the so-called gold standard of such assessment: *randomized controlled trials*, (e.g., Noice, Noice, & Staines, 2004).

A common assumption is that an acting programme would emphasise script memorisation, but our programme specifically directs participants to avoid intentional memorisation and to concentrate on understanding the scripted events in depth. This mirrors the procedure followed by most of the professional actors we employed in our early expertise studies.

The nature of acting expertise

Although rote repetition is widely assumed to be the strategy of choice for word-for-word retention, we found that our sample of professional actors rarely acquired dialogue by rote. One actor specifically addressed this:

> Most of the time I memorize by magic – and that is, I don't really memorize. There is no effort involved. There seems to be no process involved: It just happens. One day early on, I know the lines.
>
> *(Noice, 1992, 420)*

A number of experiments using both protocol analysis and empirical methodology revealed that actors (apparently unwittingly) employ most of the learning principles identified by an earlier generation of cognitive researchers (Noice & Noice, 1997a). Actors engage in extensive elaboration, perspective-taking, self-referencing, self-generation, mood congruency, distinctiveness and so forth. They also determine the goal of every utterance of the character, breaking down scripts into what they call 'beats' (small goal-directed chunks of dialogue). These beats lay out the entire role as a causal chain, a process that often entails generating multiple elaborations for a few words of dialogue. (A well-established cognitive concept is that both causal chaining and additional elaboration lead to greater recall, e.g., Graesser & Clark, 1985.)

However, our research showed that retention of the textual material is only the preparatory part of the actor's work. The actor must then *mean* the words of the script each time he or she says them so that every performance is unique even though the words are unchanged. That is, *how* the words are said depends on the spontaneous mental-physical-emotional interactions between the actors at every moment of the performance. This double process appears to be in general use by professionals; they analyse the role prior to rehearsal, but during rehearsal, they try to devote all their conscious awareness to remaining in the present moment by attending to the other actors, only glancing down at the script when necessary. Eventually, they find they are *off book* and *performance-ready*.

A vexing question not answered by the specification of actors' script-processing approaches was whether their rapid acquisition of dialogue was a product of their expertise (acquired by long practice) or was based on an explicit strategy that could be taught to non-actors. To answer this, the authors tried to train undergraduates in the actors' strategy (Noice & Noice, 1997b). At first, we assumed that the preliminary part of the learning strategy (the deep processing of the script) was responsible for an actor's highly efficient memory. Therefore, we taught students to elaborate on a text by asking goal-directed questions (e.g., 'Am I flirting with her when I say this?'). Consistent with previous findings, students who elaborated by questioning the underlying meaning remembered more than controls who read the same text purely for comprehension. (Indeed, most of the early investigations of cognitive learning principles had employed *read-only* controls.) However, we then tried to make the conditions more stringent by having the controls deliberately memorise the same material. If the actors' strategy still produced more retention, it would suggest the existence of an optimal means of studying verbal material. Unfortunately for our hypothesis, the deliberate memorisation controls outperformed the students who used the actors' analytic strategy (*M* = 57% for memorisation strategy vs. *M* = 33% for actors' analytic strategy, Noice & Noice, 1997b). Putting aside our original assumption, we turned to the other half of the actors' process, the one they use for rehearsal and performance. Instead of instructing the students to analyse the text, they were told to read the material, imagining someone they knew who needed this information. Indeed, we specifically told them *not* to try to remember the words but to put all their concentration on meaning them. Suddenly, the results were reversed—meaning the words produced greater retention than memorising them (*M* = 50% for memorisation strategy vs. *M* = 60% for actors' performance strategy). We have been able to replicate this finding using different populations, procedures and materials other than dramatic dialogue (e.g., newspaper articles). In writing up the results for publication in the cognitive literature, we coined the term *active experiencing* (AE) to refer to this process in which participants are asked to refrain from deliberate memorisation but to use all physical, mental and emotional channels to communicate the meaning of material to another person, either actually present or imagined (for a review, see Noice & Noice, 2004). This process of trying to 'live' the material rather than just memorise it appeared to make a major difference in the results.

This brief summary of some of our early studies brings us back to the chance meeting with Pasqualina and Walter Perrig, described at the beginning of this chapter. They were in the midst of a longitudinal experiment to determine what factors might boost recall and recognition performance in a group of retired office and executive employees of a large chemical corporation in Basle, Switzerland. When the results of all their interventions over the years (vitamin regimes, weight training, etc.) were analysed, the acting programme was one of the few that produced statistically significant increases on their assessment measures (Noice et al., 1999). The results of this pilot study reinforced our notion that acting instruction might provide sufficient mental stimulation to enhance various cognitive abilities. Therefore, in future experiments, we would not be restricted to showing increased retention and recognition of theatrical dialogue but might see enhancement of a number of general cognitive abilities (creativity, episodic memory, prose comprehension, problem-solving, etc.).

Experimental procedure used across our various studies

To supply evidence of effectiveness, the experimental group receives our four-week acting class, one of the control groups receives an alternate course of instruction (e.g., visual or performing arts) and the second control group continues with their usual daily activities (no-treatment condition). The experimental group and the active controls meet for two 70-minute sessions (60 minutes of training and 10 minutes for a break to allow for socialising) each week for four weeks. All three groups are simultaneously pre-tested the day before classes start and post-tested on the day following the final class sessions. An important feature of the testing is that none of the assessment instruments are targeted to the training but simply test abilities necessary or helpful for independent living. This procedure is relatively rare. In many cognitive interventions, the instruction is specifically targeted, such as training in the method of loci and testing the result with a list-learning task, teaching of test-taking strategies or practice runs on tasks similar to those on the pre-post tests. Because we did not engage in any such tactics, the improvements following our acting classes would appear to be the result of the intense activation experienced by imaginatively becoming immersed in a fictional situation and performing it in front of the group. To minimise attrition, all participants receive a small gift of $40.00 at the end of the course, but only if they attend every session, including the pre- and post-tests.

In sum, our theatre classes are based on a widely accepted notion that good acting consists of *spontaneous* and *truthful* performance.[1] Therefore, if the scene calls for Character A to plead with Character B, then Actor A pleads with Actor B. The actor does not *pretend* to plead or try to duplicate the external behaviour of someone engaged in pleading; he or she simply pleads with the other actor *for real*. Although this duplicates what we do in our everyday lives, the constraints of playing a theatrical scene turn this into a highly complex cognitive task. In life, we make up our dialogue as we go along, depending on our needs of the moment, but the actor has already learnt and rehearsed the scene yet must render it spontaneously (*not* just apparently spontaneously) at the moment of utterance. Countless theatrical theorists have struggled with ways to truly live *moment to moment* while simultaneously retrieving the exact words and movements from memory. One well-known acting guru, Sanford Meisner, described it as 'living truthfully under imaginary circumstances' (Meisner & Longwell, 2008, 15). The majority of contemporary acting teachers would probably agree with this definition, but there such agreement would end. Meisner had his own approach to get his students to accomplish this tricky task. Many other highly regarded acting teachers and their disciples (e.g., Benedetti, 2008; Chekhov, 1993; Cohen, 2007; Hagen, 1973; Lewis, 1958; Strasberg, 1988) have their own original roadmaps to get students to engage in this truthful, spontaneous performance of well-learnt material. One rationale for our own eclectic intervention is that when non-actors attempt to engage in such professional techniques, the process is even more challenging (and hence, more stimulating) than when trained actors perform the same tasks.

Therefore, we embarked on a series of studies to find empirical evidence on the validity of our rationale. As usual with such a research programme, each study not only answered a specific research question but prompted additional ones that required additional experimentation. The first large, controlled intervention in the USA (Noice, Noice, & Staines, 2004) used 124 community-dwelling participants (aged 60–86) recruited from senior centres or retirement homes.

Overview of that first US intervention

Following the previously described experimental procedure, these participants were randomly divided into three groups: the experimental group (theatre), the group to control for non-content-specific effects (art appreciation) and the no-treatment control group. It should be noted that the art appreciation group was, however, very active. For example, in one session, participants arrived to find reproductions of various art works displayed on easels throughout the large training room. They were given tokens with bold titles such as 'I love it!' or 'Fine for a museum but not in my house!' or simply 'Yuck!' They had to post the tokens on the various art works and defend their choices during subsequent discussion with their peers.

As previously explained, the main experimental group (theatre) was given a beginning acting course, with each session emphasising genuine involvement in the dramatic situation. The instructor (a professional actor/director and professor of theatre) emphasised that he would be correcting people from time to time in order to illustrate common mistakes that are part of the process. However, to put the participants at ease, he emphasised that even highly experienced professional actors receive the same sort of corrections from directors. At each session, every participant did every exercise, with the rest of the group serving as an audience. The exercises became much longer and increasingly more complex, with new 'acting tools' frequently being added. In the beginning, the participants held the scripts during performance so that the emphasis was always on genuinely experiencing the essence of truthful acting and never on remembering the literal words. Thus, participants were trained to become cognitively, emotionally and physiologically involved in dramatic situations by practicing a wide range of theatrical intentions such as 'To confront,' 'To apologize,' 'To demand attention' or 'To tease.' The goal throughout was to give participants the ability to become so engrossed in the character's inner life that obvious behavioural changes occurred without intentional manipulation. These changes included facial expressions, body language, tones of voice, affect states and so forth. It must be stressed again, although no intentional memorisation of dialogue was ever required, the repeated rehearsals of these short scenes resulted in retention of the exact words as a concomitant of making the strong mental effort necessary to genuinely affect their acting partner by *meaning what they were saying*, using all channels of communication.

Both before and after the intervention, all participants were pre-tested on three cognitive measures (recall, working memory, problem-solving), and two affective measures: self-esteem (Rosenberg, 1965) and a quality of life scale consisting of self-acceptance, positive relations and personal growth (Ryff, 1989; Ryff & Keyes, 1995). Results for the cognitive/affective measures showed that, compared to the no-treatment controls, the theatre group improved significantly on quality of life ($p = 0.002$), list recall ($p = 0.007$), problem-solving ($p = 0.015$) and barely missed significance on working memory ($p = 0.056$).[2] In general, although the visual art group made some gains, such gains were far weaker, both in number and degree. Because the no-treatment controls were given a courtesy course following the testing, they could no longer be used for comparison purposes. However, when the theatre group was retested in four months without reinstatement of the training, no declines were found in any of the measures.

Some open questions to be answered

The 2004 experiment left open the question of why the art appreciation group did not improve as dramatically as the theatre group. Perhaps the reason was the lack of public performance. Therefore, the authors designed an experiment to compare acting to an art form requiring such performance: singing (Noice & Noice, 2009). The crucial question was whether singing training would produce results comparable to acting training, or if the specific nature of acting was responsible for the observed gains. Indeed, the reason the singing group was employed this time was that our previous choice, art appreciation, was designed to rule out the notion that *any* stimulating course given in a social situation might produce these benefits. We considered using other performing arts as controls but chose singing because of its practicality; dancing would eliminate many potential participants due to physical requirements, and learning to play a musical instrument involved both a longer time commitment plus the availability of fairly expensive instruments. However, our voice teacher, in addition to being a soloist with the Chicago Symphony, had a great deal of experience training beginners and verified that noticeable vocal improvement can occur within four weeks. For this experiment, we expanded our test battery to include eight cognitive instruments, plus an affective instrument, a memory controllability index and an activity scale. All tests were individually administered, and were experimenter-timed or self-timed, as called for in the instructions for each particular test.[3] The rationale for the choice of the additional instruments was three-fold:

a Each ability tested was important for healthy, independent living,

b All instruments were valid, reliable and standard in the field, and previously observed declines in these measures had been shown to have implications for dementia (e.g., Bennett et al., 2006).

c The tests could be administered in less than 90 minutes to avoid fatiguing the participants or subjecting them to multiple test sessions.

Overall rationale of experiment

In this iteration of our theatre intervention, we predicted that the acting group would improve significantly relative to the singing and no-treatment groups, because our former research led us to believe that acting contains a unique combination of various elements found in a number of successful ageing studies. It is novel, effortful, enjoyable, multi-modal/multi-factorial and mentally and physically stimulating. It requires participants to truthfully and spontaneously react to fictional situations, an experience that is at the core of the acting process. Moreover, it is emotionally activating and it encourages bonding in a social situation. An ever-increasing body of evidence indicates that such social interaction correlates positively with healthy cognitive ageing. Litwin and Shiovitz-Ezra (2010) found that the more social ties older adults had, the greater the well-being and the less anxiety they experienced. We are not aware of any other type of leisure activity that encompasses all these elements in such concentrated form. Consequently, we expected to find significant increases in the test scores of the acting group relative to the singing and no-treatment groups.

Results

The results were positive as predicted, showing greater improvement for the acting group over both the singing group and the no-treatment group on immediate and delayed word recall, immediate and delayed story recall, means-end problem solving, verbal fluency and personal growth (all $p < 0.05$). An interesting finding emerged from the data. Even though the acting group outperformed the singing group on all the cognitive measures, the acting and singing groups did equally well on the affective measure, personal growth. This suggests that, even in the absence of any demonstrated cognitive enhancement, the singing group still felt better about their lives after training.

This new experiment was also designed to investigate many other open questions. Are the benefits of theatre training restricted to those relatively healthy older adults who still function well in their own homes (as in the two former experiments) or could seniors (primarily low-income seniors) living in retirement homes make similar gains? Would participants almost a full decade older than those in the former study benefit? Would physical problems that inhibited mobility lessen or eliminate the positive effects? Would the fact that the majority of the current participants were less well educated affect the benefits? Our data showed that participants in the acting group prevailed over all these possible roadblocks.

Another question answered

As usual, still other important issues had to be addressed by future experimentation. Up to this point, all iterations of the theatre intervention had been carried out by the same professional actor, director and professor of theatre who devised the performance aspects of the programme. Perhaps the results were due to the specific qualities of this one individual acting teacher. Therefore, in the next study (Noice & Noice, 2013), retirement home activity directors without any background or experience in theatre were taught to administer the programme. Along with personal instruction, the activity directors attended a complete intervention given by the original instructor and took notes.

Yet another question was answered by this two-part study. In addition to the activity directors, one newly recruited professional actor/director/acting teacher was also engaged to administer the same intervention. However, she received only minimal instruction (mainly e-mailed descriptions of the training and short follow-up phone conversations before each session). If this new instructor produced pre-post results similar to those produced by the originator of the intervention, that finding would suggest that other qualified acting teachers could successfully duplicate the programme throughout the country.

Since the design necessitated two types of instructors, it obviously required two interventions, one conducted by activity directors and the other by the outside professional. With this one exception, the programme remained the same (i.e., number and length of sessions, content of the acting intervention, payment of participants, the cognitive/affective test battery). However, we added one functional test for both the activity directors and the professional instructor: The Observed Tasks of Daily Living, Revised (Diehl et al., 2005). This complex instrument contains 28 tasks, and uses actual objects such as prescription bottles, utility directories and check registers. Participants are required to perform such tasks as comparing prescription ingredient labels, paying bills and looking up and dialing telephone numbers. Three areas of functioning are examined: medical matters, telephone use and handling finances. This test in its original form was designed for the ACTIVE experiment (Ball et al., 2002), but, thanks to the creators' generosity, the revised version was released to the wider community of researchers a few years later.

The results of Experiment 1 (activity directors) showed that a few significant gains could be achieved when the intervention is conducted by persons without formal theatre training. However, the size and variety of those gains are certainly not comparable to those achieved by the professional (immediate word recall, verbal fluency and problem solving $p > 0.01$, East Boston Memory Test, $p = 0.007$). Our demographic analysis in Experiment 1 produced one quite atypical result: amount of education was not correlated with the observed gains. This unusual finding suggests that, unlike other training or correlation studies (e.g., Tun & Lachman, 2008; van Hooren et al., 2007), improvements from this programme may be independent of prior education provided the differences are not too great. (In our current research, using a very highly educated sample that included many PhDs, the usual education advantage was observed.)

Of course, researchers have examined a variety of other approaches for enhancing healthy cognitive ageing. For example, Basak et al. (2008) investigated commercially produced computer games and found improved executive function in participants who played for 23.5 hours; other positive results were found by Smith et al. (2009) and Nouchi et al. (2012). However, as pointed out in a review article on technologically delivered cognitive improvement programmes, 'objective evidence that a particular game or computer intervention can improve an individual's quality of life outside of the laboratory is sparse' (Charness & Boot, 2009, 256). Moreover, such games are obviously solitary undertakings and lack the strong benefits associated with social interactions described earlier.

Another advantage of our intervention is that written instructions and scripts are left in the libraries of the retirement homes. Some administrators have already reported integrating theatrical activities into their regularly scheduled programmes. As of this writing, in one facility, six productions, complete with costumes and sets made by the residents, have already been staged. Furthermore, in the majority of the retirement homes, some intervention participants performed scenes or sketches in the facility's subsequent talent shows. Thus, this programme sets up the possibility of cognitively enhancing lifelong learning for some of the older adults who partake in it.

The search for empirical evidence of benefits continues. We have recently concluded a four-year study using fMRI brain scans and electronically delivered cognitive tests to document any pre-post intervention gains. The control group received a wide-ranging course about theatre that did not include any acting instruction. In this way we tried to rule out the possibility that any previously observed behavioural advantages were due to the inherent appeal of theatre itself rather than the intense mental activity involved in learning and applying acting techniques. The data are still being analysed, but at least one important positive result has emerged: a significant increase in episodic memory (Banducci et al., 2017). This is an important finding because prior research has shown that a decline in episodic memory is one of the first and most salient signs of incipient Alzheimer's disease (e.g., Bennett et al., 2006; Wilson & Bennett, 2003).

We would like to conclude this chapter with a brief summary of the advantages of acting classes and their unique place in the small group of healthy ageing interventions not requiring intense physical activity. The efficacy of acting classes has been demonstrated by multiple investigations that were supported by major government organisations (e.g., National Institute of Aging). The acting intervention involves a very high degree of social interaction, shown to be an important factor in healthy ageing (e.g., Litwin & Shiovitz-Ezra, 2010). It is *novel* for all the participants (no former amateur or professional actor is eligible to participate in our studies), and novelty is another prime ingredient shown to enhance healthy ageing (e.g., Hultsch et al., 1999). Furthermore, acting is generally fulfilling and enjoyable, thus encouraging future involvement; in fact, in all our studies, the attrition rate for the theatre groups was, more often than not, *zero*, compared to fairly frequent dropouts in the control groups. (Control group attrition was of course analysed and accounted for in the statistical reports.) Additionally, the theatre intervention is brief and relatively low-cost, requiring only one teacher and no expensive equipment. Finally, we hope the experiments discussed here encourage further inquiry into theatre and other participatory arts by many additional investigators. There is a great need for this type of leisure activity for older adults. We certainly recognise the demonstrated value of the exercise and dance programmes, but for those unwilling to push the limits of their perceived physical stamina, we still believe 'The play's the thing.'

Notes

1 The emphasis on truth and spontaneity during the performance of a fictional situation comes from the teachings of Stanislavski (for an up-to-date, one-volume translation of these teachings, see Stanislavski (2008). As of today, the dominant approach in the majority of college acting programmes is still Stanislavski-based, with many, many individual variations. This is true of our own approach, both in the older adult programme and at Elmhurst College. We acknowledge that still other approaches (some, but certainly not all, integrating Stanislavski-like components) are taught in secondary and post-secondary education. These include the approaches of Lecoq, Grotowski, Brecht, Meyerhold and others. Followers of these later theorists testify to the efficacy of their techniques. Nevertheless, we opted to base our own eclectic intervention on the acting principles most frequently taught in the USA.

2 For readers unacquainted with this statistical reporting convention, the p value indicates the number of times (out of 100) that the finding might have occurred by chance. Thus a p value of 0.05 means that, if the experiment were performed 100 times, in 95 of those trials, the result would have been due to manipulation (in our case, participating in the acting course). The other five times might be due to coincidence. In general, a p value of 0.05 or below is considered 'statistically significant' and can be relied upon. Some of our measures scored as low as 0.001, indicating that, 999 of 1000 times, improvements following the acting course could not have been due to chance.

3 For those interested in the details of specific tests, see the Reference section under the following names: Morris et al. (1989), Wechsler (1987), Albert et al. (1991), Platt and Spivack (1975), Ryff (1989), Ryff and Keys (1995), Lachman et al. (1995), Wilson et al. (2002).

References

Albert, M. S., Smith, L., Scherr, P., Taylor, J., Evans, D. A., & Funkenstein, H. (1991). Use of brief cognitive tests to identify individuals in the community with clinically diagnosed Alzheimer's disease. *International Journal of Neuroscience, 57*, 167–78.

Ball, K., Berch. D. B., Helmers, K. F., Jobe, J. B., Leveck, M. D., Marsiske, M., Morris, J. N., ... Willis, S. L. (2002). Effects of cognitive training interventions with older adults. *Jama, 288*, 2271–81.

Banducci, S., Daugherty, A. R., Biggan, J., Cooke, G., Voss, M. W., Noice, T., Noice, H., Kramer, A. F. (2017). Active experiencing training improves episodic memory recall in older adults. Frontiers in Aging Neuroscience, 9. doi.org/10.3389/fnagi.2017.00133.

Basak, C., Boot, W. R., Voss, M. Q., & Kramer, A. (2008). Can training in a real-time strategy video game attenuate cognitive decline in older adults? *Psychology and Aging, 23*, 765–77.

Benedetti, R. (2008). *The actor in you*. New York: Pearson.

Bennett, D. A., Schneider, J. A., Arvanitakis, Z., Kelly, J. F., Aggarwal, N. T., Shah, R. C., & Wilson, R. S. (2006). Neuropathology of older persons without cognitive impairment from two community-based studies. *Neurology, 66*, 1837–44.

Charness, N., & Boot, W. R. (2009). Aging and information technology use: Potential and barriers. *Current Directions in Psychological Science, 18*, 253–58.

Chekhov, M. (1993). *On the technique of acting*. New York: Harper Paperbacks.

Cohen, R. (2007). *Acting one*. New York: McGraw-Hill.

Diehl, M., Marsiske, M., Horgas, A. L., Rosenberg, A., Saczynski, J. S., & Willis, S. L. (2005). The revised observed tasks of daily living: A performance-based assessment of everyday problem solving in older adults. *The Journal of Applied Gerontology, 24*(3), 211–30.

Graesser, A. C., & Clark, L. F. (1985). *Structures and procedures of implicit knowledge*. Norwood, NJ: Ablex.

Hagen, U. (1973). *Respect for acting*. New York: Macmillan.

Hertzog, C., Kramer A. F., Wilson R. S., Lindenberger, U. (2009). Enrichment effects on adult cognitive development: Can the functional capacity of older adults be preserved and enhanced? *Psychological Science in the Public Interest, 9*, 1–65.

Hultsch, D. F., Hertzog, C., Small, B. J., & Dixon, R. A. (1999). Use it or lose it: Engaged lifestyle as a buffer of cognitive decline in aging? *Psychology and Aging, 14*, 245–63.

Kattenstroth, J. C., Kalisch, T., Holt, S., Tegenthoff, M., & Dinse, H. R. (2013). Six months of dance intervention enhances postural, sensorimotor, and cognitive performance in elderly without affecting cardio-respiratory functions. *Frontiers in Aging Neuroscience, 5*, 5.

Lachman, M. E., Bandura, M., Weaver, S. L., & Elliott, E. (1995). Assessing memory control beliefs: The memory controllability inventory. *Aging and Cognition, 2*(1), 67–84.

Lewis, R. (1958). *Method—Or madness?* New York: Samuel French.

Litwin, H., & Shiovitz-Ezra, S. (2010). Social network type and subjective well-being in a national sample of older Americans. *The Gerontologist, 51*, 379–88.

Meisner, S., & Longwell, D. (1987). *Sanford Meisner on acting*. New York: Vintage Books.

Morris, J., Heyman, A., Mohs, R., Hughes, J. P., van Belle, G., Fillenbaum, G., (1989). The consortium to establish a registry for Alzheimer's disease (CERAD). Part I. Clinical and neuro-psychological assessment of Alzheimer's disease. *Neurology, 39*, 1159–65.

Noice, H. (1992). Elaborative memory strategies of professional actors. *Applied Cognitive Psychology, 6*, 417–27.

Noice, H., & Noice, T. (1993). The effects of segmentation on the recall of theatrical material. *Poetics, 22*, 51–67.

Noice, H. & Noice, T. (1994). Analyzing the role preparation of a professional actor. Discourse Processes, 18, 345–69.

Noice, H. & Noice, T. (1996). Two approaches to learning a theatrical script. Memory, 4(1), 1–17.

Noice, T., & Noice, H. (1997a). *The nature of expertise in professional acting. A cognitive view*. Mahvah: N.J. Erlbaum.

Noice, T., & Noice, H. (1997b). Effort and active experiencing as factors in verbatim recall. *Discourse Processes, 23*, 51–69.

Noice, T., & Noice, H. (2004). A cognitive learning principle derived from the role acquisition strategies of professional actors. *Cognitive Technology, 9*, 1–24.

Noice, H., & Noice, T. (2007). The non-literal enactment effect: filling in the blanks. *Discourse Processes, 44*(2), 73–89.

Noice, H., & Noice, T. (2009). An arts intervention for older adults living in subsidized retirement homes. *Aging, Neuropsychology and Cognition, 1,* 1–24.
Noice, H., & Noice, T. (2013). Extending the reach of an evidence-based theatrical intervention. *Experimental Aging Research, 39,* 398–418.
Noice, T. Noice, H., & Kramer, A. (2014). Participatory arts for older adults: a review of benefits and challenges. *The Gerontologist, 54,* 741–53.
Noice, H., Noice, T., Perrig-Chiello, P., & Perrig, W. (1999). Improving memory in older adults by instructing them in professional actors' learning strategies. *Applied Cognitive Psychology, 13,* 315–28.
Noice, H., Noice, T., & Staines, G. (2004). A short-term intervention to enhance cognitive and affective functioning in older adults. *Journal of Aging and Health, 16,* 1–24.
Nouchi, R., Taki, Y., Takeuchi, H., Hashizume, H., Atkitsukie, Y., Shigemune, Y., & Kawashima, R. (2012). Brain training game improves executive functions and processing speed in the elderly: A randomized controlled trial. *PLoS ONE,* 7, e29676.
Platt, J., & Spivack, G. (1975). *The MEPS procedure manual.* Philadelphia, PA: Community Mental Health/Mental Retardation Center, Hahnemann Medical College and Hospital.
Rebok, G. W., Ball, K., Guey, L. T., Jones, R. N., Kim, H.-Y., King, J. W., ... Willis, S. L. (2014). Ten-year effects of the advanced cognitive training for independent and vital elderly cognitive training trial on cognition and everyday functioning in older adults. *Journal of the American Geriatric Society, 62,* 16–24.
Robertson-Tchabo, E. A., Hausman, C. P., & Arenberg, D. (1976). A classical mnemonic for older learners: A trip that works. *Educational Gerontology, 1,* 215–26.
Rosenberg, M. (1965). *Society and the adolescent self-image.* Princeton, NJ: Princeton University Press.
Ryff, C. D. (1989). Happiness is everything, or is it? Explorations on the meaning of psychological well-being. *Journal of Personality and Social Psychology, 57,* 1069–81.
Ryff, C. D., & Keyes, C. L. M. (1995). The structure of psychological well-being revisited. *Journal of Personality and Social Psychology, 69*(4), 719–7.
Smith, G. E., Housen, P., Yaffe, K., Ruff, T., Kennison, R.F., Mahncke, H. W., & Zelinski, E. M. (2009). A cognitive training program based on principles of brain plasticity: Results from the Improvement in Memory with Plasticity-based Adaptive Cognitive Training (IMPACT) study. *Journal of the American Geriatric Society, 57,* 594–603.
Stanislavski, K. (2008). *An actor's work.* J. Benedetti (Translated and edited). New York: Routledge.
Strasberg, L. (1988). *A dream of passion.* N.Y. Penguin Random House: Plume.
Tun, P. A., & Lachman, M. E. (2008). Age differences in reaction time and attention in a national telephone sample of adults: Education, sex, and task complexity matter. *Developmental Psychology, 44,* 1421–9.
van Hooren, S. A. H., Valentijn, A. M., Bosma, H., Ponds, R. W. H. M., van Boxtel, M. P. J., & Jolles, J. (2007). Cognitive functioning in healthy older adults aged 64–81: A cohort study into the effects of age, sex, and education. *Aging Neuropsychology, and Cognition, 14,* 40–54.
Verhaeghen, P., Marcoen, A., & Goossens, L. (1992). Improving memory performance in the aged through mnemonic training: A meta-analytic study. *Psychology and Aging,* 7, 242–51.
Verghese, J., LeValley, A., Derby, C. A., Kuslansky, G., Katz, M., Hall, M., ... Lipton, R. B. (2006). Leisure activities and the risk of dementia in the elderly. *Neurology, 66,* 821–7.
Wechsler, D. (1987). *Wechsler memory scale – revised manual.* San Antonio, TX: Psychological Corporation.
Wilson, R. S., & Bennett, D. A. (2003). Cognitive activity and risk of Alzheimer's disease. *Current Directions in Psychological Science, 12,* 87–91.
Wilson, R. S., Mendes de Leon, C. F., Barnes, L. L., Schneider, J. A., Bienias, J. L., Evans, D. A., & Bennett, D. A. (2002). Participation in cognitively stimulating activities and risk of incident Alzheimer disease. *Journal American Medical Association, 287*(6), 742–8.

27

THE PERFORMANCE OF CARING

Theatre, empathetic communication and healthcare

Rick Kemp and Rachel DeSoto-Jackson

Introduction

Empathy has been the subject of much interest in recent years, both in public discourse and in the fields of neuroscientific research. A popular understanding of empathy focuses on the idea of feeling another's emotion. In neuroscience, the term and the concept is much debated, with a recognition that it is complex and multi-modal in having cognitive, emotional, behavioural and cultural components. This chapter uses a case study to examine the application of theatrical expertise to training in empathetic communication for students in healthcare professions. We use the term 'expertise' advisedly and emphasise that we conceive of theatrical expertise as embodied, situated and enactive, with a tendency to inclusivity since it is always collaborative in its application. The case study is contextualised by information on concepts of empathy in the cognitive sciences, which suggest potential explanations for the self-reported benefits of the training by its participants. Despite the contested nature of the concept of empathy in the cognitive sciences, numerous behavioural studies over the last three decades suggest that empathy (with its varied definitions) is desirable in the interactions between patients and healthcare professionals (e.g. Bas-Sarmiento et al. 2017; Charon 2001; Herbek and Yammarino 1990; Petrucci et al. 2016; Roter et al. 1997; Suchman et al. 1997; Vanderford et al. 2001;). Patients' perception of a lack of empathy by health professionals has been shown to negatively influence outcomes in treatment of a variety of conditions, ranging from the common cold (Rakel et al. 2009) to cancer (Back et al. 2003). Furthermore, technological advances in healthcare have also prompted a trend towards using information technology to drive patient-provider relationships (Weiner and Biondich 2006). This technology can further reduce the interpersonal interaction in encounters between patients and healthcare professionals, and thus the means through which empathy is both communicated and experienced. Significantly, it has been shown that primary-care physicians who spend less time with patients are more likely to experience malpractice claims (Levinson et al. 1997). While there have been numerous studies on the importance of empathy in healthcare, less attention has been focused on developing effective models of teaching empathy and, more specifically, empathetic communication in healthcare education.

Theatrical expertise offers one approach to developing a method of teaching empathetic communication within healthcare. Actors and directors are experts in the execution, recognition and analysis of features like posture, gesture, eye movements, facial expressions, spatial proxemics and the tone and cadence of voice in interpersonal communication. These abilities in themselves are features of communication; we are considering them within a conceptual context of enactivism, which leads us to suggest that the conscious awareness and guided practice of these behavioural activities can stimulate both enhanced awareness of emotion in others and a responsive affective state in oneself. This proposal is made using a recognition of communicative empathy as a two-stage process, involving both the recognition and the reflection of affective state (Coll et al. 2017). This conceptual framework of empathy is integrated with practical theatrical expertise within an undergraduate course offered by our Department of Theater and Dance at Indiana University of Pennsylvania (IUP). Titled 'The Performance of Caring,' the course was originally intended for nursing students, but has since been expanded to include majors in multiple disciplines.

Cognitive science and empathy

We focus on a cognitive definition of empathy that is frequently used in clinical studies, rather than seeking to provide a comprehensive review of neuroscientific understandings of empathy. This focused definition facilitates the operationalisation of a behavioural training. As the authors of one neuroscientific study of empathy state, 'Empathy involves experiencing emotion vicariously, and understanding the reasons for those emotions. It may be served partly by a motor simulation function, and therefore share a neural basis with imitation (as opposed to mimicry), as both involve sensorimotor representations of intentions based on perceptions of others' actions' (Braadbart et al. 2014, 367). This description correlates with the concept of 'embodied simulation' developed by neuroscientist Vittorio Gallese to describe the neural mechanisms underlying social cognition: 'The results of neurocognitive research suggest that in the brain of primates, mirror neurons, and more generally the premotor system, play a major role in several aspects of social cognition, from action and intention understanding to language processing' (Gallese 2007, 659). Gallese has demonstrated in several studies that 'embodied simulation can underpin basic forms of social cognition like the capacity of empathizing with others' emotions and sensations' [Gallese 2003a, 2003b, 2005a, 2005b] (ibid.). While Gallese focuses on what is known as 'bottom up' processing (the information received from sensory input), recent research reviewed by Andy Clark in *Surfing Uncertainty: Prediction, Action and the Embodied Mind* (2016) suggests that cognition is the result of an interaction between sensory input and information coming 'top down' from memory, beliefs and hypotheses. These are more abstract mental concepts that, to the degree in which they influence action, result in what is called 'top down' processing. The 'bottom' and 'top' in the metaphor reflect a hypothesis that the brain's activity is hierarchical, with more abstract mental concepts at a 'higher' level than sensory input. It is important to remember that this description is a metaphor, since the dominant conceptualisation of the brain's mechanisms in embodied cognition is that of networks of patterns of neural activity that are constantly varying the ways in which neurons connect through synapses. Clark's 'predictive processing' model suggests that as sensory input is received, it is compared with 'predictions' arising from mental models formed by past experience. An example of a 'match' between the sensory input and a 'prediction' could be correlating the visual information gained from looking at a tree with the mental concept of 'tree.' If the sensory information does not match the concept (for example, if one determines that the branches and trunks of the tree are made of metal), then this 'unpredicted input' is compared with higher level, more abstract concepts until a match is found (a metal sculpture of a tree). Of course, this process is happening at extraordinary speed and at a level of cognition of which we are not consciously aware. The predictive processing model of cognition has relevance for the concept of empathy because top-down predictions interact with bottom-up perception. Thus, pre-formed beliefs can affect one's understanding of another person's affective state. One neuroscientific study on empathy has demonstrated that one is less likely to feel empathy for a person who is perceived to be in some way 'different' than oneself, with one of the most significant factors in perception of difference being race (Xu et al. 2009). This bias happens below the level of conscious awareness, so the challenges of empathy training include developing activities that can not only bring unconscious biases to the level of awareness, but also have an effect on these biases. A pathway towards accomplishing these goals involves linking guided behavioural activities with neuronal 'embodied simulation,' which is what we explore in this chapter.

We have chosen to use Gallese's term 'embodied simulation' to describe behavioural interaction in healthcare education. In the training of healthcare professionals, 'simulation' describes clinical practices in education involving realistic scenarios in an environment that poses no risk to patient safety. The formats of simulations range from clinical assessment of high fidelity manikins (the term used in healthcare to denote realistic mannequins) that have biometric features such as a heart rate and respiratory sounds, to procedural task-training through human interactions with 'Simulated Patients' (SPs) portrayed by trained actors. The Association for Standardized Patient Educators (ASPE) defines an SP as 'a person trained to portray a patient scenario...for the instruction, assessment, or practice of communication and/or examining skills of a health care provider' (ASPE 2017, n.p.). Our use of the term 'embodied simulation' describes the process of learning through behavioural interaction. We use this term for two reasons. Firstly, it distinguishes this approach from types of training that rely solely on the use of manikins and/or video. As theatre practitioners, we believe that embodied learning is necessary for significant behavioural change to occur. This opinion is supported by studies demonstrating the success of 'learning-by-doing' approaches in this field (e.g. Gracía-Pérez and Marta Gil-Lacruz 2017). Secondly, it consciously echoes Gallese's term for the neuronal behaviour involved in social cognition. The two are linked because observable behaviour stimulates neurological embodied simulation – our term is intended to describe the learning processes that stimulate the neurological process. This approach is compatible with a core principle of coordination dynamics, which is that the coordination of neurons in the brain and the coordinated actions of people and animals share a common mathematical or dynamical structure. As indicated in Chapter 3, this principle suggests that, as cognitive psychologists Keith Sawyer and Stacy DeZutter state, 'When cognitive processes are distributed across groups, they become visible, and scientists can observe them by analyzing the verbal and gestural interactions among the participants' (2009, 81).

The field of embodied cognition also strongly proposes that a sense of self is 'situated' in that it arises intersubjectively among humans and also from interaction with the environment. By linking embodied simulation in its observable, behavioural sense with neurological embodied simulation, we propose that the actions involved in embodied training can be used to guide the neurological processes that occur below the level of awareness. In a previous publication, one of us (Kemp) proposed that the actor's persona in rehearsal and performance can be thought of as a 'situational self' (2012). Within the theatrical approach to empathetic communication training that we examine, the same concept is used to encourage healthcare professionals to recognise that their behaviour and communication is affected by, tied to and affects the interpersonal context (situation) within which they are operating.

Simulated patients (SPs)

The importance of embodied simulation in healthcare education is supported by a study that indicates 'empathy can be improved and successfully taught at medical school especially if it is embedded in the students' actual experiences with patients' (Mercer and Reynolds 2002). This study further defines clinical empathy as 'an ability to: (a) understand the patient's situation, perspective, and feelings (and their attached meanings); (b) to communicate that understanding and check its accuracy; and (c) to act on that understanding with the patient in a helpful (therapeutic) way' (Mercer and Reynolds 2002, n.p.). While the use of manikins in simulation can be useful for invasive training procedures such as performing a tracheotomy, what is absent are the tiny movements of non-verbal communication that occur in response to others' behaviour and with which patients express affect. The use of SPs in healthcare simulation has grown since their initial use in 1963 when neurologist Howard Barrows began utilising non-medical students to portray illness and give feedback to medical students. With the success of this model of interactive learning, he termed these individuals 'Simulated Patients,' which he defined as 'a person who has been carefully coached to simulate an actual patient so accurately that the simulation cannot be detected by a skilled clinician' (ASPE 2017 n.p.). Barrows recognised that the impact of using an SP in medical education lies in their ability to perform realistic reactions, 'not just the history, but the body language, the physical findings, and the emotional and personality characteristics as well' (ibid.).

With the emergence of SP use in healthcare education, the terms 'Simulated' and 'Standardized' have often been used interchangeably. As methods of SP training have progressed, the two terms have become distinct from one another (Lewis et al. 2017). 'Simulated Patients' are trained to portray the roles of patients and family members within realistic medical scenarios. SPs act and react to the healthcare professional based on the interaction rather than a prescribed scripted response (Allegheny Health Network 2017; Lewis et al. 2017). While there are learning objectives to be met and contextual information conveyed, SP encounters allow healthcare students to learn in an environment in which responses mimic real interactions. The SP is trained to facilitate a realistic encounter that presents not only the history and physical findings, but also uses body language with emotional and personality characteristics. 'Standardized Patients' are used to portray specific medical conditions when consistent and standard responses are necessary. They are used extensively in testing healthcare students' clinical skills, usually as a part of an Objective Structured Clinical Examination, a form of performance-based testing used to measure candidates' clinical competence (OSCE 2015). Standardized Patients are more likely to provide reliable reactions to questions asked by healthcare professionals. The Standardized Patient will respond to questions and situations in a consistent manner, regardless of how the healthcare professional interacts with the Standardized Patient (Allegheny Health Network 2017). Within an educational model, the SP is preferable, as a range of reactions can be experienced without adhering to a scripted response. This allows for variations in communication based on the verbal and non-verbal stimuli from the healthcare professional. For example, if the healthcare professional neglects to greet the patient and give their name when entering the room, this may prompt the SP to respond in a way that indicates a lack of trust for the healthcare professional, which in turn affects aspects of the interaction such as full and honest responses during health history intake. Observations of these (and other) reactive behaviours would later be shared by the SP during a debrief with the healthcare professional at the conclusion of the simulation. Studies demonstrating the effectiveness of debriefing as a learning tool and the development of models of debriefing strategies have been prolific in recent years (e.g. Cantrell 2008; Decker et al. 2013; Sawyer et al. 2016).

The increase of simulation use in healthcare education has created a growing need for SPs to perform not only the roles of patients, but also of their family members. Reflecting this, the term 'Simulated Patient' has been updated to 'Simulated Participant' to acknowledge the various roles an SP may perform as well as 'all human role players in any simulation context' (Lewis et al. 2017, 2). In this chapter, the term 'Simulated Patient' refers specifically to the trained actors in a simulation and is not synonymous with healthcare student participants. The identification of best practices and the generation of research in the area of SP training have grown since the inception of the ASPE in 2001. A study published in the journal *BMC Medical Education* cites the importance of 'the authenticity of role play and quality of feedback provided by SPs' and recognises that 'the available literature on SP training mostly addresses instructor-led training where the SPs are given direction on their roles' (Perera et al. 2009, 1). Historically, SPs often come from diverse backgrounds, and many are without theatrical training as actors. Likewise, SP educators and trainers often emerge from healthcare and related fields rather than theatre programmes. As the use of SPs in healthcare education expands, the inclusion of theatrical experts offers an approach to SP training that is grounded in a rich tradition of actor training. While there are many schools of actor training methodology within the field, an Applied Theatre approach presents specific methodology transferable to SP training through which more effective simulations might be achieved.

The field of Applied Theatre has gained momentum in recent decades as a form of theatrical work that occurs in non-traditional spaces such as prisons, health and therapy settings, community centres and museums (e.g. Landy and Montgomery 2012; Prendergast and Saxton 2010; Prentki and Preston 2008; Rohd 1998; Taylor 2003). While the methodologies within Applied Theatre are wide ranging, they are connected by an overarching goal of using participatory theatrical practice to create communities, embrace human interconnections and generate social change through the use of ensemble practice. One such methodology is Theatre of the Oppressed, which is a system of interactive theatre focused on rehearsing actions for change developed by Brazilian director and activist Augusto Boal (1985). Theatre of the Oppressed, like many Applied Theatre methods, is grounded in ensemble practice. An Applied Theatre, ensemble-led approach to healthcare simulation results in the creation of original scenarios that reflect greater depth of patient identity, history and life circumstances in the development of an SP role. For example, an individual actor creating a character history of a patient with diabetes is limited to their own perception of this disease. In an ensemble-led approach, the actors create a character history as a group by offering various perspectives, encounters and experiences that might shape a patient's history. This in turn offers a simulation experience that better portrays the diversity of patients, thereby more effectively training healthcare professionals to meet patient needs. An Applied Theatre approach also offers the ability to address ineffective communication and collaboration among healthcare professionals. This phenomenon has been identified as a barrier to patient safety and quality of care (Institute of Medicine 2003). As medical disciplines have become increasingly specialised, specialists have had to focus on developing skills in their specific area of expertise resulting in an unintended outcome of compromising patient care due to failures of interprofessional collaboration (Palaganas, Epps, and Raemer 2014). (The unhyphenated form of 'interproffesional' is standard in medical fields). The emergence of interprofessional healthcare simulation has evolved in response to these collaboration failures in patient care. Currently, interprofessional simulations are utilised in a variety of professions and settings including emergency medicine, obstetrics, operating rooms, rapid response teams, psychiatry, pharmacology, radiology, occupational therapy and allied health professions (Titzer, Swenty, and Hoehen 2012; Zhang, Thompson, and Miller, 2011). However, to improve patient safety, the Institute of Medicine (IOM) recommends that healthcare professional educational programmes 'design and implement early and continuous interprofessional collaboration through joint classroom and clinical training opportunities' (Institute of Medicine 2010, 6). Interprofessional simulations allow healthcare students to gain a deeper understanding of the various healthcare professional roles that serve to provide quality patient care in a safe, controlled environment.

Case study: the performance of caring

'There is a change in myself that I have realized from the beginning of the semester until now. There is growth in my communication, body language, empathy, and mindfulness skills. My goal is take this information I have learned and practiced from this class and continue to use them [sic] throughout my career and life in general. These skills will help me to be a better nurse that [sic] is attentive, understanding, empathetic, and kind' (DeSoto-Jackson 2016, n.p.). This reflection, written by a student in The Performance of Caring course, highlights the perceived value of training in empathetic communication. Effective empathetic communication goes beyond existing 'SBAR' protocols for interprofessional engagement. SBAR is an acronym for Situation, Background, Assessment, Recommendation, a technique that is used to facilitate prompt communication in healthcare settings (Thomas, Bertram, and Johnson 2009). While empathy in clinical settings is viewed as a desirable behaviour in healthcare professionals, dedicated practice to embodying the behavioural characteristics of empathy in practice is seldom part of the pedagogy. In response to this need, the Performance of Caring course, originating in 2013, was developed at IUP to offer training for healthcare education students in empathetic communication skills.

The Department of Theater and Dance began its collaboration with the College of Health and Human Services to provide SP for clinical simulations in the Department of Nursing. With the success of this early collaboration, the Performance of Caring was created to serve nursing students in developing empathetic communication skills. Since then, the department has successfully implemented and broadened a number of Applied Theatre interdisciplinary collaborations designed to enhance the education and training of current and future healthcare professionals, both on campus and in the region. These efforts take a variety of forms, yet all incorporate methods from the field of Applied Theatre for embodying core principles that are learnt and practiced in order to improve healthcare communication. While this approach originates in theatre training, it is remarkably consonant with a set of recommendations for teaching communication skills to oncologists:

> (1) lecture-style methods alone are ineffective; (2) adult learning principles should be used; (3) teaching must include skills practice; (4) teaching must attend to learner attitudes and emotions; (5) the learning environment should integrate knowledge, skills, and attitudes; and (6) reinforcement is critical for the learning process.
>
> *(Back et al. 2003)*

The Performance of Caring course, currently taught by DeSoto-Jackson, offers students theatrical practice-based exercises and skill-building that teach the 'language' of empathy and non-verbal communication so as to positively affect patient/healthcare professional interaction. The course has since been expanded to include a focus on interprofessional communication prompted by the changing healthcare landscape. In its current formation, the course includes students from the following majors and fields of study: Nursing, Nutrition, Dietetics, Sports Medicine, Criminology, Hospitality Management, Psychology and Early Childhood Development. Students enrolled in the Performance of Caring often report at the beginning of the semester that they struggle with knowing how to talk to patients and how to demonstrate caring, empathetic communication (DeSoto-Jackson 2016). Applied Theatre offers many tools for practical approaches to preparing students with enhanced skills in interpersonal, empathetic communication using methods common to the training of performing artists. Using exercises and training techniques derived from Augusto Boal's Theatre of the Oppressed – a method within the field of Applied Theatre – students engage in collaborative learning models with specific attention to the ways in which culture and structures of power affect communication (Boal 2002).

The course is organised within three overarching units: Basic Communication, Building Empathetic Relationships, and Overcoming Challenges. Students are assessed through diagnostic, mid-term and final simulations that apply communication strategies learnt in each unit. Students' grades are largely determined using a self-assessment and peer feedback model which has been shown to be effective in simulation practice (Perera et al. 2009). SPs, rather than Standardized Patients, are utilised within this course to allow for a variety of interactions rather than standardised responses. Attention is given to performing diverse simulation scenarios within each skill-building unit including cultural communication, de-escalation, confrontation and communicating major life changes. As expressed by a student in the course, 'This class has helped me to become more confident when communicating my needs. I really feel one area I excel in is empathy. I always felt I was able to connect with people when in conversation, but sometimes feel uncomfortable doing so in a professional environment. This class has taught me the importance of listening and opening up when appropriate. There is something very special that happens when two people connect' (DeSoto-Jackson 2016, n.p.).

In contrast to other theatre-based forms of SP training, the Performance of Caring course incorporates many of Boal's Theatre of the Oppressed techniques. These include Forum Theatre, in which 'spectactors' move from observing enacted situations to proposing actions for change by participating in the enactment, and Image Theatre, in which students use static embodied poses to create images of social interaction. This process enhances participants' perceptions of the ways in which posture and spatial arrangement communicate meaning (Boal 2002). Using these techniques, participants develop a depth of understanding of the circumstances facing patients in diverse populations and learn best practices for improving their communication with these patients. These methods enhance the understanding and embodiment of verbal and non-verbal cues which are essential to empathetic communication. Ensemble-based characterisation exercises serve to further enhance the understanding of the variety of circumstances impacting patients within diverse groups. An example of this is an ensemble-created scenario in which a patient identifies as transgender and multi-racial. Through this scenario, students learn the complexities of gender identity and how misidentification can hinder the building of trust with a patient. Student nurses also learn to manage their own anxiety when encountering unfamiliar situations and patients with whom they don't initially identify with. Within the scenario, students apply physicalised practices learned in the unit such as eye contact, lower vocal resonances and relaxed body postures, which are conducive to building trust with a patient. This in turn may elicit more forthcoming responses during patient intake, resulting in more effective patient care. Students also learn how to adapt their mannerisms to a variety of situations in which they work with patients. Often students in the course exhibit reluctance to change or adjust their body posture or spatial proximity during an interaction, which results in an apparently 'frozen' state. Using improvisational theatrical practices, students learn how to 'prepare to be unprepared,' which allows them the flexibility to adjust to their circumstances and react confidently during encounters which are unexpected. This method of Applied Theatre training offers students concrete skills in de-escalation, assertiveness and confrontation in addition to empathetic communication. As one student reports, 'I am now prepared to meet difficult situations in the workplace through the interactive scenarios we have learned in class. My confidence has also increased exponentially through this class as well as learning to be mindful of my body language' (DeSoto-Jackson 2017a, n.p.).

The course units and specific exercises are structured around DeSoto-Jackson's original model of an empathetic communication process termed the L.O.V.E.™ approach: **L**isten, **O**bserve, **V**alue, **E**mpathize. This model expands to include specific skills for each step in the L.O.V.E. approach to empathetic communication: **Listen** actively, **Observe** verbal/non-verbal cues, **Value** other perspectives and **Empathize** through verbal/non-verbal response. What follows are specific exercises within each area that target skill-building.

Listen: 'Minute Conversations' is an early exercise adapted from Peggy McIntosh's approach to unpacking privilege and is used to teach students active listening, which many students report struggling with in our technology-saturated society (McIntosh 2003). Students begin moving around the room, greeting each other as they pass by. When they hear 'stop' from the facilitator, students form pairs with another student close to them. Within these pairs, students are asked a targeted question of personal significance such as 'describe a time when you felt powerless.' While the question itself is valuable, the focus in this exercise is on students' listening to their partner's story. The listening students are instructed to not respond verbally and to minimise non-verbal reactions to their partner's story. The students are given one minute to tell their story before they alternate with their partners, who then get to share their own stories based on the original question. This exercise is repeated three times with three different questions and with different partners in each rotation, so students practice actively listening with a variety of other people. During the debrief of this exercise, students often express surprise at how difficult it was for them to remain silent and focus on listening. They recognise how often they prematurely prepare responses in their mind rather than focusing on what the people in front of them are saying when they are speaking, or how often they interrupt other people when they are speaking. Following this exercise, students also begin to understand how important embodied communication is to feeling that they have been heard. This forms the basis of recognising verbal and non-verbal cues in others as well as themselves.

Observe: 'Habits Interview' is an exercise that builds an awareness of individual and others' verbal and non-verbal cues. This exercise begins by teaching students the basic components of verbal and non-verbal communication from actor training such as posture, gestures, facial expressions, tone, volume, diction, dialect, repetition, tempo. Space is also identified as a contributor to effective communication, and students explore the myriad ways spatial proximity can be employed in communication, such as sitting face-to-face or standing at a distance. Within pairs, one student begins the exercise by creating a spatial arrangement of the environment to conduct an interview. While this student is creating this space, his or her partner receives the interview questions which include instructions for conducting the interview. These have an unexpected instruction; rather than recording the verbal content of their partner's answers to the listed questions, they are asked instead to record the vocal and non-verbal cues that their partners demonstrate. At the end of the exercise, there is always astonishment when partners reveal that they have been recording these cues rather than their responses. This leads to a detailed debrief in which the partners and the instructor reveal the observed vocal and non-verbal cues. The exercise is then repeated for the other partner; while they are now aware of the observation, students have found it to be just as informative. Two skills are targeted in this exercise: observation of self and observation of others. Both are key to demonstrating empathetic communication through observation of patient behaviours and self-control of the health professionals' own behaviour.

Value: 'Culture Box' is an assignment that initially appears to be an ice-breaker activity. However, what is discovered through the activity is a deeper understanding of the experiences of others and how those experiences shape individual perspectives. This exercise is adapted from the Intergroup Dialogue approach to communicating across diverse groups (Dessel 2008). A key premise of Intergroup Dialogue is the distinction between discussion, debate and dialogue, with dialogue being preferable, since it is a communication mode that seeks to understand another person's perspective without judgement and without trying to further a position (discussion) or win an argument (debate) (Maxwell et al. 2011). The method's approach is intended to teach communicators how to value other perspectives; this does not imply that one must adopt or agree with the other person's perspective. In the Culture Box activity, students are asked to create a 'box' that contains three physical objects that represent their cultural identity. This is paired with learning the many aspects of cultural identity recognised through social categories such as race, gender, religion or belief system. The box can take any form that best represents a students' unique identity. For one student, this meant putting their items inside of a pizza box to represent their college lifestyle. During the activity, students share the stories behind their objects, and respond to questions from their peers. In the debrief following this activity, students unanimously express increased feelings of connection to their peers and appreciation for the various identities represented. This leads to a more in-depth analysis of the ways in which personal connections can be made with those from seemingly different identities (by asking questions, finding commonality, relating to an emotional response) which leads to valuing diverse perspectives. This is an important skill for empathetic communication in healthcare, given the diversity of patients who seek medical care.

Empathize: 'Forum Theatre' and 'Image Theatre' are techniques that are used throughout the course as methods of rehearsing the practical, embodied skills of empathetic communication, which apply the definition of empathy as a two-stage process involving both recognition and the reflection of affective state (Coll et al. 2017). In traditionally structured simulations, students engage in the action of the scenario, and then receive feedback on their performance during the debrief. However, this single-stage process limits the students' ability to practice embodied skills that create behavioural changes over time. Within theatrical performance, it is widely recognised that rehearsal is a critical means of learning embodied action. In traditional theatre models, rehearsals are often orchestrated by a director who ensures a unifying process from which a performance is presented to a spectating audience. Forum Theatre uses a different process that favours an ensemble model of collective creation and employs an interactive performance format that seeks direct audience engagement through interaction. By utilising a Forum Theatre method, diverse voices and perspectives can be introduced. This offers an expanded view of the simulation scenario. Using this method within the course, students begin by creating scenarios within small groups based on their lived experiences. These are later expanded to generate additional scenarios of interactions they expect to encounter in their professional careers. Students then use 'Image Theatre' to create an embodied tableau of the scenario. In this method, students produce still images using their bodies to represent the scene. By freezing the scene, other students are able to offer specific feedback on observed non-verbal cues represented in the image. After analysis of the non-verbal cues, students perform their scenarios with a SP (the SP is portrayed by other students during class sessions and by trained actors during the assessment simulations). During these embodied simulations, students call out 'stop' to halt the action and make direct interventions in that moment. This is a method specific to Forum Theatre and offers students the opportunity to make immediate changes to their verbal and non-verbal communication. The use of this rehearsal method has resulted in students' increased ability to make adjustments in the moment and emphasises the recognition and reflection stages of empathy through the L.O.V.E.™ approach to empathetic communication.

Conclusion

The Performance of Caring course has yielded overwhelmingly positive responses from students with each iteration. One student writes, 'This is one of the most challenging yet rewarding classes I have ever taken and worth all of the money I spent on tuition' (DeSoto-Jackson 2017b, n.p.). Upon completion of the course, students have self-reported increased confidence, reduced performance anxiety in simulations, the development of practice-based empathetic communication skills and a heightened awareness of verbal and non-verbal behaviours. The importance of this course for healthcare professionals is encapsulated in one senior nursing student's response; 'I wish I had taken this course earlier since the skills I have learned in this class have made me better at communicating with patients. This should be a required course for all nursing majors' (DeSoto-Jackson 2016, n.p.).

These innovative methods, applied within the framework of cognitive science, are intended to increase empathetic communication skills for healthcare professionals, which will lead to greater patient safety and quality of care. As we continue to explore the use of Applied

Theatre approaches, we anticipate that the course will continue to evolve. Our next step is to generate empirical evidence of the efficacy of DeSoto-Jackson's L.O.V.E.™ model. We are also developing a condensed model of this training that can be used to train healthcare professionals who are currently in the field. The potential benefits of this approach are supported by psychologists Thirioux, Birault and Jaafari, who address the issue of burnout in physicians: 'Burnout is a multidimensional work-related syndrome that is characterized by emotional exhaustion, depersonalization – or cynicism – and diminution of personal accomplishment' (2016, n.p.). They point out that while burnout and empathy are closely linked, the nature of the relationship between the two remains poorly understood, in part because of a lack of distinction between empathy and sympathy. They propose that 'clarifying the link between burnout, empathy and sympathy would enable developing specific training in medical students and continuous professional formation in senior physicians and would potentially contribute to the prevention of burnout in medical care' (ibid.). We hope that the approach we have outlined in this chapter can contribute to the development of this type of training for healthcare professionals and prompt further research in embodied empathetic communication.

References

Allegheny Health Network. 2017. 'Standardized and simulated patient program.' Accessed October 15, 2017. www.ahn.org/education/star-center-standardized-patient-program/patients.

Association of Standardized Patient Educators. 2017. 'About ASPE.' Accessed October 15, 2017. www.aspeducators.org/about-aspe.

Back, Anthony L., Robert Arnold, James A. Tulsky, Walter F. Baile, and Kelly A. Fryer-Edwards. 2003. 'Teaching communication skills to medical oncology fellows.' *Journal of Clinical Oncology*, 21, no. 12: 2433–6. doi:10.1200/JCO.2003.09.073.

Bas-Sarmiento, Pilar, Martina Fernández-Gutiérrez, María Baena-Baños, and Jose Manuel Romero-Sánchez. 2017. 'Efficacy of empathy training in nursing student: A quasi-experimental study.' *Nurse Education Today*, 59: 59–65.

Boal, Augusto. 1985. *Theatre of the Oppressed*. New York: Theatre Communications Group.

Boal, Augusto. 2002. *Games for Actors and Non-Actors*. 2nd edition. Translated by Adrian Jackson. New York: Routledge.

Braadbart, L., H. de Grauw, D. I. Perret, G. D. Waiter, and J. H. Williams. 2014. 'The shared neural basis of empathy and facial imitation accuracy.' *Neuroimage*, 84: 367–75. doi:10.1016/j.neuroimage.2013.08.061. Epub 2013 September 3.

Cantrell, Mary Ann. 2008. 'The importance of debriefing in clinical simulations.' *Clinical Simulation in Nursing*, 4, no. 2: 19–23.

Coll, Michel-Pierre, Essi Viding, Markus Rütgen, Giorgia Silani, Claus Lamm, Caroline Catmur and Geoffrey Bird. 'Are we really measuring empathy? Proposal for a new measurement framework.' Neuroscience & Biobehavioral Reviews. 2017 Dec; 83:132–39. doi.org/10.1016/j.neubiorev.2017.10.009

Charon, R. 2001. 'Narrative medicine: A model for empathy, reflection, profession and trust.' *JAMA*, 286: 1897–902.

Clark, Andy. 2016. *Surfing Uncertainty: Prediction, Action and the Embodied Mind*. New York: Oxford University Press.

Decker, Sharon, Mary Fey, Stephanie Sideras, Sandra Caballero, Leland Rockstraw, Teri Boese, Ashley Franklin, Donna Gloe, Lori Lioce, Carol Sando, Coleen Meakim, and Jimmie Borum. 2013. 'Standards of best practice: Simulation standard VI: The debriefing process.' *Clinical Simulation in Nursing*, 9, no. 6: 26–29.

DeSoto-Jackson, Rachel. 2016. 'The performance of caring-fall 2016.' Unpublished notes. Indiana University of Pennsylvania.

DeSoto-Jackson, Rachel. 2017a. 'The performance of caring-spring 2017.' Unpublished notes. Indiana University of Pennsylvania.

DeSoto-Jackson, Rachel. 2017b. 'The performance of caring-fall 2017.' Unpublished notes. Indiana University of Pennsylvania.

Dessel, A. 2008. 'Intergroup dialogue.' In A. Gitterman and R. Salmon (Eds.), *Encyclopedia of Social Work with Groups*. Binghamton, NY: Routledge.

Gallese, Vittorio. 2001. 'The "Shared Manifold" hypothesis: From mirror neurons to empathy.' *Journal Consciousness Studies*, 8: 33–50.

Gallese, Vittorio. 2003a. 'The manifold nature of interpersonal relations: The quest for a common mechanism.' *Philosophical Transactions of the Royal Society B*, 358: 517–28. doi:10.1098/rstb.2002.1234.

Gallese, Vittorio. 2003b. 'The roots of empathy: The shared manifold hypothesis and the neural basis of intersubjectivity.' *Psychopathology* 36: 171–80. doi:10.1159/000072786.

Gallese, Vittorio. 2005a. 'Embodied simulation: From neurons to phenomenal experience.' *Phenomenology and the Cognitive Sciences*, 4: 23–48. doi:10.1007/s11097-005-4737-z.

Gallese, Vittorio. 2005b. '"Being like me": Self-other identity, mirror neurons and empathy.' In S. Hurley and N. Chater (Eds.), *Perspectives on Imitation: From Cognitive Neuroscience to Social Science*, vol. 1, pp. 101–118. Cambridge, MA: MIT Press.

Gallese, Vittorio. 2007. 'Before and below 'theory of mind': Embodied simulation and the neural correlates of social cognition.' *Philosophical Transaction of the Royal Society B*,362: 659–69. doi:10.1098/rstb.2006.2002. Published online 13 February 2007.

Gracía-Pérez, Maria-Louisa, and Marta Gil-Lacruz. 2017. 'The impact of a continuing training program on the perceived improvement of quality health care delivered by health care professionals.' *Evaluation and Program Planning*, September, 2017.

Herbek, Thomas A., and Yammarino, Francis J. 1990. 'Empathy training for hospital staff nurses.' *Group Organization Management*, 15, no. 3: 279–95.

Institute of Medicine. 2010. *The future of nursing – Focus on education*. Washington DC: The National Academies of Sciences, Engineering, Medicine. Accessed October 3, 2017. www.nationalacademies.org.

Kemp, Rick. 2012. *Embodied Acting: What neuroscience tells us about performance*. Abingdon and New York: Routledge.

Landy, Robert, and David T. Montgomery. 2012. *Theatre for Change: Education, Social Action and Therapy*. Palgrave Macmillan.

Levinson, Wendy, Debra L. Roter, John P. Mullooly, Valerie T. Dull, Richard M. Frankel. 1997. 'Physician-patient communication. The relationship with malpractice claims among primary care physicians and surgeons.' *JAMA*, 277, no. 7: 553–59.

Lewis, Karen, Carrie Bohnert, Wendy Gammon, Henrike Hölzer, Lorraine Lyman, Cathy Smith, Tonya Thompson, Amelia Wallace, and Gayle Gliva-McConvey. 2017. 'The Association of Standardized Patient Educators (ASPE) Standards of Best Practice (SOBP).' *Advances in Simulation*, 2, no. 10.

Maxwell, Kelly E., Roger B. Fisher, Monita C. Thompson, and Charlie Beling. 2011. In Kelly E. Maxwell, Biren (Ratnesh) A. Nagda, and Monita C. Thompson (Eds.), *Facilitating Intergroup Dialogues: Bridging Differences, Catalyzing Change*. Sterling, VA: Stylus Publishing, Inc.

McIntosh, Peggy. 2003. 'White privilege: Unpacking the invisible knapsack.' In S. Plous (Ed.), *Understanding Prejudice and Discrimination*, pp. 191–6. New York: McGraw-Hill.

Mercer, Stewart, and William Reynolds. 2002. 'Empathy and quality care.' *The British Journal of General Practice: The Journal of the Royal College of General Practitioners*, 52: S9–12. doi:10.1007/0-387-33608-7.

OSCE. 2015. 'Clinical skills assessment medical OSCE exam.' Accessed March 9, 2018. www.oscehome.com/What_is_Objective-Structured-Clinical-Examination_OSCE.html.

Palaganas, J. C., C. Eppsand, and D. B. Raemer. 2014. 'A history of simulation-enhanced interprofessional education." *Journal of Interprofessional Care*, 28, no. 2: 110–15.

Prendergast, Monica, and Juliana Saxton. 2010. *Applied Theatre: International Case Studies and Challenges for Practice*. Intellect ltd.

Prentki, Tim, and Sheila Preston. 2008. *The Applied Theatre Reader*. New York: Routledge.

Perera, Jennifer, Joachim Perera, Juriah Abdullah, and Nagarajah Lee. 2009. 'Training simulated patients: Evaluation of a training approach using self-assessment and peer/tutor feedback to improve performance.' *BMC Medical Education* 2009, 9, no. 37: 1–8.

Petrucci, Cristina, Carmen La Cerra, Federica Aloisio, Paola Montanari, and Loreto Lancia. 2016. 'Empathy in health professional students: A comparative cross-sectional study.' *Nurse Education Today*, 41: 1–5.

Rakel, David P., Theresa J. Hoeft, Bruce P. Barrett, Betty A. Chewning, Benjamin M. Craig, and Min Niu. 2009. 'Practitioner empathy and the duration of the common cold.' *Family Medicine,* 41, no. 7: 494–501.

Rohd, Michael. 1998. *Theatre for Community Conflict and Dialogue: The Hope Is Vital Training Manual.* Heinemann Drama.

Roter, D., S. Stewart, N. Putnam, and M. Lipkin. 1997. 'Communication patterns of primary care physicians.' *JAMA*, 277: 350–6.

Sawyer, R. K., and Stacy De Sutter. 2009. 'Distributed creativity: How collective creations emerge from collaboration.' *Psychology of Aesthetics, Creativity, and the Arts*, 3, no. 2: 81–92.

Sawyer, Taylor, Walter Eppich, Marissa Brett-Fleegler, Vincent Grant, and Adam Cheng. 2016. 'More than one way to debrief: A critical review of healthcare simulation debriefing methods.' *Simulation in Healthcare*, 11, no. 11: 209–17.

Suchman, A., K. Markakis, H. Beckman, and R. Frankel. 1997. 'A model of empathic communication in the medical interview.' *JAMA*, 277: 678–82.

Taylor, Philip. 2003. *Applied Theatre: Creating Transformative Encounters in the Community.* Heinemann Drama.

Thirioux, B., F. Birault, and N. Jaafari. 2016. 'Empathy is a protective factor of burnout in physicians: New neuro-phenomenological hypotheses regarding empathy and sympathy in care relationship.' *Frontiers in Psychology.* doi:10.3389/fpsyg.2016.00763.

Thomas, Cynthia, Evelyn Bertram, and Doreen Johnson. 2009. 'The SBAR communication technique: Teaching nursing students professional communication skills.' *Nurse Educator*, 34, no. 4: 176–80.

Titzer J. L., C. F. Swenty, W. G. Hoehn. 2012. 'An interprofessional simulation promoting collaboration and problem solving among nursing and allied health students.' *Clinical Simulation in Nursing,* 8: 325–33.

Vanderford, M., T. Stein, R. Sheeler, and S. Skochelak. 2001. 'Communication challenges for experienced clinicians: Topics for an advanced communication curriculum.' *Health Communications,* 13: 261–84.

Weiner, Michael, and Paul Biondich. 2006. 'The influence of information technology on patient-physician relationships.' *Journal of General Internal Medicine.* doi:10.1111/j.1525-1497.2006.00307.x.

Xu, Xiaojing, Xiangyu Zuo, Xiaoying Wang, and Shihui Han. 2009. 'Do you feel my pain? Racial group membership modulates empathic neural responses.' *The Journal of Neuroscience*, 29, no. 26: 8525–9.

Zhang, Chao, Sarah Thompson, Connie Miller. 2011. 'A review of simulation-based interprofessional education.' *Clinical Simulation in Nursing*, 7, no. 4: e117–e126.

28
AWARENESS PERFORMING
Practice to protocol

Experience Bryon

Situated awareness

Situated Awareness (SA) is said to have been identified during World War I by Oswald Boelke to help predict the strategy and position of the enemy. The field of aviation fervently took it up first during the 1980s, resulting in a flurry of research and scholarship that laid much of the groundwork still in operation. Among those who currently use it are nuclear power plant operators, fire fighters, certain police units, nurses, medical emergency responders and military command personnel. It is employed by those who 'must ascertain the critical features in widely varying situations to determine the best course of action' since 'Inaccurate or incomplete SA in these environments can lead to devastating loss of life' (Endsley 1995, 32–33).

This chapter offers a summary of SA and surveys key contributions from cognitive science that problematise the status of an agent's 'mind' in current SA theory. It then draws from performance theory to question the positioning of the agent/operator's mind as a *custodian* of knowledge, introducing the notion of an *active middle field* whereby awareness occurs as *practice*. In the tradition of social science literature, through theory and hypothesis, this chapter identifies knowledge from performance practice, selecting transferable principles that could potentially be valuable for SA. A translational proposition drawing from the *Integrative Performance Practice* is shared as a way of exploring *modes* of awareness, as *practice*, towards the possibility of more effective SA.

SA has multiple definitions. Pew and Mavor, working within applications of SA to military simulations, because of its recognised linkage to effective combat decision-making in the tactical environment, offer a good rundown of many of the approaches, starting with a simple one: 'the up-to-the-minute cognizance or awareness required to operate equipment, or maintain a system' (Adams et al. 1995, 85). The definition of SA by one of the most prolific theorists on the subject, generally considered the most comprehensive, is as follows: 'the perception of the elements in the environment within a volume of time and space, the comprehension of their meaning and the projection of their status in the near future' (Endsley & Garland 2000; Endsley 1988, 1995, 1999, 2001). Another, taking into account the possibility of subjective 'mental models of reality,' says that SA can be viewed 'as one important function mechanism of *reflective-orientational activity,* which provides a conscious and dynamic orientation in the situation' (Bedny & Meister 1999).

Smith and Hancock, claiming a more ecological approach, describe SA 'as adaptive externally directed consciousness. [it is] the invariant in the agent-environment system that generates the momentary knowledge and behaviour required to attain the goals specified by an arbiter of performance in the environment.' Smith and Hancock stress that 'SA is not in the agent but exists in the invariant interaction of the agent and his or her environment' (1995, 145–6).

Stanton, Chambers and Piggott's analysis of these three approaches is that Endsley describes an 'information processing framework,' Bedny and Meister a 'reflective quality' or an 'activity approach' and Smith, Hancock and Gilson offer an 'embedded world view' (Stanton et al. 2001). They argue that these three approaches evidence a dualistic framing, rendering SA either a *product* or a *process,* with the product approach as 'the resultant state of situational awareness in the mind of the human operator' and the process as 'the acts of acquiring situational awareness by the human operator.' Endsley embraces this product/process distinction and states that 'situated assessment, as an active process of seeking information from the environment, is defined separately from situation awareness as the resultant of that process' (2000, 16).

Adams, Tenny, Pew and Gill further elaborate on the product/process distinction, exploring the psychological components of SA. '[*P*]*roduct* refers to the state of awareness with respect to the information and knowledge. Whereas *process* refers to the various perceptual and cognitive activities involved in constructing, updating, and revising the state of awareness' (1995, 88). This is important, because where *product* represents a 'continually updated schema' and *process* a 'continuous sampling of the environment' (Salmon et al. 2008), in both of these we get a distinct subject (agent/operator) – object (task/action) dualism. This dualism is problematic in that it often puts the onus of SA in the *mind* of the operating agent. Further, it often neglects to fully consider the *ways* in which the *practice(s)* of that agent/operator works across the agent/operator – task/action divide.

SA failure

This chapter posits that the *ways* in which agents *practice* the *acts* of conditioning, training and executions of SA is important, since a compromised *practice* of awareness has the potential to adversely affect even the most strategically designed supporting protocols, plans and applications of technology.

In examining language around error types, one can see how many aspects of failure may be connected to the *way* one engages with SA. I have italicised the aspects of failure that can be attributed to *processual* elements and thus could also benefit from possible *practice(s)* that would enable and condition the agent towards more optimum performance of the processes of perception, comprehension and prediction.

Endsley's 'error taxonomy' includes:

Level 1; **Failure to correctly perceive information**: data not available, data hard to discriminate or detect, *failure to monitor or observe data, misperception of data,* and *memory loss.*

Level 2; **Failure to correctly integrate or comprehend information**: *lack of or poor mental model, use of incorrect mental model, over-reliance on default values*, and other.

Level 3; **Failure to project future actions or state of the system**: *lack of, or poor mental model, over projection of current trends*, and other.

Other general considerations: *failure to maintain multiple goals* and *habitual schema.*

(Endsley 1999, 268 emphasis added)

SA failures include compromised or conflicted 'shared mental models,' 'attentional narrowing,' where information and tasks can be neglected (Endsley 1997, 272), 'automaticity' within the habitual schema, where people can 'fall into the habit of executing a task in a habitual way, making them less receptive to incoming environmental cues where information that is outside the routinized sequence is not attended to' (Endsley 2001, 15, 1997, 270), and data and technology overload. To understand the vocabularies used around this, a study on air traffic controllers illuminates the usage of the terminologies 'If the person misread or misheard the information, the error was assigned to the "perception" category. If the person was distracted or not attentive, the error was assigned to the "attention" category. If the person could not recall, or confused, the information then the error was assigned to the "memory" category. Finally, if there was poor interpretation, poor understanding, poor judgement, poor reasoning, or poor planning then the error was assigned to "decision making"' (Durso in Stanton et al. 2001, 12).

Crucially, what defines the agent/operator who is successful is a somewhat romantic notion of 'the right stuff,' which implies a 'complete and natural adherence to task goals and to criteria for performance' (Smith et al. 1995, 139):

> To possess SA – to have the right stuff – the agent must have developed a level of *adaptive capability* sufficient to match the specification of task goals and the criteria for assessing performance variables.
>
> *(ibid, 139 emphasis added)*

Typically, however, advice for training in SA often boils down to 'how to employ a system to best achieve SA (when to look for what where), appropriate scan patterns, or techniques for making the most of limited information' and supplements to traditional technology-oriented training 'that concentrates mainly on the mechanics of how a system operates' (Endsley 1999, 273).

While there are studies dedicated to making technology more user-friendly (Rahman 2012; Rahman et al. 2012), and studies towards helpful models such as PAYE (plan-as-you-execute) which aim to improve the OODA (observe-orient-decide-act) protocols necessary under conditions of a NTA (non-traditional adversaries) (Boyd, 1987 in Ntuen 2006), until we shift our understanding towards awareness as *active*, rather than as an object residing within an agent's mind, we will have an incomplete picture of SA theory. Further, many of these tools fail to directly address the agent/operator's aforementioned 'adaptive capability.'

Mind, knowledge and cognition in SA

Endsley asserts that 'true situation awareness only *exists in the mind of the human operator*. Therefore, presenting a ton of data will do no good unless it is successfully transmitted, absorbed and assimilated in a timely manner by the human to form situation awareness' (Endsley 2001, 4 emphasis added). This can bring up more questions than it answers, especially when one looks at how awareness *operates* in relation to *mind*. Croft et al. identify that Endsley's account relies heavily on an operator's 'explicit conscious knowledge' (2004, 82), and in doing so fails to account for the *implicit*. For them, implicit memory 'is defined as the non-intentional, non-conscious retrieval of previously acquired information, and is demonstrated by performance of tasks that do not require conscious recollection of past experiences' (2004, 84). Implicit learning 'is implicit when we acquire new information without intending to do so, and in such a way that the resulting knowledge is difficult to express' (Berry & Dienes in Croft et al. 2004, 82). They successfully argue that if the aims of SA measurements are to assess what an operator knows and what knowledge is available to them in the act of complex tasks under unpredictable situations, then 'future research efforts will need to concentrate on developing implicit techniques to supplement existing measures' (Croft et al. 2004, 99).

Another cognition approach to the 'mind' problem draws attention to the relationships between selective attention and short-term memory (STM), sometimes referred to as working memory. This becomes particularly important in situations when prediction and anticipation are required. Banbury et al., in adopting a 'cognitive streaming account of situated awareness' (see Jones 1993; Jones et al. 1996), conclude that it is the *processing of information,* rather than the storage of information, that speaks to effectiveness, and further that skill level also changes things. 'STM capacity was critical for novice pilots, while memory skill was important for expert pilots' (Banbury et al. 2004, 119). When we take into account that much of the research around anticipation connects *pattern matching* with the unfolding experience, we get somewhere interesting. Exactly *how* the *practice* of pattern matching occurs, and what the ramifications for memory and action are, is still being explored; however, they agree that unconscious, or *implicit,* memory contributes to the operator's overall knowledge (ibid 131) and that cognitive streaming includes a certain skill of *connecting* things through *threads* between what they call *pointers.* 'Rehearsal is by the use of threads to trace back through the pointers and objects (i.e., the cognitive stream). By retracing the cognitive stream, the pointers between the objects are rehearsed and not the objects themselves' (ibid 125).

And returning to the question of whether SA resides in the mind, the technology-led view, particularly prominent in defense domains, conversely proposes that 'SA is imbued in the technology and displays somehow,' suggesting that 'SA is in the device, not the person' (Stanton et al. 2010). Stanton and his colleagues posit that SA is an emergent property of a system. Drawing from Hutchins' theory of distributed cognition (1995a, 1995b), they conclude that 'viewing the system as a whole, it does not matter if humans or technology own this information, just that the right information is activated and passed to the right agent at the right time' (Stanton et al. 2010, 34). Here, the agent can mean either a human and/or non-human actor.

In these examples we see a leaning towards embodied cognition, a research area that performance practice has been increasingly employing to explore the dynamics of knowledge, meaning and action (see: McConachie & Hart 2006; Kemp 2012; Shaughnessy 2012; Bryon et al. 2017).

Embodied approaches to cognition propose that 'thinking and behavior are properties of the whole human organism, not the brain alone, and that body, brain and cognition are "situated" – engaged with the surrounding environment' (Kemp 2017, 95). When the brain, body and world are viewed as interdependent and often even as constituting each other, the term 'situated' can be 'used to draw attention to the fact that the body is embedded in and dynamically interacting with an environment' (Shapiro 2010, 124). Varela, Thompson and Rosch, pioneers of the theory, emphasise 'the growing conviction that cognition is not the representation of a pre-given world by a pre-given mind but is rather the enactment of a world and a mind on the basis of a history of the variety of actions that a being in the world performs' (2017, 9). Andy Clark's work on the notion of predictive processing suggests that [brains] like ours:

> are predictive engines, constantly trying to guess at the structure and shape of incoming sensory array. Such brains are incessantly pro-active, restlessly seeking to generate the sensory data for themselves using the incoming signal (in a surprising inversion of much traditional wisdom) mostly as a means of checking and correcting their best top-down guessing.
>
> *(Clark 2016, 3)*

Clark emphasises that this 'involves learning to predict our own evolving sensory states – states that are responding both to the body-in-action and to the world' (Clark 2016, xv). This leads to questions about *how* the *performance* of reception, perception, reflection and action *operates* in *practice*.

Performance theory and practice

These questions are addressed by the analytic conceptual model of the *active aesthetic,* the key premise behind the Integrative Performance Practice (IPP) (Bryon 2014). IPP was developed to allow optimal ability while acting, singing/sounding and dancing/moving, providing one practice from which plural aesthetic outcomes could emerge simultaneously, without compromising the technical excellence of any one genre. IPP differs from many other performance practices in that its primary function is to minimise the double or triple tasking that occurs when working across techniques and disciplines. This could be useful for SA, as multi-tasking is disorienting, stressful and can compromise the effectiveness of discrete tasks. IPP is currently taught as part of performance curricula in schools/conservatoires throughout the UK, USA and Australia. It prioritises the practitioner (for the purposes of this translational proposition, the SA operator), and works with *awareness* in specific ways to best integrate chosen aspects of one's technique, instrument, task and environment.

The analytic and conceptual framework of the *active aesthetic*, as currently employed in IPP and beyond, is used to measure and/or capture knowledge as it engages across disciplines (for examples see Campos 2013; Scott 2016; Moran 2017; Kaufman 2017; Bryon 2018). It proposes that knowledge, information and meaning are *active,* and as such, within a performance/event/operation, do not reside in a codification, plan or protocol (for SA) or text, choreography, blocking or score (for performance). Knowledge, information and meaning *occur* in *ways* of doing – in the *practice* of a practice. It is a matter of *manner* rather than of method. The *active aesthetic* offers a way to 'witness or evaluate the *way* of *doing* in process, the qualities within the *activation* of *generating* a *process*' (Bryon 2014, 61).

A tool for delineating a moment or event, to witness its *active aesthetic*, is the metaphor of a *middle field*, which helps to collapse the aforementioned subject (agent/operator) – object (task/action) dualism. The notion of the *middle field* posits that a performer/agent does not do performance, but rather does something (performing) from which the performer – subject – (agent/operator) and the performance object – (task/action) *emerge*. A dancer does not dance, a singer does not sing, a pilot does not fly, a soldier does not fight/protect, but rather they do things, are done onto and interact with things in fields of dancing, singing, flying, fighting/protecting, from which they as practitioner/operator and task/action emerge, often simultaneously. The status of the agent/operator in this sense aligns with cognitive neuroscience's concept of enactivism.

> The agent is embodied and dynamically-coupled to the world in which she is embedded; thus, agent, world and action are necessarily intricately interwoven, and the agent's body, experience, action, and world shape the way in which she deals with her everyday pragmatic concerns. There is an inseparability of mind and world, and it is embodied practice rather than cognitive deliberation that marks the agent's engagement with its world.
>
> *(Stuart 2008, 256)*

This *middle field* is not linear and encompasses some things that are not of our bodies and some things that are. It also encompasses many aspects that are not in the realm of representation.

What lives within this *middle field* can include felt senses, highly technical physical patterns of learnt behaviour, muscles groups firing in exact ways, surprises, productive tensions, conscious/explicit and unconscious/implicit actions and knowledge, STM and long-term memory, the room, the floor, technologies, one's nervous system, one's history, the day of the week, muscle and also emotional memory…and so much else. It's there to help bracket occurrences within an *event*, be that an entire performance/operation or a single dance move, or step in weapon assemblage.

To think of *awareness* as a *practice*, we can consider how, rather than being a measurable object, it lives in *activation*. Here the agent/operator and/or task/action simultaneously emerge from this dynamic rather than serve as custodians of awareness. The 'right stuff' may just become accessible through carefully chosen *practices* of awareness derived from performance.

Awareness as practice

Before applying 'technique' or 'skill sets' to texts, blocking, choreography or any pre-given representative sequence (which for the purposes of this application could be likened to a drill, plan or protocol), practitioners learn to engage in various *modes* of awareness as part of their conditioning towards skills training. In performance, these often operate parallel to performance technique and other operational skills, as *ways of doing,* the actions of tasks.

Awareness is a slippery word. It can mean modes or sites of consciousness and/or spheres of knowing, including perception awareness, motor awareness, awareness of action, mindfulness as awareness, pre-reflective awareness and qualities of states of being or feeling as an embodied self. In IPP awareness is active, and includes exterocepetive, proprioceptive and interoceptive sensory dynamics.

As expected from any practice called 'integrative,' there are crossovers to be found with other performance practices; however, as a comprehensive compare and contrast of performance practices is beyond the scope of this chapter, as are full descriptions of the many exercises which sit within the categories mentioned later, I will speak primarily from IPP to share specific *modes* of awareness that could aid in the enhancement of 'adaptive ability,' in addition to possibly decreasing the previously noted *failures* in SA such as 'attentional narrowing,' 'automaticity,' 'over-reliance on default values,' 'lack of, or poor mental models,' 'failure to maintain multiple goals and habitual schema' and 'compromised or conflicted shared mental models.'

Physiology of awareness

Those who rely on SA often work in conditions of emergency, and with emergency comes stress and fear. Theories of stress observe that with stress comes a 'perceptual tunneling' and 'cognitive tunneling' (Ritter et al. 2017), along with 'perceptual distortions and biases and the inability to process symbolic information and alter[ed] motor abilities' (Rahman et al. 2012, 358). Although performers do not generally work in conditions of emergency, they do have in common with emergency workers that they need to be exacting under conditions of stress while multi-tasking and dealing with external and internal stimuli.

In IPP, practitioners condition their bodies to initiate *every* movement, task and breath from a very low point of centre, at the perineum, which resides at the bottom of the pelvic floor (Bryon 2014, 97–165). When done correctly, this supports an exacting engagement with the iliopsoas muscle system (psoas), which in turn allows the release of superfluous tensions and an economical use of the skeletal, muscular and fascial systems (Koch 1997).

Practitioners learn this in coordination with a three-part breathing practice that engages the perineum on the exhale and releases it on the inhale. The feeling is that of length, balance, poise and readiness for action. Time seems to slow down. After a few weeks of training, practitioners find that responses are faster and clearer, and memory of texts and blocking/choreography is more readily available, even when nervous (or in fear). The reason for this is that the perineal contraction in line with the three-part breath affects not only the physicality of the agent/performer but also their entire psychophysical-emotional system, including how they perceive and receive the environment. The psoas, the fight or flight centre of the body, directly connects to the parasympathetic and sympathetic nervous systems. The practice of perineal contraction is also found in yoga, called *moola banda*; however, it differs slightly in IPP in that it is not a lock but rather a directional contraction upwards from the pelvic floor on the exhale. It is known to regulate nervous activity, 'lowering the breathing rate, heart rate, blood pressure and stabilizing the brain waves,' helping to stimulate 'the unconscious mind – so that suppressed mental energy is allowed to surface into conscious awareness' (Buddhananda 2000, 36–38). The practice of the three-part breath in line with a conditioning of the psoas system for agents of SA could help to minimise stress responses and compromises to reception, perception and decision-making. As it was specifically developed to work across disciplines, it does not compete but rather integrates across knowledge platforms and various physical conditionings. It is the primary supporting mechanism that makes all the following modes of awareness accessible.

Agential awareness

In IPP, awareness is a *practice* rather than a 'state' as often articulated in theories of SA. Awareness is a dynamic from which all tasks are executed. IPP practitioners begin by distinguishing between thoughts, feelings and awareness, engaging in progressively challenging mental and physical exercises designed to reveal that while (agents) *have* feelings and *have* thoughts, they are *not* those thoughts or feelings. Those familiar with Buddhist teachings or 'mindfulness' practices will find the language familiar. However, in IPP, the goal is NOT peace, or therapy, but rather the harnessing of a productive tension supporting an exactitude of action.

In a practice of *witnessing* thoughts and feelings rather than judging, diagnosing, boxing or acting on them, practitioners render these less 'sticky' and less able to influence or highjack a task. O'Regan, Myin and Noë's articulation of the phenomenon of *grabbiness* correlates nicely: 'Grabbiness is the fact that sensory stimulation can grab your attention away from what you were previously doing' (2004). In IPP, counter to many defense and sports practices, thoughts and feelings are not seen as *interfering* or *blocking* a *focused* action. This departs from Green and Gallwey's oft-cited equation: P(Performance) = p(potential) – I(interference) (1987, 23). Rather, interference is *witnessed* and integrated.

Montero's findings concur, revealing 'that in expert-level [activities] thinking is generally better than not thinking' (2011, 2012, 2015). With IPP, to come from *awareness* allows the *self* to know that one is thinking/feeling/sensing, and can allow the agent/operator *choices* regarding the role that certain mental processes might take in the act of executing a task while in the flow of doing.

Further, a marked distinction between *focus* and *awareness* is important. Focus, more aligned with hypervigilance and alertness, differs from awareness, which is more peripheral but also directed. Practitioners learn that if they *focus* on a left foot, for instance, it is directed, yes, but the harder one does it, the more proprioceptive and kinesthetic spheres tend to diminish, tensions arise, breathing shallows and attention narrows. In more challenging tasks, feeling and thoughts are more likely to 'stick.' Drawing one's *awareness* to an object (or even a task) allows 'for an attending to that object while having access to the entire environment and psycho/physical-vocal-emotive self' (Bryon 2014, 70).

In this way, an agent's practice of being aware that they are aware is important, as is the execution of tasks *from* this awareness. Agent/performers in this way can *act* from a self-space that is non-verbal, inclusive, receptive and directed. From the cognition perspective, one might say this mode of awareness lives in the realm of *pre-reflectivity*.

> I can, of course, reflect on and attend to my experience, I can make it the theme or object of my attention, but prior to reflecting on it, I wasn't 'mind- or self-blind'. The experience was already present to me, it was already something *for me,* and in that sense it counts as being *pre-reflectively* conscious.
>
> *(Gallagher & Zahavi 2008, 46 emphasis added)*

Legrand points out, 'Self-relative information is not information about the self, but information about the world relative to the self.' Further, crucially for SA, at 'the sensorimotor level, this self-relativity is given by the reciprocal modulation of perceptual afference and motor efference. This provides the basis for the functioning of the body schema' (2008, 513). Afference refers to the ways our nervous system receives and senses, while efference refers to motor function and how we act on stimuli. Together, they create a system of sensations, decisions and reactions.

A practice of returning to this mode of agential awareness could help with such diagnosed SA 'failures' as 'poor mental model,.' 'failure to monitor or observe data' and 'misperceptions of data.'

Agency and the task

In IPP, presence is determined in the act of doing. *Presence-ing* is non-trancelike; it is active, not passive; it is not fuzzy or pleasurable, nor team affirming and often not rewarding. It is distinct from an experience of *being in the flow.* There is no 'getting it' or working automatically.

Practitioners learn that there is no such thing as a good habit, but rather a doing of a doing *anew* each time. Musical scales, barre work or for SA repetitive exercises/drills are never executed by rote. A drill/exercise does not do a practitioner, but rather the practitioner engages in it, *from* awareness *with* awareness. This helps to avoid a tunnel vision, dissociative type of focus where a task can get hijacked by thoughts or feelings.

Practitioners observe that what one *actually* practices is what one *actually* gets good at, so if one practices 'trying to do something,' 'blocking out,' 'serving a collective,' 'repeating' or 'trying not to do something,' that is *exactly* what they will get good at. These activities are rarely the goal of the required task.

Like many performance trainings, IPP intersects with select Eastern meditative practices; however, IPP does not work towards achieving 'states' of presence, inner peace, acceptance or Zen, but rather towards establishing pathways of attending to chosen tasks.

IPP employs a modified version of the Vippassna Walking Meditation (Bryon 2014, 77). Evan Thompson, working from a combination of meditation, neurobiology and phenomenology, reveals the benefits of such a practice. In his description of attending to a task/target while *witnessing* other information, one can see how this might apply towards effective SA.

> ... there was less mental "clinging" or "sticking" to the first target, so that attention was open and ready for the second target, making it easier to detect. This reduction in mental clinging or stickiness was reflected in the EEG brain waves, which showed that fewer attentional processes were devoted to the first target after intensive meditation training, making more attentional processes available for the second target. Furthermore, the individuals who showed the largest decrease over time in the neural activity they required to detect the first target also showed the greatest improvement in detecting the second target. Thus, a more efficient neural response to the first target seems to facilitate detecting the second one.
>
> *(Thompson 2015, 53)*

A practice of returning to this mode of *presence-ing,* or working from awareness in action, could help the agent/operator remain *adaptive* and at the same time more discerning of stimuli that they allow to influence a task.

Awareness with others

In thinking about collective SA, returning to the notion of distributed cognition theory is helpful. Drawing once again from Hutchins' seminal work on distributed cognition (1995a, 1995b), Salmon et al. propose that a Distributed Situation Awareness 'assume[s] that collaborative systems possess cognitive properties (such as SA) that are not part of individual cognition. No single member has the overall SA, rather it is distributed around the system' (2007, 413).

Orchestras, ensembles, choruses, companies, groups and troupes all engage in dynamics of awareness between selves and other. Recent enquiries into group dynamics and performance, specifically those of musicians, evoke terminology such as 'we-agency,' 'plural self-awareness,' 'collective agency,' 'we intentionality' and 'ensemble cohesion,' situating the phenomenon at the level of pre-reflexive embodiment (Høffding 2015; Schiavio & Høffding 2015; Saliceet al. 2017).

Where this work resonates most with IPP is in the discourse around the individual/group – intentionality/behaviour dynamic. Once again, following the enactive approach, which considers 'behaviour and intentions as two domains of the same embodied subjectivity...instantiating the self-regulative processes at the basis of the circular interplay with the environment (Chemero 2009), Schiavio and Høffding ask, 'Is it possible to describe the musicians' experience of playing together without positing fixed categories such as "goals," "attention," "mental states," and "communication?"' (2015, 381) With IPP, working through an *active aesthetic,* it is not only possible but imperative, and as such could offer new solutions to the difficulties around compromised or conflicted 'shared mental models' mentioned earlier with regard to SA. In IPP, awareness is both individual and collective simultaneously.

In IPP, since it was designed specifically to work across disciplines and skill sets, group work is specific in terms of its practices of awareness. Practitioners build incrementally from partners to groups *after* mastering the previously outlined modes of awareness. Group work in IPP expands the field of awareness; following the pre-reflexive awareness it moves beyond the self/body schema, while maintaining the integrity of *presence-ing* supported by the three-part breath with an exacting isolation of the psoas system as integral to the task. In a practice called 'Meeting in the Middle,' practitioners become aware of activities around and behind them, things out of sightlines, along with unplanned occurrences such as a missed step, a partner's late entry or a falling set piece. It is particularly useful when it takes more than one performer/operator to accurately execute a task, such as a catch in mid-air, or a tricky entry in a particular place in the score. The exercise involves 'manipulating the field of attention in relation to *another*; exploring the *crossover* of *contact*; finding a point of Meeting in the Middle; and cultivating a common contact of awareness that includes other and self' (Bryon 2014, 87).

Potentially useful for SA is that this practice is not about positive troupe or team building, but rather about engaging in a task with exactitude from agential awareness and in cooperation with others.

Conclusion

Whether one is a nuclear power plant operator, fire fighter, police officer, nurse, medical emergency responder or a military commander, the *manners* or *ways* in which one engages with SA differ. Further, even when two operators are in the same vocation and using similar plans and protocols or helpful models, such as the aforementioned PAYE or OODA, there is also a difference.

As Endsley indicated, 'situational awareness is influenced by task factors and individual factors,' explaining 'why two people faced with different task factors might arrive at different conclusions, as might people with different abilities, experience, and training' (Stanton & Chambers 2001, 5). This individual factor was explored in this chapter as a *symptom* of *practice* rather than as an inherent quality in the agent/operator.

After surveying SA theory and exposing the dangers of an agent/operator – task/action divide, this chapter introduced more recent theories from a cognitive perspective to expand the notion of awareness, countering the idea that awareness *resides* in the mind. Awareness as a dynamic, living between and in relationship to the environment, technology, group and body, was explored. Through a combination of performance theory and enactivism, the agent/operator and task/action were re-considered as *emergent properties* of a *practice* of *awareness* rather than a linear progression of *one* derived from the *other*. Finally, four transferable principles from performance practice were offered as select modes of awareness: *physiology of awareness*, *agential awareness*, *agency and the task* and *awareness with others*.

IPP, like any performance practice, can hardly be summarised in so short a contribution. What I hope is that in this brief introduction of transferable principles, I have made a good case for the consideration of awareness as a *practice* to enhance SA, to empower and protect agent/operators who often work in environments of emergency and to minimise the types of failures identified through SA theory. Importantly, incorporating select practises from performance that correlate with the four modes of awareness discussed could offer complementary tools that in no way compete with or compromise existing techniques and models of SA, but could radically enhance them.

References

Adams, M. J., Tenney, Y. J., and Pew, R. W. (1995). Situation awareness and the cognitive management of complex systems. *Human Factors*. *37*, 1. pp. 85–104. doi:10.1518/001872095779049462.

Banbury, S. P., Croft, D. G., Macken, W. J., and Jones, D. M. (2004). A cognitive streaming account of situation awareness. In S. Banbury & S. Tremblay (Eds.), *A cognitive approach to situation awareness: Theory and application* (pp. 117–34). Farnham: Ashgate.

Banbury, S., & Tremblay, S. (Eds.) (2004). *A cognitive approach to situation awareness: Theory and application*. Farnham: Ashgate.

Bedny, G., & Meister, D. (1999). Theory of activity and situation awareness. *International Journal of Cognitive Ergonomics*. *3*, 1. pp. 63–72. doi:10.1207/s15327566ijce0301_5.

Berry, D., & Dienes, Z. P. (1993). *Implicit learning: Theoretical and empirical issues*. Hove: Erlbaum.

Boyd, J. R. (1987). *A discourse on winning and losing*. Maxwell AFB: Air Defence University.

Bryon, E. (2014). *Integrative performance: Practice and theory for the interdisciplinary performer.* Abingdon: Routledge.

Bryon, E., Bishop, M., & McLaughlin, D. (Eds.). (2017). *Embodied cognition, acting and performance.* Connection Science. 29, 1.

Bryon, E. (Ed.) (2018). *Performing interdisciplinarity: Working across disciplinary boundaries through an active aesthetic.* Abingdon: Routledge.

Buddhananda, S. (2000). *Moola Bandha, the master key,* New Delhi: Yoga publications Trust.

Campos, L. (2013). *Action research project the application of practice as research pedagogical principles in traditionally designed performance modules.* Retrieved from http://theatrefutures.org.uk/centrefor-learningandteachingintheperformingarts/files/2014/02/Campos_Action_Research_2013.pdf.

Chemero, A. (2009). *Radical embodied cognitive science.* Cambridge, MA. MIT Press.

Clark, A. (2016). *Surfing uncertainty: Prediction, action, and the embodied mind.* Oxford: University Press.

Croft, D. G., Banbury, S. P., Butler, L. T., & Berry, D. C., (2004). The role of awareness in situation awareness. In S. Banbury & S. Tremblay (Eds.), *A cognitive approach to situation awareness: Theory and application.* (pp. 82–103). Farnham: Ashgate.

Durso, F. T., Truitt, T. R., Hackworth, C. A., Crutchfield, J. M., & Manning, C. A. (1998). En route operational errors and situational awareness. *The International Journal of Aviation Psychology.* 8, 2. pp. 177–94. doi:10.1207/s15327108ijap0802_6.

Endsley, M. R. (1988). October. Design and evaluation for situation awareness enhancement. In *Proceedings of the human factors and ergonomics society annual meeting* (Vol. 32, 2. pp. 97–101). Los Angeles, CA: Sage.

Endsley, M. R. (1995). Toward a theory of situation awareness in dynamic systems. *Human Factors: The Journal of the Human Factors and Ergonomics Society. 37,* 1. pp. 32–64. doi:10.1518/001872095 779049543.

Endsley, M. R. (1999). Situation awareness in aviation systems. In J. A. Wise, V. D. Hopkin, D. J. Garland (Eds.), *Handbook of aviation human factors* (4th ed., pp. 12–13 to pp. 12–18). Boca Raton, FL: Taylor and Francis.

Endsley, M. R., & Garland, D. J. (2000). Theoretical underpinnings of situation awareness: A critical review. In M. R. Endsley & D. J. Garland (Eds.), *Situation awareness analysis and measurement* (pp. 3–32). Mahwah, NJ: Erlbaum.

Endsley, M. R. (2001, November). Designing for situation awareness in complex systems. In *Proceedings of the Second International Workshop on symbiosis of humans, artifacts and environment.*

Gallagher, S., & Zahavi, D. (2008). *The phenomenological mind. An introduction to philosophy of mind and cognitive science.* Abingdon: Routledge.

Green, B., & Gallwey, W. T. (1987). *The inner game of music.* London: Pan Macmillan.

Høffding, S. (2015). *A phenomenology of expert musicianship.* PhD Thesis. Department of Philosophy. University of Copenhagen.

Hutchins, E. (1995a). *Cognition in the wild.* Cambridge, MA: MIT Press.

Hutchins, E., (1995b). *How a cockpit remembers its speeds. Cognitive Science. 19,* pp. 265–88. Retrieved from http://onlinelibrary.wiley.com/store/10.1207/s15516709cog1903_1/asset/s15516709cog1903_1. pdf;jsessionid=807DDA1451C02AAF2EC3E7CB4A774DF1.f03t02?v=1&t=j9edaiix&s= cf79d8582b3d98b07a52281cd92937359603e526.

Jones, D. M., Beaman, C. P., & Macken, W. J. (1996). The object-oriented episodic record model. In S. E. Gathercole (Ed.), *Models of short-term memory* (pp. 209–38). Hove: Erlbaum.

Jones, D. (1993). Objects, streams, and threads of auditory attention. In A. D. Baddeley & L. Weiskrantz (Eds.), *Attention: Selection, awareness, and control: A tribute to Donald Broadbent* (pp. 87–104). New York, NY: Clarendon Press/Oxford University Press.

Kaufman, J. M. (2017). Noise as queer dramaturgy: Towards a reflexive dramaturgy-as-research praxis in devised theatre for young audiences. *Arts Praxis.* Retrieved from http://steinhardt.nyu.edu/ scmsAdmin/media/users/cl1097/J_Kaufman_-_Noise_as_Queer_Dramaturgy.pdf.

Kemp, R. (2017). The embodied performance pedagogy of Jacques Lecoq. *Connection Science, Embodied Cognition, Acting and Performance. 29,* 1. pp. 94–105. doi:10.1080/09540091.2016.1233521.

Kemp, R. (2012). *Embodied acting: What neuroscience tells us about performance.* Abingdon and New York, NY: Routledge.

Koch, L. (1997). *The Psoas book.* Felton, CA: Guinea Pig Publications.

Legrand, D. P. M. (2007). Pre-reflective selfconsciousness: On being bodily in the world. *Janus Head, 9.* pp. 493–519. Retrieved from www.janushead.org/9-2/Legrand.pdf.

McConachie, B., & Hart, F. E. (Eds.) (2006). *Performance and cognition: Theatre studies and the cognitive turn*. Abingdon: Routledge.

Montero, B. G. (2011). Effortless bodily movement. *Philosophical Topics, 39*, 1. pp. 67–79. Retrieved from https://barbaramontero.files.wordpress.com/2012/09/effortless-bodily-movement.pdf.

Montero, B. (2012). Practice makes perfect: The effect of dance training on the aesthetic judge. *Phenomenology and the Cognitive Sciences. 11*, 1. pp. 59–68. doi:10.1007/s11097-011-9236-9.

Montero, B. G. (2015). Thinking in the zone: The expert mind in action. *The Southern Journal of Philosophy. 53*, S1. pp. 126–40. doi:10.1111/sjp.12119.

Moran, N. (2017). *The right light: Interviews with contemporary lighting designers*. London: Palgrave.

Ntuen, C. A. (2006). *Cognitive constructs and the sensemaking process*. North Carolina agriculture and technical State Univ Greensboro NC center for human machine studies. Retrieved from www.dtic.mil/get-tr-doc/pdf?AD=ADA463415.

O'Regan, J. K., Myin, E., & Noë, A. (2004). Towards an analytic phenomenology: The concepts of "bodiliness" and "grabbiness". In: A. Carsetti (Ed.), *Seeing, thinking and knowing*. Theory and Decision Library A. 38. Dordrecht: Springer.

Rahman, M. (2012), September. Direct perception-action coupling: A neo-gibsonian model for critical human-machine interactions under stress. In *Proceedings of the Human Factors and Ergonomics Society Annual Meeting* (Vol. 56(1). pp. 1401–5). Los Angeles, CA: SAGE Publications.

Rahman, M., Balakrishnan, G., & Bergin, T. (2012). Designing human–machine interfaces for naturalistic perceptions, decisions and actions occurring in emergency situations. Theoretical Issues in *Ergonomics Science, 13*, 3. pp. 358–79.

Ritter, F. E., Reifers, A. L., Klein, L. C., & Schoelles, M. J. (2007). Lessons from defining theories of stress for cognitive architectures. In W. D. Gray (Ed.), *Integrated models of cognitive systems* (Vol 1., pp. 254–62). New York, NY: Oxford University Press.

Salice, A., Høffding, S., & Gallagher, S. (2017). Putting plural self-awareness into practice: The phenomenology of expert musicianship. In *Topoi* (pp. 1–13), The Netherlands: Springer. doi:10.1007/s11245-017-9451-2.

Salmon, P. M., Stanton, N. A., Jenkins, D. P., Walker, G. H., Young, M. S. & Aujla, A. (2007) July. What really is going on? Review, critique and extension of situation awareness theory. In *International Conference on Engineering Psychology and Cognitive Ergonomics* (pp. 407–16). Berlin, Heidelberg: Springer.

Schiavio, A., & Høffding, S. (2015). Playing together without communicating? A pre-reflective and enactive account of joint musical performance. *Musicae Scientiae. 19*, 4. pp. 366–88. doi:org/10.1177/1029864915593333.

Scott, J., 2016. 'Affective encounters': live intermedial spaces in sites of trauma. *Research in Drama Education: The Journal of Applied Theatre and Performance. 21*, 3. pp. 332–36.

Shapiro, L. (2010). *Embodied cognition*. Abingdon: Routledge.

Shaughnessy, N. (2012). *Applying performance: Live art, socially engaged theatre and affective practice*. London: Palgrave.

Smith, K., & Hancock, P. A. (1995). Situation awareness is adaptive, externally directed consciousness. *Human Factors: The Journal of the Human Factors and Ergonomics Society. 37*, 1. pp. 137–148. doi:10.1518/001872095779049444.

Stanton, N. A., Chambers, P. R., & Piggott, J. (2001). Situational awareness and safety. *Safety Science. 39*, 3. pp. 189–204. doi:10.1016/S0925-7535(01)00010-8.

Stanton, N. A., Salmon, P. M., Walker, G. H., & Jenkins, D. P. (2010). Is situation awareness all in the mind? Theoretical Issues in *Ergonomics Science. 11*, 1–2. pp. 29–40. doi:10.1080/14639220903009938.

Stuart, S. A. (2008). From agency to apperception: Through kinaesthesia to cognition and creation. *Ethics and Information Technology, 10*, 4. pp. 255–64. DOI:10.1007/s10676-008-9175-5.

Thompson, E. (2015). *Waking, dreaming, being: Self and consciousness in neuroscience, meditation, and philosophy*. New York, NY: Columbia University Press.

Varela, F. J., Thompson, E., & Rosch, E. (2017). *The embodied mind: Cognitive science and human experience*. Cambridge, MA: MIT Press.

29

IMAGINING THE ECOLOGIES OF AUTISM[1]

Melissa Trimingham and Nicola Shaughnessy

Autism is a prevalent theme in both literary and cognitive studies. There are numerous fictional narratives of the condition, many of which are written by non-autistic authors as well as autobiographical accounts of the experience, poetry by autistic writers, television, film and theatre representations (Grandin 1995; Hacking 2010; Semino 2014). The condition is also discussed in connection with key topics in cognitive science that include theory of mind, empathy, attention, intersubjectivity and emotion. This chapter draws upon insights from cognitive neuroscience, performance theory and practice, as well as our personal and professional experience of living and working with autistic people in order to better understand the characteristics of the autistic imagination.

Autism and the imagination

Disability studies in the West have challenged the prevalent discourses on autism that stress deficits in need of remediation or cure, emerging from the medical model that underpins diagnosis. These new thinkers, often autistic themselves, demonstrate that autistic people are, in fact, 'neurodiverse' and communicate in distinctive ways that 'neuro-typicals' may not understand. Whilst social imagination has been identified as an area of difficulty in diagnostic criteria, there is increasing recognition of imagination as an area of 'difference,' as is evident in Bruce Mills's account:

> With the increased incidence of autism and the insights arising from autists' self-reporting and artistic work...we might begin to re-think past paradigms that oppose typical/normal with atypical/abnormal creative processes. In the continuum that marks the different cognitive processes that produce 'art' we might begin to refine an understanding of the imagination in relation to autism...The nature of play- and its symbolic and imaginative dimensions- might vary in relation to the particular manner in which the 'player' processes the world.
>
> *(Mills 2005, 131)*

Indeed, research by Bonnie Evans (2013) draws attention to changing definitions of autism through the history of diagnosis and the shift from excess to deficit in descriptions of the imagination in autism.[2] A more nuanced and variable picture of imagination in autism is now emerging, with research on autistic creativity and studies of artistic self-expression, particularly in visual art, poetry and music (Roth 2008; Happé and Frith 2010; Shaughnessy 2013). However, more work is needed to understand and engage with the different languages of communication within the autistic population (see, for example, Savarese 2015, and the testimony of artist Amanda Balls[3]). Moreover, there has been recognition recently that autistic people physically perceive the world in ways that neurotypicals do not. Differently inflected senses can derive both intense pleasure and fear from objects, material form, light and sound. Naoki Higashida, in *The Reason I Jump: one boy's voice from the silence of autism,* describes this: 'Sometimes I actually pity you for not being able to see the beauty of the world in the same way we do. Really, our vision of the world can be incredible, just incredible' (2013, no pagination). The 'beautiful otherness of the autistic mind' (Happé and Frith 2010) can be infused with pain and confusion. As Higashida describes it, ' the detail…claims our attention, and then our hearts kind of drown in it, and we can't concentrate on anything else.' Higashida describes sensory confusion: 'It's as if my limbs are a mermaid's rubbery tail'; and when noise surrounds him, 'it feels as if the ground is shaking and the landscape around us starts coming to get us, and it's absolutely terrifying.' This confusion comes about because autistic individuals may have 'hyper-' (over) or 'hypo-' (under) sensitive modalities of touch, hearing, sight, smell and taste (Bogdashina 2003). They also have an eye for detail that is far more sensitive than neurotypical perception, leading to joy in repetition and fascination with tiny details ('stimming'). As Happé and Frith point out, 'repetition is not repetition…if you have expert levels of discrimination' (2010, xvii).

The authors of this chapter have themselves undergone a journey of changing understanding through our lived experience of bringing up our own autistic children, and also through our work as theatre researchers. Using creative, practical and embodied performance approaches, informed by cognitive neuroscience and in collaboration with psychologists, we have discovered more about the characteristics of the autistic imagination, realising how rich and full that inner life might be, and becoming sensitive to the problems that everyday encounters can cause for this population. This interdisciplinary research takes place in the context of an increasing recognition of the potential of the arts as a key to exploring thinking and feeling states (perception, engagement, emotion), and the encouragement of individual agency in those with autism in order to improve health and well-being. Psychologists are using creative methods as research tools, acknowledging the value of arts practices for developmental science: 'a current wave of methodologically rigorous studies shows the depth of arts learning, as well as how arts engagement can be harnessed for transfer to other skills' (Goldstein, Lerner and Winner 2016, 1505). Hence, it is stated that 'developmental psychologists cannot afford to ignore such a developmentally entwined experience as the arts' (1505).

Imagining Autism: a novel approach

The case study for this chapter, 'Imagining Autism Drama, Performance and Intermediality as Interventions for Autistic Spectrum Conditions' arose in the context of collaboration between the arts and sciences.[4] The use of the term 'intervention' in the title is itself a contested term between the communities of practice engaged with the project. As Scott Kaufman says, 'perhaps the problem isn't with the interventions but with the whole model that presupposes a normal' (2017 http://behavioralscientist.org/rethinking-autism-social-awkwardness-social-creativity/). Matthew Lerner, a psychologist who uses drama as a research tool (Socio-Dramatic-Autism Research Interventions), has developed skills-based methods using drama to enhance social communication and social interaction in working with students with Asperger syndrome. Lerner suggests a potential paradigm shift:

> ...what emerges from the research is that rather than think about the task of understanding social cognition of people on the spectrum and also how to treat them, teaching them to behave in a way that better approximates the way other people are doing it, it becomes more incumbent on us to understand people on the spectrum, and meet them where they are, and really try to understand how they are experiencing a situation. What are the skills, features, tendencies, passions they are bringing to the table, and how can we use those to make this social world they seem to want to access? How can we make it accessible and rewarding?
>
> *(Lerner in Kaufman 2017, no pagination)*

Crucially, Lerner suggests, in a statement indicative of the shift in autism research from a medical to a social model of disability: 'Instead of viewing people with ASD as "socially awkward" individuals who need to be "fixed," we should instead conceptualize them as *socially creative*. They may not do things the "right" way, but they do them their way' (Lerner in Kaufman 2017, no pagination).

Imagining Autism was a project defined in the context of the field of applied and socially engaged theatre, referring to theatre adapted for work in social, educational and community contexts intended to have a therapeutic or educational purpose. Within this field, the terms of reference are also being contested in relation to the 'social turn' (Nicholson 2011; Shaughnessy 2013). In this case, the original aim was to use drama as a means of producing efficacious and beneficial effects for autistic children through their participation in activities that engaged them socially, physically and creatively. Psychologists and theatre practitioners worked together, initially with psychologists evaluating the work, and theatre practitioners designing and delivering the exercises. As the work developed, however, the research design was adapted, moving towards a 'transdisciplinary' practice, engaging bi-lingual vocabularies whereby the disciplinary perspectives informed and impacted on each other to create new approaches (Shaughnessy 2017b). The team worked in three schools with three groups of six to eight children, aged 7–11, with a diagnosis of autism. The intervention involved participants in weekly drama sessions (45 minutes) in a portable tent (described as the 'pod'). Each week the participating children experienced a different themed visual and sensory immersive environment: forest, underwater, outer space, the Arctic and under the city. The environments offered imaginary worlds presenting a magical multisensory playground of light, colour, sound and moving images, where trained practitioner/performers interacted with the children either as themselves (as friendly guides) or using puppets and masked characters; they conjured storms (sound, light and a soft plastic sheet rising and falling lightly over the children), sunny woodlands (birds singing and dappled light) or a moon-scape (ultra violet light). They shared a loose narrative with the children (structured as a journey), encouraging speech, movement and creative expression (including humour, building on autistic children's interests in slapstick). Whilst the pod functioned as a container, the 'way out' was always clear, and practitioners could follow a child into the school hall and play there if preferred. There were several features distinguishing our approach as 'novel,' a term used by the psychologists within the project and by subsequent commentators responding to its outcomes (O'Sullivan 2015). The project may be described as a 'counterpoint to mimesis'; its methods draw upon contemporary performance practices and 'post-dramatic' paradigms which are non-illusionist and in which practitioners and materials draw attention to the performative nature of the activity. This means the work is based on a particular mode of pretence with participants and practitioners aware of themselves as co-creators of fictional constructs. This contrasts with drama practices based on social scripts and skills-based learning. It also flattens the hierarchies between performance creators, facilitating practitioners and participants. The pod is a space in which we were able to work together through a process of mutual discovery. The awareness of the 'not real' also involved an 'in the moment' mode of engagement, requiring spontaneity to improvise, following the children's cues without recourse to scripts or habits as a process of 'becoming' in the making of meaning. The environments also stimulated a high level of sensory arousal, in contrast to the low arousal learning environments recommended for teaching autistic children. Whilst this caused some concern with teachers about whether the participants would cope with this level of sensory stimulation, the positive outcomes indicated the potential value of short-term exposure to highly stimulating sensory approaches.

As drama specialists and applied theatre practitioners, we might be charged with 'applying' drama to a perceived 'problem' (the autistic inability to communicate); interacting with the children as the other, from our neurotypical cognitive base; and intending to bring about an improvement in the children's 'condition,' which psychologists would test and prove.[5] We suggest that the project's sub-title 'Drama, Performance and Intermediality as Interventions for Autistic Spectrum Conditions' reveals the quantum shift in our own understanding since the small pilot project in 2010 when we first began to build our immersive and interactive environments – the calming green forest, white Arctic landscape and blue underwater worlds. The word 'interventions' was troubling for us at the time, but the psychologists we were working with persuaded us that this was the terminology recognised in research contexts for the participatory performance practices we were planning. The intervention, however, has stimulated a transformation of understanding, affecting the researchers as much as the researched.

The holistic process that subsequently took place inside – and outside – the pod or performance space was a changing ecology of material form, people and affect that changed child, practitioner, teacher, sibling and parent. Vibrant, responsive materials and objects in our environments changed all whom they touched physically – and those whom they touched more indirectly. The autistic community has its own culture that meshed inextricably into the neurotypical cultures that we as practitioners and researchers brought into the pod; far from this being a discrete aesthetic experience contained within the pod, wider cultures of education and the family were similarly enmeshed and permanently recast. The sensory and interactive elements were designed to help participants develop felt understanding through experiential, physical and immersive media. The training methods emphasised the importance of play, turn-taking, liveness, open space, physicality, improvisation, shared attention, responding to the other, reading non-verbal cues and working as an ensemble. The participatory and process-based approaches emphasised autonomy and authorship, offering a license to play creatively (the importance of play is an aspect which is often overlooked post-diagnosis). The psychologists established proof of concept that the methods positively impacted upon language, social interaction, empathy and imagination (Beadle Brown et al. 2017). Although the intervention was of a low intensity and duration (45 minutes per week for one term), statistically significant changes were recorded in several areas of deficit and across the spectrum. The biggest changes were in reciprocal social interaction, emotion recognition and the severity of autistic symptoms as rated by parents and teaching staff. However, significant improvements were also found for at least some of the children in socialisation, communication, imagination and play, with at least some children from all three schools showing improvements in at least one area. The research collaboration also led to unexpected insights into the imagination in autistic children, demonstrating how it is differently inflected from the typically developing child, particularly in terms of visual and auditory perceptual processes, awareness of time and space, physical and verbal modes of creative expression and responses to objects and interactive media.

A developing body of work is now building evidence of the effectiveness of drama as a means of engaging with and improving outcomes for autistic people (O'Sullivan 2015). Examples include the UK-based actor and director Kelly Hunter's 'Hunter Heartbeat Method,' which uses Shakespeare's language to interact with autistic children across the spectrum, an approach that has been positively evaluated by Ohio State University (www.kellyhunter.co.uk/ohio.php). This is also a sensory and play-based method, emphasising creativity and being 'in the moment' as co-producers of participatory performance: 'These games are derived from Shakespeare's poetic exploration of how it feels to be alive, specifically through his obsession with the eyes and the mind and with reason and love; how we see, think and feel, which forms the spine of his poetry throughout the whole canon' (http://kellyhunter.co.uk/shakespeares-heartbeat/the-hunter-heartbeat-method/).

Cognitive perspectives

Because engaging in practical drama activities benefits participants, a key term in connection with drama and autism is experiential. This learning through doing is an embodied practice, and the cognitive turn in theatre and performance was one of the core theoretical perspectives informing the development of the Imagining Autism approach (McConachie and Hart 2006; Lutterbie 2011; Kemp 2012). Cognitive theory spoke to our experience as drama practitioners and as parents; we knew that using the multimodalities of participatory performance and play-based approaches (working with puppets and objects), physical interaction, auditory stimuli (soundscapes, musical triggers, microphones) and visual and haptic materials (e.g. costumes, hats, torches) could elicit communication, shared attention and sustained interaction. Participatory performance (with interacting auditory, bodily, temporal and spatial elements) is a means of accessing cognitive processing, helping us to understand how autistic children perceive and connect to the world around them. Play-based practices within multisensory creative environments are a scaffolding for meaning-making and learning. The approach can be conceived in terms of cognitive ecologies, through its engagement with three ecologies, those of the environment, social relations and human subjectivities. The idea of ecologies, encapsulated in the title of this chapter, as interconnected landscapes of cognition, is taken from Edward Hutchins (1995), but Trimingham has also used this metaphor in connection with new materialist interpretations of the pod environment in Imagining Autism (Trimingham 2017). The ecological framework is also the conceptual underpinning for participatory performance praxis (Shaughnessy 2012; Harpin and Nicholson 2017): Harpin and Nicholson use terms that resonate with the practice-based research methodologies of Imagining Autism, referring to participatory performance as 'affective encounters, [that] bring together the sentient with the spatial and environmental' in a 'dance between the affective agency of environments, social relations and subjectivities' (7). Trimingham's background in professional theatre and interests in phenomenology complemented Shaughnessy's work on language, gender identity and autobiographical performance. Shaughnessy's initial interest in mirror neurons via Rizzolatti (2009) was in the context of her son's diagnosis, leading to her earliest writing on theatre, cognition and affective neuroscience, whilst Trimingham came to this body of work through a perceptual route in her studies of

Merleau-Ponty in relation to the Bauhaus. Significantly, modernism was also a shared interest for both authors as this was to become a feature of what we conceptualise as an aesthetics of neurodiversity (Shaughnessy 2017a).

As our work has developed and in dialogue with other theatre and performance scholars, we use '4 E' cognition as a conceptual framework. Rhonda Blair and Amy Cook (2016) offer a particularly useful summary of this model which provides a way of encompassing some key aspects of how to connect performance and theatre to cognitive science. Cognition, they explain, can usefully be seen as embodied, embedded, extended and enacted, each stage claiming slightly more for the imbrication of selfhood and environment.

Embodied – cognition isn't separable from our physicality.

Embedded – cognition depends heavily on off-loading cognitive work and taking advantage of affordances, or potentials, in the environment.

Extended – cognition extends beyond the boundaries of the individual organism, encompassing aspects of the social, interpersonal environment.

Enacted – cognition is inseparable from action (and is often an outgrowth or even an attribute of action).

In Imagining Autism, we described our approach in terms that refer to the interactions between participants and the following core elements that map onto the 4 Es:

1 Physical (embodied) through non-verbal communication using movement and clowning, informed by Richard Hayhow's 'mimetics, a psycho-physical practice (Trowsdale and Heyhoe 2015), and Phoebe Caldwell's intensive interaction.[6]
2 Immersive (embedded) through the multisensory 'pod,' the tent-like container housing the themed scenic environments, with loose material elements acting as triggers or affordances.
3 Participatory (extended) through interaction between practitioners and peers using improvisation, turn-taking, copying, call and response.
4 Play (enacted) through the playing out (action) of loose narratives in which autistic children pretend they are experiencing the various journeys to the forest, arctic, outer space, under the sea and under the city.

However, there is a related feature at play here, situated within the 'inters' of the interdisciplinary, interactive and intermedial, a third space beyond dualisms in which there is shared understanding of cultures, languages, perspectives. Through drama, we enter into an empathic and relational understanding of how self and other are interdependent. As Harpin and Nicholson have summarised in their introduction to *Performance and Participation*, these practices are informed by an understanding of 'how different forms of participation are reshaping questions of agency as relational practices rather than individualised acts' (Harpin and Nicholson 2016, 13).

Fundamental to the project's approach has been a focus on empathic engagement with autism as difference and the use of drama as a means of tapping into the experience of perceiving differently. This is a basis for our training methods, the 'walking in another's shoes' exercises that draw upon our experience of working with the autistic community to enable professionals, families and carers to 'imagine' the experience of another through creative practice. Playing with puppets, finding your clown and experiencing the world from different perceptual perspectives are some of the practical approaches developed by the project team for training practitioners as well as being disseminated through workshops for teachers and care-workers. The team have worked with as many as 60–70 staff in a single workshop, exploring different ways of engaging with space, objects and people. In developing our practice-based training methods, we aimed to do three things:

- To develop performance training systems and vocabularies appropriate for practitioners to work with autistic people
- To free practitioners from habit, enabling them to respond in new and original ways to stimuli and to be open to play
- To engage in imaginative and empathic dialogue with autists.

Just as trained musicians can find jazz improvisation difficult, so the devising methods used in contemporary performance practice involve making material with new vocabularies that can be challenging to practitioners experienced in traditional approaches to training. In many respects, the practitioners needed to be self-abnegating, to free themselves to respond in new, non-typical ways to these encounters with difference.

Relational practices and new materialisms

For an autistic person whose senses are differently tuned to the world from neurotypicals, the pod was a space of improvisation, liberated amidst a wealth of different materials, where participants were free to play and reveal their intense sensory worlds to those of us willing to learn. Let us take Matthew, one of the participants at one of the schools, as a case study. He was cautious in his first encounter with the pod, running in and out of the forest in week 1, but playing happily with the mole character outside the pod in the school hall. Matthew had very limited speech when the project began; in week 2 (outer space), he discovered the microphone and began to play with his voice, exploring different registers (his voice was breaking) through a form of sound and word painting. By week 5, the Arctic, he had progressed rapidly:

> The microphone, we have speculated, changed Matthew's relationship to his voice, making him aware of its potential as an instrument for self-expression. His voice was breaking and he experimented with the different registers, playing with the sounds of words through a form of onomatopoeic sound painting that sounded like descriptive scene setting, even though the language was emergent as he produced a range of speech sounds rather than words, with a story telling intonation. We could trace a developing grammar as the changing intonation (rising and falling) of repeated singles words (most frequently "space") created holophrastic sentences. Matthew's demonstration of communicative intent (rarely seen before according to parents and teachers), and the sense making processes that are articulated through language give some insight into his cognitive and perceptual processing as well as his developing [...]awareness of those around him. Meaning is being constructed through the integration of visual, acoustic and bodily modalities. We see him thinking physically as he moves around the environment, in between the episodes on the microphone, using a torch and a moon rod puppet as an extension of his arms. On the microphone, his voice is the exploratory instrument and as the project progressed his language developed from a rhythmic rhyming rap in the Arctic setting to a poetic meditation in one of the final sessions, uttered from a cardboard tube that functioned as a pretend microphone in the underwater

> environment: 'Now I see the world, don't let it change the past.' This was adopted as the subtitle for the film, as Matthew became the project's performance poet.
>
> *(Shaughnessy 2017b, 499)*

Matthew also enjoyed wrapping himself up protectively in the blankets, covers and cloths in the pod; he was gradually coaxed out of these wraps by a mischievous puppet pulling at his blanket, a naughty husky dog running away with it and a friendly snowman taking his side and defending him against the 'thieves.' Together, Matthew and the snowmen chased them away. Through simple material means (puppets, blankets, a microphone) inside the pod, Matthew was led into interaction with others, so that outside the pod, in the classroom and especially at home, he began to speak. He made observations on the weather to his mother, for example, and commented that the car was driving unusually slowly in the snow ('Mummy, the car is *dead*!'), communicating to his mother and to his teachers for the first time that he was noticing the world around him.

In one school, the children we worked with were diagnosed with autism but were exceptionally able verbally. It was obvious, however, that they had problems relating to their peer group and making friends, whether this took the form of excessive withdrawal or over-exuberance and controlling behaviours. It was this group that showed the greatest gain in empathic skills following the project. Watching the filmed documentary footage of the sessions gives the clue as to why this might be so. These children were constantly playing together, for example, in the chaos of storms at sea, freezing to death in the Arctic (whether by bad weather or the evil of the Ice Queen), in the forest launching spontaneously into the game 'What's the Time, Mr Wolf?' (with fully costumed Mr Fox instead of a wolf) or protecting the woodpecker in the Arctic (who had been sadly blown off course when migrating to Africa!) from being eaten by a hungry Inuit. There was absolute freedom to play, even to the extent that Greg (noisy, excitable and often in rivalry with his peer John for control) burst a hole through the cloth roof of the pod and announced he could see a 'sabre toothed tiger!' There was not much left of this roof section by the end of the session, but there was no harm done, just the release of an extended fantasy that all the children then joined in. Later, Greg took over the job of protecting the woodpecker, holding him closely and taking him to places of safety, and Joseph, a very quiet child, in his session also took over this role. Two other children, before even entering the pod, dressed up as two polar bear cubs, 'blood brothers' (they announced), and they sustained this role throughout, both getting up to rough and loud mischief, but controlled from their worst excesses by the Inuit. In the other group, two children were also dressed up as mischievous furry animals throughout the piece; the Inuit called them 'puppies' and trained them by tempting them with fish.

A sense of humour pervaded much of the improvisation in every session of the pod, usually slapstick physical humour that often provoked loud laughter. The snowman, for example, lay 'sunning' himself on a lounger in the Arctic and when he got up, Joseph slipped into his chair. The snowman, of course, pretended not to notice, immediately covered Joseph with a white cloth – and 'sat' down. He sprang up in surprise, much to Joseph's delight; and of course, in true clown spirit, the snowman immediately repeated the joke several times over, to even more delighted laughter from Joseph. This is a space of improvisation, of learning through play and interaction, a space that is vital to cognitive development. These spaces of play and learning have been characterised by anthropologists Tim Ingold and Elizabeth Hallam as spaces whose aim is not to 'project future states, but to follow the paths along which such projections take shape' (Hallam and Ingold 2007, 15).

In the words of Katherine Hayles, 'Materiality, like the object itself, is not a pre-given entity but rather a dynamic process that changes as the focus of attention shifts' (Hayles 2012, 14). The focus of attention or consciousness is moreover only a fraction of neurotypical thought. The unconscious is a 'perceptive capacity that catches the abundant overflow too varied, rich and deep to make it through the bottleneck of attention' (ibid.). But the abundant overflow is not too varied, rich and deep to escape the attention of an autistic child whose eye for detail is often sharp and defined, and their 'bottleneck of attention' is often much larger than that of other children's. Researchers and practitioners in Imagining Autism were drawn into the shifting ecology of autism through the objects, materials and environments they inhabited temporarily alongside the autistic children. So began a 'cultural' exchange between the participants, that is the children and the practitioners. The 'abundant overflow' also began to seep out from the defined place of performance into the surround, that is, the '[s]paces produced through networks of social interactions' (ibid.), those spaces of classroom, school, family and community.

The impact of these relatively short but intense encounters is evident in the following extracts, which are taken from parental reports (interviews and questionnaires) by the psychologists in the post-project assessment which took place three to six months after the project finished. (All children's names have been changed.)

> He started saying things he never said before. I am flabbergasted by the amount of language. Every time I wrote down something was the day after Imagining Autism and it continues.... The big changes came from Imagining Autism and not from school. The biggest change is that he now comments – e.g. "medicine's empty". He loved the sessions. School said he would skip along the corridor to go there.... He has gained confidence and the ability to communicate more. He is now having a conversation. Before I would ask and get a minimal answer.

Another parent wrote:

> He said things like "car was taking alien eyes off", "bell was ringing the alien was crying" and started to make expressions on his face. He commented on feelings which he has never said about...For the first time in his life when he plays figures are talking to each other and he is making up a story. Imaginative play with toys is a breakthrough. He started to play with related toys after sessions e.g. space toys...He has gained in his imagination, he is talking more, commenting on everything. He is identifying emotions, and naming them. He gave me a kiss and a cuddle which is very rare. He is reasoning things out – we had a conversation for 15 minutes for the first time.

In this anti-Cartesian geography of play, cultures do not collide but they intermingle and imperceptibly transform. In the cultural and social context of the pod, performing objects, puppets and materiality itself served to cross boundaries and articulate new realities.

Participatory futures

Imagining Autism was developed at a time of change in autism awareness and diagnosis, with the autistic community increasingly giving voice, stressing the importance of community-based participatory research models, advocating research that has relevance to the lived experience of autistic children. Whilst autism research outputs have doubled in the twenty-first century, with similar increases in funding, the priority has been 'basic science,' with only 5% of research reported to have been dedicated to support and education (Pellicano 2017). As stated in *A Future Made Together* (2013), for the most part 'advances in research fail to impact upon those who need them most: autistic people, their parents and carers and those who help support them' (Pellicano, Dinsmore and Charman 2013, 4). Historically, decision-making in research has rarely involved the autism community.

The importance of engaging autistic artists as actors in media representations has been foregrounded through productions such as the National Theatre's stage version of Mark Haddon's novel *The Curious Incident of the Dog in the Night-time* (first performed 2012). Engaging autistic practitioners in the design and delivery of Imagining Autism has been critical to the project's development. A series of outreach events in educational and community settings have been informed and supported by a developing network of autistic advocates. At a 2017 residency at The Atkinson, a leading arts centre in Southport, UK, the team were joined by an autistic self-advocate and practitioner, Annette Foster, who immeasurably enriched both the encounters with participants and the training offered to parents and teachers. Her experience concludes this chapter.

The future for Imagining Autism is imbricated in these communities of change. Karan Barad's terms 'intra-action' or 'complex manifold of connections' (Barad 2007, 388) can be used to describe both the 'pod' environment of the 4 Es in action and the tangled debates within the autistic community of which we are necessarily a part. In our understanding of autistic children, we have travelled, we hope, a very long way from our earliest interdisciplinary endeavours. We hope that the project, and those outside it, will be able to respond flexibly and imaginatively to the challenges that lie ahead, as understanding of autistic people as different, imaginative and with their own strengths continues to grow, enriching and transforming an increasingly neurodiverse world.

We conclude with the words of Annette Foster, a live artist whose PhD explores autism and gender identity through performance.[7] As an autistic artist associated with the project, she offers her perspective on Imagining Autism as an advocate of the autistic community, in dialogue with the authors of this chapter:

NS/MT: You have described the Imagining Autism pod as an 'autistic space.' As an autistic person, can you explain how you perceive the environment?

The Imagining Autism pod is an Autistic Space because the children who are invited into the pod are allowed to be their autistic selves. The oppressive nature of living in a neurotypical world is lifted for those 45 minutes they are in the sensorial immersive environments. They are encouraged to be themselves and relate to the world in their own way. No one tells them not to stim, not to touch that or not to sit there. They are free to explore the space from their own unique perspective and develop relationships with objects, puppets and people in the space. It is a safe, all-encompassing space, an opportunity to play without judgement. Practitioners are discouraged from correcting and directing and are trained to follow the children's direction of play. This might be by copying the words or sounds they make, by playing with objects in a certain way or encouraging an exploration of touch. The pod is a multiple sensorial space geared towards people who are drawn to sensory stimuli, for example, glow in the dark rocks, large space rocks covered in bubble wrap, a microphone, shadow puppets screen, video projection and various puppets. Characters in this space are verbal or non-verbal, encouraging children to communicate in whatever way they feel comfortable. The make-believe world becomes a space where autistic children can be free.

I was involved in Imagining Autism as a practitioner in 2017, and for two weeks I played the role of the alien, operating a Bunraku puppet. It felt oddly appropriate as I have always felt like an alien, having been diagnosed with Autism six years ago, at the age of 39. As the alien, I have large white eyes and a globular head that I move around a lot trying to reach the children in some way. They seem to like me. I touch the world with white scratchy bulbous fingers that are very good at pointing and waving. I have long stretchy legs that can extend to make me a tall alien (if I like), but most of the time I sit on the ground so that I am on the human children's level, as they are the ones I am most interested in. I hide in outer space, behind the shadow wall, in the glowing dark, waiting for them to arrive. I am nervous, scratching my head with my white gloved hands and checking that I have my moon rocks in place. I rearrange my spaceship, various planets and stars ready for take-off. I hear the children board the rocket ship and take off. I wait in anticipation with my sparkly bag in tow, in this dark extravaganza of an intergalactic cosmos filled with space rocks that I love to eat.

The human children arrive and I am excited as well as nervous about meeting them, wondering whether I will be able to relate to them, in this strange world we create together. I slowly emerge as the alien on the shadow screen, first waving at the children. My main mission on this planet is to find and eat rocks. I like the greeny yellow ones and hate the orange ones. I am obsessed and rope the children into finding rocks for me. If I don't like one I let them know with a loud 'Yuck.' I don't speak human only alien but most of the children understand my language fluently in a few minutes.

Although I've never done puppetry before, I've been a performance artist for the last 20 years and knew what was needed to do this work. I use my skills of being in the moment, taking cues from the children to interact, react, make mischief and be playful. I feel free to be my autistic self when I follow the children's lead. At parties, I was always the person drawn to children or animals and children rather than doing the 'small talk' with adults. Children are easier to read and also more genuine, they say what they mean and they do what they say. I like that. I find children much more open and accepting of difference than most adults. With most children, you are either instantly accepted for who you are or they will be direct with you and ask you questions about your difference, which I don't mind as once you answer their questions they accept it.

MT/NS: In your encounters with autistic participants when you worked as a practitioner on the project, do you feel your interactions were distinctive? How far did the project's training inform your approach and how far were you drawing upon personal experience to co-produce the activities?

I think my interactions with the autistic participants were unique. I am autistic, I relate to these children, I understand them in a way that possibly neurotypical people don't. This is instinctive. I see stimming as something that is needed to survive in a neurotypical world rather than something to be corrected. I understand from my own life experiences that autistic children are often misunderstood or not listened to. They are told that their way is wrong most of the time. Or at least this was my experience as an undiagnosed autistic child. Autistic people, especially late diagnosed people, are very good at copying, figuring out the rules of the neurotypical world. I understand intuitively how to connect with autistic people: stop, look, listen, pay attention to detail, copy and then, once you understand how, to, play with them, show you understand their world.

The project training informed my approach to an extent as it reminded me what I do instinctively as an autistic person and it allowed me to be myself during the training experience. I felt as if the training was quite freeing; people were training to take a journey into the autistic universes, the individual universes of autistic people, all unique and complicated, as all autistic people are different. The training gave validity and legitimacy to the autistic experience for the first time in my life. It was a very unusual experience to be with neuro-typical people learning about the autistic experience and how to communicate with autistic children. I felt I was in a room of people learning how to speak a language close to my own. I was being taught things I already knew instinctively, but perhaps, had never been articulated in this way. I was aware of my position as a spokesperson for the autistic community, even though I am only one perspective and can't speak for all. However, if I felt something didn't feel right, I could call it into question and help the team to understand what to tweak to make it more authentic or inclusive.

NS/MT: You have a professional background as an artist and as a performer; do you think that being an autistic person is a factor in your creativity?

I am an artist first and foremost. My work over the years has been an eclectic mix of performance, live art and visual art. It has always been autobiographical informed by feminism, identity, gender, sexuality and difference. However, for the last two years I've been working on a creative project as an autistic self-advocate, and this led to my PhD and performance practice becoming research as means of enquiry into and articulation of autistic identity. There's a misconception that Autistic people don't have imagination. As a late diagnosed autistic person who is also an artist, this is something I am challenging in all that I do! If you look at my website (nettypage.com), for example, it clearly shows I have an imagination and that I have built a whole career around my creativity as a multidisciplinary visual and performance artist. Autism researchers often refer to difficulties with 'social imagination.' I like to think of it as just a different social imagination from an autistic perspective, which is how you imagine other people and social situations. One of the problems with the stereotypes of autistic people lacking imagination, being maths and science wizards or male brained, is that it reinforces our obsession with binaries and categories: male/female, typical/atypical, normal/abnormal systemising/empathising. This doesn't take into account the complexities of neurodiversity. I resist being defined in terms of the medical model and want autistic people to be understood as different, rather than deficit. As a late diagnosed person on the spectrum, I have a unique experience of the world and this has been conveyed through my creative work throughout the years. So, I would say that I am an artist first, and then an autistic person.

Notes

1 Part of this chapter was first published in *Applied Theatre Crossings,* an annual journal in Chinese and English edited by Cariad Astles and Xiaoxin Wang.

2 The first reference to autism is attributed to the Swiss Psychologist Eugen Bleuler who used the term in 1911 in his study of schizophrenia to describe social isolation and withdrawal into a fantasy world (its etymology is from the Greek eaftismos, meaning 'self-enclosed'). Leo Kanner and Hans Asperger developed new terminologies and understanding of the condition. Both use the term autism but refer to different characteristics. Kanner's work is associated with what came to be regarded as 'classic' autism (in which individuals are profoundly affected by what was later defined as the 'triad of impairments' in communication, social interaction and social imagination). Asperger, however, describes the higher functioning manifestation of the condition, associated with individuals who may be highly verbal but have difficulties in social interaction due to what has been referred to as a 'systemising' brain. In the 1970s, Lorna Wing's revolutionary work replaced the black and white paradigms of Kanner and Asperger with the multi-coloured concept of the condition as a spectrum of behaviours and features. See Wing (1996) and Frith (1991).

3 See 'In my Language' by Amanda Balls, www.youtube.com/watch?v=JnylM1hI2jc.

4 Imagining Autism was funded by the UK's Arts and Humanities Research Council. Based at the University of Kent, the researchers were Professor Nicola Shaughnessy (Principal Investigator, Drama), Dr Melissa Trimingham, (Co-Investigator, Drama), Professor Julie Beadle-Brown (Co-investigator, Tizard Centre for Learning Disability) and Professor David Wilkinson (Co- investigator, Psychology).

5 Children were assessed using the Autism Diagnostic Observation Schedule (ADOS), Vineland Adaptive Behaviour (VABS), cognitive measures, parental and teacher ratings, observation, interviews and practitioner ratings. The results were summarised thus by the psychologists:

At the immediate follow-up, all 22 children improved on at least one measure. Of the 6 who improved on ADOS social interaction scores, 5 also improved on three or more other measures. Among the other measures, the most significant changes were seen in the number of facial expressions the children recognised. 4 out of six children maintained the changes in social interaction on the ADOS at 3 months post intervention with three showing maintenance at 9 months. All children maintained or showed increased changes in emotion recognition at follow up – between 5 months and 1 year after post intervention. The majority of children maintained these improvements at follow-up (between 5 and 12 months post intervention)

Julie Beadle Brown/David Wilkinson speaking at *Imagining Autism: Exploding the Myths*, 21 March 2014, held at The Gulbenkian Theatre, Canterbury.

6 See www.phoebecaldwell.co.uk and Caldwell film (2010), '*Autism and Intensive Interaction,*' Jessica Kingsley.

7 Annette Foster is a professional live artist undertaking a practice-based PhD at the University of Kent: 'Autism, Performance and Identity: Articulating Women and trans/nonbinary people experience of Autism through live art and performance.'

References

Barad, Karan (2007) *Meeting the Universe Halfway: Quantum Physics and the Entanglement of Matter and Meaning*. Durham, NC: Duke University Press.

Beadle-Brown, J. et al. (2017), 'Imagining Autism: Feasibility of a Drama Based Intervention on the Social, Communicative and Imaginative behaviour of Children with Autism.' *Autism: International Journal of Research and Practice*. September 1–13.

Blair, Rhonda and Amy Cook, eds (2016) *Theatre, Performance and Cognition: Languages, Bodies and Ecologies*. London: Bloomsbury Methuen.

Bogdashina, Olga (2003) *Sensory Perceptual Issues in Autism and Asperger Syndrome: Different Perceptual Worlds*. London: Jessica Kingsley.

Evans, Bonnie (2013) 'How autism became autism: the radical transformation of a central concept of child development in Britain.' *History of the Human Sciences*. 26.3. 3–31.

Frith, Uta, ed. (1991) *Autism and Asperger Syndrome*. Cambridge: Cambridge University Press.

Goldstein, Thalia R., Matthew Lerner and Ellen Winner (2017) 'The arts as a venue for developmental science: realizing a latent opportunity'. *Child Development*. 88.5. 1505–12.

Grandin, Temple (1995) *Thinking in Pictures: And Other Reports from My Life in Autism*. New York: Doubleday.

Hacking, Ian (2010) 'Autistic autobiography'. In *Autism and Talent*, edited by Francesca Happé and Uta Frith, 195–208. Oxford: Oxford University Press.

Hallam, Elizabeth and Tim Ingold, eds (2007) *Creativity and Cultural Improvisation*. Oxford and New York: Berg.

Happé, Francesca and Uta Frith, eds (2010) *Autism and Talent*. Oxford: Oxford University Press.

Harpin, Anna and Helen Nicholson, eds (2016) *Performance and Participation: Practices, Audiences, Politics*. London: Palgrave Macmillan.

Hayles, N. Katherine (2012) *How We Think: Digital Media and Contemporary Technologies*. Chicago, IL and London: Chicago University Press.

Higashida, Naoki (2013) *The Reason I Jump: One Boy's Voice from the Silence of Autism*. Trans. David Mitchell and Keiko Yoshida. London: Hodder and Stoughton.

Hutchins, Edwin (1995) *Cognition in the Wild*. Cambridge, MA: Massachusetts Institute of Technology.

Kaufman, Scott Barry (2017) '*Rethinking Autism: from social awkwardness to social creativity*' www.scottbarrykaufman.com/rethinking-autism-social awkwardness-social-creativity/.

Kemp, Rick (2012) *Embodied Acting: What Neuroscience Tells Us about Performance*. London: Routledge.

Lutterbie, John (2011) *Towards a General Theory of Acting*. Basingstoke: Palgrave Macmillan.

McConachie, Bruce and Lynda Hart, eds (2006) *Performance and the Cognitive Turn*. London: Routledge.

Mills, Bruce (2005) 'Autism and the imagination'. In *Autism and Representation*, edited by Mark Osteen, 118–32. London and New York: Routledge.

Nicholson, Helen (2011) *Theatre, Education and Performance: The Map and the Story*. London: Routledge.

O'Sullivan, Carmel (2015) 'Drama and Autism'. In *Encyclopaedia of Autism*, edited by R. F. Volkmar, 1–13. New York: Springer.

Pellicano, Elizabeth (2017) 'Knowing Autism'. *8th Annual Centre for Research in Autism and Education Lecture*, University of Cambridge, 26 September 2017.

Pellicano, Elizabeth, Adam Dinsmore and Tony Charman (2013) *A Future Made Together: shaping autism research in the UK*. London: Institute of Education.

Rizzolatti, Giacomo, Maddalena Fabbri-Destro and Luigi Catteneo, 2009 'Mirror Neurons and their clinical relevance.' *Nature Clinical Practice Neurology*, 5. 24–34.

Roth, Ilona (2008) *Imaginative Minds*. Proceedings of the British Academy, 167. Oxford: Oxford University Press.

Savarese, Ralph James (2015) 'What Some Autistics Can Teach Us About Poetry: A Neurocosmopolitan Approach.' In *The Oxford Handbook of Cognitive Literary Studies*, edited by Lisa Zunshine, 393–420, Oxford Handbooks, Oxford, England.

Semino, Elena (2014) 'Language, mind and autism in Mark Haddon's *The Curious Incident of the Dog in the Night-time*.' In *Linguistics and Literary Studies*, edited by Monika Fludernik and Daniel Jacob, 279–304. Berlin: de Gruyter.

Shaughnessy, Nicola (2012) *Applying Performance: Live Art, Socially Engaged Theatre and Affective Practice*. Basingstoke: Palgrave Macmillan.

Shaughnessy, Nicola (2013) 'Imagining otherwise: Autism, neuroaesthetics and contemporary performance.' *Interdisciplinary Science Reviews*. 38.4. 321–34.

Shaughnessy, Nicola (2017a) 'Do you see what I see? Art, science and evidence in autism research'. In *Applied Practice: Evidence and Impact in Music, Theatre and Art*, edited by Nick Roe and Matthew Reason, 80–94. London: Bloomsbury Methuen.

Shaughnessy, Nicola (2017b) 'Valuing performance: purposes at play in participatory theatre practice.' In *Artistic Citizenship: Artistry, Social Responsibility and Ethical Praxis*, edited by David Elliott, Marissa Silverman and Wayne Bowman, 480–509. New York: Oxford University Press.

Trimingham, Melissa (2017) 'Ecologies of Autism: vibrant space in Imagining Autism.' In *Scenography Expanded an Introduction to Contemporary Stage Design*, edited by Joslin McKinney and Scott Palmer, 183–96. London: Bloomsbury.

Trowsdale, Jo, and Richard Hayhow (2015) 'Psycho-physical theatre practice as embodied learning for young people with learning disabilities'. *International Journal of Inclusive Education*. 19.10. 1022–36.

Wing, Lorna (1996) *The Autistic Spectrum: A Guide for Parents and Professionals*. London: Robinson.

30

TOWARDS CONSILIENCE

Integrating performance history with the co-evolution of our species

Bruce McConachie

Two phenomena are 'consilient' when they 'jump together.' Biologist E.O. Wilson chose *Consilience: The Unity of Knowledge* as the title of his 1998 book because he is encouraging scholars to use rigorous empirical methods in order to create 'a common groundwork of explanation' that bridges the sciences and humanities (1998, 8). I am investigating potential areas of consilience between history and evolution because of my interest in the politics of climate change and resource depletion. Wilson, also an advocate for radical political reform to alter our increasingly dangerous future, understands that the unity of knowledge he seeks can directly improve democratic citizenship and governance. 'Most of the issues that vex humanity' – among which Wilson includes ethnic conflict, endemic poverty and the environment – 'cannot be solved without integrating knowledge from the natural sciences with that of the social sciences and humanities,' he states (13).

The chapters in this part of our book aim to 'translate' the knowledge learnt through our work in applying cognitive science to theatre and performance into other areas of useful endeavour; my chapter here is no exception. How to take what we know about political performances – both in political arenas and on the stage – and apply that evolutionary and cognitive knowledge to the challenges of global climate change is my ultimate goal. Of course, if we lived in Wilson's future of epistemological consilience, translation would not be necessary; we could all use the same language, at least at a foundational level, to analyse and describe what we need to understand from the arts, the sciences and the social sciences. In fact, in a consilient world of knowledge, we could eliminate Part IV altogether from this book. Although our various disciplinary epistemologies are still a long way from 'jumping together,' history and biology are closer now than they were 20 years ago when Wilson sounded his clarion call.

Behind Wilson's preference for empirical methodology is the epistemology of naturalism. 'Naturalism in philosophy,' states philosopher Peter Godfrey-Smith, 'requires that we begin our philosophical investigations from the standpoint provided by our best current scientific picture of human beings and their place in the universe.... The questions we try to answer, however, need not be derived from the sciences' (2003, 154). I will shortly outline the material reasons behind the theory of co-evolution and then use it as a framework to explain significant aspects of political and performance history. Naturalism will also guide my discussion of our ecological problems in the Anthropocene, the recent geological epoch we earthlings have entered. Following Godfrey-Smith, my essay will be based upon empirical science and then ask questions that require historical and social scientific answers.

First, though, we need to understand why history and science remained so far apart for most of the twentieth century. Carl Hempel's essay, 'The Function of General Laws in History,' was the first modern attempt to model historical causation on scientific, cause-and-effect explanations. Written in 1942, the essay sustained withering fire from several postwar historians in the West, whose chief objection was that historians' chief aim was to shape persuasive and plausible narratives about the past. Whether such narratives 'jumped with' Hempel's (or anybody else's) notion of general scientific laws was not a necessary criterion for their validity, said Hempel's opponents. Most historians accepted their criticisms and endorsed conventional notions of evidence, narrative and rhetoric over scientific explanation. In retrospect, like most of their generation, these historians helped to broaden the gulf between C.P. Snow's 'two cultures' – the sciences and the humanities – in the academy.

Some historians reconsidered the need for cause-and-effect explanations in the 1990s, however, after the excesses of the poststructuralist turn threatened to reduce good historical narratives to little more than subjective tales, garnished with a sprinkling of 'facts' (always in scare quotes) and served with rhetorical flourish. Objecting to the epistemological problems of poststructuralism, historian Thomas Haskell argued in 1998 that 'the time is ripe for second thoughts about causation' (14): 'Whatever else history may be, it cannot but be about the ways things come into being and go out of being, which is to say about cause and effect' (14). For Haskell, there was no necessary contradiction between 'beginning-middle-end' narrative and scientific explanation; narrative was simply 'an especially supple form of causal reasoning' (14). Haskell's insight with regard to historical explanation, now sustained by most cognitive linguists, began to close the gap between the two cultures in historiography.

One implicit claim behind Haskell's assertion was that some historical narratives might be better than others not just because their analysis of the evidence was more astute or because their narratives were more comprehensive. It could also be because the causal reasoning they contained was more attuned to human realities than the reasoning behind other narratives. Were there bedrock human realities that operated below the levels of culture and history? While working on *Cognition, Evolution, and Performance*, published in 2015, I began to suspect that the co-evolution of genetics and culture might provide such an approach to the answers.

Co-evolution and deep history

As it happens, E.O. Wilson was one of the first modern biologists to bring evolution and history closer together by arguing for co-evolution. Darwin had discussed the possibility of group evolution, but the notion fell into disfavour until 1981 when Wilson, with Charles Lumsden, published *Genes, Mind, and Culture*. Other biologists contributed models and ideas for co-evolution, and in 1985, Robert Boyd and Peter J. Richerson published the book that offered the most comprehensive explanation for the process, *Culture and the Evolutionary Process*. Based primarily on mathematical models that demonstrated how cultural processes might have worked with genetic changes to enhance the evolutionary success of some groups in our species, Boyd and Richerson's book provided substantial empirical and theoretical support for the probability of group, as distinct from individual evolution. They capped their success with *Not by Genes Alone: How Culture Transformed Human Evolution* (2005), which applied their theory directly to current knowledge about the evolution of *Homo sapiens*.

The co-evolutionary scenario for our species resulting from our ancestors' ability to pass on survivable traits through both genes and culture is fairly straightforward. By the time of the early Pleistocene epoch, probably around 2 million years ago, the genetic evolution of the *Homo* genus had outfitted hominins with psychological and behavioural adaptations that enabled them to practice the rudiments of culture and teach them to their children. Cultures of cooperation within some hominid bands began to flourish and played a significant role in their survival. Later, when similar enculturated groups grew numerous enough to begin competing for territory, those hominid bands with a more cooperative culture than other groups had an advantage in this competition. Because they could hunt and gather more effectively, they survived difficult times to produce more children. Over many generations, then, those genes that helped to shape a culture of cooperation in some hominid bands were expressed more often in the band's offspring. At the beginning of our co-evolution, genetics induced our ancestors to invent and propagate cooperative cultures. After several hundred thousand years, however, culture was in the driver's seat; it was putting pressure on our genes to expand our capabilities for learning more culture. The result was a dynamic feedback loop between hominin cultures and genes that substantially improved our intersubjective dynamics to shape what eventually became our ultrasocial species.

As Richerson and Boyd put it in *Not by Genes Alone*:

> [C]ulture must have influenced the reproductive success of our ancestors; otherwise, the features of our brain that make culture possible would not have evolved. The operational products of this evolution are innate predispositions and organic constraints that influence the ideas that we find attractive, the skills that we can learn, the emotions that we can experience, and the very way we see the world.
>
> *(2005, 13)*

Consequently, by the time of *H. heidelbergensis*, the common ancestor of both our species and the Neanderthals that flourished around 600,000 years ago, the confluence of genes, culture and the environment 'favored the evolution of a suite of new social instincts suited to life in such groups, including a psychology which 'expects' to be structured by moral norms...; new emotions such as shame and guilt... and a psychology which 'expects' the social world to be divided into symbolically marked groups' (214). And these attributes of ultrasociality would only increase over time.

Not by Genes Alone turns the usual either-or, nature-or-nurture question about behaviour into a both-and situation. 'To ask whether behavior is determined by genes or environment does not make sense,' say Richerson and Boyd. '*Every* bit of the behavior (or physiology or morphology, for that matter) of every single organism living on the face of the earth results from the interaction of genetic information stored in the developing organism and the properties of its environment' (ital. in original) (266–67). In this instance, Richerson and Boyd's 'environment' includes epigenetics (external constraints on genetic expression), foundational intersubjectivity (social behaviour not dependent upon symbolic communication) and symbolic activity (cultural relations dependent on symbols), the 'four dimensions' discussed in Eva Jablonka and M.J. Lamb's *Evolution in Four Dimensions: Genetic, Epigenetic, Behavioral, and Symbolic Variation in the History of Life* (2005). Because a significant and important part of the hominin 'environment' during the later stages of our evolution was social and symbolic, it should not be surprising that culture would eventually become a co-equal factor in our evolution.

The validity and persuasiveness of Boyd and Richerson's work in co-evolution has spawned a minor industry of new scholarship. In ethics and philosophy, humanists and scientists are asking new questions about cooperation, altruism and the origins of human morality. Two economists, Samuel Bowles and Herbert Gintis, explode the myth of *Homo economicus* from an evolutionary point of view in *A Cooperative Species*. Evolutionary anthropologist Joseph Henrich underlines the key role played by prosocial emotions and norms in *The Secret of Our Success*. In his *Ultrasociety: How 10,000 Years of War Made Humans the Greatest Cooperators on Earth*, anthropologist Peter Turchin comes to the surprising conclusion that man-unkind applied the lessons of cooperation to constitute ever larger forms of social organisation primarily to wage war successfully. Peter Richerson teamed up with a cultural psychologist in 2013 to co-edit *Cultural Evolution: Society, Technology, Language, and Religion*, an anthology of 20 essays that ranges broadly among the four areas of its subtitle to determine some of the major results of co-evolution in human cultures.

That the early cultures of our genus were important for our evolutionary success may be good news for anthropologists, economists, ethicists and others, but unless historians decide to investigate hominin bands during the Pleistocene epoch, co-evolution opens few obvious doors for them. Several historians, however, are now investigating what Daniel Lord Smail and others have called 'deep history.' Smail and Andrew Shryock published a multi-authored book in 2011, entitled *Deep History: The Architecture of Past and Present*, urging the investigation of history before the invention of writing. In this 2011 book, the co-editors laid out their historiographical project and its rationale for fully integrating deep history into conventional, 'shallow' history – that is, narratives about the past that rely primarily on written evidence for their factual basis. As the authors of *Deep History* point out, narrowing the purview of history in this way is fairly recent; historians considered the entire past to be grist for their mills until the publication of Darwin's *On the Origin of Species* in 1859. Traditionally constrained by the *Book of Genesis*, 'history' for Jewish, Christian and Muslim writers up to the mid-nineteenth century began with the Creation, and included the stories of human, animal and geological elements from that time to the present. After Darwin's revelation of deep time, however, the fields of archaeology, geology and physical anthropology developed new techniques for understanding the bones, rocks and tools that were beginning to be uncovered and that required interpretation.

This left Western historians with a choice: they could either join their new colleagues and begin to integrate deep history with the rest of the past or they could stick to what they knew and continue to elaborate historical trends from the civilisation of ancient Sumer forward, when written sources were available. They chose the latter course. According to Shryock and Smail, 'The decision to truncate history was a deliberate intellectual and epistemological move, bound up with the fate of the discipline itself' (2011, 7). Eager to display their rigour as a discipline in the new universities that were emerging at the end of the nineteenth century, historians adopted the analysis of written documents as their chief methodology. 'No documents, no history' (7), wrote Charles Langlois and Charles Seignobos, in a widely influential handbook on historical study. Despite fitful attempts after that to extend historical investigation into deep time, written documents and shallow history have prevailed until very recently.

Smail and Shryock aim to reverse that fateful decision foolishly made 150 years ago. As they point out, many Victorian historians believed in human exceptionalism, despite Darwin's insights joining humans to the rest of the animal kingdom. While the notion that past human actions could be studied as a part of evolution slowly gained ground in the biological and anthropological sciences, it rarely penetrated historical investigation. Instead, many historians drew on Hegelian ideas to posit that mankind had to gradually free itself from nature before it could gain political awareness and agency. Hegelian thinking, for example, was behind French historian Jules Michelet's bogus claim that 'when the world was born there began a war that will last until the world's end, and that is the war of man against nature, of the spirit against the flesh, of liberty against determinism. History is nothing but the story of this endless conflict' (9).

When historians and humanists turned to evolution at all, it was usually to embrace Social Darwinism, the doctrine of individual 'survival of the fittest,' which Darwin despised. As Shryock and Smail remark, such attempts to bring 'evolutionary' models into humanistic scholarship 'produced Victorian disasters' that often trumpeted racist and xenophobic conclusions. Consequently, they add, 'the soft social sciences and the humanities have never really come to terms intellectually with human evolution' (12). They also note, however, that contemporary historians, while still turning to written documents, are no longer disciplinarily bound by them. In the 'ongoing merger of history and social science,' they state, 'histories can be written from every type of trace, from the memoir to the bone fragment and the blood type' (13). Smail and Shryock's invitation to explore 'deep history' nicely complements Boyd and Richerson's interest in joining history and co-evolution.

The historical project I am working on involves tracing the origins and consequences of two foundational socio-political dynamics and the performances that regularly accompanied them in deep and shallow history. The first is the tension between domination and prestige as the basis for the exercise of power and authority in human cultures. Alpha male power and the all-male coalitions it facilitated derived from the genetics of previous great apes on our family tree and dominated the early politics of hominin hunter-gatherers. Around a million years ago, however, alpha male domination gave way to what evolutionary biologist Joseph Henrich and others have called the exercise of prestige, a different mode of behaviour and a new source of political legitimacy that could also allow for the authority of women and mixed-sex coalitions within hunter-gatherer bands. This type of governance involved a more egalitarian mode of sharing power among the sexes and deployed strategies of cooperation that enabled the band's survival and success in the context of increasing competition from outsiders. Nonetheless, tensions between the politics of domination and prestige continued during the Pleistocene, changed substantially during the centuries of shallow history and remain a part of human history today.

This political tension was complemented (and later complicated) by homophily, our species' predisposition to separate 'us' from 'them,' which also has deep roots in the social behaviour of our primate ancestors. Initially a simple matter of distinguishing 'our' band from those 'other' hominins in the area, homophily became more complex as bands grew in size, individuals within them divided into subgroups and the band as a whole made alliances with other groups for marriages and trading. The result was an increasing tension between exclusivity and expansion, which sought friendly relations with outsiders, where possible. These differences within homophily led, over time, to what we can now see as a host of tensions between narrow tribalism (in religion, ethnicity, nationalism, etc.) and cosmopolitanism. My forthcoming book, tentatively entitled *Politics, Performances, and the Anthropocene in Coevolutionary Perspective*, will explore the origins of human authority and homophily and the major historical shifts in these modes of governance and affiliation from the Pleistocene to the present.

Enter the Anthropocene

I am interested in the long history of politics and performances primarily because I want to understand more about our options for corrective political action on climate change and resource depletion in our current geological epoch, now known to many as the Anthropocene.

This term entered general scientific discourse in 2002, when chemist-geologist Paul Crutzen published an announcement in *Nature* advocating that his colleagues adopt it to emphasise the central role of humankind in shaping the earth's biosphere and geology. Anthropocence derives from the Greek and means, roughly, 'the human era.' As I write this essay in 2018, the members of the International Union of Geological Sciences have yet to determine the beginning of the Anthropocene epoch, but many geologists now favour a date during the 1950s. This dating accords with what Crutzen and others have termed the 'great acceleration' of carbon emissions into the atmosphere and radioactive Plutonium fallout around the world from the testing of thermonuclear bombs.

For authors Clive Hamilton, Christophe Bonneuil and François Gemenne, writing in the introduction to their anthology, *The Anthropocene and the Global Environmental Crisis*, the realities of the Anthropocene make two compelling claims on our attention. First, the new epoch 'claims that humans have become a telluric force, changing the functioning of the Earth as much as volcanism, tectonics, the cyclic fluctuations of solar activity, or changes in the Earth's orbital movements around the sun' (2015, 3). As a result, natural history and human history are now thoroughly interwoven. This recognition is important because 'modern humanities and social sciences have pictured society as if they were above material and energy cycles.... Now they must come back down to Earth' (4). 'The second claim is that the human inhabitants of our planet will face, in a time lapse of just a few decades, global environmental shifts of an unprecedented scale and speed, not [seen] since the emergence of the genus *Homo* some 2.5 million years ago.... It means inhabiting an impoverished and artificialised biosphere in a hotter world increasingly characterized by catastrophic events and new risks' (4). The authors add: 'Reinventing a life of dignity for all humans in a finite and disrupted Earth has become the master issue of our time' (5).

Significantly, Hamilton, Bonneuil and Gemenne choose the word 'reinventing' when considering how all of humanity might live 'a life of dignity' in the Anthropocene. Although recognising that statesman and philosophers have invented and tried to practice modes of ecological, political and economic justice in the past, they do not anticipate that we can wipe the slate clean of our past predispositions, which would include past tensions in our orientations towards authority and homophily, and craft new modes of governance and affiliation from scratch. And they point to the calamities that await us if we do not, as a species, once again use our capability to imagine and radically reshape our social and political lives together very soon. Much of what might have worked during the Holocene, the geologic epoch that ended in the 1950s, is no longer tenable as the Anthropocene tightens its grip on our biosphere.

The Anthropocene and the Global Environmental Crisis includes scientist Bruno Latour's essay, 'Telling Friends from Foes in the Time of the Anthropocene,' which calls upon readers to recognise that 'the old division of labor between science and politics is totally ill-equipped to handle the conflicts we have to navigate' in our new epoch. The 'friends' and 'foes' of Latour's title are not the usual suspects that a progressive citizen of the Holocene would have chosen. To be sure, Latour recognises that climate change deniers and capitalist despoilers of the Earth remain enemies of humankind, which he inventively renames 'the Earthbound.' His primary foes, however, are those who wish to perpetuate the modernist division between science and politics. States Latour:

> 'There are two sides: those who stick to a traditional science-versus-politics version and those who have understood that this older political epistemology (to give it the more accurate label)... is what renders both politics and science weak when the issues at stake are too large for too many interested people directly affected by their decisions.

> This is where there is a real distinction to be made between a Holocene and an Anthropocene settlement. What might have been good for humans [in the past]... has lost any sense for the Earthbound.'
>
> *(2015, 148)*

Latour urges us to recognise that there is currently a 'war' going on between the modernists stuck in Holocene assumptions and those who understand the new realities of the Anthropocene. Consequently, his foes include scientists who refuse to climb off their pedestal of apolitical objectivity and small 'd' democrats who so cherish the institutions of the Holocene that they would continue to allow ignorant politicians to make policy decisions. Presumably, one way to reach Latour's new 'settlement' and to preserve a significant measure of democratic government would be to guarantee all Earth-bound citizens clean air, water, sufficient food and other necessities and to enforce these universal rights with a world state that empowered scientific policy-makers. Even without planetary-wide laws, however, it is clear that more governance, not less, will be necessary for all humans to attain a 'life of dignity' in the new epoch. But, of course, we are a long way from such a political solution.

Latour recognises that such a transformation of global politics will not occur in the immediate future and may never occur at all. From my co-evolutionary perspective, Latour's hope for building workable political institutions that bring together scientists and politicians rests upon possible alliances across the world among those who favour a politics of prestige over domination and those who embrace social affiliations that move the Earth's cultures away from tribalist loyalties and towards a version of cosmopolitanism. Such alliances, however, always difficult at any time, now seem out of reach for most of the Earthbound. Fears of terrorism, starvation, political oppression and warfare – much of it caused or exacerbated by climate change and resource depletion – stalk many populations, and most nation-states remain preoccupied with economic growth as the cure-all for their problems. These conditions have led to the rise of strong-man rule and an increase in us-vs-them policies around the world. Even in better times, few politicians in democratic countries would willingly hand over part of their power to scientists they cannot control. While it almost goes without saying that a consilient approach to the truths of politics and biology would be an enormous benefit to the future inhabitants of the Anthropocene, such a radical shift in practice and epistemology is presently beyond reach.

All the more reason, then, for those of us concerned about the coming devastations to write wake-up calls for action to the Earthbound. Although the dominant economic institutions of the last 500 years are largely responsible for the present and coming disasters of our epoch, our fate as a species has primarily to do with the politics of the next 80 years. The tradition of alpha male domination – evident in the many dictators, political bosses, presidential bullies, religious terrorists and other strong men who stir up tribalistic animosities for their own benefit – has already led to situations in which populations have been ravaged by climate change and resource depletion. Think, for example, of Russia, Nigeria and Puerto Rico. Nurturing and supporting coalitions of prestige that seek to open borders rather than close them cannot guarantee the necessary global effort to protect a diverse biosphere, but at least they have the potential to shape very different institutions than those that now imprison many of the Earthbound institutions with norms and goals that can respond to genuine human needs in the fierce ecology of the Anthropocene.

Performing our predispositions

I expect that my narrative for *Politics, Performances, and the Anthropocene in Coevolutionary Perspective* will emphasise the twists and turns that our predilections for modes of authority and group affiliation have taken in the last 2 million years. My two foundational dichotomies, domination-prestige and tribalism-cosmopolitanism, are best thought of as spectrums of practice and belief, which began as clear choices in hunter-gatherer times, but became muddied and multiple after our species turned to agriculture and invented more complex societies. In modern cultures today, some fields of activity swing towards the domination and tribalism ends of both spectrums, while others generally work as modes of prestige and cosmopolitanism. These alignments of authority and homophily are far from inevitable, however. In the field of politics, parliamentary democracies (such as the UK) often function primarily in terms of prestige, but may also embrace tribalism. And some authoritarian regimes (e.g., China) favour domination at home, but practice a guarded cosmopolitanism abroad.

With the caveat of complexity in mind, I will summarise what I presently take to be the first major turning point in my historical narrative. It occurred around a million years ago during *Homo erectus* times when, according to several anthropologists, our female ancestors, probably aligned with some males in a few hunter-gatherer bands, asserted the right to control their own sexuality and overthrew the domination of the reigning alpha males and their allies to establish a governing coalition for their band that claimed authority on the basis of prestige.

I will discuss this turning point more thoroughly than the others because, without it, our co-evolution as a species could not have moved our genus (and, later, our species) much beyond the sociality and politics of our primate cousins. The genus *Homo* began to flourish on the plains of Africa primarily because of their ability to cooperate with each other. The proto-chimpanzees from whom our genus descended could also cooperate in small ways, but lacked a sense of fairness and group collaboration, according to evolutionary anthropologist Michael Tomasello. Chimps and other great apes were (and are) primarily built for competition, not for the cooperation of reciprocity or for acts of altruism that go beyond great ape mother-child bonding. Hominins of both sexes in the mid-Pleistocene could use empathy, sympathy and other social emotions to share parenting and improvise communication for many collaborative tasks that benefitted their hunter-gatherer band. Given the difficulties of survival for *H. erectus* bands in the middle Pleistocene, these small groups needed leaders who could make or guide decisions about where to hunt, how to protect the band from predators and when to move camp – leadership roles never played by great ape males. Some alpha males probably exercised more command and control than others, but those who could encourage cooperation through empathy and example might be rewarded with gratitude and loyalty, social emotions absent among great apes. Most alpha male hominins probably earned their right to political leadership through hunting, a cooperative part of group survival. Certainly alpha male violence continued as a last resort, but the importance of group collaboration in child care and hunting would have worked against its constant deployment.

Did the dominance of alpha males in hominin groups also allow them sexual access to all of the fertile women of the band, however, as was the case with bands of proto-chimps and gorillas? Most evolutionary biologists believe that it did. According to anthropologist Terence Deacon, sometime during the middle years of the Pleistocene, the *Homo* genus experienced an evolutionary bottleneck in the growth of their cooperative sociality. Without the ability to legitimate the social rights and obligations of long-term sex partners, he believes, the unstable cohesion of these *H. erectus* bands could easily fragment through male rivalries and sexual jealousy.

There is evidence, however, that, over time, relations between the sexes became more egalitarian. Although the sexual division of labour likely changed very little – women did most of the gathering and child-caring while men did the hunting – Pleistocene women gained increasingly equal power within their bands. Some reasons for this shift likely included the growing importance and control of fire for cooking, environmental factors limiting the success of male hunting and innovations in hominin weaponry, which meant that, for the first time, groups of women could potentially gang up on an abusive alpha male in the band and kill him. Whether such gender revolts occurred often cannot be known, but most adults in every band probably understood that they were possible; by itself, this knowledge likely helped to level the field of power relations between the sexes. The first victory of female adults, whether achieved violently or peacefully, probably had to do with restraining male physical threats in order to gain greater control over their sexuality. This 'reverse-dominance coalition,' as anthropologist Camilla Power and others have termed it, allowed a female collective to deter and dominate individual males who sought to dominate them. Hominin rituals of proto-marriage probably provided another restraint on the sexual appetites of adult males, alpha or not.

Perhaps the best evidence for governance by coalitions organised according to prestige rather than dominance comes from ethnographers and anthropologists who have studied contemporary hunter-gatherer bands over the last 60 years. For his *Moral Origins: The Evolution of Virtue, Altruism, and Shame* (2011), evolutionary anthropologist Chris Boehm collated 150 ethnographic studies of surviving hunter-gatherer societies. Boehm calls these groups 'Late Pleistocene Appropriates' (LPAs) because he believes that many of their cultural practices date from the time of *H. heidelbergensis*, about 600,000 years ago. In their essay 'Zoon Politicon' in *Cultural Evolution*, Herbert Gintis and Carel van Schaik underline the dislike of coercion central to the politics of these LPA bands: 'Hunter-gatherers share with other primates the striving for hierarchical power, but social dominance aspirations are successfully countered because individuals do not accept being controlled by an alpha male and are extremely sensitive to attempts of group members to accumulate power through coercion' (2013, 36). Arrogant bullies and scheming Machiavels will be warned, punished and occasionally ostracised from the band. If these strategies are unsuccessful, note Gintis and van Schaik, 'the group will delegate one or more members, usually including one close relative of the offender, to kill him' (36). In the absence of traditional forms of hierarchy, these small bands generally govern themselves through persuasion and coalition building. Building and governing through mixed-sex coalitions remained the predominant form of politics among our species for perhaps half a million years, until several thousand years into the Holocene epoch.

The gender-based politics of prestige and coalition-building are evident in some of the rituals practised by contemporary hunter-gatherer bands and tribes. We can see one example of this in the contemporary rituals of the BaYaka Pygmy tribes in Central Africa. Deep historical conflicts between the genders are on display and open for resolution in the initiation ceremonies of the Mbendjele tribe. Based in tribal creation myths that narrate the initial separation and eventual synthesis of male and female tribes, these playful rituals feature performances of war-like battles of the sexes that are resolved in the end. They begin with the Mbendjele men pretending to ambush the women using pig-hunting techniques, which is then countered by a female 'Ngoku charge,' as the women lock arms and rampage through the camp, threatening to trample the men. The reign of 'Ngoku,' meant to embody a primordial time when women ruled over men, then encourages the women (often led by the grandmothers) to mock male sexual prowess and to openly flaunt their own sexual desires. Such teasing occasionally angers some of the men, but most accept it. As anthropologists Chris Knight and Jerome Lewis explain, 'They usually join in good-humoredly, eventually laughing at their wives hilarious impersonations of themselves. These reenactments are displayed with such exaggeration and parody as to provoke helpless laughter' (Knight and Lewis 2014, 308).

This playful resolution parallels the narrative ending of the origin myth of the Mbendjele. In the story, the mythic battle of the sexes culminates in play fights that are alternately won and lost by both genders, until husbands and wives relax into lovemaking. Knight and Lewis remark that 'ritualized play pervades the very arena which, in other primates – chimpanzees, for example – leads recurrently to sexual violence' (2014, 309). They speculate that these initiation rituals may echo and comically resolve actual, painful events in the Pleistocene past of the tribe, when the women overthrew the rule of an alpha male and established governance by a reverse-dominance coalition, a polity that mostly remains in place among the BaYaka today.

Although prestige and cooperation, reinforced psychologically and socially, had provided the basis of our species' success in the Pleistocene, our ancestors mostly squandered this advantage for much of the Holocene. As most hunter-gatherers gradually transformed themselves into farmers and herders to seek survival, the size of their language-based cultures grew through warfare from bands into tribes and eventually into archaic states, which tightened the us-them bonds of homophily. Egalitarian politics mostly disappeared from these increasingly complex societies, however; dominating despots with superior weapons eager to pursue wars of conquest for land and slaves took over the archaic states. The first major break in the dynamics of warfaring tribalism occurred in China at the end of the Period of Warring States (433–221 BCE), when successive emperors began to incorporate and enculturate all of their subjects as Han Chinese, using culture to achieve political stability. The second was the advent of universalistic religious cultures, such as Buddhism, Christianity and Islam, across much of the Eurasian land mass. Mass conversions to the same religion facilitated trust, commerce and cosmopolitanism among those cultures that shared the same god and, for the first time, placed religious limits on political rule. The Islamic Caliphate in northern Africa and Spain, for example, practised religious tolerance and encouraged trade among different religious groups for much of the early medieval period. I will discuss Chinese emperor worship during the Han period and Islamic rituals in northern Africa and Spain to explore relevant aspects of authority and homophily in those cultures.

The universalistic religions had established laws that were separate from the power of dominating kings and emperors, but it took the Enlightenment of the eighteenth century to begin cross-cultural conversations that would seek to free the individual from traditional forms of religious loyalty as well as from the power of the modernising state. Dramatic theatre and opera played an important political role during the eighteenth century, as the works of Voltaire, Lessing, Diderot, Mozart and others led many Europeans to imagine the possibility of equal rights for all humans, regardless of state or religion. I will look closely at the politics of authority, especially the shift towards more egalitarian social relations, in three of Mozart's late operas, *The Marriage of Figaro, Don Giovanni* and *The Magic Flute.*

Partly because of the wars of the next 200 years, aspirations for individual rights emerged as a significant expectation for most of the world's populations after World War II and were affirmed by the United Nations in 1948. The Universal Declaration of Human Rights celebrated human rights for all *H. sapiens*, barred discrimination on the basis of race and other identifications based in tribalistic homophily, and elevated the protection of these rights above the sovereignty of any nation. As the heirs of Enlightenment humanism, the theatres of Bertolt Brecht, Arthur Miller, Jean Giraudoux and several other late modernist playwrights helped to embody these commitments during the postwar era. In particular, I will probe Brecht's *Mother Courage and Her Children* and *Life of Galileo* for their interest in advancing egalitarian and cosmopolitan values.

Although many of the cultures in Africa and Asia gained national freedom and a measure of democracy in the mid-1950s near the beginning of the Anthropocene, the capitalist revolution and its expansion into colonialism and imperialism begun during the previous epoch had already ensnared them in webs of exploitation that soon made a mockery of many of their hopes for democracy and a politics based in prestige instead of domination. As Immanuel Wallerstein and others have made clear, pre-1945 world systems had been dominated by European city-states and nations, but the United States, preeminent following two world wars mostly on European soil, quickly established itself as the global hegemonic power after World War II. Through banking, trade and defense agreements, plus infusions of development aid, the United States quickly solidified its leadership in western Europe and around the world. Motivated to 'develop' Third World countries to 'save' them from communism, the United States encouraged its corporations to drain them of mineral and renewable resources in the name of free trade in exchange for U.S. oil and agricultural products. Even with competition from the so-called Communist Bloc, the U.S. world system – reliant on the 'soft power' of Hollywood and other image makers as much as on its corporate clout and outsized military – imposed its cultural hegemony on the rest of the globe from 1945 to 2005. For this reason, the chapter on the Anthropocene will focus on the culture of the United States and its relations with the rest of the Earth during these years.

American hegemony and its struggle with the Soviet Union opened a space for expanding individual rights throughout the world, which began to break the hold of traditional religious loyalties and opened up cosmopolitan orientations to many more of the Earthbound than ever before. Simultaneously, the lust for territorial gain through warfare that had animated our species for most of the Holocene subsided, quelled partly by the reality of UN peacekeeping and the possibility of nuclear warfare. Stanley Kubrick parodied Cold War fears and desires in *Dr. Strangelove: Or How I Learned to Stop Worrying and Love the Bomb (1964)*, a film that revealed that both sides had largely abdicated leadership in the race for technological superiority. A similar hollowness at the core of national resolve haunted Francis Ford Coppola's *Apocalypse Now* and other films about the U.S. war in Vietnam. The United States 'won' the Cold War when the USSR fell apart in 1989, but Hollywood had shown that few Americans believed that our democratic values had triumphed. In part, this was also because racism and inequality continued to fester, after most of the (white) population turned against full equality for black citizens at the end of the 1960s. The implicit demands for equality in the stand-up and filmic career of Richard Pryor in the 1970s stand in sharp contrast to the domesticated performances of Bill Cosby – 'America's favorite Dad' – on stage and TV in the 1980s and early 1990s.

The Great Acceleration of postwar petrocultures continued to speed up, however, as world capitalism, in its neoliberal phase after 1990, kept its foot on the gas. Western neoliberal oligarchs increased trade with the Chinese communists, whose politics of domination was transforming China into a new world power. Meanwhile what international political historian Francis Fukuyama calls 'political decay' (2014, 455) – the erosion of accountability and legislative paralysis in nominally democratic nation-states – had allowed the oligarchs to gain excessive power in the European Union and the United States. The crash of 2008 might have been a wake-up call for the elites that dominated world capitalism and government, but shallow economic recovery, bureaucratic straightjackets, political fragmentation, widening inequality and continuing racism distracted them and most Americans from the decay of their national political culture. I will end the book chapter with brief analyses of Steven Sondheim's *Lincoln* (2012) and Lin Manuel Miranda's *Hamilton* (2015), two performances set in American history that both celebrated an expansive vision of U.S. democracy and revealed some of its chronic difficulties with regard to authority and affiliation.

I will focus on what the Trump presidency reveals (so far) about the likely future of democratic governance in the United States in the context of the continuing bad news about climate change and resource depletion. I expect I will end my narrative with suggestions for future performative actions that may help us to deal more equitably and justly with the disasters to come. Not surprisingly, my prognosis for the future will be pessimistic.

Looking toward the next 80 years, as the Anthropocene tightens its grip, two things are certain: the co-evolution of our predispositions for organising politics around domination or prestige and affiliating socially in tribal or cosmopolitan ways will continue to play a significant role in our actions; they are not going away and should not be ignored. The notion that we can somehow escape our co-evolutionary predilections and reinvent human nature through old religions or new technologies has always been a utopian fantasy. Nor will my history endorse notions of apocalypse or complete despair; such fears are not only unrealistic, they empower the narcissists and myth-makers. Humanity will not perish. Although many of us will certainly die in greater numbers from climate- and resource-related problems, our species will survive. Moving to consilient discourse and effective governance in the Anthropocene will be difficult, but it is not impossible. Our co-evolutionary past matters because it can help us to understand and navigate the parameters of our possible futures. But the good news here is that humanity has the potential to energise its evolutionary political capabilities to enable us to ride out the ravages of climate change and depleted resources and to begin to manage a fairer life for the survivors in the epoch of the Anthropocene. Political performances – both those that comment on the exercise of governmental power and those that directly embody and practise versions of authority and solidarity to shape our norms, laws and institutions – can help us to reach that potential.

References

Boehm, Chris. 2011. *Moral Origins: The Evolution of Virtue, Altruism, and Shame.* New York: Basic Books.

Bowles, Samuel and Gintis, Herbert. 2011. *A Cooperative Species: Human Reciprocity and Its Evolution.* Princeton, NJ: Princeton University Press.

Boyd, Robert and Richerson, Peter J. 1985. *Culture and the Evolutionary Process.* Chicago, IL: University of Chicago Press.

Crutzen, Paul J. 2002. "Geology of mankind," *Nature* 415: 23.

Deacon, Terence W. 1997. *The Symbolic Species: The Co-evolution of Language and the Brain.* New York: W.W. Norton.

Fukuyama, Francis. 2014. *Political Order and Political Decay: From the Industrial Revolution to the Globalization of Democracy.* New York: Farrar, Strauss, and Giroux.

Gintis, Herbert and van Schaik, Carel. 2013. "*Zoon Politicon*: the evolutionary roots of human sociopolitical systems" in *Cultural Evolution: Society, Technology, Language, and Religion*, eds. Peter J. Richerson and Morten H. Christiansen, 25–44. Cambridge, MA: MIT Press.

Godfrey-Smith, Peter. 2002. *Theory and Reality: An Introduction to the Philosophy of Science.* Chicago, IL: University of Chicago Press.

Haskell, Thomas L. 1998. *Objectivity Is Not Neutrality: Explanatory Schemes in History.* Baltimore, MD and London: Johns Hopkins University Press.

Hempel, Carl G. 1959. "The function of general laws in history [1942]" in *Theories of History*, ed. Patrick Gardiner, 344–55. New York: Free Press.

Henrich, Joseph. 2016. *The Secret of our Success: How Culture Is Driving Human Evolution, Domesticating Our Species, and Making Us Smarter.* Princeton, NJ: Princeton University Press.

Jablonka, Eva and Marion J. Lamb. 2005. *Evolution in Four Dimensions: Genetic, Epigenetic, Behavioral, and Symbolic Variation in the History of Life.* Cambridge, MA: MIT Press.

Knight, Chris and Lewis, Jerome. 2014. "Vocal deception, laughter, and the linguistic significance of reverse dominance" in *The Social Origins of Language*, eds. Daniel Dor, Chris Knight, and Jerome Lewis, 297–314. Oxford: Oxford University Press.

Power, Camilla. 2014. "The evolution of ritual as a process of sexual selection" in *The Social Origins of Language*, in *The Social Origins of Language*, eds. Daniel Dor, Chris Knight, and Jerome Lewis, 196–207. Oxford: Oxford University Press.

Richerson, Peter J. and Boyd, Robert. 2005. *Not by Genes Alone: How Culture Transformed Human Evolution*. Chicago, IL: University of Chicago Press.

Smail, Daniel L. and Shryock, Andrew. 2011. *Deep History: The Architecture of Past and Present*. Berkeley: University of California Press.

Snow, C. P. 1969. *The Two Cultures and a Second Look: An Expanded Version of the Two Cultures and the Scientific Revolution*. Cambridge: Cambridge University Press.

Tomasello, Michael. 2016. *A Natural History of Human Morality*. Cambridge, MA: Harvard University Press.

Turchin, Peter. 2016. *Ultrasociety: How 10,000 Years of War Made Humans the Greatest Cooperators on Earth*. Chaplin, CT: Beresta Books.

Wallerstein, Immanuel. 2004. *World-Systems Analysis: An Introduction*. Durham, NC: Duke University Press.

Wilson, Edward O. 1998. *Consilience: The Unity of Knowledge*. New York: Knopf.

GLOSSARY

Shelby Brewster

Affect

Like the closely related concept of emotion, affect has been subject to a wide variety of theoretical explanations (see 'Emotion'). These include definitions supported by empirical evidence and philosophical definitions without empirical confirmation, such as the affect theory of Gilles Deleuze and Félix Guattari. In the 1960s, psychologist Silvan Tomkins argued that *affect* implied the biological component of human emotions that are genetically transmitted. He described nine primary affects that can be distinguished by their expression and intensity. However, others have developed differing definitions of the term. For example, Giovanna Colombetti uses 'affect' as a broad umbrella term which encompasses emotion, moods and feelings. She argues that the human mind and body are always already affective. Because Colombetti ties affect to meaning and argues that all living systems encounter the world as meaningful, then all living systems are affective.

Colombetti, Giovanna. 2014. *The Feeling Body: Affective Science Meets the Enactive Mind.* Cambridge, MA: MIT Press.

Tomkins, Silvan S. 2008. *Affect, Imagery, Consciousness: The Complete Edition.* New York: Springer Publishing Company.

Autonomy

Autonomy is a significant concept for the paradigm of enaction (see 'Autopoiesis' and 'Dynamic Systems Theory'). The general meaning of autonomy is self-governed, as opposed to heteronomy, or other-governed. In terms of systems, as Francisco Varela theorised, autonomous systems are characterised by organisational closure: a set of self-referential processes and relations determines the system. For example, the human body is an autonomous system composed of a number of self-referential processes, including blood circulation. Autonomy does not imply, however, that systems are completely closed off from their environments. As Evan Thompson explains in *Mind in Life,* an autonomous system is always structurally coupled to its environment and continuously reacts with it, as well as other systems. In contrast to the computational theory of mind, an autonomous system does not have inputs and outputs, but instead work through an ongoing coupling with their environments. Organisational closure and structural coupling also mean that autonomous systems determine their own boundaries, which can be material (as in a cellular membrane) or not.

Varela, Francisco. 1979. *Principles of Biological Autonomy*. New York: Elsevier North Holland.
Thompson, Evan. 2007. *Mind in Life: Biology, Phenomenology, and the Sciences of the Mind*. Cambridge, MA: Harvard University Press.

Autopoiesis

Biologists Humberto Maturana and Francisco Varela coined the term 'autopoiesis' as part of their 1973 work *Autopoiesis and Cognition*. It is the biological expression of autonomy, in which a materially-bounded system has organisational closure, meaning it is self-referential and internally produces conditions for its survival (see 'Autonomy'). Maturana and Varela claimed that autopoietic systems have several characteristics: they are a network of processes that continuously regenerate themselves through interactions with their environment, and they are self-contained systems. Evan Thompson, in *Mind in Life: Biology, Phenomenology, and the Sciences of the Mind*, built on Maturana and Varela's work, specifically the connection between autopoiesis and cognition. Thompson takes autopoiesis to entail internal self-production in order to interact with the environment, what Varela called structural coupling. With this definition, autopoiesis always includes cognition, a characteristic that Thompson attributes to all living systems.

Maturana, H. R., and F. J. Varela. 1980. *Autopoiesis and Cognition: The Realization of the Living*. Boston, MA: D. Reidel Publishing Company.
Thompson, Evan. 2007. *Mind in Life: Biology, Phenomenology, and the Sciences of the Mind*. Cambridge, MA: Harvard University Press.

Behaviourism

Behaviourism is an approach to analysing human behaviour that generally focuses on environmental factors rather than on evolutionary characteristics. This approach dominated psychology in the middle of the twentieth century. Although they may have different methodologies, behaviourists pursue a general science of human behaviour based on observable phenomena. In this view, the brain responds to external stimuli, resulting in particular social or individual behaviours; the brain could not do too much to resist any kind of conditioning or direction. Belief in these claims made cognitive research essentially unnecessary, until cognitive theories of behaviour began emerging in the 1960s.

Baum, William M. 2017. *Understanding Behaviorism: Behavior, Culture, and Evolution*. 3rd ed. Oxford: Wiley.

Bodymind

For researchers who seek to challenge Cartesian mind-body dualism, the use of the term *bodymind* emphasises the complete integration of the brain and body. The body is not separate from the mind, and vice versa.

Damasio, Antonio. 1999. *The Feeling of What Happens: Body, Emotion, and the Making of Consciousness*. London: Vintage.

Body schema

In order to move through space, navigate obstacles and interact with object, the brain maintains a neural representation of the body in space. This is called the body schema. Through largely unconscious sensory processing, the brain maintains a schema of the body's position in space (see 'Proprioception'). As Holmes and Spence point out, the body schema might also include auditory, visual and other sensory information. Using both the body schema and the perception of peripersonal space, the space immediately surrounding the body, we can maintain our place in space and avoid potential obstacles that might suddenly appear.

Holmes, Nicholas P., and Charles Spence. 2004. 'The body schema and multisensory representation(s) of peripersonal space.' *Cognitive Processing* 5: 94–105.

Bottom-up and top-down processing

These are two broad explanations of the connection between perception and cognition. Bottom-up, or data-driven, cognition is driven by sensory information, such as when you see a bear moving towards you in the woods. In this approach, processing occurs via the input of sensory information, generally without a higher level of direction. Examples of bottom-up processing include rapid visual identification or quick reaction to a stimulus. Top-down processing, in contrast, involves the application of prior knowledge or conceptual data when interpreting sensory information. Andy Clark's predictive processing model, for example, includes both modes of processing but emphasises the importance of top-down processing for human prediction (see 'Predictive Processing').

Goldstein, E. Bruce. 2005. *Cognitive Psychology: Connecting Mind, Research, and Everyday Experience.* Belmont, CA: Thomson.

Cognition

Cognition includes any and all mental actions and processes, including thought, knowledge and sensory input. In humans, cognition includes things like memory, visual processing, attention, use of concepts, pattern recognition, mental imagery and language. There are a wide variety of approaches to studying human cognition from several scientific fields: biology, psychology, computer science, linguistics and philosophy, for example.

Weisberg, Robert W., and Lauretta M. Reeves. 2013. *Cognition: From Memory to Creativity.* Hoboken, NJ: John Wiley & Sons, Inc.

Cognitivism

Cognitivism emerged within the field of psychology in the 1950s, particularly as a response to behaviourism (see 'Behaviourism'). Rather than privileging the role of the environment in directing human behaviour, cognitivism supports a computational view of the mind. Under traditional cognitivism, as exemplified by the work of Jerry Fodor, Allen Newell and Herbert A. Simon, mental processes *are* computational processes. Mental objects, including thoughts, memories and beliefs, are symbolic structures which are then manipulated. In other words, cognitivists believe there are a set of algorithms which order the brain's activity, like a mathematical equation. Certain inputs (stimuli from outside the brain) result in certain outputs (emotions, bodily responses, etc.).

Newell, Allen, and Herbert A. Simon. 1961. 'Computer Simulation of Human Thinking.' *Science* 134: 2011–17.

Computational theory of mind

Computational Theory of Mind (CTM) is a collection of views that composed the dominant cognitive paradigm beginning in the 1950s. It is also called computationalism or the classical computational theory of mind. CTM holds that the mind is a computing system, and mental processes such as reasoning or decision-making are executed like programmes within a computer. The brain uses a symbolic language to carry out these processes, which are completely non-conscious, essentially inaccessible to the mind. One of the most influential versions of CTM was first put forth by Jerry Fodor in the 1970s. Fodor argues that mental states are created through relations between individual minds and mental representations. In this paradigm, the mind didn't function like a computer; the mind *was* a computer. The key problems with CTM, especially for enactivists, are the separation of the mind from the materiality of the body and its environment, and the explanatory gap CTM creates by separating cognition from subjective mental experience. CTM was followed by several other cognitive paradigms, including connectionism and enaction.

Fodor, Jerry. 2010. *LOT2: The Language of Thought Revisited*. Oxford: Oxford University Press.
Pylyshn, Zenon W. 1986. *Computation and Cognition: Toward a Foundation for Cognitive Science*. Cambridge, MA: MIT Press.

Conceptual integration

Conceptual integration, sometimes called blending, is a vital process for human cognition. As Gilles Fauconnier and Mark Turner explain in *The Way We Think,* this theory of cognition holds that blending is a ubiquitous subconscious process necessary for both thought and language. Concepts from two or more spaces or domains are integrated into another space, the space of the blend. Theatrical characters are a useful exemplar of how conceptual blending works. Audiences watching Kenneth Branagh's performance of Hamlet construct a partial match of concepts of 'Kenneth Branagh' with concepts of 'Hamlet' in a third mental space. This space then dynamically develops an emergent 'Branagh-Hamlet' construct. Blending allows humans to compress any number of distinct behaviours and events into a single moment. As such, blending is an essential part of both viewing and creating a performance. While they are vital to cognition, the processes of conceptual blending take place unconsciously.

Fauconnier, Gilles, and Mark Turner. 2002. *The Way We Think: Conceptual Blending and the Mind's Hidden Complexities*. New York: Basic Books.
Turner, Mark. 1996. *The Literary Mind: The Origins of Thought and Language*. Oxford: Oxford University Press.

Conceptual metaphor

In 1980, George Lakoff and Mark Johnson first published *Metaphors We Live By*. Their work was a departure from the representationalism of CTM, which holds that language involves a relationship between words and an objectively real world. Instead, Lakoff and Johnson argued that language emerges from embodied experience, especially through metaphors. They maintain that language does not exist separate from human conceptual systems, but is based on those very systems. And, for them, most of our conceptual structures come from metaphors. Lakoff identified a number of image schemas from which metaphors are created; these enable understanding of something through our embodied experience of something else. For example, one of the most common kinds of conceptual metaphors involve spatial organization or orientation, such as understanding future events as 'up' or 'ahead.' 'What's coming up tomorrow?' exemplifies one version of this conceptual metaphor. Like all conceptual metaphors, it is grounded in embodied experience, namely, that humans look forward as they move forward, and objects appear larger as one moves towards them.

Lakoff, George, and Mark Johnson. 1999. *Philosophy in the Flesh: The Embodied Mind and Its Challenge to Western Thought*. New York: Basic Books.

Lakoff, George, and Mark Johnson. 2003. *Metaphors We Live By*. Chicago, IL: University of Chicago Press.

Connectionism

Connectionism is a theory of information processing. In cognitive science, it is also called parallel distributed processing or neural network modelling. Connectionism emerged in the 1980s as a reaction to earlier symbolic systems of cognition. Connectionists sought to model cognitive function in a model similar to the network of the nervous system. Under this paradigm, cognition is explained through neural networks modelled on the changing organisation of neurons in the brain. The more synapses among the neurons, the more 'weight' each of them carry in the model. Unlike symbolic models, in connectionism cognitive processing occurs in parallel, or in many units simultaneously. Mental processes emerge from this shifting dynamic of networks and weights. In this model, neural networks are dynamic systems, as the network as a whole responds and adapts to its environment (see 'Dynamic Systems').

Bechtel, William, and Adele Abrahamsen. 1991. *Connectionism and the Mind: An Introduction to Parallel Processing in Networks*. Cambridge, MA: Blackwell.

Rumelhart, D. E., and J. L. McClelland. 1986. *Parallel Distributed Processing: Explorations in the Microstructure of Cognition. Volume 1: Foundations*. Cambridge, MA: MIT Press.

Rumelhart, D. E., and J. L. McClelland. 1986. *Parallel Distributed Processing: Explorations in the Microstructure of Cognition. Volume 2: Psychological and Biological Models*. Cambridge, MA: MIT Press.

Consciousness

The challenge of defining and describing consciousness remains one of the central preoccupations of psychology and cognitive science. Generally, consciousness can be identified either as *state consciousness* (the ways in which particular mental processes or emotions might be described as conscious) or *creature consciousness* (the consciousness of a whole organism). Robert Van Gulick describes five types of creature consciousness. The first is *sentience*, the capacity to sense and perceive the environment, and respond to it. The second is *wakefulness*, the active use of the capacity for sense and perception. The third is *self-awareness*: a creature is conscious if it can self-reflect on its mental processes and states. The fourth is *qualitative experience*, meaning that an organism has consciousness if it has an experiential life, even if we as humans cannot comprehend it. Finally, *transitive consciousness* is a relation between an organism and some object in its environment of which it is aware. There are a number of theories that seek to explain consciousness, including philosophical, cognitive and neurobiological theories.

Edelman, Gerald M., and Giulio Tononi. 2000. *A Universe of Consciousness: How Matter Becomes Imagination*. New York: Basic Books.
Van Gulick, Robert. 2012. 'Consciousness and Cognition.' In *The Oxford Handbook of Philosophy of Cognitive Science,* edited by Eric Margolis, Richard Samuels, and Stephen P. Stich, 19–40. Oxford: Oxford University Press.

Distributed cognition

Distributed cognition is a specific theoretical approach to human cognition that relies on the idea of the extended mind, which holds that mind and environment are continuously interacting through a series of feedback loops. In other words, according to philosopher Andy Clark, the mind is *extended* into the environment. One of Clark's classic examples of extended mind involves the completion of a math problem. If one uses a paper and pencil to complete the calculation, in Clark's view those tools are integral to the cognitive process and so should be considered part of the system's 'mind.' Edwin Hutchins takes a slightly different approach to distributed cognition, exploring the cognitive processes of social and cultural systems, which may also involve the use of tools. Hutchins argues that human cognition is always both cultural and social; it does not occur in a void separated from the social and physical environment. He uses the case study of a group of navy sailors to demonstrate how social and cultural groups can be considered cognitive systems, as cognition is distributed across and among individual minds.

Clark, Andy. 2008. *Supersizing the Mind: Embodiment, Action, and Cognitive Extension*. New York: Oxford University Press.
Hutchins, Edwin. 1995. *Cognition in the Wild*. Cambridge, MA: MIT Press.

Dynamic systems theory

Dynamic Systems Theory (DST) is a mathematical approach used to define and analyse the behaviours of complex systems. DST has been applied in a variety of fields, including cognitive science. As Robert F. Port and Timothy Van Gelder outline in *Minds as Motion,* DST solves the problem of time that computational or representational theories of cognition do not answer for. Because human cognition occurs in real time, DST offers tools to understand the process as it occurs. J.A. Scott Kelso used dynamic systems to explore how human brains and behaviour are self-organizing. At any given moment, dynamic systems reveal an overall state, and the behaviour of the system can be defined as the change in that state over time.

Kelso, J. A. Scott. 1995. *Dynamic Patterns: The Self-Organization of Brain and Behaviour.* Cambridge, MA: MIT Press.
Port, Robert F., and Timothy Van Gelder. 1995. *Minds as Motion: Explorations in the Dynamics of Cognition*. Cambridge, MA: A Bradford Book.

Embodied cognition

Embodied cognition is a theory of cognition reaching across a number of different fields. The goal of embodied cognition is to account for the way human minds, bodies and environments interact together. The framework maintains that embodiment is necessary for cognition. Cognitive processes, then, emerge from the coupled systems of organisms and their environments. (See 'Autonomy,' 'Autopoiesis,' and 'Conceptual Metaphor.')

Lakoff, George, and Mark Johnson. 1999. *Philosophy in the Flesh: The Embodied Mind and Its Challenge to Western Thought*. New York: Basic Books.

Shapiro, Lawrence A. 2004. *The Mind Incarnate*. Cambridge, MA: MIT Press.

Varela, Francisco, Evan Thompson, and Eleanor Rosch. 1991. *The Embodied Mind: Cognitive Science and Human Experience*. Cambridge, MA: MIT Press.

Emergence

Emergence is a property of a system in which some characteristics or phenomena are not part of the system's individual components but emerge from the interactions of components within the system as a whole (see 'Dynamic Systems Theory'). This concept has long been common in other branches of science, including linguistics and biology, and has recently been explored in cognitive science. Emergence has become especially important to the enaction paradigm (see 'Enaction'). As Evan Thompson and Francisco Varela argue, the processes essential for human consciousness are not restricted to the brain, but instead spread across the boundaries between the brain, the body, and environment. As such, the neural processes *emerge* from the interactions between and among the brain-body-environment system.

McClelland, James L. 2010. 'Emergence in Cognitive Science.' *Topics in Cognitive Science* 2: 751–70.

Thompson, Evan, and Francisco J. Varela. 2001. 'Radical Embodiment: Neural Dynamics and Consciousness.' *Trends in Cognitive Science* 5 (10): 418–25.

Emotion

Emotion is one of the most contested concepts in cognitive science. There remain a number of theories of emotion with significant disagreement among them. The Theory of Basic Emotions (BET) was developed in the 1960s in the work of Sylvan Tomkins, Paul Ekman, Wallace V. Friesen and others. BET holds that emotions are a result of genetically predetermined sets of instructions which create changes in the brain, expression and behaviour. There are seven biologically basic emotions that developed as evolutionary adaptations: fear, anger, happiness, sadness, disgust, surprise and contempt. Under BET, once these basic emotions are activated, their expression is predetermined and cannot be interrupted. Further, these emotions are the same for all humans, regardless of cultural difference. Enactivists take a different approach to emotion, rejecting the Cartesian theory of emotion, which views emotions as physical or bodily sensations that only serve to inform the mind about the body. In this view, which is still common in emotion science, the information sent from the body needs to be cognitively evaluated, which generates a bodily response. In the enaction paradigm, however, emotion is an integral part of both autopoiesis and cognition. Emotion cannot be separated from meaning, as Giovanna Colombetti argues. She explains that emotion is a capacity of a whole organism, not of a separated mind or body. The organism generates meaning as it encounters the world; no evaluative process or cognitive appraisal is necessary to determine a bodily response to stimuli.

Colombetti, Giovanna. 2014. *The Feeling Body: Affective Science Meets the Enactive Mind* Cambridge, MA: MIT Press.

Damasio, Antonio. 1999. *The Feeling of What Happens: Body, Emotion, and the Making of Consciousness*. London: Vintage.

Empathy

C. Daniel Batson identifies eight different ways that cognitive scientists and philosophers have defined empathy. They include the knowledge of another's internal state, including thoughts and feelings; taking on the posture or physical expression of another; feeling the same emotion that another feels; projecting oneself into another's place; imagining the feelings and thoughts of another; imagining how one would feel or think in another's place; feeling distress by another's suffering and feeling for another's suffering. Evan Thompson elaborates on the enactivist conception of empathy, which is rooted in phenomenology. Integral to this theory of emotion is the idea of subjectivity, which Thompson argues is always *inter*subjective. In other words, 'no mind is an island,' and human subjectivity necessitates empathetic recognition of both self and other. Therefore, empathy is fundamental to subjectivity and cognition. Drawing on the work of Edith Stein, Thompson explains that empathy is a particular type of open intentionality, one directed towards another's experience. It includes perception and inference, but empathy cannot be reduced to these functions; instead, empathy entails directly experiencing the other as a person, an intentional being. Thompson identifies four types of empathy: passive or involuntary coupling, imagination or transposition of self into the place of the other, intersubjective understanding and the moral perception of the other as a person.

Batson, C. Daniel. 2009. 'These Things Called Empathy: Eight Related but Distinct Phenomena.' In *The Social Neuroscience of Empathy*, edited by Jean Decety and William Ickes, 3–15. Cambridge, MA: MIT Press.

Stein, Edith. 1989. *On the Problem of Empathy*. Translated by Waltraut Stein. Washington, DC: ICS Publications.

Thompson, Evan. 2007. *Mind in Life: Biology, Phenomenology, and the Sciences of Mind*. Cambridge, MA: Harvard University Press.

Enaction (or Enactivism)

The concept of enaction was first proposed by Varela, Thompson and Rosch in their 1991 book *The Embodied Mind: Cognitive Science and Human Experience*. They positioned this new cognitive framework in opposition to the traditional Computational Theory of Mind. Drawing on foundations in phenomenology and Buddhist meditation, Varela, Thompson and Rosch argued that, rather than being the result of a mind encountering a pregiven environment, cognition entails the coupling of both the environment and the mind. Building on this work, in *Enaction: Toward a New Paradigm for Cognitive Science*, Stewart, Gapenne and Di Paolo describe the enactive paradigm as composed of three themes. Firstly, enaction connects first-person experience with third-person scientific research; secondly, as a framework it can unite the variety of organisation involved in cognitive science; thirdly, this paradigm focuses on reflexivity, particularly recognising that cognitive scientists themselves engage in cognition when they do their research. While as a paradigm enaction does not preclude other, more specific approaches to cognition, its foundational claim is as follows: the embodied action of a living organism is in a continual reciprocal relationship with the world in which it lives, constituting the organism's perception and grounding its cognition.

Stewart, John, Olivier Gapenne, and Ezequiel Di Paolo. 2010. *Enaction: Toward a New Paradigm for Cognitive Science*. Cambridge, MA: MIT Press.

Varela, Francisco, Evan Thompson, and Eleanor Rosch. 1991. *The Embodied Mind: Cognitive Science and the Human Experience*. Cambridge, MA: MIT Press.

Entrainment

Entrainment is the process in which two or more rhythms become synchronised with one another. There are a number of natural occurrences of entrainment, such as crickets chirping in unison or the circadian rhythm of any organism. In humans, entrainment can be either unconscious (the rhythmic coordination of the body's internal processes) or conscious (dancing or playing music).

Phillips-Silver, Jessica, C. Athena Aktipis, and Gregory A. Bryant. 2010. 'The Ecology of Entrainment: Foundations of Coordinated Rhythmic Movement.' *Music Perception: An Interdisciplinary Journal* 28 (1): 3–14.

Exteroception

Exteroception denotes the processes involved in sensing information from the environment or stimuli external to the body. This includes the five traditional senses, as well as organs like the skin that receive sensory information from the environment.

fMRI

fMRI, or functional magnetic resonance imaging, is a non-invasive technique that measures brain activity based on changes in oxygenated blood flow. Because blood flow is directed towards regions of the brain when they are active, an fMRI can detect particular areas of the brain used in particular tasks. It has been one of the most common methods of brain mapping and measuring brain activity since the 1990s.

Imagination

Imagination entails a mental representation of a thing that is not necessarily present or real. Radu Bogdan defines imagination as 'the capacity to form and deploy thoughts about non-actual, possible, future, or counterfactual scenarios in a deliberate, self-conscious, effortful, reflective, and introspectively active form of offline processing of information' (5). Humans typically develop this capacity around the age of four. At this stage, human children can begin to create situations, ideas and concepts that are unconnected to their present. Imagination is closely connected to memory (see 'Memory').

Bogdan, Radu J. 2013. *Mindvaults: Sociocultural Grounds for Pretending and Imagining*. Cambridge, MA: MIT Press.
Johnson, Mark. 1987. *The Body in the Mind: The Bodily Basis of Meaning, Imagination, and Reason*.

Interoception

Interoception involves sensory receptors within the body that enable one to perceive internal stimuli. This includes the movement of organs or muscles, such as stretching, as well as sensations like pain and hunger.

Memory

The standard taxonomy of memory distinguishes between declarative and non-declarative memory. Declarative memory entails information that can be consciously recalled or retrieved, and includes both *episodic memory* (specific episodes that are part of one's own experience, such as the name of a childhood pet) and *semantic memory* (information concerned with the world in general, such as the difference between a cat and a dog). Non-declarative or procedural memory, on the other hand, involves remembering how to do something, such as ride a bicycle or use a computer. Recent research within the paradigm of enaction, however, indicates that memory may not function in this way (see 'Enaction'). For example, Lawrence Barsalou argues that because humans perceive the world as situations rather than individual entities, memory involves creating and recalling simulations, or concepts. New experiences change stored simulations; when encountering a new experience, then, humans retrieve a simulation which helps them determine how to behave in this new situation.

Barsalou, Lawrence W. 2009. 'Situating Concepts.' In *Cambridge Handbook of Situated Cognition*, edited by Philip Robbins and Murat Aydede, 236–63. Cambridge: Cambridge University Press.

Squire, Larry R. 2009. 'Memory and Brain Systems: 1969–2009.' *Journal of Neuroscience* 29 (41): 12711–16.

Mirror neuron system

Mirror neurons were first discovered in macaque monkeys by a team of neuroscientists in 1996. Their experimental data indicated a set of neurons, which they termed mirror neurons, that were active both when a monkey performed a particular action and when it observed the same action being performed by another. Since their discovery, many posited the existence of a similar system of mirror neurons in humans, though experimental data on this issue are difficult to maintain due to the complexities of the human brain. However, several scientists have argued for the importance of mirror neurons systems (MNS) in the operations of human emotions and empathy.

Keysers, Christian. 2011. *The Empathetic Brain: How the Discovery of Mirror Neurons Changes Our Understanding of Human Nature.*

Rizzolatti, Giacomo. 2008. *Mirrors in the Brain: How Our Minds Share Actions and Emotions.* Oxford: Oxford University Press.

Play

Many cognitive psychologies and evolutionary biologists have recognised the importance of play in human cognitive development. A small number of animal species, humans included, engage in play. Typically, play behaviour is marked as a semi-distinct or distinct event separate from other activity. Even when these behaviours mimic actual behaviour, such as young wolves imitating fighting without drawing blood, they serve important evolutionary functions. In humans, play serves to develop social behaviour and increase cognitive flexibility, as Brian Boyd has demonstrated. Boyd concludes that the arts, including performance, developed from the human capacity for play, a characteristic that was (and is) necessary for our evolutionary survival.

Boyd, Brian. 2009. *On the Origin of Stories: Evolution, Cognition, and Fiction.* Cambridge, MA: Harvard University Press.

Fagen, Robert. 1995. 'Animal Play, Games of Angels, Biology, and the Brain.' In *The Future of Play Theory: A Multidisciplinary Inquiry into the Contributions of Brian Sutton-Smith,* edited by Anthony D. Pelligrini, 23–44. Albany: State University of New York Press.

Predictive processing

Predictive processing (PP) is a model of cognitive activity recently proposed by Andy Clark. Drawing from recent scholarship in embodied cognition, Clark argues that the human brain is engaged in a continuous process of predicting. Perception, then, is a process in which the brain guesses or predicts what is in the environment, refining predictions based on incoming sensory information and memory. Predictive processing joins bottom-up and top-down processing (see 'Bottom-up processing'). Predictive processing can explain, for example, the jarring experience of sipping a hot drink and tasting coffee instead of tea. If this is the case, then much of what we perceive is determined by the structure of our expectations about it. While Clark believes predictive processing is an extremely significant strategy for cognition, he maintains that it is just one among many.

Clark, Andy. 2016. *Surfing Uncertainty: Prediction, Action, and the Embodied Mind.* Oxford: Oxford University Press.

Proprioception

Proprioception is the sense of one's body, including the position of body parts and the sense of effort in movement. In humans, proprioceptors are located in muscles, tendons and joints. Information from these proprioceptors is sent to the brain and integrated into the overall sense of the body's position and movement. For example, sobriety tests often use proprioception to test for intoxication, as impaired proprioception causes an inability to touch one's own nose with a finger (because the sense of the body's position in space is affected).

Representationalism

Representationalism is a general approach in cognitive science and philosophy, closely related to traditional cognitivism (see 'Cognitivism'). There are many specific representationalist theories that seek to explain cognition. Their basic premise is that humans process external reality via mental representations. In this view, mental states, including thoughts, desires, beliefs and perceptions, are about something external to the mind. All mental processes, such as thinking and imagining, are composed of a series of mental images or representations. The mind can only perceive representations of objects outside of it, never the objects themselves. Critics of this approach suggest that it is an inadequate explanation of consciousness, since it separates mind from perception in a 'Cartesian theatre.'

Fodor, Jerry A. 1981. *Representations: Philosophical Essays on the Foundations of Cognitive Science.* Cambridge, MA: MIT Press.

Theory of mind

Theory of Mind (ToM) is an approach in cognition that emphasises the ability of humans to recognise and ascribe mental states to others, and use that information to understand and predict behaviour. Humans develop ToM quite early, as a process that includes developing perception and language. ToM as a branch of cognitive science includes a wide range of approaches and methodologies. Generally, ToM includes two broad groups of theories: theory-theory and simulation-theory. Under theory-theory, or 'folk psychology,' a person uses a basic theory of psychology to deduce the mental states of others. Simulation theory, on the other hand, holds that ToM occurs when people mentally simulate the action of another. ToM is distinct from empathy. ToM holds that there is always a separation between the mind of the self and the mind of another; therefore, some activity (either folk psychology or simulation) must be performed in order to 'read' the other's mind.

Goldman, Alvin I. 2006. *Simulating Minds: The Philosophy, Psychology, and Neuroscience of Mindreading*. Oxford: Oxford University Press.

Wellman, Henry M. 2015. *Making Minds: How Theory of Mind Develops*. Oxford: Oxford University Press.

INDEX

Made in the USA
Monee, IL
13 September 2026

62977684R10212